FORCE

3-150 HORSEPOWER, AL ...INDER

SELOC®

Managing Partners	Dean F. Morgantini, S.A.E.
	Barry L. Beck
Executive Editor	Kevin M. G. Maher, A.S.E.
Manager-Marine/Recreation	James R. Marotta, A.S.E.
Production Specialists	Melinda Possinger
	Ronald Webb
Author	George Jansen

Manufactured in USA
© 2000 Seloc Publishing
104 Willowbrook Lane
West Chester, PA 19382
ISBN 0-89330-055-1
Library Of Congress Catalog Card No. 00-132-144
4567890123 4321098765

www.selocmarine.com
1-866-SELOC55

Contents

1 GENERAL INFORMATION AND BOATING SAFETY
- 1-2 HOW TO USE THIS MANUAL
- 1-3 BOATING SAFETY
- 1-11 SAFETY IN SERVICE

2 TOOLS AND EQUIPMENT
- 2-2 TOOLS AND EQUIPMENT
- 2-4 TOOLS
- 2-11 FASTENERS, MEASUREMENTS AND CONVERSIONS

3 MAINTENANCE AND TUNE-UP
- 3-2 ENGINE MAINTENANCE
- 3-10 BOAT MAINTENANCE
- 3-12 TUNE-UP
- 3-28 WINTER STORAGE CHECKLIST
- 3-28 SPRING COMMISSIONING CHECKLIST

4 FUEL SYSTEM
- 4-2 FUEL AND COMBUSTION
- 4-3 CARBURETED FUEL SYSTEM

5 IGNITION AND ELECTRICAL SYSTEMS
- 5-2 UNDERSTANDING AND TROUBLESHOOTING ELECTRICAL SYSTEMS
- 5-8 BREAKER POINTS (MAGNETO IGNITION)
- 5-13 MOTOROLA® DISTRIBUTOR IGNITION
- 5-22 ELECTRONIC IGNITON SYSTEMS UP TO 25 HP
- 5-24 PRESTOLITE® ELECTRONIC IGNITION
- 5-32 THUNDERBOLT® ELECTRONIC IGNITION SYSTEM
- 5-38 CAPACITOR DISCHARGE MODULE (CDM) SYSTEM
- 5-45 CHARGING CIRCUIT
- 5-50 STARTING CIRCUIT
- 5-59 IGNITION AND ELECTRICAL WIRING DIAGRAMS

6 LUBRICATION AND COOLING
- 6-2 COOLING SYSTEM
- 6-10 WARNING SYSTEMS

Contents

7 POWERHEAD AND POWERHEAD OVERHAUL

- **7-2** ENGINE MECHANICAL
- **7-8** POWERHEAD RECONDITIONING
- **7-22** POWERHEAD EXPLODED VIEWS
- **7-29** TORQUE SEQUENCE DIAGRAMS

8 LOWER UNIT

- **8-2** LOWER UNIT
- **8-7** LOWER UNIT OVERHAUL

9 TRIM AND TILT

- **9-2** MANUAL TILT
- **9-3** CENTER MOUNTED PISTON TRIM/TILT SYSTEM
- **9-11** SINGLE RAM INTEGRAL POWER TILT/TRIM SYSTEM

10 REMOTE CONTROLS

- **10-2** REMOTE CONTROL BOX
- **10-11** TILLER HANDLE

11 HAND REWIND STARTER

- **11-2** HAND REWIND STARTER
- **11-2** OVERHEAD TYPE STARTER
- **11-11** BENDIX TYPE STARTER

GLOSSARY

- **11-17** GLOSSARY

MASTER INDEX

- **11-19** INDEX

IN A SPORT THAT DEMANDS NERVES BE MADE OF STEEL, YOU CAN IMAGINE HOW TOUGH THE TOOLS HAVE TO BE.

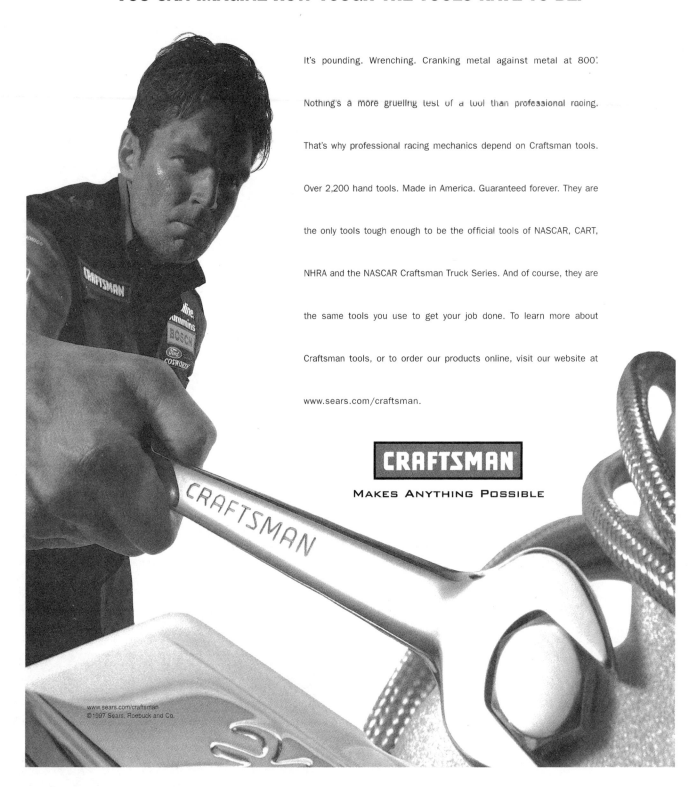

It's pounding. Wrenching. Cranking metal against metal at 800°. Nothing's a more gruelling test of a tool than professional racing. That's why professional racing mechanics depend on Craftsman tools. Over 2,200 hand tools. Made in America. Guaranteed forever. They are the only tools tough enough to be the official tools of NASCAR, CART, NHRA and the NASCAR Craftsman Truck Series. And of course, they are the same tools you use to get your job done. To learn more about Craftsman tools, or to order our products online, visit our website at www.sears.com/craftsman.

CRAFTSMAN

MAKES ANYTHING POSSIBLE

www.sears.com/craftsman
©1997 Sears, Roebuck and Co.

SAFETY NOTICE

Proper service and repair procedures are vital to the safe, reliable operation of all marine engines, as well as the personal safety of those performing repairs. This manual outlines procedures for servicing and repairing outboards using safe, effective methods. The procedures contain many NOTES, CAUTIONS and WARNINGS which should be followed, along with standard procedures, to eliminate the possibility of personal injury or improper service which could damage the vessel or compromise its safety.

It is important to note that repair procedures and techniques, tools and parts for servicing marine engines, as well as the skill and experience of the individual performing the work, vary widely. It is not possible to anticipate all of the conceivable ways or conditions under which these engines may be serviced, or to provide cautions as to all possible hazards that may result. Standard and accepted safety precautions and equipment should be used during cutting, grinding, chiseling, prying, or any other process that can cause material removal or projectiles.

Some procedures require the use of tools specially designed for a specific purpose. Before substituting another tool or procedure, you must be completely satisfied that neither your personal safety, nor the performance of the marine engine, will be compromised.

Although information in this manual is based on industry sources and is complete as possible at the time of publication, the possibility exists that some vehicle manufacturers made later changes which could not be included here. While striving for total accuracy, Nichols Publishing cannot assume responsibility for any errors, changes or omissions that may occur in the compilation of this data.

PART NUMBERS

Part numbers listed in this reference are not recommendations by Nichols Publishing for any product brand name. They are references that can be used with interchange manuals and aftermarket supplier catalogs to locate each brand supplier's discrete part number.

SPECIAL TOOLS

Special tools are recommended by the marine manufacturer to perform a specific task. Use has been kept to a minimum, but, where absolutely necessary, they are referred to in the text by the part number of the tool manufacturer. These tools can be purchased, under the appropriate part number, from your local dealer or regional distributor, or an equivalent tool can be purchased locally from a tool supplier or parts outlet. Before substituting any tool for the one recommended, read the SAFETY NOTICE at the top of this page.

ALL RIGHTS RESERVED

No part of this publication may be reproduced, transmitted or stored in any form or by any means, electronic or mechanical, including photocopy, recording, or by information storage or retrieval system, without prior written permission from the publisher.

ACKNOWLEDGMENTS

Nichols Publishing expresses appreciation to the following companies who supported the production of this book.

- Mercury Marine—Fod du Lac, WI
- Belk's Marine—Holmes, PA
- Rapair/CDI Electronics—Madison, AL

Thanks to John Hartung and Judy Belk of Belk's Marine for allowing us full access to their dealership for a portion of our photoshoot and to Clark Beard of CDI Electronics for teaching us everything he knows about troubleshooting outboards.

Nichols Publishing would like to express thanks to all of the fine companies who participate in the production of our books:
- Hand tools supplied by Craftsman are used during all phases of our vehicle teardown and photography.
- Many of the fine specialty tools used in our procedures were provided courtesy of Lisle Corporation.
- Lincoln Automotive Products (1 Lincoln Way, St. Louis, MO 63120) has provided their industrial shop equipment, including jacks (engine, transmission and floor), engine stands, fluid and lubrication tools, as well as shop presses.
- Rotary Lifts (1-800-640-5438 or www.Rotary-Lift.com), the largest automobile lift manufacturer in the world, offering the biggest variety of surface and in-ground lifts available, has fulfilled our shop's lift needs.
- Much of our shop's electronic testing equipment was supplied by Universal Enterprises Inc. (UEI).
- Safety-Kleen Systems Inc. has provided parts cleaning stations and assistance with environmentally sound disposal of residual wastes.
- United Gilsonite Laboratories (UGL), manufacturer of Drylok® concrete floor paint, has provided materials and expertise for the coating and protection of our shop floor.

MARINE TECHNICIAN TRAINING

INDUSTRY SUPPORTED PROGRAMS
OUTBOARD, STERNDRIVE & PERSONAL WATERCRAFT

- Dyno Testing • Boat & Trailer Rigging • Electrical & Fuel System Diagnostics
- Powerhead, Lower Unit & Drive Rebuilds • Powertrim & Tilt Rebuilds
- Instrument & Accessories Installation

TRAIN IN SUNNY FLORIDA!

For information regarding housing, financial aid and employment opportunities in the marine industry, contact us today:

CALL TOLL FREE
1-800-528-7995

An Accredited Institution

SM
Name
Address
City   State   Zip
Phone

MARINE MECHANICS INSTITUTE
A Division of CTI
9751 Delegates Drive • Orlando, Florida 32837
2844 W. Deer Valley Rd. • Phoenix, AZ 85027

MEMBER NMMA

FINANCIAL ASSISTANCE AVAILABLE FOR THOSE WHO QUALIFY!

HOW TO USE THIS MANUAL 1-2
CAN YOU DO IT? 1-2
WHERE TO BEGIN 1-2
AVOIDING TROUBLE 1-2
MAINTENANCE OR REPAIR? 1-2
DIRECTIONS AND LOCATIONS 1-2
PROFESSIONAL HELP 1-2
PURCHASING PARTS 1-3
AVOIDING THE MOST COMMON
  MISTAKES 1-3
BOATING SAFETY 1-3
REGULATIONS FOR YOUR BOAT 1-3
  DOCUMENTING OF VESSELS 1-4
  REGISTRATION OF BOATS 1-4
  NUMBERING OF VESSELS 1-4
  SALES & TRANSFERS 1-4
  HULL IDENTIFICATION
    NUMBER 1-4
  LENGTH OF BOATS 1-4
  CAPACITY INFORMATION 1-4
  CERTIFICATE OF COMPLIANCE 1-4
  VENTILATION 1-4
  VENTILATION SYSTEMS 1-5
REQUIRED SAFETY EQUIPMENT 1-5
  TYPES OF FIRES 1-5
  FIRE EXTINGUISHERS 1-5
  WARNING SYSTEM 1-6
  PERSONAL FLOTATION
    DEVICES 1-6
  SOUND PRODUCING DEVICES 1-8
  VISUAL DISTRESS SIGNALS 1-8
EQUIPMENT NOT REQUIRED BUT
  RECOMMENDED 1-10
  SECOND MEANS OF
    PROPULSION 1-10
  BAILING DEVICES 1-10
  FIRST AID KIT 1-10
  ANCHORS 1-10
  VHF-FM RADIO 1-11
  TOOLS & SPARE PARTS 1-11
COURTESY MARINE
  EXAMINATIONS 1-11
SAFETY IN SERVICE 1-11
DO'S 1-11
DON'TS 1-12

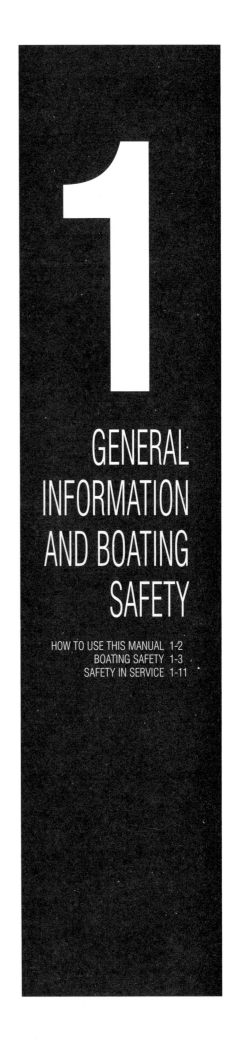

1

GENERAL INFORMATION AND BOATING SAFETY

HOW TO USE THIS MANUAL 1-2
BOATING SAFETY 1-3
SAFETY IN SERVICE 1-11

1-2 GENERAL INFORMATION AND BOATING SAFETY

HOW TO USE THIS MANUAL

This manual is designed to be a handy reference guide to maintaining and repairing your Force outboard. We strongly believe that regardless of how many or how few years experience you may have, there is something new waiting here for you.

This manual covers the topics that a factory service manual (designed for factory trained mechanics) and a manufacturer owner's manual (designed more by lawyers these days) covers. It will take you through the basics of maintaining and repairing your Force outboard, step-by-step, to help you understand what the factory trained mechanics already know by heart. By using the information in this manual, any boat owner should be able to make better informed decisions about what they need to do to maintain and enjoy their Force outboard.

Even if you never plan on touching a wrench (and if so, we hope that you will change your mind), this manual will still help you understand what a mechanic needs to do in order to maintain your engine.

Can You Do It?

If you are not the type who is prone to taking a wrench to something, NEVER FEAR. The procedures in this manual cover topics at a level virtually anyone will be able to handle. And just the fact that you purchased this manual shows your interest in better understanding your Force outboard.

You may find that maintaining your Force outboard yourself is preferable in most cases. From a monetary standpoint, it could also be beneficial. The money spent on hauling your boat to a marina and paying a tech to service the engine could buy you fuel for a whole weekend's boating. If you are unsure of your own mechanical abilities, at the very least you should fully understand what a marine mechanic does to your boat. You may decide that anything other than maintenance and adjustments should be performed by a mechanic (and that's your call), but know that every time you board your boat, you are placing faith in the mechanic's work and trusting him or her with your well-being, and maybe your life.

It should also be noted that in most areas a factory trained mechanic would command a hefty hourly rate for off site service. This hourly rate is charged from the time they leave their shop to the time they return home. The cost savings in doing the job yourself should be readily apparent at this point.

Where to Begin

Before spending any money on parts, and before removing any nuts or bolts, read through the entire procedure or topic. This will give you the overall view of what tools and supplies will be required to perform the procedure or what questions need to be answered before purchasing parts. So read ahead and plan ahead. Each operation should be approached logically and all procedures thoroughly understood before attempting any work.

Avoiding Trouble

Some procedures in this manual may require you to "label and disconnect . . ." a group of lines, hoses or wires. Don't be lulled into thinking you can remember where everything goes — you won't. If you reconnect or install a part incorrectly, things may operate poorly, if at all. If you hook up electrical wiring incorrectly, you may instantly learn a very, very expensive lesson.

A piece of masking tape, for example, placed on a hose and another on its fitting will allow you to assign your own label such as the letter "A", or a short name. As long as you remember your own code, the lines can be reconnected by matching letters or names. Do remember that tape will dissolve when saturated in fluids. If a component is to be washed or cleaned, use another method of identification. A permanent felt-tipped marker can be very handy for marking metal parts; but remember that fluids will remove permanent marker.

SAFETY is the most important thing to remember when performing maintenance or repairs. Be sure to read the information on safety in this manual.

Maintenance or Repair?

Proper maintenance is the key to long and trouble-free engine life, and the work can yield its own rewards. A properly maintained engine performs better than one that is neglected. As a conscientious boat owner, set aside a Saturday morning, at least once a month, to perform a thorough check of items which could cause problems. Keep your own personal log to jot down which services you performed, how much the parts cost you, the date, and the amount of hours on the engine at the time. Keep all receipts for parts purchased, so that they may be referred to in case of related problems or to determine operating expenses. As a do-it-yourselfer, these receipts are the only proof you have that the required maintenance was performed. In the event of a warranty problem, these receipts will be invaluable.

It's necessary to mention the difference between maintenance and repair. Maintenance includes routine inspections, adjustments, and replacement of parts that show signs of normal wear. Maintenance compensates for wear or deterioration. Repair implies that something has broken or is not working. A need for repair is often caused by lack of maintenance.

For example: draining and refilling the engine oil is maintenance recommended by all manufacturers at specific intervals. Failure to do this can allow internal corrosion or damage and impair the operation of the engine, requiring expensive repairs. While no maintenance program can prevent items from breaking or wearing out, a general rule can be stated: MAINTENANCE IS CHEAPER THAN REPAIR.

Directions and Locations

♦ See Figure 1

Two basic rules should be mentioned here. First, whenever the Port side of the engine (or boat) is referred to, it is meant to specify the left side of the engine when you are sitting at the helm. Conversely, the Starboard means your right side. The Bow is the front of the boat and the Stern is the rear.

Most screws and bolts are removed by turning counterclockwise, and tightened by turning clockwise. An easy way to remember this is: righty-tighty; lefty-loosey. Corny, but effective. And if you are really dense (and we have all been so at one time or another), buy a ratchet that is marked ON and OFF, or mark your own.

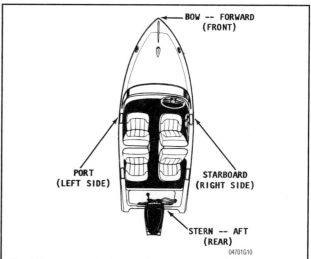

Fig. 1 Common terminology used for reference designation on boats of all size. These terms are used though out the manual

Professional Help

Occasionally, there are some things when working on a Force outboard that are beyond the capabilities or tools of the average Do-It-Yourselfer (DIYer). This shouldn't include most of the topics of this manual, but you will have to be the judge. Some engines require special tools or a selection of special parts, even for basic maintenance.

Talk to other boaters who use the same model of engine and speak with a trusted marina to find if there is a particular system or component on your engine that is difficult to maintain. For example, although the technique of valve adjustment on some engines may be easily understood and even performed by a DIYer, it might require a handy assortment of shims in various sizes and a few hours of disassembly to get to that point. Not having the assortment of shims handy might mean multiple trips back and forth to the parts store, and this might not be worth your time.

You will have to decide for yourself where basic maintenance ends and where professional service should begin. Take your time and do your research first (starting with the information in this manual) and then make your own decision. If you really don't feel comfortable with attempting a procedure, DON'T DO IT. If you've gotten into something that may be over your head, don't panic. Tuck your tail between your legs and call a marine mechanic. Marinas and independent shops will be able to finish a job for you. Your ego may be damaged, but your boat will be properly restored to its full running order. So, as long as you approach jobs slowly and carefully, you really have nothing to lose and everything to gain by doing it yourself.

Purchasing Parts

▶ See Figure 2

When purchasing parts there are two things to consider. The first is quality and the second is to be sure to get the correct part for your engine. To get quality parts, always deal directly with a reputable retailer. To get the proper parts always refer to the information tag on your engine prior to calling the parts counter. An incorrect part can adversely affect your engine performance and fuel economy, and will cost you more money and aggravation in the end.

Just remember, a tow back to shore will cost plenty. That charge is per hour from the time the towboat leaves their home port, to the time they return to their homeport. Get the picture. . . .$$$?

So whom should you call for parts? Well, there are many sources for the parts you will need. Where you shop for parts will be determined by what kind of parts you need, how much you want to pay, and the types of stores in your neighborhood.

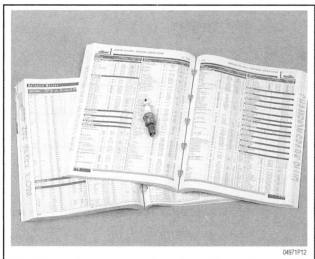

Fig. 2 Manufacturers provide parts catalogs, giving application and part number information, for most replacement parts

Your marina can supply you with many of the common parts you require. Using a marina for as your parts supplier may be hand because of location (just walk right down the dock) or because the marina specializes in your particular brand of engine. In addition, it is always a good idea to get to know the marina staff (especially the marine mechanic).

The marine parts jobber, who is usually listed in the yellow pages or whose name can be obtained from the marina, is another excellent source for parts. In addition to supplying local marinas, they also do a sizeable business in over-the-counter parts sales for the do-it-yourselfer.

Almost every community has one or more convenient marine chain stores. These stores often offer the best retail prices and the convenience of one-stop shopping for all your needs. Since they cater to the do-it-yourselfer, these stores are almost always open weeknights, Saturdays, and Sundays, when the jobbers are usually closed.

The lowest prices for parts are most often found in discount stores or the auto department of mass merchandisers. Parts sold here are name and private brand parts bought in huge quantities, so they can offer a competitive price. Private brand parts are made by major manufacturers and sold to large chains under a store label.

Avoiding the Most Common Mistakes

There are 3 common mistakes in mechanical work:

1. Incorrect order of assembly, disassembly or adjustment. When taking something apart or putting it together, performing steps in the wrong order usually just costs you extra time; however, it CAN break something. Read the entire procedure before beginning disassembly. Perform everything in the order in which the instructions say you should, even if you can't immediately see a reason for it. When you're taking apart something that is very intricate, you might want to draw a picture of how it looks when assembled at one point in order to make sure you get everything back in its proper position. When making adjustments, perform them in the proper order; often, one adjustment affects another, and you cannot expect satisfactory results unless each adjustment is made only when another cannot change it.

2. Overtorquing (or under torquing). While it is more common for over torquing to cause damage, under torquing may allow a fastener to vibrate loose causing serious damage. Especially when dealing with aluminum parts, pay attention to torque specifications and utilize a torque wrench in assembly. If a torque figure is not available, remember that if you are using the right tool to perform the job, you will probably not have to strain yourself to get a fastener tight enough. The pitch of most threads is so slight that the tension you put on the wrench will be multiplied many times in actual force on what you are tightening.

3. Cross threading. This occurs when a part such as a bolt is screwed into a nut or casting at the wrong angle and forced. Cross threading is more likely to occur if access is difficult. It helps to clean and lubricate fasteners, then to start threading with the part to be installed positioned straight in. Always start a fastener, etc. with your fingers. If you encounter resistance, unscrew the part and start over again at a different angle until it can be inserted and turned several times without much effort. Keep in mind that some parts may have tapered threads, so that gentle turning will automatically bring the part you're threading to the proper angle, but only if you don't force it or resist a change in angle. Don't put a wrench on the part until it has been tightened a couple of turns by hand. If you suddenly encounter resistance, and the part has not seated fully, don't force it. Pull it back out to make sure it's clean and threading properly.

BOATING SAFETY

In 1971 Congress ordered the U.S. Coast Guard to improve recreational boating safety. In response, the Coast Guard drew up a set of regulations.

Beside these federal regulations, there are state and local laws you must follow. These sometimes exceed the Coast Guard requirements. This section discusses only the federal laws. State and local laws are available from your local Coast Guard. As with other laws, "Ignorance of the boating laws is no excuse." The rules fall into two groups: regulations for your boat and required safety equipment on your boat.

Regulations For Your Boat

Most boats on waters within Federal jurisdiction must be registered or documented. These waters are those that provide a means of transportation between two or more states or to the sea. They also include the territorial waters of the United States.

1-4 GENERAL INFORMATION AND BOATING SAFETY

DOCUMENTING OF VESSELS

A vessel of five or more net tons may be documented as a yacht. In this process, the U.S. Coast Guard issues papers as they are for large ships. Documentation is a form of national registration. The boat must be used solely for pleasure. Its owner must be a U.S. citizen, a partnership of U.S. citizens, or a corporation controlled by U.S. citizens. The captain and other officers must also be U.S. citizens. The crew need not be.

If you document your yacht, you have the legal authority to fly the yacht ensign. You also may record bills of sale, mortgages, and other papers of title with federal authorities. Doing so gives legal notice that such instruments exist. Documentation also permits preferred status for mortgages. This gives you additional security and aids financing and transfers of title. You must carry the original documentation papers aboard your vessel. Copies will not suffice.

REGISTRATION OF BOATS

If your boat is not documented, registration in the state of its principal use is probably required. If you use it mainly on an ocean, a gulf, or other similar water, register it in the state where you moor it.

If you use your boat solely for racing, it may be exempt from the requirement in your state. States may also exclude dinghies. Some require registration of documented vessels and non-power driven boats.

All states, except Alaska, register boats. In Alaska, the U.S. Coast Guard issues the registration numbers. If you move your vessel to a new state of principal use, a valid registration certificate is good for 60 days. You must have the registration certificate (certificate of number) aboard your vessel when it is in use. A copy will not suffice. You may be cited if you do not have the original on board.

NUMBERING OF VESSELS

A registration number is on your registration certificate. You must paint or permanently attach this number to both sides of the forward half of your boat. Do not display any other number there.

The registration number must be clearly visible. It must not be placed on the obscured underside of a flared bow. If you can't place the number on the bow, place it on the forward half of the hull. If that doesn't work, put it on the superstructure. Put the number for an inflatable boat on a bracket or fixture. Then, firmly attach it to the forward half of the boat. The letters and numbers must be plain block characters and must read from left to right. Use a space or a hyphen to separate the prefix and suffix letters from the numerals. The color of the characters must contrast with that of the background, and they must be at least three inches high.

In some states your registration is good for only one year. In others, it is good for as long as three years. Renew your registration before it expires. At that time you will receive a new decal or decals. Place them as required by state law. You should remove old decals before putting on the new ones. Some states require that you show only the current decal or decals. If your vessel is moored, it must have a current decal even if it is not in use.

If your vessel is lost, destroyed, abandoned, stolen, or transferred, you must inform the issuing authority. If you lose your certificate of number or your address changes, notify the issuing authority as soon as possible.

SALES & TRANSFERS

Your registration number is not transferable to another boat. The number stays with the boat unless its state of principal use is changed.

HULL IDENTIFICATION NUMBER

A Hull Identification Number (HIN) is like the Vehicle Identification Number (VIN) on your car. Boats built between November 1, 1972 and July 31, 1984 have old format HINs. Since August 1, 1984 a new format has been used.

Your boat's HIN must appear in two places. If it has a transom, the primary number is on its starboard side within two inches of its top. If it does not have a transom or if it was not practical to use the transom, the number is on the starboard side. In this case, it must be within one foot of the stern and within two inches of the top of the hull side. On pontoon boats, it is on the aft crossbeam within one foot of the starboard hull attachment. Your boat also has a duplicate number in an unexposed location. This is on the boat's interior or under a fitting or item of hardware.

LENGTH OF BOATS

For some purposes, boats are classed by length. Required equipment, for example, differs with boat size. Manufacturers may measure a boat's length in several ways. Officially, though, your boat is measured along a straight line from its bow to its stern. This line is parallel to its keel.

The length does not include bowsprits, boomkins, or pulpits. Nor does it include rudders, brackets, outboard motors, outdrives, diving platforms, or other attachments.

CAPACITY INFORMATION

♦ See Figure 3

Manufacturers must put capacity plates on most recreational boats less than 20 feet long. Sailboats, canoes, kayaks, and inflatable boats are usually exempt. Outboard boats must display the maximum permitted horsepower of their engines. The plates must also show the allowable maximum weights of the people on board. And they must show the allowable maximum combined weights of people, engines, and gear. Inboards and stern drives need not show the weight of their engines on their capacity plates. The capacity plate must appear where it is clearly visible to the operator when underway. This information serves to remind you of the capacity of your boat under normal circumstances. You should ask yourself, "Is my boat loaded above its recommended capacity" and, "Is my boat overloaded for the present sea and wind conditions?" If you are stopped by a legal authority, you may be cited if you are overloaded.

Fig. 3 A U.S. Coast Guard certification plate indicates the amount of occupants and gear appropriate for safe operation of the vessel

CERTIFICATE OF COMPLIANCE

Manufacturers are required to put compliance plates on motorboats greater than 20 feet in length. The plates must say, "This boat," or "This equipment complies with the U. S. Coast Guard Safety Standards in effect on the date of certification." Letters and numbers can be no less than one-eighth of an inch high. At the manufacturer's option, the capacity and compliance plates may be combined.

VENTILATION

A cup of gasoline spilled in the bilge has the potential explosive power of 15 sticks of dynamite. This statement, commonly quoted over 20 years ago, may be an exaggeration, however, it illustrates a fact. Gasoline fumes in the bilge of a boat are highly explosive and a serious danger. They are heavier than air and will stay in the bilge until they are vented out.

Because of this danger, Coast Guard regulations require ventilation on many powerboats. There are several ways to supply fresh air to engine and gasoline tank compartments and to remove dangerous vapors. Whatever the choice, it must meet Coast Guard standards.

➥**The following is not intended to be a complete discussion of the regulations. It is limited to the majority of recreational vessels. Contact your local Coast Guard office for further information.**

GENERAL INFORMATION AND BOATING SAFETY

General Precautions

Ventilation systems will not remove raw gasoline that leaks from tanks or fuel lines. If you smell gasoline fumes, you need immediate repairs. The best device for sensing gasoline fumes is your nose. Use it! If you smell gasoline in an engine compartment or elsewhere, don't start your engine or operate any electrical accessories. The smaller the compartment, the less gasoline it takes to make an explosive mixture.

Ventilation for Open Boats

In open boats, the air that moves through them disperses gasoline vapors. So they are exempt from ventilation requirements.

To be "open," a boat must meet certain conditions. Engine and fuel tank compartments and long narrow compartments that join them must be open to the atmosphere." This means they must have at least 15 square inches of open area for each cubic foot of net compartment volume. The open area must be in direct contact with the atmosphere. There must also be no long, unventilated spaces open to engine and fuel tank compartments into which flames could extend.

Ventilation for All Other Boats

Powered and natural ventilation is required in an enclosed compartment with a permanently installed gasoline engine that has a cranking motor. A compartment is exempt if its engine is open to the atmosphere. Diesel powered boats are also exempt.

VENTILATION SYSTEMS

There are two types of ventilation systems. One is "natural ventilation." In it, air circulates through closed spaces due to the boat's motion. The other type is "powered ventilation." In it, a motor driven fan or fans circulate air.

Natural Ventilation System Requirements

A natural ventilation system has an air supply from outside the boat. The air supply may also be from a ventilated compartment or a compartment open to the atmosphere. Intake openings are required. In addition, intake ducts may be required to direct the air to appropriate compartments.

The system must also have an exhaust duct that starts in the lower third of the compartment. The exhaust opening must be into another ventilated compartment or into the atmosphere. Each supply opening and supply duct, if there is one, must be above the usual level of water in the bilge. Exhaust openings and ducts must also be above the bilge water. Openings and ducts must be at least three square inches in area or two inches in diameter. Openings should be placed so exhaust gasses do not enter the fresh air intake. Exhaust fumes must not enter cabins or other enclosed, non-ventilated spaces. The carbon monoxide gas in them is deadly.

Cowls or similar devices must cover intake and exhaust openings. These registers keep out rainwater and water from breaking seas. Most often, intake registers face forward and exhaust openings aft. This aids the flow of air when the boat is moving or at anchor since most boats face into the wind when anchored.

Power Ventilation System Requirements

♦ See Figure 4

Powered ventilation systems must meet the standards of a natural system. They must also have one or more exhaust blowers. The blower duct can serve as the exhaust duct for natural ventilation if fan blades do not obstruct the airflow when not powered. Openings in engine compartment, for carburetion are in addition to ventilation system requirements.

Required Safety Equipment

Coast Guard regulations require that your boat have certain equipment aboard. These requirements are minimums. Exceed them whenever you can.

TYPES OF FIRES

There are four common classes of fires:
- Class A—fires are in ordinary combustible materials such as paper or wood.
- Class B—fires involves gasoline, oil and grease.

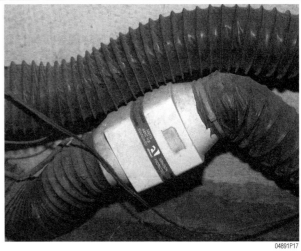

Fig. 4 Typical blower and duct system to vent fumes from the engine compartment

- Class C—fires are electrical.
- Class D—fires involve ferrous metals

One of the greatest risks to boaters is fire. This is why it is so important to carry the correct number and type of extinguishers onboard.

The best fire extinguisher for most boats is a Class B extinguisher. Never use water on Class B or Class C fires, as water spreads these types of fires. You should never use water on a Class C fire as it may cause you to be electrocuted.

FIRE EXTINGUISHERS

♦ See Figure 5

If your boat meets one or more of the following conditions, you must have at least one fire extinguisher aboard. The conditions are:
- Inboard or stern drive engines
- Closed compartments under seats where portable fuel tanks can be stored
- Double bottoms not sealed together or not completely filled with flotation materials
- Closed living spaces
- Closed stowage compartments in which combustible or flammable materials are stored
- Permanently installed fuel tanks
- Boat is 26 feet or more in length.

Contents of Extinguishers

Fire extinguishers use a variety of materials. Those used on boats usually contain dry chemicals, Halon, or Carbon Dioxide (CO_3). Dry chemical extinguishers contain chemical powders such as Sodium Bicarbonate—baking soda.

Carbon dioxide is a colorless and odorless gas when released from an extinguisher. It is not poisonous but caution must be used in entering compartments filled with it. It will not support life and keeps oxygen from reaching your lungs. A fire-killing concentration of Carbon Dioxide is lethal. If you are in a compartment with a high concentration of CO_3, you will have no difficulty breathing. But the air does not contain enough oxygen to support life. Unconsciousness or death can result.

HALON EXTINGUISHERS

Some fire extinguishers and 'built-in' or 'fixed' automatic fire extinguishing systems contain a gas called Halon. Like carbon dioxide it is colorless and odorless and will not support life. Some Halons may be toxic if inhaled.

To be accepted to the Coast Guard, a fixed Halon system must have an indicator light at the vessel's helm. A green light shows the system is ready. Red means it is being discharged or has been discharged. Warning horns are available to let you know the system has been activated. If your fixed Halon system discharges, ventilate the space thoroughly before you enter it. There are no residues from Halon but it will not support life.

Although Halon has excellent fire fighting properties, it is thought to deplete

Fig. 5 An approved fire extinguisher should be mounted close to the operator for emergency use

the earth's ozone layer and has not been manufactured since January 1, 1994. Halon extinguishers can be refilled from existing stocks of the gas until they are used up, but high federal excise taxes are being charged for the service. If you discontinue using your Halon extinguisher, take it to a recovery station rather than releasing the gas into the atmosphere. Compounds such as FE 241, designed to replace Halon, are now available.

Fire Extinguisher Approval

Fire extinguishers must be Coast Guard approved. Look for the approval number on the nameplate. Approved extinguishers have the following on their labels: "Marine Type USCG Approved, Size . . . , Type . . . , 162.208/," etc. In addition, to be acceptable by the Coast Guard, an extinguisher must be in serviceable condition and mounted in its bracket. An extinguisher not properly mounted in its bracket will not be considered serviceable during a Coast Guard inspection.

Care and Treatment

Make certain your extinguishers are in their stowage brackets and are not damaged. Replace cracked or broken hoses. Nozzles should be free of obstructions. Sometimes, wasps and other insects nest inside nozzles and make them inoperable. Check your extinguishers frequently. If they have pressure gauges, is the pressure within acceptable limits? Do the locking pins and sealing wires show they have not been used since recharging?

Don't try an extinguisher to test it. Its valves will not reseat properly and the remaining gas will leak out. When this happens, the extinguisher is useless.

Weigh and tag carbon dioxide and Halon extinguishers twice a year. If their weight loss exceeds 10 percent of the weight of the charge, recharge them. Check to see that they have not been used. They should have been inspected by a qualified person within the past six months, and they should have tags showing all inspection and service dates. The problem is that they can be partially discharged while appearing to be fully charged.

Some Halon extinguishers have pressure gauges the same as dry chemical extinguishers. Don't rely too heavily on the gauge. The extinguisher can be partially discharged and still show a good gauge reading. Weighing a Halon extinguisher is the only accurate way to assess its contents.

If your dry chemical extinguisher has a pressure indicator, check it frequently. Check the nozzle to see if there is powder in it. If there is, recharge it. Occasionally invert your dry chemical extinguisher and hit the base with the palm of your hand. The chemical in these extinguishers packs and cakes due to the boat's vibration and pounding. There is a difference of opinion about whether hitting the base helps, but it can't hurt. It is known that caking of the chemical powder is a major cause of failure of dry chemical extinguishers. Carry spares in excess of the minimum requirement. If you have guests aboard, make certain they know where the extinguishers are and how to use them.

Using a Fire Extinguisher

A fire extinguisher usually has a device to keep it from being discharged accidentally. This is a metal or plastic pin or loop. If you need to use your extinguisher, take it from its bracket. Remove the pin or the loop and point the nozzle at the base of the flames. Now, squeeze the handle, and discharge the extinguisher's contents while sweeping from side to side. Recharge a used extinguisher as soon as possible.

If you are using a Halon or carbon dioxide extinguisher, keep your hands away from the discharge. The rapidly expanding gas will freeze them. If your fire extinguisher has a horn, hold it by its handle.

Legal Requirements for Extinguishers

You must carry fire extinguishers as defined by Coast Guard regulations. They must be firmly mounted in their brackets and immediately accessible.

A motorboat less than 26 feet long must have at least one approved hand-portable, Type B-1 extinguisher. If the boat has an approved fixed fire extinguishing system, you are not required to have the Type B-1 extinguisher. Also, if your boat is less than 26 feet long, is propelled by an outboard motor, or motors, and does not have any of the first six conditions described at the beginning of this section, it is not required to have an extinguisher. Even so, it's a good idea to have one, especially if a nearby boat catches fire, or if a fire occurs at a fuel dock.

A motorboat 26 feet to less than 40 feet long, must have at least two Type B-1 approved hand-portable extinguishers. It can, instead, have at least one Coast Guard approved Type B-2. If you have an approved fire extinguishing system, only one Type B-1 is required.

A motorboat 40 to 65 feet long must have at least three Type B-1 approved portable extinguishers . It may have, instead, at least one Type B-1 plus a Type B-2. If there is an approved fixed fire extinguishing system, two Type B-1 or one Type B-2 is required.

WARNING SYSTEM

Various devices are available to alert you to danger. These include fire, smoke, gasoline fumes, and carbon monoxide detectors. If your boat has a galley, it should have a smoke detector. Where possible, use wired detectors. Household batteries often corrode rapidly on a boat.

You can't see, smell, nor taste carbon monoxide gas, but it is lethal. As little as one part in 10,000 parts of air can bring on a headache. The symptoms of carbon monoxide poisoning—headaches, dizziness, and nausea—are like seasickness. By the time you realize what is happening to you, it may be too late to take action. If you have enclosed living spaces on your boat, protect yourself with a detector. There are many ways in which carbon monoxide can enter your boat.

PERSONAL FLOTATION DEVICES

Personal Flotation Devices (PFDs) are commonly called life preservers or life jackets. You can get them in a variety of types and sizes. They vary with their intended uses. To be acceptable, the Coast Guard must approve them.

Type I PFDs

A Type I life jacket is also called an offshore life jacket. Type I life jackets will turn most unconscious people from facedown to a vertical or slightly backward position. The adult size gives a minimum of 22 pounds of buoyancy. The child size has at least 11 pounds. Type I jackets provide more protection to their wearers than any other type of life jacket. Type I life jackets are bulkier and less comfortable than other types. Furthermore, there are only two sizes, one for children and one for adults.

Type I life jackets will keep their wearers afloat for extended periods in rough water. They are recommended for offshore cruising where a delayed rescue is probable.

Type II PFDs

♦ See Figure 6

A Type II life jacket is also called a near-shore buoyant vest. It is an approved, wearable device. Type II life jackets will turn some unconscious people from facedown to vertical or slightly backward positions. The adult size gives at least 15.5 pounds of buoyancy. The medium child size has a minimum of 11 pounds. And the small child and infant sizes give seven pounds. A Type II life jacket is more comfortable than a Type I but it does not have as much buoyancy. It is not recommended for long hours in rough water. Because of this, Type IIs are recommended for inshore and inland cruising on calm water. Use them where there is a good chance of fast rescue.

Type III PFDs

Type III life jackets or marine buoyant devices are also known as flotation aids. Like Type IIs, they are designed for calm inland or close offshore water where there is a good chance of fast rescue. Their minimum buoyancy is 15.5 pounds. They will not turn their wearers' face up.

Type III devices are usually worn where freedom of movement is necessary. Thus, they are used for water skiing, small boat sailing, and fishing among other activities. They are available as vests and flotation coats. Flotation coats are useful in cold weather. Type IIIs come in many sizes from small child through large adult.

Life jackets come in a variety of colors and patterns—red, blue, green, camouflage, and cartoon characters. From a safety standpoint, the best color is bright orange. It is easier to see in the water, especially if the water is rough.

Type IV PFDs

♦ See Figures 7 and 8

Type IV ring life buoys, buoyant cushions and horseshoe buoys are Coast Guard approved devices called throwables. They are made to be thrown to people in the water, and should not be worn. Type IV cushions are often used as seat cushions. Cushions are hard to hold onto in the water. Thus, they do not afford as much protection as wearable life jackets.

The straps on buoyant cushions are for you to hold onto either in the water or when throwing them. A cushion should never be worn on your back. It will turn you face down in the water.

Type IV throwables are not designed as personal flotation devices for unconscious people, non-swimmers, or children. Use them only in emergencies. They should not be used for, long periods in rough water.

Ring life buoys come in 18, 20, 24, and 30 inch diameter sizes. They have grab lines. You should attach about 60 feet of polypropylene line to the grab rope to aid in retrieving someone in the water. If you throw a ring, be careful not to hit the person. Ring buoys can knock people unconscious

Type V PFDs

Type V PFDs are of two kinds, special use devices and hybrids. Special use devices include boardsailing vests, deck suits, work vests, and others. They are approved only for the special uses or conditions indicated on their labels. Each is designed and intended for the particular application shown on its label. They do not meet legal requirements for general use aboard recreational boats.

Hybrid life jackets are inflatable devices with some built-in buoyancy provided by plastic foam or kapok. They can be inflated orally or by cylinders of compressed gas to give additional buoyancy. In some hybrids the gas is released manually. In others it is released automatically when the life jacket is immersed in water.

The inherent buoyancy of a hybrid may be insufficient to float a person unless it is inflated. The only way to find this out is for the user to try it in the water. Because of its limited buoyancy when deflated, a hybrid is recommended for use by anon-swimmer only if it is worn with enough inflation to float the wearer.

If they are to count against the legal requirement for the number of life jackets you must carry on your vessel, hybrids manufactured before February 8,

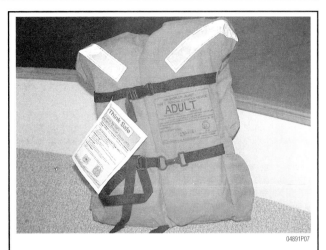

Fig. 6 Type II approved flotation devices are recommended for inshore and inland cruising on calm water. Use them where there is a good chance of fast rescue

Fig. 7 Type IV buoyant cushions are made to be thrown to people in the water. If you can squeeze air out of the cushion, it is faulty and should be replaced

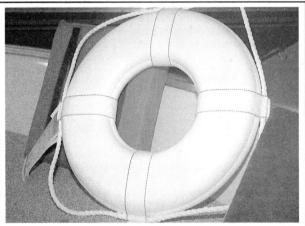

Fig. 8 Type IV throwables, such as this ring life buoy, are not designed as personal flotation devices for unconscious people, non-swimmers, or children

1995 must be worn whenever a boat is underway and the wearer is not below decks or in an enclosed space. To find out if your Type V hybrid must be worn to satisfy the legal requirement, read its label. If its use is restricted it will say, "REQUIRED TO BE WORN" in capital letters.

Hybrids cost more than other life jackets, but this factor must be weighed against the fact that they are more comfortable than Type I, II, or III life jackets. Because of their greater comfort, their owners are more likely to wear them than are the owners of Type I, II, or III life jackets.

The Coast Guard has determined that improved, less costly hybrids can save lives since they will be bought and used more frequently. For these reasons a new federal regulation was adopted effective February 8, 1995. The regulation increases both the deflated and inflated buoyancy of hybrids, makes them available in a greater variety of sizes and types, and reduces their costs by reducing production costs.

Even though it may not be required, the wearing of a hybrid or a life jacket is encouraged whenever a vessel is underway. Like life jackets, hybrids are now available in three types. To meet legal requirements, a Type I hybrid can be substituted for a Type I life jacket. Similarly Type II and III hybrids can be substituted for Type II and Type III life jackets. A Type I hybrid, when inflated, will turn most unconscious people from facedown to vertical or slightly backward positions just like a Type I life jacket. Type I and III hybrids function like Type II and III life jackets. If you purchase a new hybrid, it should have an owner's manual attached, which describes its life jacket type and its deflated and inflated buoyancy. It warns you that it may have to be inflated to float you. The manual also tells you how to don the life jacket and how to inflate it. It also tells you how to change its inflation mechanism, recommended testing exercises, and inspection and maintenance procedures. The manual also tells you why you need a life jacket and why you should wear it. A new hybrid must be packaged with at least three gas cartridges. One of these may already be loaded into the inflation mechanism. Likewise, if it has an automatic inflation mechanism, it must be packaged with at least three of these water sensitive elements. One of these elements may be installed.

Legal Requirements

A Coast Guard approved life jacket must show the manufacturer's name and approval number. Most are marked as Type I, II, III, IV, or V. All of the newer hybrids are marked for type.

You are required to carry at least one wearable life jacket or hybrid for each person on board your recreational vessel. If your vessel is 16 feet or more in length and is not a canoe or a kayak, you must also have at least one Type IV on board. These requirements apply to all recreational vessels that are propelled or controlled by machinery, sails, oars, paddles, poles, or another vessel. Sailboards are not required to carry life jackets.

You can substitute an older Type V hybrid for any required Type I, II, or III life jacket provided that its approval label shows it is approved for the activity the vessel is engaged in, approved as a substitute for a life jacket of the type required on the vessel, used as required on the labels, and used in accordance with any requirements in its owner's manual, if the approval label makes reference to such a manual.

A water skier being towed is considered to be on board the vessel when judging compliance with legal requirements.

You are required to keep your Type I, II, or III life jackets or equivalent hybrids readily accessible, which means you must be able to reach out and get them when needed. All life jackets must be in good, serviceable condition.

General Considerations

The proper use of a life jacket requires the wearer to know how it will perform. You can gain this knowledge only through experience. Each person on your boat should be assigned a life jacket. Next, it should be fitted to the person who will wear it. Only then can you be sure that it will be ready for use in an emergency.

Boats can sink fast. There may be no time to look around for a life jacket. Fitting one on you in the water is almost impossible. This advice is good even if the water is calm, and you intend to boat near shore. Most drownings occur in inland waters within a few feet of safety. Most victims had life jackets, but they weren't wearing them.

Keeping life jackets in the plastic covers they came wrapped in and in a cabin assures that they will stay clean and unfaded. But this is no way to keep them when you are on the water. When you need a life jacket it must be readily accessible and adjusted to fit you. You can't spend time hunting for it or learning how to fit it.

There is no substitute for the experience of entering the water while wearing a life jacket. Children, especially, need practice. If possible, give, your guests this experience. Tell them they should keep their arms to their sides when jumping in to keep the life jacket from riding up. Let them jump in and see how the life jacket responds. Is it adjusted so it does not ride up? Is it the proper size? Are all straps snug? Are children's life jackets the right sizes for them? Are they adjusted properly? If a child's life jacket fits correctly, you can lift the child by the jacket's shoulder straps and the child's chin and ears will not slip through. Non-swimmers, children, handicapped persons, elderly persons and even pets should always wear life jackets when they are aboard. Many states require that everyone aboard wear them in hazardous waters.

Inspect your lifesaving equipment from time to time. Leave any questionable or unsatisfactory equipment on shore. An emergency is no time for you to conduct an inspection.

Indelibly mark your life jackets with your vessel's name, number, and calling port. This can be important in a search and rescue effort. It could help concentrate effort where it will do the most good.

Care of Life Jackets

Given reasonable care, life jackets last many years. Thoroughly dry them before putting them away. Stow them in dry, well ventilated places. Avoid the bottoms of lockers and deck storage boxes where moisture may collect. Air and dry them frequently.

Life jackets should not be tossed about or used as fenders or cushions. Many contain kapok or fibrous glass material enclosed in plastic bags. The bags can rupture and are then unserviceable. Squeeze your life jacket gently. Does air leak out? If so, water can leak in and it will no longer be safe to use. Cut it up so no one will use it, and throw it away. The covers of some life jackets are made of nylon or polyester. These materials are plastics. Like many plastics, they break down after extended exposure to the ultraviolet light in sunlight. This process may be more rapid when the materials are dyed with bright dyes such as "neon" shades.

Ripped and badly faded fabrics are clues that the covering of your life jacket is deteriorating. A simple test is to pinch the fabric between your thumbs and forefingers. Now try to tear the fabric. If it can be torn, it should definitely be destroyed and discarded. Compare the colors in protected places to those exposed to the sun. If the colors have faded, the materials have been weakened. A fabric covered life jacket should ordinarily last several boating seasons with normal use. A life jacket used every day in direct sunlight should probably be replaced more often.

SOUND PRODUCING DEVICES

All boats are required to carry some means of making an efficient sound signal. Devices for making the whistle or horn noises required by the Navigation Rules must be capable of a four second blast. The blast should be audible for at least one-half mile. Athletic whistles are not acceptable on boats 12 meters or longer. Use caution with athletic whistles. When wet, some of them come apart and loose their "pea." When this happens, they are useless.

If your vessel is 12 meters long and less than 20 meters, you must have a power whistle (or power horn) and a bell on board. The bell must be in operating condition and have a minimum diameter of at least 200 mm (7.9 inches) at its mouth.

VISUAL DISTRESS SIGNALS

▶ See Figure 9

Visual Distress Signals (VDS) attract attention to your vessel if you need help. They also help to guide searchers in search and rescue situations. Be sure you have the right types, and learn how to use them properly.

It is illegal to fire flares improperly. In addition, they cost the Coast Guard and its Auxiliary many wasted hours in fruitless searches. If you signal a distress with flares and then someone helps you, please let the Coast Guard or the appropriate Search And Rescue Agency (SAR) know so the distress report will be canceled.

Recreational boats less than 16 feet long must carry visual distress signals on coastal waters at night. Coastal waters are:
- The ocean (territorial sea)
- The Great Lakes
- Bays or sounds that empty into oceans
- Rivers over two miles across at their mouths upstream to where they narrow to two miles.

GENERAL INFORMATION AND BOATING SAFETY   1-9

Fig. 9 Internationally accepted distress signals

Recreational boats 16 feet or longer must carry VDS at all times on coastal waters. The same requirement applies to boats carrying six or fewer passengers for hire. Open sailboats less than 26 feet long without engines are exempt in the daytime as are manually propelled boats. Also exempt are boats in organized races, regattas, parades, etc. Boats owned in the United States and operating on the high seas must be equipped with VDS.

A wide variety of signaling devices meet Coast Guard regulations. For pyrotechnic devices, a minimum of three must be carried. Any combination can be carried as long as it adds up to at least three signals for day use and at least three signals for night use. Three day/night signals meet both requirements. If possible, carry more than the legal requirement.

➡**The American flag flying upside down is a commonly recognized distress signal. It is not recognized in the Coast Guard regulations, though. In an emergency, your efforts would probably be better used in more effective signaling methods.**

Types of VDS

VDS are divided into two groups; daytime and nighttime use. Each of these groups is subdivided into pyrotechnic and non-pyrotechnic devices.

DAYTIME NON-PYROTECHNIC SIGNALS

A bright orange flag with a black square over a black circle is the simplest VDS. It is usable, of course, only in daylight. It has the advantage of being a continuous signal. A mirror can be used to good advantage on sunny days. It can attract the attention of other boaters and of aircraft from great distances. Mirrors are available with holes in their centers to aid in "aiming." In the absence of a mirror, any shiny object can be used. When another boat is in sight, an effective VDS is to extend your arms from your sides and move them up and down. Do it slowly. If you do it too fast the other people may think you are just being friendly. This simple gesture is seldom misunderstood, and requires no equipment.

DAYTIME PYROTECHNIC DEVICES

Orange smoke is a useful daytime signal. Hand-held or floating smoke flares are very effective in attracting attention from aircraft. Smoke flares don't last long, and are not very effective in high wind or poor visibility. As with other pyrotechnic devices, use them only when you know there is a possibility that someone will see the display.

To be usable, smoke flares must be kept dry. Keep them in airtight containers and store them in dry places. If the "striker" is damp, dry it out before trying to ignite the device. Some pyrotechnic devices require a forceful "strike" to ignite them.

All hand-held pyrotechnic devices may produce hot ashes or slag when burning. Hold them over the side of your boat in such a way that they do not burn your hand or drip into your boat.

Nighttime Non-Pyrotechnic Signals

An electric distress light is available. This light automatically flashes the international morse code SOS distress signal (●●● --- ●●●). Flashed four to six times a minute, it is an unmistakable distress signal. It must show that it is approved by the Coast Guard. Be sure the batteries are fresh. Dated batteries give assurance that they are current.

Under the Inland Navigation Rules, a high intensity white light flashing 50-70 times per minute is a distress signal. Therefore, use strobe lights on inland waters only for distress signals.

Nighttime Pyrotechnic Devices

♦ See Figure 10

Aerial and hand-held flares can be used at night or in the daytime. Obviously, they are more effective at night.

Currently, the serviceable life of a pyrotechnic device is rated at 42 months from its date of manufacture. Pyrotechnic devices are expensive.

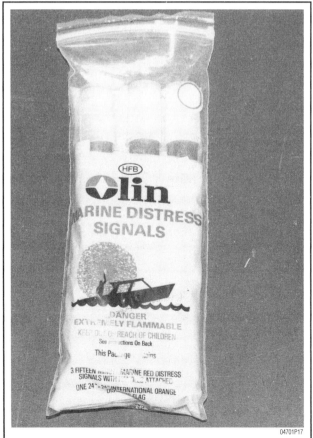

Fig. 10 Moisture protected flares should be carried onboard any vessel for use as a distress signal

1-10 GENERAL INFORMATION AND BOATING SAFETY

Look at their dates before you buy them. Buy them with as much time remaining as possible.

Like smoke flares, aerial and hand-held flares may fail to work if they have been damaged or abused. They will not function if they are or have been wet. Store them in dry, airtight containers in dry places. But store them where they are readily accessible.

Aerial VDSs, depending on their type and the conditions they are used in, may not go very high. Again, use them only when there is a good chance they will be seen.

A serious disadvantage of aerial flares is that they burn for only a short time. Most burn for less than 10 seconds. Most parachute flares burn for less than 45 seconds. If you use a VDS in an emergency, do so carefully. Hold hand-held flares over the side of the boat when in use. Never use a road hazard flare on a boat, it can easily start a fire. Marine type flares are carefully designed to lessen risk, but they still must be used carefully.

Aerial flares should be given the same respect as firearms since they are firearms! Never point them at another person. Don't allow children to play with them or around them. When you fire one, face away from the wind. Aim it downwind and upward at an angle of about 60 degrees to the horizon. If there is a strong wind, aim it somewhat more vertically. Never fire it straight up. Before you discharge a flare pistol, check for overhead obstructions. The flare might damage these. They might deflect the flare to where it will cause damage.

Disposal of VDS

Keep outdated flares when you get new ones. They do not meet legal requirements, but you might need them sometime, and they may work. It is illegal to fire a VDS on federal navigable waters unless an emergency exists. Many states have similar laws.

Emergency Position Indicating Radio Beacon (EPIRB)

There is no requirement for recreational boats to have EPIRBs. Some commercial and fishing vessels, though, must have them if they operate beyond the three mile limit. Vessels carrying six or fewer passengers for hire must have EPIRBs under some circumstances when operating beyond the three mile limit. If you boat in a remote area or offshore, you should have an EPIRB. An EPIRB is a small (about 6 to 20 inches high), battery-powered, radio transmitting buoy-like device. It is a radio transmitter and requires a license or an endorsement on your radio station license by the Federal Communications Commission (FCC). EPIRBs are activated by being immersed in water or by a manual switch.

Equipment Not Required But Recommended

Although not required by law, there are other pieces of equipment that are good to have onboard.

SECOND MEANS OF PROPULSION

♦ See Figure 11

All boats less than 16 feet long should carry a second means of propulsion. A paddle or oar can come in handy at times. For most small boats, a spare trolling or outboard motor is an excellent idea. If you carry a spare motor, it should have its own fuel tank and starting power. If you use an electric trolling motor, it should have its own battery.

Fig. 11 A typical wooden oar should be kept onboard as an auxiliary means of propulsion. It can also function as a grab hook for someone fallen overboard

BAILING DEVICES

All boats should carry at least one effective manual bailing device in addition to any installed electric bilge pump. This can be a bucket, can, scoop, hand operated pump, etc. If your battery "goes dead" it will not operate your electric pump.

FIRST AID KIT

♦ See Figure 12

All boats should carry a first aid kit. It should contain adhesive bandages, gauze, adhesive tape, antiseptic, aspirin, etc. Check your first aid kit from time to time. Replace anything that is outdated. It is to your advantage to know how to use your first aid kit. Another good idea would be to take a Red Cross first aid course.

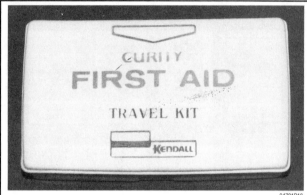

Fig. 12 Always carry an adequately stocked first aid kit on board for the safety of the crew and guests

ANCHORS

♦ See Figure 13

All boats should have anchors. Choose one of suitable size for your boat. Better still, have two anchors of different sizes. Use the smaller one in calm water or when anchoring for a short time to fish or eat. Use the larger one when the water is rougher or for overnight anchoring.

Carry enough anchor line of suitable size for your boat and the waters in which you will operate. If your engine fails you, the first thing you usually

Fig. 13 Choose and anchor of sufficient weight to secure the boat without dragging. In some cases separate anchors may be needed for different situations

GENERAL INFORMATION AND BOATING SAFETY 1-11

should do is lower your anchor. This is good advice in shallow water where you may be driven aground by the wind or water. It is also good advice in windy weather or rough water. The anchor will usually hold your bow into the waves.

VHF-FM RADIO

Your best means of summoning help in an emergency or in case of a breakdown is a VHF-FM radio. You can use it to get advice or assistance from the Coast Guard. In the event of a serious illness or injury aboard your boat, the Coast Guard can have emergency medical equipment meet you ashore.

TOOLS & SPARE PARTS

▶ See Figures 14 and 15

Carry a few tools and some spare parts, and learn how to make minor repairs. Many search and rescue cases are caused by minor breakdowns that boat operators could have repaired. If your engine is an inboard or stern drive, carry spare belts and water pump impellers and the tools to change them.

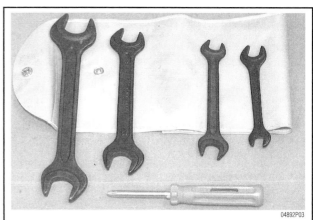

Fig. 14 A few wrenches, a screwdriver and maybe a pair of pliers can be very helpful to make emergency repairs

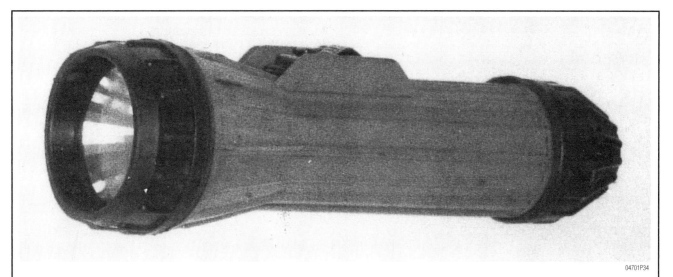

Fig. 15 A flashlight with a fresh set of batteries is handy when repairs are needed at night. It can also double as a signaling device

Courtesy Marine Examinations

One of the roles of the Coast Guard Auxiliary is to promote recreational boating safety. This is why they conduct thousands of Courtesy Marine Examinations each year. The auxiliarists who do these examinations are well trained and knowledgeable in the field.

These examinations are free and done only at the consent of boat owners. To pass the examination, a vessel must satisfy federal equipment requirements and certain additional requirements of the coast guard auxiliary. If your vessel does not pass the Courtesy Marine Examination, no report of the failure is made. Instead, you will be told what you need to correct the deficiencies. The examiner will return at your convenience to redo the examination.

If your vessel qualifies, you will be awarded a safety decal. The decal does not carry any special privileges, it simply attests to your interest in safe boating.

SAFETY IN SERVICE

It is virtually impossible to anticipate all of the hazards involved with maintenance and service, but care and common sense will prevent most accidents.

The rules of safety for mechanics range from "don't smoke around gasoline," to "use the proper tool(s) for the job." The trick to avoiding injuries is to develop safe work habits and to take every possible precaution. Whenever you are working on your boat, pay attention to what you are doing. The more you pay attention to details and what is going on around you, the less likely you will be to hurt yourself or damage your boat.

Do's

- Do keep a fire extinguisher and first aid kit handy.
- Do wear safety glasses or goggles when cutting, drilling, grinding or prying, even if you have 20–20 vision. If you wear glasses for the sake of vision, wear safety goggles over your regular glasses.
- Do shield your eyes whenever you work around the battery. Batteries contain sulfuric acid. In case of contact with the eyes or skin, flush the area with water or a mixture of water and baking soda, then seek immediate medical attention.
- Do use adequate ventilation when working with any chemicals or hazardous materials.
- Do disconnect the negative battery cable when working on the electrical system. The secondary ignition system contains EXTREMELY HIGH VOLTAGE. In some cases it can even exceed 50,000 volts.
- Do follow manufacturer's directions whenever working with potentially hazardous materials. Most chemicals and fluids are poisonous if taken internally.

1-12 GENERAL INFORMATION AND BOATING SAFETY

- Do properly maintain your tools. Loose hammerheads, mushroomed punches and chisels, frayed or poorly grounded electrical cords, excessively worn screwdrivers, spread wrenches (open end), cracked sockets, or slipping ratchets can cause accidents.
- Likewise, keep your tools clean; a greasy wrench can slip off a bolt head, ruining the bolt and often harming your knuckles in the process.
- Do use the proper size and type of tool for the job at hand. Do select a wrench or socket that fits the nut or bolt. The wrench or socket should sit straight, not cocked.
- Do, when possible, pull on a wrench handle rather than push on it, and adjust your stance to prevent a fall.
- Do be sure that adjustable wrenches are tightly closed on the nut or bolt and pulled so that the force is on the side of the fixed jaw. Better yet, avoid the use of an adjustable if you have a fixed wrench that will fit.
- Do strike squarely with a hammer; avoid glancing blows. But, we REALLY hope you won't be using a hammer much in basic maintenance.
- Do use common sense whenever you work on your boat or motor. If a situation arises that doesn't seem right, sit back and have a second look. It may save an embarrassing moment or potential damage to your beloved boat.

Don'ts

- Don't run the engine in an enclosed area or anywhere else without proper ventilation—EVER! Carbon monoxide is poisonous; it takes a long time to leave the human body and you can build up a deadly supply of it in your system by simply breathing in a little every day. You may not realize you are slowly poisoning yourself.
- Don't work around moving parts while wearing loose clothing. Short sleeves are much safer than long, loose sleeves. Hard-toed shoes with neoprene soles protect your toes and give a better grip on slippery surfaces. Jewelry, watches, large belt buckles, or body adornment of any kind is not safe working around any vehicle. Long hair should be tied back under a hat.
- Don't use pockets for toolboxes. A fall or bump can drive a screwdriver deep into your body. Even a rag hanging from your back pocket can wrap around a spinning shaft.
- Don't smoke when working around gasoline, cleaning solvent or other flammable material.
- Don't smoke when working around the battery. When the battery is being charged, it gives off explosive hydrogen gas. Actually, you shouldn't smoke anyway. Save the cigarette money and put it into your boat!
- Don't use gasoline to wash your hands; there are excellent soaps available. Gasoline contains dangerous additives that can enter the body through a cut or through your pores. Gasoline also removes all the natural oils from the skin so that bone dry hands will suck up oil and grease.
- Don't use screwdrivers for anything other than driving screws! A screwdriver used as a prying tool can snap when you least expect it, causing injuries. At the very least, you'll ruin a good screwdriver.

TOOLS AND EQUIPMENT 2-2
SAFETY TOOLS 2-2
    WORK GLOVES 2-2
    EYE & EAR PROTECTION 2-2
    WORK CLOTHES 2-3
CHEMICALS 2-3
    LUBRICANTS & PENETRANTS 2-3
    SEALANTS 2-3
    CLEANERS 2-3
TOOLS 2-4
HAND TOOLS 2-5
    SOCKET SETS 2-5
    WRENCHES 2-6
    PLIERS 2-8
    SCREWDRIVERS 2-8
    HAMMERS 2-8
    OTHER COMMON TOOLS 2-8
    SPECIAL TOOLS 2-8
    ELECTRONIC TOOLS 2-9
    GAUGES 2-10
MEASURING TOOLS 2-10
    MICROMETERS & CALIPERS 2-10
    DIAL INDICATORS 2-10
    TELESCOPING GAUGES 2-11
    DEPTH GAUGES 2-11
FASTENERS, MEASUREMENTS AND CONVERSIONS 2-11
BOLTS, NUTS AND OTHER THREADED RETAINERS 2-11
TORQUE 2-13
STANDARD AND METRIC MEASUREMENTS 2-13
SPECIFICATIONS CHART
    CONVERSION FACTORS 2-14

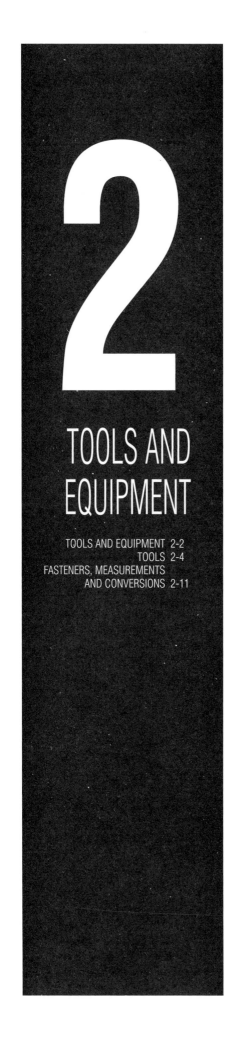

2

TOOLS AND EQUIPMENT

TOOLS AND EQUIPMENT 2-2
TOOLS 2-4
FASTENERS, MEASUREMENTS AND CONVERSIONS 2-11

2-2 TOOLS AND EQUIPMENT

TOOLS AND EQUIPMENT

Safety Tools

WORK GLOVES

▶ See Figures 1 and 2

Unless you think scars on your hands are cool, enjoy pain and like wearing bandages, get a good pair of work gloves. Canvas or leather are the best. And yes, we realize that there are some jobs involving small parts that can't be done while wearing work gloves. These jobs are not the ones usually associated with hand injuries.

A good pair of rubber gloves (such as those usually associated with dish washing) or vinyl gloves is also a great idea. There are some liquids such as solvents and penetrants that don't belong on your skin. Avoid burns and rashes. Wear these gloves.

And lastly, an option. If you're tired of being greasy and dirty all the time, go to the drug store and buy a box of disposable latex gloves like medical professionals wear. You can handle greasy parts, perform small tasks, wash parts, etc. all without getting dirty! These gloves take a surprising amount of abuse without tearing and aren't expensive. Note however, that it has been reported that some people are allergic to the latex or the powder used inside some gloves, so pay attention to what you buy.

EYE & EAR PROTECTION

▶ See Figures 3 and 4

Don't begin any job without a good pair of work goggles or impact resistant glasses! When doing any kind of work, it's all too easy to avoid eye injury through this simple precaution. And don't just buy eye protection and leave it on

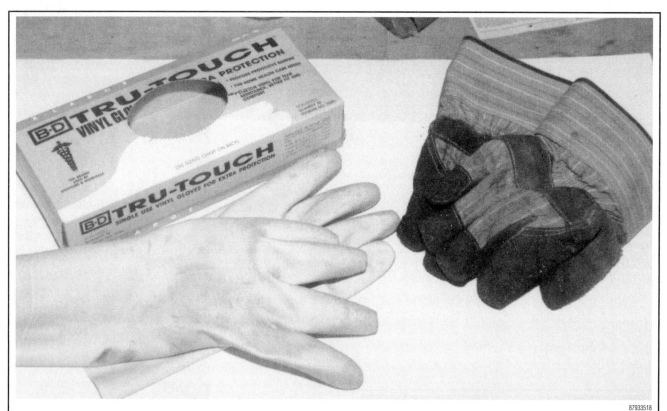

Fig. 1 Three different types of work gloves. The box contains latex gloves

Fig. 2 Latex gloves come in handy when you are doing those messy jobs, like handling an oil soaked filter

Fig. 3 Don't begin any job without a good pair of work goggles or impact resistant glasses. Also good noise reducing earmuffs are cheap insurance to protect your hearing

Fig. 4 Things have a habit of breaking, chipping, splashing, spraying, splintering and flying around. And, for some reason, your eye is always in the way

TOOLS AND EQUIPMENT 2-3

the shelf. Wear it all the time! Things have a habit of breaking, chipping, splashing, spraying, splintering and flying around. And, for some reason, your eye is always in the way!

If you wear vision correcting glasses as a matter of routine, get a pair made with polycarbonate lenses. These lenses are impact resistant and are available at any optometrist.

Often overlooked is hearing protection. Power equipment is noisy! Loud noises damage your ears. It's as simple as that! The simplest and cheapest form of ear protection is a pair of noise-reducing ear plugs. Cheap insurance for your ears. And, they may even come with their own, cute little carrying case.

More substantial, more protection and more money is a good pair of noise reducing earmuffs. They protect from all but the loudest sounds. Hopefully those are sounds that you'll never encounter since they're usually associated with disasters.

WORK CLOTHES

Everyone has "work clothes." Usually these consist of old jeans and a shirt that has seen better days. That's fine. In addition, a denim work apron is a nice accessory. It's rugged, can hold some spare bolts, and you don't feel bad wiping your hands or tools on it. That's what it's for.

When working in cold weather, a one-piece, thermal work outfit is invaluable. Most are rated to below zero (Fahrenheit) temperatures and are ruggedly constructed. Just look at what the marine mechanics are wearing and that should give you a clue as to what type of clothing is good.

Chemicals

There is a whole range of chemicals that you'll find handy for maintenance work. The most common types are, lubricants, penetrants and sealers. Keep these handy onboard. There are also many chemicals that are used for detailing or cleaning.

When a particular chemical is not being used, keep it capped, upright and in a safe place. These substances may be flammable, may be irritants or might even be caustic and should always be stored properly, used properly and handled with care. Always read and follow all label directions and be sure to wear hand and eye protection!

LUBRICANTS & PENETRANTS

▶ See Figure 5

Anti-seize is used to coat certain fasteners prior to installation. This can be especially helpful when two dissimilar metals are in contact (to help prevent corrosion that might lock the fastener in place). This is a good practice on a lot of different fasteners, BUT, NOT on any fastener which might vibrate loose causing a problem. If anti-seize is used on a fastener, it should be checked periodically for proper tightness.

Fig. 5 Antiseize, penetrating oil, lithium grease, electronic cleaner and silicone spray. These products have hundreds of uses and should be a part of your chemical tool collection

Lithium grease, chassis lube, silicone grease or a synthetic brake caliper grease can all be used pretty much interchangeably. All can be used for coating rust-prone fasteners and for facilitating the assembly of parts that are a tight fit. Silicone and synthetic greases are the most versatile.

➡**Silicone dielectric grease is a non-conductor that is often used to coat the terminals of wiring connectors before fastening them. It may sound odd to coat metal portions of a terminal with something that won't conduct electricity, but here is it how it works. When the connector is fastened the metal-to-metal contact between the terminals will displace the grease (allowing the circuit to be completed). The grease that is displaced will then coat the non-contacted surface and the cavity around the terminals, SEALING them from atmospheric moisture that could cause corrosion.**

Silicone spray is a good lubricant for hard-to-reach places and parts that shouldn't be smeared with grease.

Penetrating oil may turn out to be one of your best friends when taking something apart that has corroded fasteners. Not only can they make a job easier, they can really help to avoid broken and stripped fasteners. The most familiar penetrating oils are Liquid Wrench® and WD-40®. A newer penetrant, PB Blaster® also works well. These products have hundreds of uses. For your purposes, they are vital!

Before disassembling any part (especially on an exhaust system), check the fasteners. If any appear rusted, soak them thoroughly with the penetrant and let them stand while you do something else (for particularly rusted or frozen parts you may need to soak them a few days in advance). This simple act can save you hours of tedious work trying to extract a broken bolt or stud.

SEALANTS

▶ See Figures 6 and 7

Sealants are an indispensable part for certain tasks, especially if you are trying to avoid leaks. The purpose of sealants is to establish a leak-proof bond between or around assembled parts. Most sealers are used in conjunction with gaskets, but some are used instead of conventional gasket material.

The most common sealers are the non-hardening types such as Permatex®No.2 or its equivalents. These sealers are applied to the mating surfaces of each part to be joined, then a gasket is put in place and the parts are assembled.

➡**A sometimes overlooked use for sealants like RTV is on the threads of vibration prone fasteners.**

One very helpful type of non-hardening sealer is the "high tack" type. This type is a very sticky material that holds the gasket in place while the parts are being assembled. This stuff is really a good idea when you don't have enough hands or fingers to keep everything where it should be.

The stand-alone sealers are the Room Temperature Vulcanizing (RTV) silicone gasket makers. On some engines, this material is used instead of a gasket. In those instances, a gasket may not be available or, because of the shape of the mating surfaces, a gasket shouldn't be used. This stuff, when used in conjunction with a conventional gasket, produces the surest bonds.

RTV does have its limitations though. When using this material, you will have a time limit. It starts to set-up within 15 minutes or so, so you have to assemble the parts without delay. In addition, when squeezing the material out of the tube, don't drop any excess sealer into the engine. The stuff will form and set and travel around the oil gallery, possibly plugging up a passage. Also, most types are not fuel-proof. Check the tube for all cautions.

CLEANERS

▶ See Figures 8 and 9

There are two types of cleaners on the market today: parts cleaners and hand cleaners. The parts cleaners are for the parts; the hand cleaners are for you. They are not interchangeable.

There are many good, non-flammable, biodegradable parts cleaners on the market. These cleaning agents are safe for you, the parts and the environment. Therefore, there is no reason to use flammable, caustic or toxic substances to clean your parts or tools.

As far as hand cleaners go, the waterless types are the best. They have

2-4 TOOLS AND EQUIPMENT

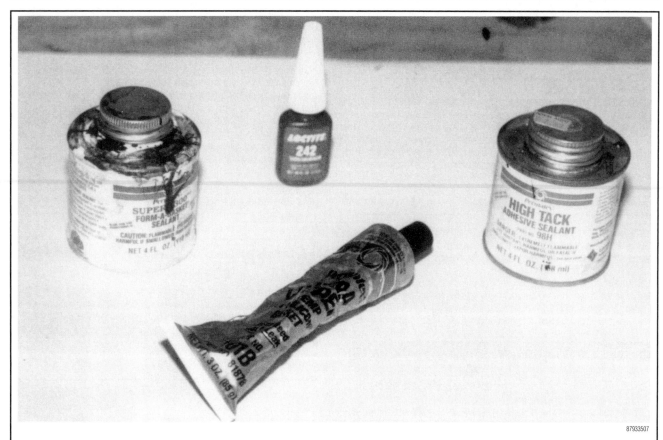

Fig. 6 Sealants are essential for preventing leaks

Fig. 7 On some engines, RTV is used instead of gasket material to seal components

Fig. 8 The new citrus hand cleaners not only work well, but they smell pretty good too. Choose one with pumice for added cleaning power

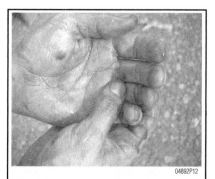

Fig. 9 The use of hand lotion seals your hands and keeps dirt and grease from sticking to your skin

always been efficient at cleaning, but leave a pretty smelly odor. Recently though, just about all of them have eliminated the odor and added stuff that actually smells good. Make sure that you pick one that contains lanolin or some other moisture-replenishing additive. Cleaners not only remove grease and oil but also skin oil.

➡ Most women will tell you to use a hand lotion when you're all cleaned up. It's okay. Real men DO use hand lotion! Believe it or not, using hand lotion before your hands are dirty will actually make them easier to clean when you're finished with a dirty job. Lotion seals your hands, and keeps dirt and grease from sticking to your skin.

TOOLS

▶ See Figure 10

Tools; this subject could fill a completely separate manual. The first thing you will need to ask yourself, is just how involved do you plan to get. If you are serious about your maintenance you will want to gather a quality set of tools to make the job easier, and more enjoyable. BESIDES, TOOLS ARE FUN!!!

Almost every do-it-yourselfer loves to accumulate tools. Though most find a way to perform jobs with only a few common tools, they tend to buy more over time, as money allows. So gathering the tools necessary for maintenance does not have to be an expensive, overnight proposition.

When buying tools, the saying "You get what you pay for . . ." is absolutely true! Don't go cheap! Any hand tool that you buy should be drop forged and/or chrome vanadium. These two qualities tell you that the tool is strong enough for the job. With any tool, go with a name that you've heard of before, or, that is

TOOLS AND EQUIPMENT

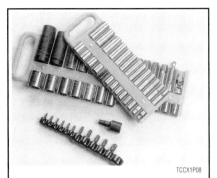

Fig. 10 Socket holders, especially the magnetic type, are handy items to keep tools in order

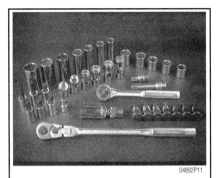

Fig. 11 A ⅜ inch socket set is probably the most versatile tool in any mechanic's tool box

Fig. 12 A swivel (U-joint) adapter (left), a ¼inch-to-⅜ inch adapter (center) and a ⅜ 8inch-to-¼inch adapter (right)

recommended buy your local professional retailer. Let's go over a list of tools that you'll need.

Most of the world uses the metric system. However, some American-built engines and aftermarket accessories use standard fasteners. So, accumulate your tools accordingly. Any good DIYer should have a decent set of both U.S. and metric measure tools.

➡ Don't be confused by terminology. Most advertising refers to "SAE and metric", or "standard and metric." Both are misnomers. The Society of Automotive Engineers (SAE) did not invent the English system of measurement; the English did. The SAE likes metrics just fine. Both English (U.S.) and metric measurements are SAE approved. Also, the current "standard" measurement IS metric. So, if it's not metric, it's U.S. measurement.

Hand Tools

SOCKET SETS

♦ See Figures 11 thru 17

Socket sets are the most basic hand tools necessary for repair and maintenance work. For our purposes, socket sets come in three drive sizes: ¼ inch, ⅜ inch and ½ inch. Drive size refers to the size of the drive lug on the ratchet, breaker bar or speed handle.

A ⅜ inch set is probably the most versatile set in any mechanic's toolbox. It allows you to get into tight places that the larger drive ratchets can't and gives you a range of larger sockets that are still strong enough for heavy duty work. The socket set that you'll need should range in sizes from ⅜ inch through 1 inch for standard fasteners, and a 6mm through 19mm for metric fasteners.

You'll need a good ½ inch set since this size drive lug assures that you won't break a ratchet or socket on large or heavy fasteners. Also, torque wrenches with a torque scale high enough for larger fasteners are usually ½ inch drive.

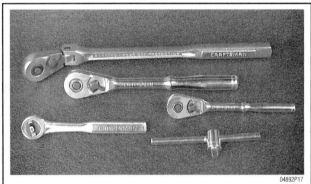

Fig. 13 Ratchets come in all sizes and configurations from rigid to swivel-headed

Fig. 14 Standard length sockets (top) are good for just about all jobs. However, some bolts may require deep sockets (bottom)

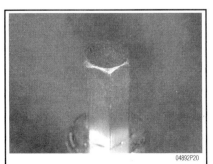

Fig. 15 Hex-head fasteners retain many components on modern powerheads. These fasteners require a socket with a hex shaped driver

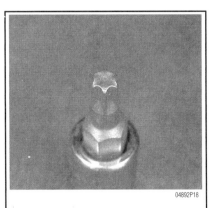

Fig. 16 Torx® drivers . . .

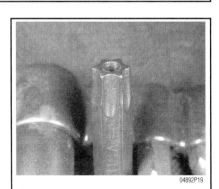

Fig. 17 . . . and tamper resistant drivers are required to remove special fasteners installed by the manufacturers

2-6 TOOLS AND EQUIPMENT

¼ inch drive sets can be very handy in tight places. Though they usually duplicate functions of the ⅜ inch set, ¼ inch drive sets are easier to use for smaller bolts and nuts.

As for the sockets themselves, they come in standard and deep lengths as well as 6 or 12 point. 6 and 12 points refers to how many sides are in the socket itself. Each has advantages. The 6 point socket is stronger and less prone to slipping which would strip a bolt head or nut. 12 point sockets are more common, usually less expensive and can operate better in tight places where the ratchet handle can't swing far.

Standard length sockets are good for just about all jobs, however, some stud-head bolts, hard-to-reach bolts, nuts on long studs, etc., require the deep sockets.

Most manufacturers use recessed hex-head fasteners to retain many of the engine parts. These fasteners require a socket with a hex shaped driver or a large sturdy hex key. To help prevent torn knuckles, we would recommend that you stick to the sockets on any tight fastener and leave the hex keys for lighter applications. Hex driver sockets are available individually or in sets just like conventional sockets.

More and more, manufacturers are using Torx® head fasteners, which were once known as tamper resistant fasteners (because many people did not have tools with the necessary odd driver shape). They are still used where the manufacturer would prefer only knowledgeable mechanics or advanced Do-It-Yourselfers (DIYers) to work.

Torque Wrenches

▶ See Figure 18

In most applications, a torque wrench can be used to assure proper installation of a fastener. Torque wrenches come in various designs and most stores will carry a variety to suit your needs. A torque wrench should be used any time you have a specific torque value for a fastener. Keep in mind that because there is no worldwide standardization of fasteners, the charts at the end of this section are a general guideline and should be used with caution. If you are using the right tool for the job, you should not have to strain to tighten a fastener.

BEAM TYPE

▶ See Figures 19 and 20

The beam type torque wrench is one of the most popular styles in use. If used properly, it can be the most accurate also. It consists of a pointer attached to the head that runs the length of the flexible beam (shaft) to a scale located near the handle. As the wrench is pulled, the beam bends and the pointer indicates the torque using the scale.

CLICK (BREAKAWAY) TYPE

▶ See Figures 21 and 22

Another popular torque wrench design is the click type. The clicking mechanism makes achieving the proper torque easy and most use ratcheting head for ease of bolt installation. To use the click type wrench you pre-adjust it to a torque setting. Once the torque is reached, the wrench has a reflex signaling feature that causes a momentary breakaway of the torque wrench body, sending an impulse to the operator's hand.

Breaker Bars

▶ See Figure 23

Breaker bars are long handles with a drive lug. Their main purpose is to provide extra turning force when breaking loose tight bolts or nuts. They come in all drive sizes and lengths. Always take extra precautions and use proper technique when using a breaker bar.

WRENCHES

▶ See Figures 24, 25, 26, 27 and 28

Basically, there are 3 kinds of fixed wrenches: open end, box end, and combination.

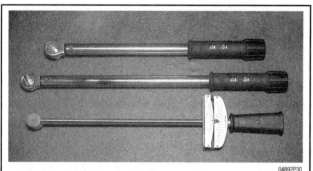

Fig. 18 Three types of torque wrenches. Top to bottom: a ⅜ inch drive beam type that reads in inch lbs., a ½ inch drive clicker type and a ½ inch drive beam type

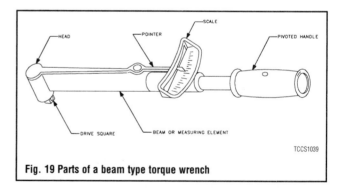

Fig. 19 Parts of a beam type torque wrench

Fig. 20 A beam type torque wrench consists of a pointer attached to the head that runs the length of the flexible beam (shaft) to a scale located near the handle

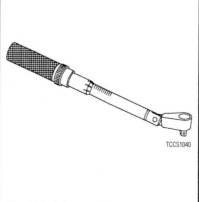

Fig. 21 A click type or breakaway torque wrench—note this one has a pivoting head

Fig. 22 Setting the proper torque on a click type torque wrench involves turning the handle until the proper torque specification appears on the dial

TOOLS AND EQUIPMENT

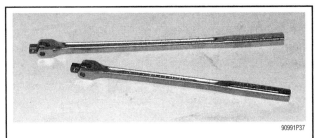
Fig. 23 Breaker bars are great for loosening large or stuck fasteners

Open end wrenches have 2-jawed openings at each end of the wrench. These wrenches are able to fit onto just about any nut or bolt. They are extremely versatile but have one major drawback. They can slip on a worn or rounded bolt head or nut, causing bleeding knuckles and a useless fastener.

Box-end wrenches have a 360° circular jaw at each end of the wrench. They come in both 6 and 12 point versions just like sockets and each type has the same advantages and disadvantages as sockets.

Combination wrenches have the best of both. They have a 2-jawed open end and a box end. These wrenches are probably the most versatile.

As for sizes, you'll probably need a range similar to that of the sockets, about ¼ inch through 1 inch for standard fasteners, or 6mm through 19mm for metric fasteners. As for numbers, you'll need 2 of each size, since, in many instances,

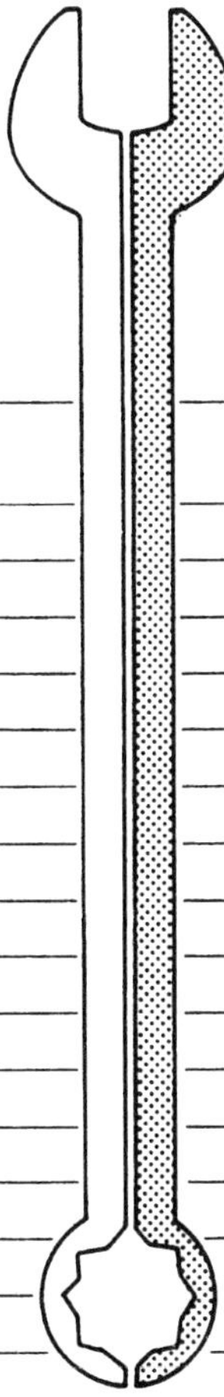

INCHES	DECIMAL	DECIMAL	MILLIMETERS
1/8"	.125	.118	3mm
3/16"	.187	.157	4mm
1/4"	.250	.236	6mm
5/16"	.312	.354	9mm
3/8"	.375	.394	10mm
7/16"	.437	.472	12mm
1/2"	.500	.512	13mm
9/16"	.562	.590	15mm
5/8"	.625	.630	16mm
11/16"	.687	.709	18mm
3/4"	.750	.748	19mm
13/16"	.812	.787	20mm
7/8"	.875	.866	22mm
15/16"	.937	.945	24mm
1"	1.00	.984	25mm

Fig. 24 Comparison of U.S. measure and metric wrench sizes

Fig. 25 Always use a backup wrench to prevent rounding flare nut fittings

Fig. 26 Note how the flare wrench sides are extended to grip the fitting tighter and prevent rounding

Fig. 27 Several types and sizes of adjustable wrenches

2-8 TOOLS AND EQUIPMENT

Fig. 28 Occasionally you will find a nut which requires a particularly large or particularly small wrench. Rest assured that the proper wrench to fit is available at your local tool store

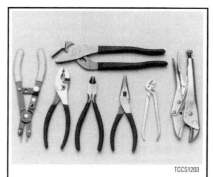

Fig. 29 Pliers and cutters come in many shapes and sizes. You should have an assortment on hand

Fig. 30 Three types of hammers. Top to bottom: ball peen, rubber dead-blow, and plastic

one wrench holds the nut while the other turns the bolt. On most fasteners, the nut and bolt are the same size so having two wrenches of the same size comes in handy.

➡ **Although you will typically just need the sizes we specified, there are some exceptions. Occasionally you will find a nut that is larger. For these, you will need to buy ONE expensive wrench or a very large adjustable. Or you can always just convince the spouse that we are talking about safety here and buy a whole (read expensive) large wrench set.**

One extremely valuable type of wrench is the adjustable wrench. An adjustable wrench has a fixed upper jaw and a moveable lower jaw. The lower jaw is moved by turning a threaded drum. The advantage of an adjustable wrench is its ability to be adjusted to just about any size fastener.

The main drawback of an adjustable wrench is the lower jaw's tendency to move slightly under heavy pressure. This can cause the wrench to slip if it is not facing the right way. Pulling on an adjustable wrench in the proper direction will cause the jaws to lock in place. Adjustable wrenches come in a large range of sizes, measured by the wrench length.

PLIERS

♦ See Figure 29

Pliers are simply mechanical fingers. They are, more than anything, an extension of your hand. At least 3 pair of pliers are an absolute necessity—standard, needle nose and channel lock.

In addition to standard pliers there are the slip-joint, multi-position pliers such as ChannelLock® pliers and locking pliers, such as Vise Grips®.

Slip joint pliers are extremely valuable in grasping oddly sized parts and fasteners. Just make sure that you don't use them instead of a wrench too often since they can easily round off a bolt head or nut.

Locking pliers are usually used for gripping bolts or studs that can't be removed conventionally. You can get locking pliers in square jawed, needle-nosed and pipe-jawed. Locking pliers can rank right up behind duct tape as the handy-man's best friend.

SCREWDRIVERS

You can't have too many screwdrivers. They come in 2 basic flavors, either standard or Phillips. Standard blades come in various sizes and thicknesses for all types of slotted fasteners. Phillips screwdrivers come in sizes with number designations from 1 on up, with the lower number designating the smaller size. Screwdrivers can be purchased separately or in sets.

HAMMERS

♦ See Figure 30

You always need a hammer for just about any kind of work. You need a ball-peen hammer for most metal work when using drivers and other like tools. A plastic hammer comes in handy for hitting things safely. A soft-faced dead-blow hammer is used for hitting things safely and hard. Hammers are also VERY useful with non air-powered impact drivers.

OTHER COMMON TOOLS

There are a lot of other tools that every DIYer will eventually need (though not all for basic maintenance). They include:
- Funnels (for adding fluid)
- Chisels
- Punches
- Files
- Hacksaw
- Portable Bench Vise
- Tap and Die Set
- Flashlight
- Magnetic Bolt Retriever
- Gasket scraper
- Putty Knife
- Screw/Bolt Extractors
- Prybar

Hacksaws have just one use—cutting things off. You may wonder why you'd need one for something as simple as maintenance, but you never know. Among other things, guide studs to ease parts installation can be made from old bolts with their heads cut off.

A tap and die set might be something you've never needed, but you will eventually. It's a good rule, when everything is apart, to clean-up all threads, on bolts, screws and threaded holes. Also, you'll likely run across a situation in which stripped threads will be encountered. The tap and die set will handle that for you.

Gasket scrapers are just what you'd think, tools made for scraping old gasket material off of parts. You don't absolutely need one. Old gasket material can be removed with a putty knife or single edge razor blade. However, putty knives may not be sharp enough for some really stubborn gaskets and razor blades have a knack of breaking just when you don't want them to, inevitably slicing the nearest body part! As the old saying goes, "always use the proper tool for the job". If you're going to use a razor to scrape a gasket, be sure to always use a blade holder.

Putty knives really do have a use in a repair shop. Just because you remove all the bolts from a component sealed with a gasket doesn't mean it's going to come off. Most of the time, the gasket and sealer will hold it tightly. Lightly driving a putty knife at various points between the two parts will break the seal without damage to the parts.

A small—8-10 inches (20-25 centimeters) long—prybar is extremely useful for removing stuck parts.

➡ **Never use a screwdriver as a prybar! Screwdrivers are not meant for prying. Screwdrivers, used for prying, can break, sending the broken shaft flying!**

Screw/bolt extractors are used for removing broken bolts or studs that have broke off flush with the surface of the part.

SPECIAL TOOLS

Almost every marine engine around today requires at least one special tool to perform a certain task. In most cases, these tools are specially designed to overcome some unique problem or to fit on some oddly sized component.

TOOLS AND EQUIPMENT 2-9

When manufacturers go through the trouble of making a special tool, it is usually necessary to use it to assure that the job will be done right. A special tool might be designed to make a job easier, or it might be used to keep you from damaging or breaking a part.

Don't worry, MOST basic maintenance procedures can either be performed without any special tools OR, because the tools must be used for such basic things, they are commonly available for a reasonable price. It is usually just the low production, highly specialized tools (like a super thin 7-point star-shaped socket capable of 150 ft. lbs. (203 Nm) of torque that is used only on the crankshaft nut of the limited production what-dya-callit engine) that tend to be outrageously expensive and hard to find. Luckily, you will probably never need such a tool.

Special tools can be as inexpensive and simple as an adjustable strap wrench or as complicated as an ignition tester. A few common specialty tools are listed here, but check with your dealer or with other boaters for help in determining if there are any special tools for YOUR particular engine. There is an added advantage in seeking advice from others, chances are they may have already found the special tool you will need, and know how to get it cheaper.

ELECTRONIC TOOLS

Battery Testers

The best way to test a non-sealed battery is using a hydrometer to check the specific gravity of the acid. Luckily, these are usually inexpensive and are available at most parts stores. Just be careful because the larger testers are usually designed for larger batteries and may require more acid than you will be able to draw from the battery cell. Smaller testers (usually a short, squeeze bulb type) will require less acid and should work on most batteries.

Electronic testers are available and are often necessary to tell if a sealed battery is usable. Luckily, many parts stores have them on hand and are willing to test your battery for you.

Battery Chargers

♦ See Figure 31

If you are a weekend boater and take your boat out every week, then you will most likely want to buy a battery charger to keep your battery fresh. There are many types available, from low amperage trickle chargers to electronically controlled battery maintenance tools that monitor the battery voltage to prevent over or undercharging. This last type is especially useful if you store your boat for any length of time (such as during the severe winter months found in many Northern climates).

Even if you use your boat on a regular basis, you will eventually need a battery charger. Remember that most batteries are shipped dry and in a partial charged state. Before a new battery can be put into service it must be filled and properly charged. Failure to properly charge a battery (which was shipped dry) before it is put into service will prevent it from ever reaching a fully charged state.

Digital Volt/Ohm Meter (DVOM)

♦ See Figure 32

Multimeters are an extremely useful tool for troubleshooting electrical problems. They can be purchased in either analog or digital form and have a price range to suit any budget. A multimeter is a voltmeter, ammeter and ohmmeter (along with other features) combined into one instrument. It is often used when testing solid state circuits because of its high input impedance (usually 10 megaohms or more). A brief description of the multimeter main test functions follows:

• Voltmeter—the voltmeter is used to measure voltage at any point in a circuit, or to measure the voltage drop across any part of a circuit. Voltmeters usually have various scales and a selector switch to allow the reading of different voltage ranges. The voltmeter has a positive and a negative lead. To avoid damage to the meter, always connect the negative lead to the negative (-) side of the circuit (to ground or nearest the groundside of the circuit) and connect the positive lead to the positive (+) side of the circuit (to the power source or the nearest power source). Note that the negative voltmeter lead will always be black and that the positive voltmeter will always be some color other than black (usually red).

• Ohmmeter—the ohmmeter is designed to read resistance (measured in ohms) in a circuit or component. Most ohmmeters will have a selector switch which permits the measurement of different ranges of resistance (usually the selector switch allows the multiplication of the meter reading by 10, 100, 1,000 and 10,000). Some ohmmeters are "auto-ranging" which means the meter itself will determine which scale to use. Since an internal battery powers the meters, the ohmmeter can be used like a self-powered test light. When the ohmmeter is connected, current from the ohmmeter flows through the circuit or component being tested. Since the ohmmeter's internal resistance and voltage are known values, the amount of current flow through the meter depends on the resistance of the circuit or component being tested. The ohmmeter can also be used to perform a continuity test for suspected open circuits. In using the meter for making continuity checks, do not be concerned with the actual resistance readings. Zero resistance, or any ohm reading, indicates continuity in the circuit. Infinite resistance indicates an opening in the circuit. A high resistance reading where there should be none indicates a problem in the circuit. Checks for short circuits are made in the same manner as checks for open circuits, except that the circuit must be isolated from both power and normal ground. Infinite resistance indicates no continuity, while zero resistance indicates a dead short.

※※ WARNING

Never use an ohmmeter to check the resistance of a component or wire while there is voltage applied to the circuit.

Fig. 31 The Battery Tender® is more than just a battery charger, when left connected, it keeps your battery fully charged

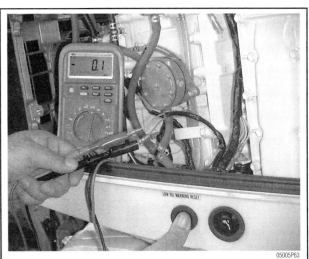

Fig. 32 Multimeters are an extremely useful tool for troubleshooting electrical problems

2-10 TOOLS AND EQUIPMENT

• **Ammeter**—an ammeter measures the amount of current flowing through a circuit in units called amperes or amps. At normal operating voltage, most circuits have a characteristic amount of amperes, called "current draw" which can be measured using an ammeter. By referring to a specified current draw rating, then measuring the amperes and comparing the two values, one can determine what is happening within the circuit to aid in diagnosis. An open circuit, for example, will not allow any current to flow, so the ammeter reading will be zero. A damaged component or circuit will have an increased current draw, so the reading will be high. The ammeter is always connected in series with the circuit being tested. All of the current that normally flows through the circuit must also flow through the ammeter; if there is any other path for the current to follow, the ammeter reading will not be accurate. The ammeter itself has very little resistance to current flow and, therefore, will not affect the circuit, but it will measure current draw only when the circuit is closed and electricity is flowing. Excessive current draw can blow fuses and drain the battery, while a reduced current draw can cause motors to run slowly, lights to dim and other components to not operate properly.

GAUGES

Compression Gauge

▶ See Figure 33

An important element in checking the overall condition of your engine is to check compression. This becomes increasingly more important on outboards with high hours. Compression gauges are available as screw-in types and hold-in types. The screw-in type is slower to use, but eliminates the possibility of a faulty reading due to escaping pressure. A compression reading will uncover many problems that can cause rough running. Normally, these are not the sort of problems that can be cured by a tune-up.

Vacuum Gauge

▶ See Figures 34 and 35

Vacuum gauges are useful for many diagnostic tasks. The gauge consists of a dial gauge calibrated in inches of mercury and a hose which allows it to be attached to various components. Some deluxe gauges are calibrated for both vacuum and pressure. Abnormal vacuum readings can be caused by any number of reasons but usually indicate the engine is not running properly.

Measuring Tools

Eventually, you are going to have to measure something. To do this, you will need at least a few precision tools in addition to the special tools mentioned earlier.

MICROMETERS & CALIPERS

Micrometers and calipers are devices used to make extremely precise measurements. The simple truth is that you really won't have the need for many of these items just for simple maintenance. You will probably want to have at least one precision tool such as an outside caliper to measure rotors or brake pads, but that should be sufficient to most basic maintenance procedures.

Should you decide on becoming more involved in boat engine mechanics, such as repair or rebuilding, then these tools will become very important. The success of any rebuild is dependent, to a great extent on the ability to check the size and fit of components as specified by the manufacturer. These measurements are made in thousandths and ten-thousandths of an inch.

Micrometers

▶ See Figure 36

A micrometer is an instrument made up of a precisely machined spindle that is rotated in a fixed nut, opening and closing the distance between the end of the spindle and a fixed anvil.

Outside micrometers can be used to check the thickness parts such shims or the outside diameter of components like the crankshaft journals. They are also used during many rebuild and repair procedures to measure the diameter of components such as the pistons. The most common type of micrometer reads in 1/1000 of an inch. Micrometers that use a vernier scale can estimate to 1/10 of an inch.

Inside micrometers are used to measure the distance between two parallel surfaces. For example, in powerhead rebuilding work, the inside mike measures cylinder bore wear and taper. Inside mikes are graduated the same way as outside mikes and are read the same way as well.

Remember that an inside mike must be absolutely perpendicular to the work being measured. When you measure with an inside mike, rock the mike gently from side to side and tip it back and forth slightly so that you span the widest part of the bore. Just to be on the safe side, take several readings. It takes a certain amount of experience to work any mike with confidence.

Metric micrometers are read in the same way as inch micrometers, except that the measurements are in millimeters. Each line on the main scale equals 1 mm. Each fifth line is stamped 5, 10, 15, and so on. Each line on the thimble scale equals 0.01 mm. It will take a little practice, but if you can read an inch mike, you can read a metric mike.

Calipers

▶ See Figures 37, 38 and 39

Inside and outside calipers are useful devices to have if you need to measure something quickly and precise measurement is not necessary. Simply take the reading and then hold the calipers on an accurate steel rule.

DIAL INDICATORS

▶ See Figure 40

A dial indicator is a gauge that utilizes a dial face and a needle to register measurements. There is a movable contact arm on the dial indicator. When the arms moves, the needle rotates on the dial. Dial indicators are calibrated to show readings in thousandths of an inch and typically, are used to measure end-play and runout on various parts.

Dial indicators are quite easy to use, although they are relatively expensive. A

Fig. 33 Cylinder compression test results are extremely valuable indicators of internal engine condition

Fig. 34 Vacuum gauges are useful for many diagnostic tasks including testing of some fuel pumps

Fig. 35 In a pinch, you can also use the vacuum gauge on a hand operated vacuum pump

TOOLS AND EQUIPMENT 2-11

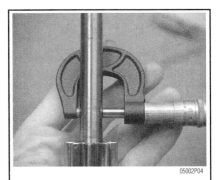

Fig. 36 Outside micrometers can be used to measure the thickness of shims or the outside diameter of a shaft

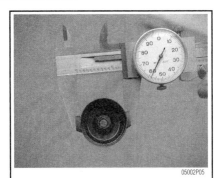

Fig. 37 Calipers, such as this dial caliper, are the fast and easy way to make precise measurements

Fig. 38 Calipers can also be used to measure depth . . .

Fig. 39 . . . and inside diameter measurements, usually to 0.001 inch accuracy

Fig. 40 Here, a dial indicator is used to measure the axial clearance (endplay) of a crankshaft during a powerhead rebuilding procedure

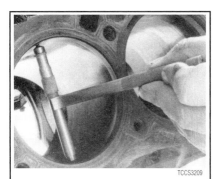

Fig. 41 Telescoping gauges are used during powerhead rebuilding procedures to measure the inside diameter of bores

variety of mounting devices are available so that the indicator can be used in a number of situations. Make certain that the contact arm is always parallel to the movement of the work being measured.

TELESCOPING GAUGES

◢ See Figure 41

A telescope gauge is used during rebuilding procedures (NOT usually basic maintenance) to measure the inside of bores. It can take the place of an inside mike for some of these jobs. Simply insert the gauge in the hole to be measured and lock the plungers after they have contacted the walls. Remove the tool and measure across the plungers with an outside micrometer.

DEPTH GAUGES

◢ See Figure 42

A depth gauge can be inserted into a bore or other small hole to determine exactly how deep it is. One common use for a depth gauge is measuring the distance the piston sits below the deck of the block at top dead center. Some outside calipers contain a built-in depth gauge so buying just one tool can save money.

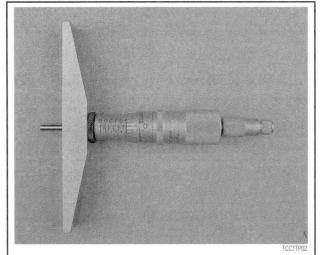

Fig. 42 Depth gauges are used to measure the depth of bore or other small holes

FASTENERS, MEASUREMENTS AND CONVERSIONS

Bolts, Nuts and Other Threaded Retainers

◢ See Figures 43, 44, 45, 46 and 47

Although there are a great variety of fasteners found in the modern boat engine, the most commonly used retainer is the threaded fastener (nuts, bolts, screws, studs, etc). Most threaded retainers may be reused, provided that they are not damaged in use or during the repair.

➡Some retainers (such as stretch bolts or torque prevailing nuts) are designed to deform when tightened or in use and should not be reused.

Whenever possible, we will note any special retainers which should be replaced during a procedure. But you should always inspect the condition of a retainer when it is removed and you should replace any that show signs of damage. Check all threads for rust or corrosion that can increase the torque necessary to achieve the desired clamp load for which that fastener was originally selected. Additionally, be sure that rounding or other damage has not

2-12 TOOLS AND EQUIPMENT

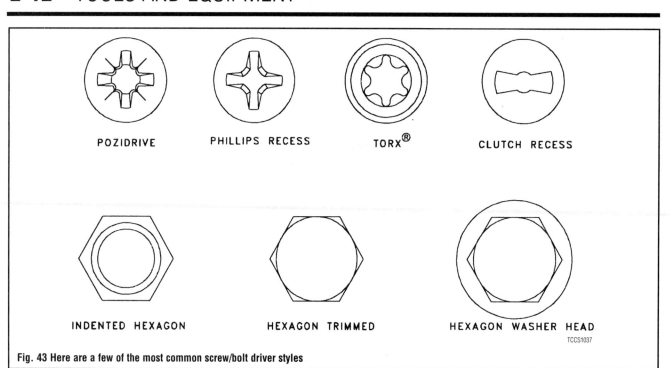

Fig. 43 Here are a few of the most common screw/bolt driver styles

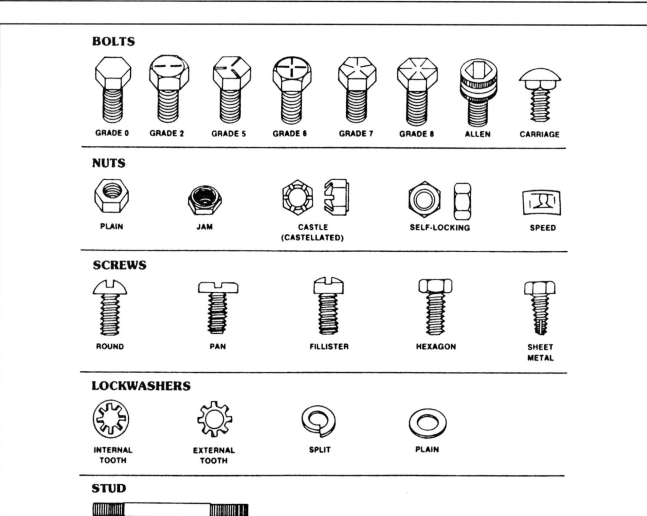

Fig. 44 There are many different types of threaded retainers

TOOLS AND EQUIPMENT 2-13

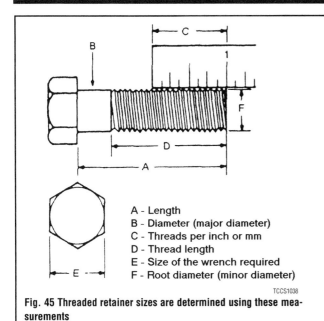

Fig. 45 Threaded retainer sizes are determined using these measurements

A - Length
B - Diameter (major diameter)
C - Threads per inch or mm
D - Thread length
E - Size of the wrench required
F - Root diameter (minor diameter)

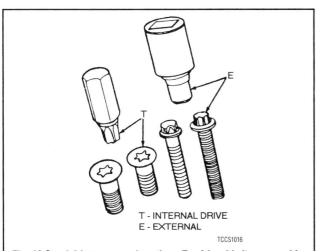

Fig. 46 Special fasteners such as these Torx® head bolts are used by manufacturers to discourage people from working on vehicles without the proper tools (and knowledge)

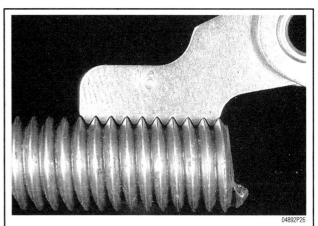

Fig. 47 Thread gauges measure the threads-per-inch and the pitch of a bolt or stud's thread

compromised the driver surface of the fastener. In some cases a driver surface may become only partially rounded, allowing the driver to catch in only one direction. In many of these occurrences, a fastener may be installed and tightened, but the driver would not be able to grip and loosen the fastener again. (This could lead to frustration down the line should that component ever need to be disassembled again).

If you must replace a fastener, whether due to design or damage, you must always be sure to use the proper replacement. In all cases, a retainer of the same design, material and strength should be used. Markings on the heads of most bolts will help determine the proper strength of the fastener. The same material, thread and pitch must be selected to assure proper installation and safe operation of the vehicle afterwards.

Thread gauges are available to help measure a bolt or stud's thread. Most part or hardware stores keep gauges available to help you select the proper size. In a pinch, you can use another nut or bolt for a thread gauge. If the bolt you are replacing is not too badly damaged, you can select a match by finding another bolt that will thread in its place. If you find a nut that threads properly onto the damaged bolt, then use that nut to help select the replacement bolt. If however, the bolt you are replacing is so badly damaged (broken or drilled out) that its threads cannot be used as a gauge, you might start by looking for another bolt (from the same assembly or a similar location) which will thread into the damaged bolt's mounting. If so, the other bolt can be used to select a nut; the nut can then be used to select the replacement bolt.

In all cases, be absolutely sure you have selected the proper replacement. Don't be shy, you can always ask the store clerk for help.

✱✱ WARNING

Be aware that when you find a bolt with damaged threads, you may also find the nut or drilled hole it was threaded into has also been damaged. If this is the case, you may have to drill and tap the hole, replace the nut or otherwise repair the threads. NEVER try to force a replacement bolt to fit into the damaged threads.

Torque

Torque is defined as the measurement of resistance to turning or rotating. It tends to twist a body about an axis of rotation. A common example of this would be tightening a threaded retainer such as a nut, bolt or screw. Measuring torque is one of the most common ways to help assure that a threaded retainer has been properly fastened.

When tightening a threaded fastener, torque is applied in three distinct areas, the head, the bearing surface and the clamp load. About 50 percent of the measured torque is used in overcoming bearing friction. This is the friction between the bearing surface of the bolt head, screw head or nut face and the base material or washer (the surface on which the fastener is rotating). Approximately 40 percent of the applied torque is used in overcoming thread friction. This leaves only about 10 percent of the applied torque to develop a useful clamp load (the force that holds a joint together). This means that friction can account for as much as 90 percent of the applied torque on a fastener.

Standard and Metric Measurements

Specifications are often used to help you determine the condition of various components, or to assist you in their installation. Some of the most common measurements include length (in. or cm/mm), torque (ft. lbs., inch lbs. or Nm) and pressure (psi, in. Hg, kPa or mm Hg).

In some cases, that value may not be conveniently measured with what is available in your toolbox. Luckily, many of the measuring devices which are available today will have two scales so Standard or Metric measurements may easily be taken. If any of the various measuring tools that are available to you do not contain the same scale as listed in your specifications, use the accompanying conversion factors to determine the proper value.

The conversion factor chart is used by taking the given specification and multiplying it by the necessary conversion factor. For instance, looking at the first line, if you have a measurement in inches such as "free-play should be 2 in." but your ruler reads only in millimeters, multiply 2 in. by the conversion factor of 25.4 to get the metric equivalent of 50.8mm. Likewise, if the specification was given only in a Metric measurement, for example in Newton Meters (Nm), then look at the center column first. If the measurement is 100 Nm, multiply it by the conversion factor of 0.738 to get 73.8 ft. lbs.

CONVERSION FACTORS

LENGTH–DISTANCE

Inches (in.)	x 25.4	= Millimeters (mm)	x .0394	= Inches
Feet (ft.)	x .305	= Meters (m)	x 3.281	= Feet
Miles	x 1.609	= Kilometers (km)	x .0621	= Miles

VOLUME

Cubic Inches (in3)	x 16.387	= Cubic Centimeters	x .061	= in3
IMP Pints (IMP pt.)	x .568	= Liters (L)	x 1.76	= IMP pt.
IMP Quarts (IMP qt.)	x 1.137	= Liters (L)	x .88	= IMP qt.
IMP Gallons (IMP gal.)	x 4.546	= Liters (L)	x .22	= IMP gal.
IMP Quarts (IMP qt.)	x 1.201	= US Quarts (US qt.)	x .833	= IMP qt.
IMP Gallons (IMP gal.)	x 1.201	= US Gallons (US gal.)	x .833	= IMP gal.
Fl. Ounces	x 29.573	= Milliliters	x .034	= Ounces
US Pints (US pt.)	x .473	= Liters (L)	x 2.113	= Pints
US Quarts (US qt.)	x .946	= Liters (L)	x 1.057	= Quarts
US Gallons (US gal.)	x 3.785	= Liters (L)	x .264	= Gallons

MASS–WEIGHT

Ounces (oz.)	x 28.35	= Grams (g)	x .035	= Ounces
Pounds (lb.)	x .454	= Kilograms (kg)	x 2.205	= Pounds

PRESSURE

Pounds Per Sq. In. (psi)	x 6.895	= Kilopascals (kPa)	x .145	= psi
Inches of Mercury (Hg)	x .4912	= psi	x 2.036	= Hg
Inches of Mercury (Hg)	x 3.377	= Kilopascals (kPa)	x .2961	= Hg
Inches of Water (H_2O)	x .07355	= Inches of Mercury	x 13.783	= H_2O
Inches of Water (H_2O)	x .03613	= psi	x 27.684	= H_2O
Inches of Water (H_2O)	x .248	= Kilopascals (kPa)	x 4.026	= H_2O

TORQUE

Pounds–Force Inches (in-lb)	x .113	= Newton Meters (N·m)	x 8.85	= in-lb
Pounds–Force Feet (ft-lb)	x 1.356	= Newton Meters (N·m)	x .738	= ft-lb

VELOCITY

Miles Per Hour (MPH)	x 1.609	= Kilometers Per Hour (KPH)	x .621	= MPH

POWER

Horsepower (Hp)	x .745	= Kilowatts	x 1.34	= Horsepower

FUEL CONSUMPTION*

Miles Per Gallon IMP (MPG)	x .354	= Kilometers Per Liter (Km/L)
Kilometers Per Liter (Km/L)	x 2.352	= IMP MPG
Miles Per Gallon US (MPG)	x .425	= Kilometers Per Liter (Km/L)
Kilometers Per Liter (Km/L)	x 2.352	= US MPG

*It is common to covert from miles per gallon (mpg) to liters/100 kilometers (1/100 km), where mpg (IMP) x 1/100 km = 282 and mpg (US) x 1/100 km = 235.

TEMPERATURE

Degree Fahrenheit (°F)	= (°C x 1.8) + 32
Degree Celsius (°C)	= (°F – 32) x .56

ENGINE MAINTENANCE 3-2
SERIAL NUMBER IDENTIFICATION 3-2
2-STROKE OIL 3-2
    OIL RECOMMENDATIONS 3-2
    PRE-MIXING FUEL 3-2
LOWER UNIT 3-3
    OIL RECOMMENDATIONS 3-3
    DRAINING LOWER UNIT 3-4
    FILLING LOWER UNIT 3-4
FUEL FILTER 3-5
    RELIEVING FUEL SYSTEM
      PRESSURE 3-6
    REMOVAL & INSTALLATION 3-6
FUEL/WATER SEPARATOR 3-7
    DRAINING 3-7
    SERVICE 3-7
LUBRICATION POINTS 3-7
    INSPECTION & LUBRICATION 3-7
PROPELLER 3-8
ANODES (ZINCS) 3-9
    SERVICING 3-9
    INSPECTION 3-9
BOAT MAINTENANCE 3-10
BATTERIES 3-10
    MAINTENANCE 3-10
    CLEANING 3-10
    TESTING 3-11
    STORAGE 3-11
FIBERGLASS HULL 3-12
TUNE-UP 3-12
INTRODUCTION 3-12
COMPRESSION TEST 3-13
    PRIMARY COMPRESSION 3-13
    SECONDARY COMPRESSION
      TEST 3-13
SPARK PLUGS 3-13
    SPARK PLUG HEAT RANGE 3-14
    SPARK PLUG SERVICE 3-14
    REMOVAL & INSTALLATION 3-14
    READING SPARK PLUGS 3-15
    INSPECTION & GAPPING 3-16
SPARK PLUG WIRES 3-16
    TESTING 3-16
    REMOVAL & INSTALLATION 3-17
IGNITION SYSTEM 3-17
SYNCHRONIZATION AND TIMING 3-17
ENGINES WITH FACE PLATE MOUNTED
  CONTROLS 3-18
    IDLE SPEED ADJUSTMENT 3-18
ENGINES WITH A SINGLE CARBURETOR
  AND TILLER HANDLE 3-18
    IDLE SPEED AND MIXTURE
      ADJUSTMENT 3-18
1984–89 ENGINES WITH A SINGLE
  CARBURETOR AND REMOTE
  CONTROL 3-19
    CHOKE ROD ADJUSTMENT 3-19
    THROTTLE ROD ADJUSTMENT 3-19
    IDLE SPEED & MIXTURE
      ADJUSTMENT 3-19
    IGNITION TIMING ADJUSTMENT 3-20
1990–99 ENGINES WITH A SINGLE
  CARBURETOR AND REMOTE
  CONTROL 3-20
    CHOKE ROD ADJUSTMENT 3-20
    THROTTLE ROD ADJUSTMENT 3-20
    IDLE SPEED ADJUSTMENT 3-21
    IGNITION TIMING
      ADJUSTMENT 3-21
MULTIPLE CARBURETOR ENGINES WITH
  MOTOROLA® DISTRIBUTOR
  IGNITION 3-21
    CHOKE ROD ADJUSTMENT 3-21
    THROTTLE ROD ADJUSTMENT 3-22
    TIMING BELT ADJUSTMENT 3-23
    IDLE SPEED ADJUSTMENT 3-24
    IGNITION TIMING
      ADJUSTMENT 3-24
MULTIPLE CARBURETOR ENGINES WITH
  NO DISTRIBUTOR 3-24
    CHOKE ROD ADJUSTMENT 3-25
    THROTTLE ROD ADJUSTMENT 3-26
    IDLE SPEED ADJUSTMENT 3-27
    IGNITION TIMING 3-27
WINTER STORAGE CHECKLIST 3-28
SPRING COMMISSIONING
  CHECKLIST 3-28
SPECIFICATIONS CHARTS
    LUBRICATION POINT/FREQUENCY 3-5
    MAINTENANCE INTERVALS 3-29
    GENERAL ENGINE
      SPECIFICATIONS 3-30
    TUNE-UP SPECIFICATIONS 3-31
    CAPACITIES 3-32
    CARBURETOR SPECIFICATIONS 3-32

3

MAINTENANCE AND TUNE-UP

ENGINE MAINTENANCE 3-2
BOAT MAINTENANCE 3-10
TUNE-UP 3-12
WINTER STORAGE CHECKLIST 3-28
SPRING COMMISSIONING
CHECKLIST 3-28

3-2 MAINTENANCE AND TUNE-UP

ENGINE MAINTENANCE

Serial Number Identification

♦ See Figure 1

The outboard serial number and the engine serial number are the manufacturer's key to engine changes. These numbers identify the year of manufacture, the qualified horsepower rating, and the parts book identification. If any correspondence or parts are required, the engine serial number must be used or proper identification is not possible. The accompanying illustration will be very helpful in locating the engine identification tag for the various models.

The outboard number is stamped on the plate usually attached to the port side of the clamp bracket.

The powerhead serial number is usually stamped on the port side of the cylinder block.

As a theft prevention measure, a special label with the outboard serial number is bonded to the starboard side of the clamp bracket. Any attempt to remove this label will result in cracks across the serial number.

On 1984–1992 engine models, the first numbers designate the horsepower and the last numbers signify the model year. Numbers from can be read as follows:

- 50F4 is a 1984 50hp engine
- 357F8 is a 1988 35hp engine
- 1201F91 is a 1991 120hp

➡ If a letter precedes the model number, this letter is very important for getting the proper parts for certain special models.

On 1993–1999 engine models the engine model numbers only give an engine model designation. The year is only available through tracking the unit's serial number at a dealership. The model number will give the horsepower and a code as follows:

- M—manual start
- E—electric start
- PT—power trim
- L—20" shaft
- X—25" shaft

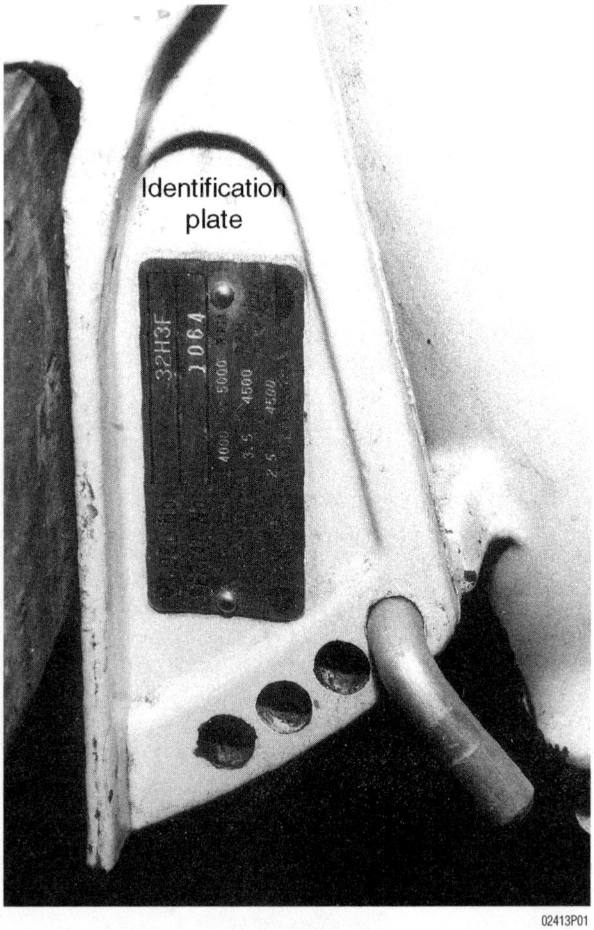

2-Stroke Oil

OIL RECOMMENDATIONS

♦ See Figures 2 and 3

Use only Quicksilver® or NMMA (National Marine Manufacturers Association) certified TCWIII 2 stroke lubricants. These oils are proprietary lubricants designed to ensure optimal engine performance and to minimize combustion chamber deposits, avoid detonation and prolong spark plug life.

➡ Remember, it is this oil, mixed with the gasoline, which lubricates the internal parts of the engine. Lack of lubrication due to the wrong mix or improper type of oil can cause catastrophic powerhead failure. Never use automotive motor oil.

PRE-MIXING FUEL

Mixing the engine lubricant with gasoline before pouring it into the tank is by far the simplest method of lubrication for 2-stroke outboards. However, this method is the messiest and has the potential to cause the most amount of harm to our environment.

The most important part of filling a pre-mix system is to determine the proper fuel/oil ratio. Most manufacturers use a 50:1 ratio (that is 50 parts of fuel to 1 part of oil) or a 100:1 ratio. Consult your owner's manual to determine what the appropriate ratio should be for your engine.

The procedure itself is uncomplicated. Simply add the correct amount of lubricant to your fuel tank and then fill the tank with gasoline. The order in which you do this is important because as the gasoline is poured into the fuel tank it will mix with and agitate the oil for a complete blending.

If you are attempting to top off your tank, here is a general guideline to determine how much oil to add. For three gallons of fuel you would add 4 ounces of oil to obtain a 100:1 ratio; 8 ounces of oil to obtain a 50:1 ratio and 16 ounces of oil to obtain a 25:1 ratio.

➡ Force outboards® all require the oil to be pre-mixed and have never had any factory produced oil injected models.

Fig. 1 Serial numbers on the identification plate of a late model outboard. Most models have the plate in this location, although some units may have it attached on the starboard support plate

Fig. 2 This scuffed piston is an example of what can happen when the proper 2-stroke oil is not used. The outboard required a complete overhaul

MAINTENANCE AND TUNE-UP 3-3

Fig. 3 2-Stroke outboard oils are proprietary lubricants designed to ensure optimal engine performance and to minimize combustion chamber deposits, avoid detonation and prolong spark plug life

Lower Unit

♦ See Figures 4 and 5

Regular maintenance and inspection of the lower unit is critical for proper operation and reliability. A lower unit can quickly fail if it becomes heavily contaminated with water or excessively low on oil. The most common cause of a lower unit failure is water contamination.

Fishing line or other foreign material, becoming entangled around the propeller shaft and damaging the seal usually causes water in the lower unit. If the line is not removed, it will eventually cut the propeller shaft seal and allow water to enter the lower unit. Fishing line has also been known to cut a groove in the propeller shaft if left neglected over time. This area should be checked frequently.

OIL RECOMMENDATIONS

♦ See Figure 6

These oils are proprietary lubricants designed to ensure optimal performance and to minimize corrosion in the lower unit.

Fig. 4 This lower unit was destroyed because the bearing carrier was frozen due to lack of lubrication

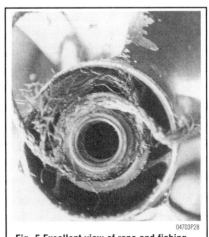

Fig. 5 Excellent view of rope and fishing line entangled behind the propeller. Entangled fishing line can actually cut through the seal, allowing water to enter the lower unit and lubricant to escape

Fig. 6 Use only Quicksilver® lower unit lube or and equivalent high quality SAE 90W hypoid gear oil

3-4 MAINTENANCE AND TUNE-UP

→Remember, it is this lower unit lubricant that prevents corrosion and lubricates the internal parts of the drive gears. Lack of lubrication due to water contamination or the improper type of oil can cause catastrophic lower unit failure.

✷✷ CAUTION

The EPA warns that prolonged contact with used engine oil may cause a number of skin disorders, including cancer! You should make every effort to minimize your exposure to used engine oil. Protective gloves should be worn when changing the oil. Wash your hands and any other exposed skin areas as soon as possible after exposure to used engine oil. Soap and water or waterless hand cleaner should be used.

DRAINING LOWER UNIT

♦ See Figures 7, 8, 9 and 10

✷✷ CAUTION

Never remove the vent or filler plugs when the lower unit is hot. Expanded lubricant will be released through the plug hole.

1. Place the lower unit in the vertical position.
2. Place a suitable container under the lower unit.
3. Loosen the oil level plug on the lower unit. This step is important! If the oil level plug cannot be loosened or removed, the complete lower unit lubricant service cannot be performed.
4. Remove the fill plug or hex key screw from the lower end of the gear housing followed by the oil level plug.
5. Allow the lubricant to completely drain from the lower unit.
6. If applicable, check the magnet end of the drain screw for metal particles. Some amount of metal is considered normal wear is to be expected but if there are signs of metal chips or excessive metal particles, the gear case needs to be disassembled and inspected.
7. Inspect the lubricant for the presence of a milky white substance, water or metallic particles. If any of these conditions are present, the lower unit should be serviced immediately.

✷✷ WARNING

The shifting mechanism of some earlier style lower units is unique. An external screw into the lower unit serves as the pivot point for a critical component of the shift mechanism. The screw actually holds the shift coupler in position inside the lower unit. If this screw should inadvertently be removed, a component of the shift mechanism would fall to the bottom of the lower unit. The lower unit would most likely need to be disassembled and properly reassembled to put the shift pivot pin back into place.

FILLING LOWER UNIT

♦ See Figure 11

✷✷ WARNING

Add only an approved gear case lubricant, such as Quicksilver® lower unit lubricant. Never use regular automotive-type grease in

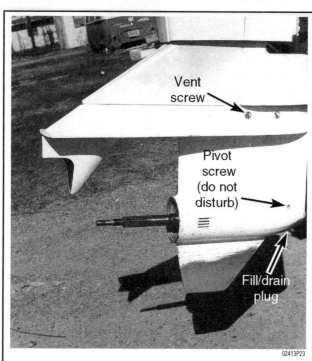

Fig. 7 The amount of lubricant in the lower unit should be checked regularly and replaced as per the maintenance interval chart. Be sure to not disturb the shift mechanism pivot screw

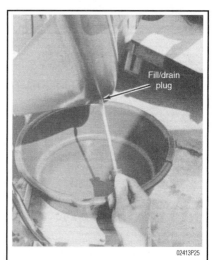

Fig. 8 A suitable container positioned under a fully raised lower unit ready to receive lubricant as it drains

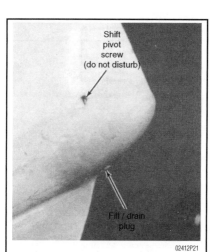

Fig. 9 The shift mechanism pivot screw should never be disturbed except during complete disassembly of the lower unit for service

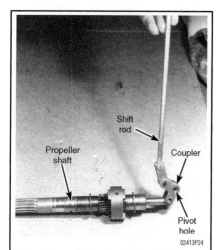

Fig. 10 Layout of some shift mechanism parts to clearly show the function of the pivot screw and to illustrate why the screw must not be disturbed

MAINTENANCE AND TUNE-UP 3-5

the lower unit because it expands and foams too much. Force Outboard lower units are not equipped to provide expansion relief.

➡ Oil should be squeezed in using a tube or with the larger quantities, by using a pump kit to fill the gear case through the drain plug.

Units Having A Drain/Fill Plug At Bottom

1. Position the drive unit approximately vertical and without a list to either port or starboard.
2. Insert the lubricant tube into the fill/drain hole at the bottom plughole, and inject lubricant until the excess begins to come out the vent hole
3. Using new gaskets, (washers) install the oil level and vent plugs first, then install the oil fill plug.

➡ The lubricant must be filled from the bottom to prevent air from being trapped in the lower unit. Air displaces lubricant and can cause a lack of lubrication or a false lubricant level in the lower unit.

4. Wipe the excess oil from the lower unit and inspect the unit for leaks.
5. Place the used lubricant in a suitable container for transportation to an authorized recycling facility.

Units Having A Hex Screw Drain At Bottom

1. Reinstall the hex drain screw into bottom of unit
2. Position the drive unit approximately vertical and without a list to either port or starboard.
3. Insert the lubricant tube into the fill hole located just above the vent plate and inject lubricant until the excess begins to come out the vent hole beside it.
4. Install the vent and fill plugs with new gaskets. Check to be sure that the gaskets are properly positioned to prevent water from entering the housing.
5. Wipe the excess oil from the lower unit and inspect the unit for leaks.
6. Place the used lubricant in a suitable container for transportation to an authorized recycling facility.

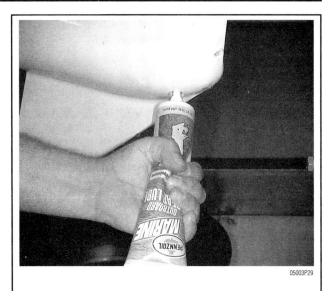

Fig. 11 Fill the lower unite with an approved lower unit lubrication. Be sure the vent screw has been removed

Fuel Filter

♦ See Figures 12, 13 and 14

The fuel filter is designed to keep particles of dirt and debris from entering the carburetor(s) and clogging the internal passages. A small speck of dirt or sand can drastically affect the ability of the carburetor(s) to deliver the proper amount of air and fuel to enter the engine. If a filter becomes clogged, the flow of gasoline will be impeded. This could cause lean fuel mixtures, hesitation and stumbling and idle problems.

Lubrication Specifications

Component	Lubricant	Frequency Fresh Water	Salt Water
Gear Housing	US Marine Gear Lube	①	①
Cranking Motor Pinion Gear	US Marine Lubricant T2961	30 Days	15 Days
Carburetor and Choke Linkage	US Marine Lubricant T2961	60 Days	30 Days
Shift Linkage	US Marine Lubricant T2961	60 Days	30 Days
Clamp Screws	US Marine Lubricant T2961	60 Days	30 Days
Swivel Bracket and Tilt Lock	US Marine Lubricant T2961	60 Days	30 Days
Stern Bracket	US Marine Lubricant T2961	60 Days	30 Days
Dcriveshaft Spline	US Marine Lubricant T2961	Every Season	Every Season
Distributor	US Marine Lubricant T2961	60 Days	30 Days
Propeller Shaft	US Marine Lubricant T2987-1	60 Days	60 Days

① - Check at 30 Hours. Replace every 100 Hours or 6 months. Replace at least once per season prior to storage.

3-6 MAINTENANCE AND TUNE-UP

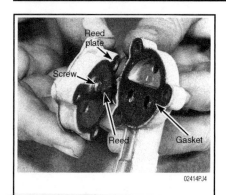

Fig. 12 Mesh filter screen built into the pump inlet nipple

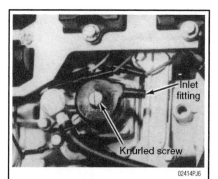

Fig. 13 Removing the knurled screw in the center allows access to the filter screen located under the inlet fitting

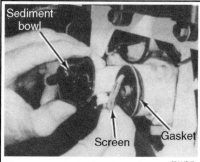

Fig. 14 Remove filter screen and inspect, clean or replace as necessary

Regular replacement of the fuel filter will decrease the risk of blocking the flow of fuel to the engine, which could leave you stranded on the water. Fuel filters are usually inexpensive and replacement is a simple task. Change your fuel filter on a regular basis to avoid fuel delivery problems to the carburetor.

In addition to the fuel filter mounted on the engine, a filter is usually found inside or near the fuel tank. Because of the large variety of differences in both portable and fixed fuel tanks, it is impossible to give a detailed procedure for removal and installation. Most in-tank filters are simply a screen on the pick-up line inside the fuel tank. Filters of this type usually only need to be cleaned and returned to service. Fuel filters on the outside of the tank are typically of the inline type and are replaced by simply removing the clamps, disconnecting the hoses and installing a new filter. When installing the new filter, make sure the arrow on the filter points in the direction of fuel flow.

RELIEVING FUEL SYSTEM PRESSURE

♦ See Figure 15

※※ CAUTION

Exercise extreme caution and always wear safety glasses whenever relieving fuel system pressure to avoid fuel spraying into the eyes and potential serious bodily injury. Please be advised that fuel under pressure may penetrate the skin or any part of the body it contacts.

To avoid the possibility of fire and personal injury, always disconnect the negative battery cable.

Always place a shop towel or rag around the fitting or connection prior to loosening to absorb any excess fuel due to spillage. Ensure that all fuel spillage is removed from engine surfaces. Ensure that all fuel soaked clothes or towels in suitable waste container.

➡ The only pressurized fuel will be at the outlet side of the engine's fuel pump.

1. Remove clamp on outlet hose.
2. Squeeze the hose close to the end of the fuel outlet nipple and gently remove it.
3. Slowly release the hose and allow the fuel in the line to safely drain into a suitable container or shop rag.
4. Replace the hose on the fuel outlet nipple and install the clamp.

REMOVAL & INSTALLATION

※※ CAUTION

Observe all applicable safety precautions when working around fuel. Whenever servicing the fuel system, always work in a well-ventilated area. Do not allow fuel spray or vapors to come in con-

Fig. 15 Relieving fuel pressure into a suitable container

tact with a spark or open flame. Do not smoke while working around gasoline. Keep a dry chemical fire extinguisher near the work area. Always keep fuel in a container specifically designed for fuel storage; also, always properly seal fuel containers to avoid the possibility of fire or explosion.

If so equipped, the engine mounted fuel filters are located in the fuel pump itself. They would be located either in the fuel pump inlet nipple or the top of the fuel filter.

1. Remove the fuel hose from the fuel pump inlet nipple.
2. Remove the inlet nipple(unthread from housing) or pump inlet housing (remove screw in center of housing) remove cover.
3. Note the installed direction of the filter screen for installation reference.

MAINTENANCE AND TUNE-UP

4. Remove filter screen and inspect/clean it.

To install:

5. Install the filter screen in the original direction of flow.
6. Install the inlet nipple or pump inlet housing.
7. Install the fuel hose on the fuel pump inlet nipple.
8. Prime the fuel system and check for leaks.

Fuel/Water Separator

▶ See Figure 16

In addition to the engine and inline fuel filters, there is usually another filter located in the fuel supply line. The fuel/water separator is used to remove water particles from the fuel prior to entering the engine or inline filter. Water can enter the fuel supply from a variety of sources and can lead to poor engine performance and ultimately, serious engine damage.

Because of the large variety of differences in both portable and fixed fuel tanks, it is impossible to give a single procedure to cover all applications. Check with the boat manufacturer or dealership who rigged the boat to get specifics on your particular fuel filtration system.

✱✱ WARNING

A neglected filter can be as harmful as it is good if not serviced regularly. The filter will over time catch water and sediment and become clogged causing a restriction in the fuel system. This could lead to a lean running condition potentially causing serious damaging the internal parts of the engine.

DRAINING

Filters equipped with a fitting on the bottom can be drained into a safe container and inspected for water and contamination.

Filters not equipped with drain fittings would need to be removed and poured into a safe container for inspection.

SERVICE

1. Drain the filter if equipped with a drain fitting.
2. Unthread the canister element or blot holding the canister in place.
3. Drain the filter if not already done.
4. Remove the element or canister and inspect it for contamination.
5. Discard filter element or canister in a safe environmentally friendly manner.
6. Reinstall a new element or canister using new gaskets as required applying a light coat of lubricant on the gasket to ease removal in the future.
7. Prime fuel system and check for leaks.

Lubrication Points

INSPECTION & LUBRICATION

▶ See Figures 17 thru 26

Keeping your Force Outboard properly lubricated will allow the engines pivot points and bearings to operate smoothly and resist premature failure.

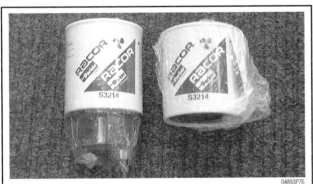

Fig. 16 Typical fuel filter (right) and a water separating fuel filter assembly (left)

Fig. 17 Use only a good quality marine grade grease for lubrication

Fig. 18 The cranking motor pinion gear and shaft should be lubricated with an approved lubricant on a regular basis

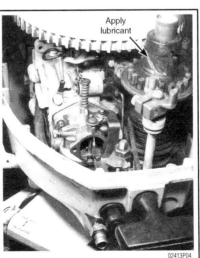

Fig. 19 The guide slot on the pinion collar should also be lubricated regularly

Fig. 20 The tower shaft bearings and ball joints are a regular lubrication point

3-8 MAINTENANCE AND TUNE-UP

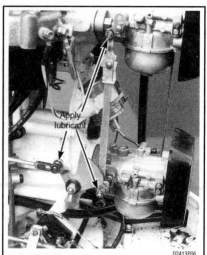

Fig. 21 The tie bar pivot points and the link rod ball joint fittings require lubrication

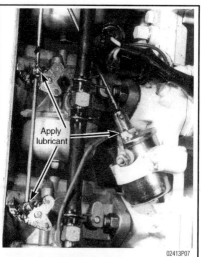

Fig. 22 The choke link swivel joints and the choke plunger surface should receive lubrication when other points are serviced

Fig. 23 An external lubrication fitting is provided to service the distributor shaft. At the same time, check the belt tension

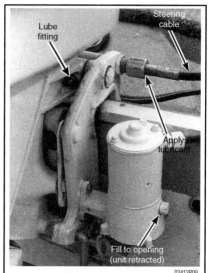

Fig. 24 The steering cable should receive lubrication at the point indicated. The swivel joint lubrication fitting is located inside the clamp bracket. Correct level for the trim/tilt hydraulic reservoir is to the bottom of the fill hole. The level should only be checked when the outboard unit is in the full down position

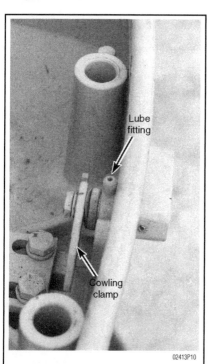

Fig. 25 Location of the lubrication fitting for the cowling clamp

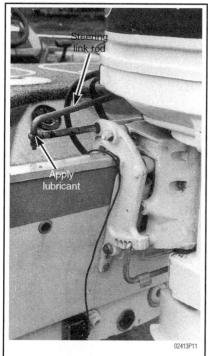

Fig. 26 Apply lubricant to the steering link rod joint, at the point indicated

If an outboard unit is operated in salt water or a corrosive environment the frequency of applying lubricant to fittings is usually cut in half for the same fitting if the unit is used in fresh water. While lubricating the engine, take a visual inspection of its general condition. This could go a long way to catch potential problems early so they can be corrected rather that just letting unknown leaks or failures lead to major breakdowns and repairs.

The "Maintenance Interval" chart can be used as a guide to periodic maintenance while the outboard is being used during the season.

1. Locate the area to be lubricated.
2. Connect the grease gun or lubricator to the fitting or area that is to be greased.
3. Inject or press lubricant into the fitting or lubrication area until the old lubricant can be seen being forced out.
4. Wipe away old lubricant with a rag or cloth

5. Use a good soap or degreaser to finish cleaning up the outboard to prevent contaminating the environment with grease.

Propeller

▶ See Figures 27 and 28

The propeller should be inspected regularly to be sure the blades are in good condition. If any of the blades become bent or nicked, this condition will set up vibrations in the motor. Remove and inspect the propeller. Use a file to trim nicks and burrs. Take care not to remove any more material than is absolutely necessary.

Also, check the rubber and splines inside the propeller hub for damage. If there is damage to either of these, take the propeller to your local marine dealer

MAINTENANCE AND TUNE-UP 3-9

Fig. 27 A propeller in good condition ready for installation

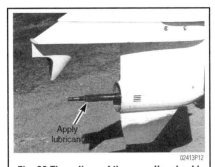

Fig. 28 The splines of the propeller should always be coated with water resistant marine grease before the propeller is installed. Failure to apply adequate lubrication could cause the propeller to seize on the shaft

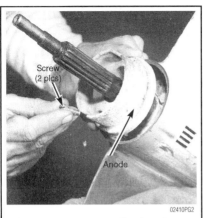

Fig. 29 Remove the propeller then the two fasteners holding the anode in place

or a prop shop. They can evaluate the damaged propeller and determine if replacing the hub can save it.

Additionally, the propeller should be removed each time the boat is hauled from the water at the end of the season. Any material entangled behind the propeller should be removed before any damage to the shaft and seals can occur. This may seem like a waste of time but the small amount of time involved in removing the propeller is returned many times by reduced maintenance and repair, including the replacement of expensive parts.

Anodes (Zincs)

The idea behind anodes is simple: When dissimilar metals are dunked in water and a small current is leaked between or amongst them, the less-noble metal (galvanically speaking) is sacrificed.

The zinc alloy that the anodes are made of is designed to be less noble than the aluminum alloy your outboard is made from. If there's any electrolysis and there almost always is, the inexpensive zinc anodes are consumed in lieu of the expensive outboard motor.

These zincs need a little attention in order to do their job. Make sure they're there, solidly attached to a clean mounting site and not covered with any kind of paint or wax.

Periodically inspect them to make sure they haven't eroded too much. At a certain point in the erosion process, the mounting holes start to enlarge, which is when the zinc might fall off. Obviously, once that happens your engine no longer has any protection.

SERVICING

▶ See Figure 29

Depending on what kind of power your boat has, you might have anywhere from one to four (or more) zincs. Regardless of the number, there are some fundamental rules to follow that will give your boat's sacrificial anodes the ability to do the best job protecting your boat's underwater hardware that they can.

The first thing to remember is that zincs are electrical components and like all electrical components, they require good clean connections. So after you've undone the mounting hardware and removed last year's zincs, you want to get the zinc mounting sites clean and shiny.

Get a piece of coarse emery cloth or some 80-grit sandpaper. Thoroughly rough up the areas where the zincs attach (there's often a bit of corrosion residue in these spots). Make sure to remove every trace of corrosion.

Zincs are attached with stainless steel machine screws that thread into the mounting for the zincs. Over the course of a season, this mounting hardware is inclined to loosen. Mount the zincs and tighten the mounting hardware securely. Tap the zincs with a hammer hitting the mounting screws squarely. This process tightens the zincs and allows the mounting hardware to become a bit loose ill the process. Now, do the final tightening. This will insure your zincs stay put for the entire season.

INSPECTION

▶ See Figures 30 and 31

For the anodes to work properly they must be bonded with the entire outboard engine.

The bonding test is a simple continuity test with a volt/ohm meter. With the meter set on the ohm scale the reading should be the same as if the meter leads

Fig. 30 Comparison of a new (left), and used (right), anode attached to the bearing carrier of larger horsepower units. The anode on the right is still acceptable for installation

3-10 MAINTENANCE AND TUNE-UP

Fig. 31 Such extensive erosion of a trim tab compared with a new tab suggests an electrolysis problem or complete disregard for periodic maintenance

were touched to each other. Just touch the black lead to the lower unit housing and the red lead to the anode.

If you use your outboard in salt water and your zincs never wear. Inspect them carefully for Paint or wax on zincs prevents them from working properly. They must be left bare. If the zincs are installed properly and not painted or waxed inspect the housing around them for signs of corrosion. If corrosion of the housing is found clean the area down to the metal, prime and repaint with a suitable paint.

On the other hand, if your zincs seem to erode in no time at all, this may be a symptom of the zincs themselves. Each manufacturer uses a specific blend of metals in their zincs. If you are using zincs with the wrong blend of metals, they may erode more quickly or leave you with diminished protection.

BOAT MAINTENANCE

Batteries

MAINTENANCE

♦ See Figure 32

Batteries require periodic servicing and a definite maintenance program will ensure extended life. If the battery should test satisfactorily but still fails to perform properly, one of four problems could be the cause.

1. An accessory might have accidentally been left on overnight or for a long period during the day. Such an oversight would result in a discharged battery.
2. Using more electrical power than the stator assembly or lighting coil can replace would result in an undercharged condition.
3. A faulty stator assembly or lighting coil, defective rectifier or high resistance somewhere in the system could cause the battery to become undercharged.
4. Failure to maintain the battery in good order. This might include a low level of electrolyte in the cells, loose or dirty cable connections at the battery terminals or possibly an excessively dirty battery top.

The most common procedure for maintaining a battery is to check the electrolyte level. Removing the cell caps and visually observing the level in the cells does this. The bottom of each cell has a split vent which will cause the surface of the electrolyte to appear distorted when it makes contact. When the distortion first appears at the bottom of the split vent, the electrolyte level is correct.

During hot weather and periods of heavy use, the electrolyte level should be checked more often than during normal operation. Add distilled water to bring the level of electrolyte in each cell to the proper level. Take care not to overfill, because adding an excessive amount of water will cause loss of electrolyte and any loss will result in poor performance, short battery life and will contribute quickly to corrosion.

➥Never add electrolyte from another battery. Use only distilled water.

CLEANING

♦ See Figures 33 and 34

Dirt and corrosion should be cleaned from the battery as soon as it is discovered. Any accumulation of acid film or dirt will permit current to flow between the terminals. Such a current flow will drain the battery over a period of time.

Clean the exterior of the battery with a solution of diluted ammonia or a paste made from baking soda and water. This neutralizes any acid that may be present. Flush the cleaning solution off with plenty of clean water.

➥Take care to prevent any of the neutralizing solution from entering the cells.

Fig. 32 Explosive hydrogen gas is normally released from the cells under a wide range of circumstances. This battery exploded when the gas ignited from someone smoking in the area when the caps were removed. Such an explosion could also be caused by a spark from the battery terminals

Fig. 33 A typical battery service tool being used to remove dirt and corrosion from the battery posts

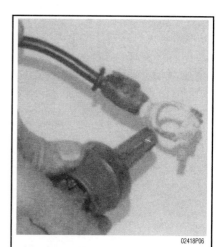

Fig. 34 A typical battery service tool being used to remove dirt and corrosion from the battery terminals

MAINTENANCE AND TUNE-UP 3-11

A poor contact at the terminals will add resistance to the charging circuit. This resistance will cause the voltage regulator to register a fully charged battery and thus cut down on the stator assembly or lighting coil output adding to the low battery charge problem.

At least once a season, the battery terminals and cable clamps should be cleaned. Loosen the clamps and remove the cables, negative cable first. On batteries with top mounted posts, the use of a puller specially made for this purpose is recommended. These are inexpensive and available in most parts stores.

Clean the cable clamps and the battery terminal with a wire brush, until all corrosion, grease, etc., is removed and the metal is shiny. It is especially important to clean the inside of the clamp thoroughly (a wire brush is useful here), since a small deposit of foreign material or oxidation there will prevent a sound electrical connection and inhibit either starting or charging. It is also a good idea to apply some dielectric grease to the terminal, as this will aid in the prevention of corrosion.

After the clamps and terminals are clean, reinstall the cables, negative cable last, do not hammer the clamps onto battery posts. Tighten the clamps securely but do not distort them. Give the clamps and terminals a thin external coating of grease after installation, to retard corrosion.

Check the cables at the same time that the terminals are cleaned. If the insulation is cracked or broken or if its end is frayed, that cable should be replaced with a new one of the same length and gauge.

TESTING

♦ See Figure 35

A hydrometer is a device to measure the percentage of sulfuric acid in the battery electrolyte in terms of specific gravity. When the condition of the battery drops from fully charged to discharge, the acid leaves the solution and enters the plates, causing the specific gravity of the electrolyte to drop.

It may not be common knowledge but hydrometer floats are calibrated for use at 80°F (27°C). If the hydrometer is used at any other temperature, hotter or colder, a correction factor must be applied.

➡ **Remember that a liquid will expand if it is heated and will contract if cooled. Such expansion and contraction will cause a definite change in the specific gravity of the liquid, in this case the electrolyte.**

A quality hydrometer will have a thermometer/temperature correction table in the lower portion, as illustrated in the accompanying illustration. By knowing the air temperature around the battery and from the table, a correction factor may be applied to the specific gravity reading of the hydrometer float. In this manner, an accurate determination may be made as to the condition of the battery.

When using a hydrometer, pay careful attention to the following points:

1. Never attempt to take a reading immediately after adding water to the battery. Allow at least ¼ hour of charging at a high rate to thoroughly mix the electrolyte with the new water. This time will also allow for the necessary gases to be created.

2. Always be sure the hydrometer is clean inside and out as a precaution against contaminating the electrolyte.

3. If a thermometer is an integral part of the hydrometer, draw liquid into it several times to ensure the correct temperature before taking a reading.

4. Be sure to hold the hydrometer vertically and suck up liquid only until the float is free and floating.

5. Always hold the hydrometer at eye level and take the reading at the surface of the liquid with the float free and floating.

6. Disregard the slight curvature appearing where the liquid rises against the float stem. This phenomenon is due to surface tension.

7. Do not drop any of the battery fluid on the boat or on your clothing, because it is extremely caustic. Use water and baking soda to neutralize any battery liquid that does accidentally drop.

8. After drawing electrolyte from the battery cell until the float is barely free, note the level of the liquid inside the hydrometer. If the level is within the Green band range for all cells, the condition of the battery is satisfactory. If the level is within the white band for all cells, the battery is in fair condition.

9. If the level is within the Green or white band for all cells except one, which registers in the red, the cell is shorted internally. No amount of charging will bring the battery back to satisfactory condition.

10. If the level in all cells is about the same, even if it falls in the Red band, the battery may be recharged and returned to service. If the level fails to rise above the Red band after charging, the only solution is to replace the battery.

STORAGE

♦ See Figure 36

If the boat is to be laid up for the winter or for more than a few weeks, special attention must be given to the battery to prevent complete discharge or possible damage to the terminals and wiring. Before putting the boat in storage, disconnect and remove the batteries. Clean them thoroughly of any dirt or corrosion and then charge them to full specific gravity reading. After they are fully charged, store them in a clean cool dry place where they will not be damaged or knocked over, preferably on a couple blocks of wood. Storing the battery up off the deck will permit air to circulate freely around and under the battery and will help to prevent condensation.

Never store the battery with anything on top of it or cover the battery in such a manner as to prevent air from circulating around the filler caps. All batteries, both new and old, will discharge during periods of storage, more so if they are hot than if they remain cool. Therefore, the electrolyte level and the specific

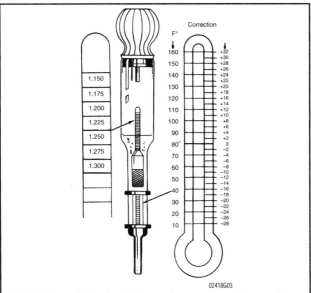

Fig. 35 The condition of a battery can be measured by the specific gravity of its cells

Fig. 36 Do not leave the cables connected to the battery while the boat is stored. The best practice is to remove the battery to a safe, dry, protected area

3-12 MAINTENANCE AND TUNE-UP

gravity should be checked at regular intervals. A drop in the specific gravity reading is cause to charge them back to a full reading.

In cold climates, care should be exercised in selecting the battery storage area. A fully charged battery will freeze at about 60 degrees below zero. A discharged battery, almost dead, will have ice forming at about 19 degrees above zero.

Fiberglass Hull

♦ See Figures 37, 38 and 39

Fiberglass reinforced plastic hulls are tough, durable, and highly resistant to impact. However, like any other material they can be damaged. One of the advantages of this type of construction is the relative ease with which it may be repaired.

A fiberglass hull has almost no internal stresses. Therefore, when the hull is broken or stove-in, it retains its true form. It will not dent to take an out-of-shape set. When the hull sustains a severe blow, the impact will be either absorbed by deflection of the laminated panel or the blow will result in a definite, localized break. In addition to hull damage, bulkheads, stringers, and other stiffening structures attached to the hull may also be affected and therefore, should be checked. Repairs are usually confined to the general area of the rupture.

➡The best way to care for a fiberglass hull is to wash it thoroughly, immediately after hauling the boat while the hull is still wet.

A foul bottom can seriously affect boat performance. This is one reason why racers, large and small, both powerboat and sail, are constantly giving attention to the condition of the hull below the waterline.

In areas where marine growth is prevalent, a coating of vinyl, anti-fouling bottom paint should be applied. If growth has developed on the bottom, it can be removed with a solution of Muriatic acid applied with a brush or swab and then rinsed with clear water.

✶✶ CAUTION

Always use rubber gloves when working with Muriatic acid and take extra care to keep it away from your face and hands. The fumes are toxic. Therefore, work in a well-ventilated area or if outside, keep your face on the windward side of the work.

Barnacles have a nasty habit of making their home on the bottom of boats that have not been treated with anti-fouling paint. Actually they will not harm the fiberglass hull, but can develop into a major nuisance.

If barnacles or other crustaceans have attached themselves to the hull, extra work will be required to bring the bottom back to a satisfactory condition. First, if practical, put the boat into a body of fresh water and allow it to remain for a few days. A large percentage of the growth can be removed in this manner. If this remedy is not possible, wash the bottom thoroughly with a high-pressure fresh water source and use a scraper. Small particles of hard shell may still hold fast. These can be removed with sandpaper.

Fig. 37 In areas where marine growth is a problem, a coating of anti-foul bottom paint should be applied

Fig. 38 The best way to care for a fiberglass hull is to wash it thoroughly

Fig. 39 Fiberglass, vinyl and rubber care products are available to protect every part of your boat

TUNE-UP

Introduction

♦ See Figure 40

A proper tune-up is the key to long and trouble-free outboard life and the work can yield its own rewards. Studies have shown that a properly tuned and maintained outboard can achieve better fuel economy than an out-of-tune engine. As a conscientious boater, set aside a Saturday morning, say once a month, to check or replace items which could cause major problems later. Keep your own personal log to jot down which services you performed, how much the parts cost you, the date and the number of hours on the engine at the time. Keep all receipts for such items as oil and filters, so that they may be referred to in case of related problems or to determine operating expenses. As a do-it-yourselfer, these receipts are the only proof you have that the required maintenance was performed. In the event of a warranty problem, these receipts will be invaluable.

The efficiency, reliability, fuel economy and enjoyment available from boating are all directly dependent on having your outboard tuned properly. The importance of performing service work in the proper sequence cannot be over emphasized. Before making any adjustments, check the specifications. Never rely on memory when making critical adjustments.

Before tuning any outboard, insure it has satisfactory compression. An outboard with worn or broken piston rings burned pistons or scored cylinder walls, will not perform properly no matter how much time and expense is spent on the tune-up. Poor compression must be corrected or the tune-up will not give the desired results.

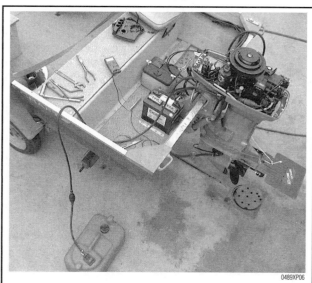
Fig. 40 A proper tune-up is the key to long and trouble-free outboard life and the work can yield its own rewards

MAINTENANCE AND TUNE-UP 3-13

The extent of the engine tune-up is usually dependent on the time lapse since the last service. In this section, a logical sequence of tune-up steps will be presented in general terms. If additional information or detailed service work is required, refer to the section of this manual containing the appropriate instructions.

Compression Test

PRIMARY COMPRESSION

Because the 2-stroke powerhead is a pump, the crankcase must be sealed against pressure created on the down stroke of the piston and vacuum created when the piston moves toward top dead center. If there are air leaks into the crankcase, insufficient fuel will be brought into the crankcase and into the cylinder for normal combustion.

➡If it is a very small leak, the powerhead will run poorly, because the fuel mixture will be lean and cylinder temperatures will be hotter than normal.

Air leaks are possible around any seal, O-ring, cylinder block mating surface or gasket. Always replace O-rings, gaskets and seals when service work is performed.

If the powerhead is running, soapy water can be sprayed onto the suspected sealing areas. If bubbles develop, there is a leak at that point. Oil around sealing points and on ignition parts under the flywheel indicates a crankcase leak.

The base of the powerhead and lower crankshaft seal is impossible to check on an installed powerhead. When every test and system have been checked out and the bottom cylinder seems to be effecting performance, then the lower seal should be tested.

SECONDARY COMPRESSION TEST

♦ See Figure 41

The actual pressure measured during a secondary compression test is not as important as the variation from cylinder to cylinder. On multi-cylinder powerhead, a variation of 15 psi or more is considered questionable. On single cylinder powerhead, a drop of 15 psi from the normal compression pressure you established when it was new is cause for concern (you did do a compression test on it when it was new, didn't you?).

➡If the powerhead been in storage for an extended period, the piston rings may have relaxed. This will often lead to initially low and misleading readings. Always run an engine to operating temperature to ensure that the reading you get is accurate.

1. Disable the ignition system by removing the lanyard clip. If you do not have a lanyard, take a wire jumper lead and connect one end to a good engine ground and the other end to the metal connector inside the spark plug boot, using one jumper for each plug wire. Never simply disconnect all the plug wires.

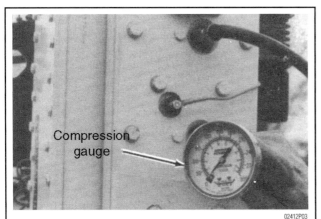

Fig. 41 A compression test should done on each cylinder as the first step in tuning or servicing the engine. Poor compression is definite sign of a problem and would need to be addressed before continuing

※※ CAUTION

Removing all the spark plugs and cranking over the powerhead can lead to an explosion if raw fuel/oil sprays out of the spark plug holes. A plug wire could spark and ignite this mix outside of the combustion chamber if it is not grounded to the engine.

2. Remove all the spark plugs and be sure to keep them in order. Carefully inspect the plugs, looking for any inconsistency in coloration and for any sign of water or rust near the tip.
3. Thread the compression gauge into the No. 1 spark-plug hole, taking care not to cross thread the fitting.
4. Open the throttle to the wide-open throttle position and hold it there.

➡Some engines allow only minimal opening if the gearshift is in neutral, to guard against over-revving.

5. Crank over the engine an equal number of times for each cylinder you test, zeroing the gauge for each cylinder.
6. If you have electric start, count the number of seconds you count. On manual start, pull the starter rope four to five times for each cylinder you are testing.
7. Record your readings from each cylinder. When all cylinders are tested, compare the readings and determine if pressures are within the 15-psi criterion.
8. If compression readings are lower than normal for any cylinders, try a "wet" compression test, which will temporarily seal the piston rings and determine if they are the cause of the low reading.
9. Using a can of fogging oil, fog the cylinder with a circular motion to distribute oil spray all around the perimeter of the piston. Retest the cylinder.
    a. If the compression rises noticeably, the piston rings are sticking. You may be able to cure the problem by decarboning the powerhead.
    b. If the dry compression was really low and no change is evident during the wet test, there is evidence that there is internal damage or excessive wear to the cylinder or piston. The cylinder head needs to be removed for further inspection
10. If two adjacent cylinders on a multi-cylinder engine give a similarly low reading then the problem may be a faulty head gasket. This should be suspected if there was evidence of water or rust on the spark plugs from these cylinders.

Spark Plugs

The spark plug performs four main functions:
- It fills a hole in the cylinder head.
- It acts as a dielectric insulator for the ignition system.
- It provides spark for the combustion process to occur.
- It removes heat from the combustion chamber.

It is important to remember that spark plugs do not create heat, they help remove it. Anything that prevents a spark plug from removing the proper amount of heat can lead to pre-ignition, detonation, premature spark plug failure and even internal engine damage, especially in a two-stroke engine.

In the simplest of terms, the spark plug acts as the thermometer of the engine. Much like a doctor examining a patient, this "thermometer" can be used to effectively diagnose the amount of heat present in each combustion chamber.

Spark plugs are valuable tuning tools, when interpreted correctly. They will show symptoms of other problems and can reveal a great deal about the engine's overall condition. By evaluating the appearance of the spark plug's firing tip, visual cues can be seen to accurately determine the engine's overall operating condition, get a feel for air/fuel ratios and even diagnose running problems.

As spark plugs grow older, they lose their sharp edges and material from the center and ground electrodes is slowly eroded away. As the gap between these two points grows, the voltage required to bridge this gap increases proportionately. The ignition system must work harder to compensate for this higher voltage requirement and hence there are a greater rate of misfires or incomplete combustion cycles. Each misfire means lost horsepower, reduced fuel economy and higher emissions. Replacing worn out spark plugs with new ones effectively restores the ignition system's efficiency and reduces the percentage of misfires, restoring power, economy and reducing emissions.

How long spark plugs last will depend on a variety of factors, including engine compression, fuel used, gap, center/ground electrode material and the conditions in which the outboard is operated.

3-14 MAINTENANCE AND TUNE-UP

SPARK PLUG HEAT RANGE

♦ See Figure 42

Spark plug heat range is the ability of the plug to dissipate heat from the combustion chamber. The longer the insulator (or the farther it extends into the engine), the hotter the plug will operate; the shorter the insulator (the closer the electrode is to the block's cooling passages) the cooler it will operate.

Selecting a spark plug with the proper heat range will ensure that the tip will maintain a temperature high enough to prevent fouling, yet be cool enough to prevent pre-ignition. A plug that absorbs little heat and remains too cool will quickly accumulate deposits of oil and carbon since it is not hot enough to burn them off. This leads to plug fouling and consequently to misfiring. A plug that absorbs too much heat will have no deposits but, due to the excessive heat, the electrodes will burn away quickly and might possibly lead to pre-ignition or other ignition problems.

Pre-ignition takes place when plug tips get so hot that they glow sufficiently to ignite the air/fuel mixture before the actual spark occurs. This early ignition will usually cause a pinging during heavy loads and if not corrected will result in severe engine damage. While there are many other things that can cause pre-ignition, selecting the proper heat range spark plug will ensure that the spark plug itself is not a hot-spot source.

SPARK PLUG SERVICE

➡ New technologies in spark plug and ignition system design have pushed the recommended replacement interval to every 100 hours of operation (6 months). However, this depends on usage and conditions.

Spark plugs should only require replacement once a season. The electrode on a new spark plug has a sharp edge but with use, this edge becomes rounded by wear, causing the plug gap to increase. As the gap increases, the plug's voltage requirement also increases. It requires a greater voltage to jump the wider gap and about two to three times as much voltage to fire a plug at high speeds than at idle.

Tools needed for spark plug replacement include: a ratchet, short extension, spark plug socket $^{13}/_{16}$ inch, a combination spark plug gauge and gapping tool.

REMOVAL & INSTALLATION

♦ See Figures 43 and 44

※※ CAUTION

. Never touch the spark plug boot or wire with your bare hands while the engine is running.

1. When removing spark plugs, work on one at a time. Don't start by removing the plug wires all at once, because unless you number them, they may become mixed up. Take a minute before you begin and number the wires with tape.
2. Disconnect the negative battery cable or turn the battery switch **OFF**.
3. If the engine has been run recently, allow the engine to thoroughly cool.

Attempting to remove plugs from a hot cylinder head could cause the plugs to seize and damage the threads in the cylinder head, especially on aluminum heads!

4. Carefully twist the spark plug wire boot to loosen it, then pull the boot using a twisting motion to remove it from the plug. Be sure to pull on the boot and not on the wire, otherwise the connector located inside the boot may become separated from the high-tension wire.

➡ A spark plug wire removal tool is recommended as it will make removal easier and help prevent damage to the boot and wire assembly.

5. Using compressed air (and safety glasses) blow debris from the spark plug area to assure that no harmful contaminants are allowed to enter the combustion chamber when the spark plug is removed. If compressed air is not available, use a rag or a brush to clean the area. Compressed air is available from both an air compressor or from compressed air in cans available at photography stores.

➡ Remove the spark plugs when the engine is cold, if possible, to prevent damage to the threads. If plug removal is difficult, apply a few drops of penetrating oil to the area around the base of the plug and allow it a few minutes to work.

6. Using a spark plug socket that is equipped with a rubber insert to properly hold the plug, turn the spark plug counterclockwise to loosen and remove the spark plug from the bore.

※※ WARNING

Avoid the use of a flexible extension on the socket. Use of a flexible extension may allow a shear force to be applied to the plug. A shear force could break the plug off in the cylinder head, leading to costly and frustrating repairs. In addition, be sure to support the ratchet with your other hand—this will also help prevent the socket from damaging the plug.

7. Evaluate each cylinder's performance by comparing the spark condition. Check each spark plug to be sure they are all the same manufacturer and have the same heat range rating. Inspect the threads in the spark plug opening of the block and clean the threads before installing the plug.
8. When purchasing new spark plugs, always ask the dealer if there has been a spark plug change for the engine being serviced. Many times manufacturers will update the type of spark plug used in an engine to offer better efficiency or performance.
9. Crank the engine through several revolutions to blow out any material which might have become dislodged during cleaning. The gasket must be fully compressed on a clean seating area to complete the heat transfer process and to provide a gas tight seal in the cylinder.
10. Inspect the spark plug boot for tears or damage. If a damaged spark plug boot is found it needs to be replaced. If the high-tension spark plug wire is a separate part and not molded in the ignition coil it should also be replaced.
11. Check the spark plug gap prior to installing the plug. Most spark plugs do not come gapped to the proper specification.
12. Carefully thread the plug into the bore by hand. If resistance is felt before the plug completely bottomed, back the plug out and begin threading again.

Fig. 42 Spark Plug heat range

Fig. 43 Firmly grasp the spark plug boot and gently twist and pull to remove. Do not pull on the spark plug wire and never touch the spark plug boot or wire with your bare hands while the engine is running

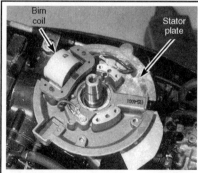

Fig. 44 Use only proper tools to remove and install spark plugs. The wrong tool could damage the spark plug, resulting in a poor running engine

MAINTENANCE AND TUNE-UP 3-15

⚠️ WARNING

Do not use the spark plug socket to thread the plugs. Always carefully thread the plug by hand or using an old plug wire to prevent the possibility of cross threading and damaging the cylinder head bore.

13. Carefully tighten the spark plug to 12.5 ft. lbs. (17Nm). If the plug you are installing is equipped with a crush washer, seat the plug, then tighten or about ¼ turn to crush the washer. Whenever possible, spark plugs should be tightened to the factory torque specification.
14. Apply a small amount of silicone dielectric grease to the end of the spark plug lead or inside the spark plug boot to prevent sticking, then install the boot to the spark plug and push until it clicks into place. The click may be felt or heard. Gently pull back on the boot to assure proper contact.
15. Connect the negative battery cable or turn the battery switch **ON**.
16. Start the outboard and insure proper operation.

READING SPARK PLUGS

▶ See Figures 45 thru 50

Reading spark plugs can be a valuable tuning aid. By examining the insulator firing nose color, you can determine much about the engine's overall operating condition.

In general, a light tan/gray color tells you that the spark plug is at the optimum temperature and that the engine is in good operating condition.

Dark coloring, such as heavy black wet or dry deposits usually indicate a fouling problem. Heavy, dry deposits can indicate an overly rich condition, too cold a heat range spark plug, possible vacuum leak, low compression, overly retarded timing or too large a plug gap.

If the deposits are wet, it can be an indication of a breached head gasket, or an extremely rich condition, depending on what liquid is present at the firing tip.

Also look for signs of detonation, such as silver specs, black specs or melting or breakage at the firing tip.

Compare your plugs to the illustrations shown to identify the most common plug conditions.

Fouled Spark Plugs

A spark plug is fouled when the insulator nose at the firing tip becomes coated with a foreign substance, such as fuel, oil or carbon. This coating makes it easier for the voltage to follow along the insulator nose and leach back down into the metal shell, grounding out, rather than bridging the gap normally.

Fig. 45 A normally worn spark plug should have light tan or gray deposits on the firing tip (electrode)

Fig. 46 A carbon-fouled plug, identified by soft, sooty black deposits, may indicate an improperly tuned powerhead

Fig. 47 A physically damaged spark plug may be evidence of severe detonation in that cylinder. Watch the cylinder carefully between services, as a continued detonation will not only damage the plug but will most likely damage the powerhead

Fig. 48 This spark plug has been left in the powerhead too long, as evidenced by the extreme gap. Plugs with such an extreme gap can cause misfiring and stumbling accompanied by a noticeable lack of power

3-16 MAINTENANCE AND TUNE-UP

Fig. 49 A bridged or almost bridged spark plug, identified by the build-up between the electrodes caused by excessive carbon or oil build up on the plug

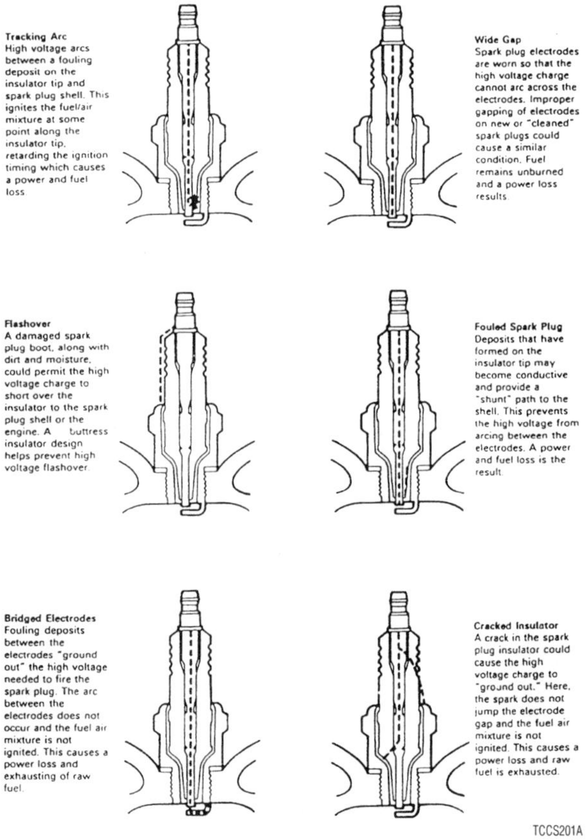

Fig. 50 Typical spark plug problems showing damage which may indicate engine problems

Different things can cause all fuel, oil and carbon fouling but in any case, once a spark plug is fouled, it will not provide voltage to the firing tip and that cylinder will not fire properly. In many cases, the spark plug cannot be cleaned sufficiently to restore normal operation. It is therefore recommended that fouled plugs be replaced.

Signs of fouling or excessive heat must be traced quickly to prevent further deterioration of performance and to prevent possible engine damage.

Overheated Spark Plugs

When a spark plug tip shows signs of melting or is broken, it usually means that excessive heat and/or detonation was present in that particular combustion chamber or that the spark plug was suffering from thermal shock.

Since spark plugs do not create heat by themselves, one must use this visual clue to track down the root cause of the problem. In any case, damaged firing tips most often indicate that cylinder pressures or temperatures were too high.

Left unresolved, this condition usually results in more serious engine damage. Detonation refers to a type of abnormal combustion that is usually preceded by pre-ignition. It is most often caused by a hot spot formed in the combustion chamber.

As air and fuel is drawn into the combustion chamber during the intake stroke, this hot spot will "pre-ignite" the air fuel mixture without any spark from the spark plugs.

Detonation

Detonation exerts a great deal of downward force on the pistons, as they are being forced upward by the mechanical action of the connecting rods. When this occurs, the resulting concussion, shock waves and heat can be severe. Spark plug tips can be broken or melted and other internal engine components such as the pistons or connecting rods themselves can be damaged.

Left unresolved, engine damage is almost certain to occur, with the spark plug usually suffering the first signs of damage.

➙When signs of detonation or pre-ignition are observed, they are symptom of another problem. You must determine and correct the situation that caused the hot spot to form in the first place.

INSPECTION & GAPPING

A particular spark plug might fit several engines and although the factory will typically set the gap to a pre-selected setting, this gap may not be the right one for your particular powerhead.

Insufficient spark plug gap can cause pre-ignition, detonation, and even engine damage. Too large a gap can result in a higher rate of misfires, noticeable loss of power, plug fouling and poor economy.

Check spark plug gap before installation. The ground electrode (the L-shaped one connected to the body of the plug) must be parallel to the center electrode and the specified size wire gauge must pass between the electrodes with a slight drag.

Do not use a flat feeler gauge when measuring the gap on a used plug, because the reading may be inaccurate. A round-wire gapping tool is the best way to check the gap. The correct gauge should pass through the electrode gap with a slight drag. If you're in doubt, try a wire that is one size smaller and one larger. The smaller gauge should go through easily, while the larger one should not go through at all.

Wire gapping tools usually have a bending tool attached. Use this tool to adjust the side electrode until the proper distance is obtained. Never attempt to bend the center electrode. Also, be careful not to bend the side electrode too far or too often as it may weaken and break off within the engine, requiring removal of the cylinder head to retrieve it.

Spark Plug Wires

TESTING

▸ See Figure 51

Each time you remove the engine cover, visually inspect the spark plug wires for burns, cuts or breaks in the insulation. Check the boots on the coil and at the spark plug end. Replace any wire that is damaged.

MAINTENANCE AND TUNE-UP 3-17

Fig. 51 Test the wires by connecting one lead of an ohmmeter to the coil end of the wire and the other lead to the spark plug end of the wire

Once a year, usually when you change your spark plugs, check the resistance of the spark plug wires with an ohmmeter. Wires with excessive resistance will cause misfiring and may make the engine difficult to start. In addition worn wires will allow arcing and misfiring in humid conditions.

Remove the spark plug wire from the engine. Test the wires by connecting one lead of an ohmmeter to the coil end of the wire and the other lead to the spark plug end of the wire. Resistance should measure approximately 7000 ohms per foot of wire. If a spark plug wire is found to have excessive (high) resistance, the entire set should be replaced.

➥Most spark plug wires in Force outboard engines are molded into the ignition coil and must be tested as part of the coil.

REMOVAL & INSTALLATION

When installing a new set of spark plug wires, replace the wires one at a time so there will be no confusion. Coat the inside of the boots with dielectric grease to prevent sticking. Install the boot firmly over the spark plug until it clicks into place. The click may be felt or heard. Gently pull back on the boot to assure proper contact. Repeat the process for each wire.

➥It is important to route the new spark plug wire the same as the original and install it in a similar manner on the powerhead. Improper routing of spark plug wires may cause powerhead performance problems.

Ignition System

Some engine models have a conventional magneto ignition system. This system utilizes contact points, condenser and an ignition coil under the flywheel, triggered by a cam keyed to the crankshaft.

Early models that have a belt driven distributor also have contact points, condenser and an externally mounted ignition coil.

Later model engines all have electronic CDI ignition systems that have become one of the most reliable components on an outboard. There is very little maintenance involved in the operation of these ignition systems and even less to repair if they fail. Most systems are sealed and there is no option other than to replace failed components.

It is very important to narrow down the problem to the ignition system and replace the correct components rather than just replace parts hoping to solve the problem. Electronic components can be very expensive and are usually not returnable.

Refer to the "Ignition and Electrical" section for more information on troubleshooting and repairing the ignition system.

Synchronization and Timing

Synchronization and timing on an outboard engine is extremely important to obtain maximum efficiency. The powerhead cannot perform properly and produce its designed horsepower output if the fuel and ignition systems have not been precisely adjusted.

Outboards equipped with a mechanical advance type Capacitor Discharge Ignition (CDI) system use a series of link rods between the carburetor and the ignition base plate assembly. At the time the throttle is opened, the ignition base plate assembly is rotated by means of the link rod, thus advancing the timing.

Many models have timing marks on the flywheel and CDI base. A timing light is used to check the ignition timing with the powerhead operating (dynamically). An alternate method is to check the timing with the powerhead not operating (static timing). This method has the timing being checked at cranking speed only and the engine not running.

Various models have unique methods of checking ignition timing. As appropriate, these differences will be explained in detail in the text.

In simple terms, synchronization is timing the fuel system to the ignition. As the throttle is advanced to increase powerhead rpm, the fuel and the ignition systems are both advanced equally and at the same rate.

Any time the fuel system or the ignition system on a powerhead is serviced to replace a faulty part or any adjustments are made for any reason, powerhead timing and synchronization must be carefully checked and verified.

On the breaker point ignitions, synchronization is automatic once the point gap and the piston travel or timing mark alignments is correct.

Models equipped with electronic ignitions are set to the appropriate advance and/or throttle are lengths and statically timed.

➥Before making any adjustments to the ignition timing or synchronizing the ignition to the fuel system, both systems should be verified to be in good working order.

Synchronizing and timing the ignition and fuel systems on an outboard motor are critical adjustments. The following equipment is essential and is called out repeatedly in this section. This equipment must be used as described, unless otherwise instructed by the equipment manufacturer. Naturally, the equipment is removed following completion of the adjustments.

Some manufacturers recommend the use of a test wheel instead of a normal propeller in order to put a load on the engine and propeller shaft. The use of the test wheel prevents the engine from excessive rpm.

• Dial Indicator—Top dead center (TDC) of the No. 1 (top) piston must be precisely known before the timing adjustment can be made. TDC can only be determined through installation of a dial indicator into the No. 1 spark plug opening

• Timing Light—During many procedures in this section, the timing mark on the flywheel must be aligned with a stationary timing mark on the engine while the powerhead is being cranked or is running. Only through use of a timing light connected to the No. 1 spark plug lead, can the timing mark on the flywheel be observed while the engine is operating

• Tachometer—A tachometer connected to the powerhead must be used to accurately determine engine speed during idle and high-speed adjustment. Engine speed-readings range from 0 to 6,000 rpm in increments of 100 rpm. Choose a tachometer with solid state electronic circuits, which eliminates the need for relays or batteries and contribute to their accuracy. For maximum performance, the idle rpm should be adjusted under actual operating conditions. Under such conditions it might be necessary to attach a tachometer closer to the powerhead than the one installed on the control panel.

• Flywheel Rotation—The instructions may call for rotating the flywheel until certain marks are aligned with the timing pointer. When the flywheel must be rotated, always move the flywheel in the indicated direction. If the flywheel should be rotated in the opposite direction, the water pump impeller vanes would be twisted. Should the powerhead be started with the pump tangs bent back in the wrong direction, the tangs may not have time to bend in the correct direction before they are damaged. The least amount of damage to the water pump will affect cooling of the powerhead

• Test Tank—Since the engine must be operated at various times and engine speeds during some procedures, a test tank or moving the boat into a body of water, is necessary. If installing the engine in a test tank, outfit the engine with an appropriate test propeller

✲✲ WARNING

Water must circulate through the lower unit to the powerhead anytime the powerhead is operating to prevent damage to the water pump in the lower unit. Just five seconds without water will damage the water pump impeller.

3-18 MAINTENANCE AND TUNE-UP

➡Remember the powerhead will not start without the emergency tether in place behind the kill switch knob.

✱✱ WARNING

Never operate the powerhead above a fast idle with a flush attachment connected to the lower unit. Operating the powerhead at a high rpm with no load on the propeller shaft could cause the powerhead to runaway causing extensive damage to the unit.

✱✱ CAUTION

Synchronization and timing procedures are done with the engine cover removed. Be sure to keep all loose clothing, hair and your hands away from the flywheel while the engine is running. Serious injury can occur if this warning were to be ignored.

Engines with Face Plate Mounted Controls

IDLE SPEED ADJUSTMENT

▸ See Figure 52

➡There is no timing adjustment on these model engines.

1. Remove engine cover to attain easy access to carburetor.
2. Install flush device or submerse lower unit in water.
3. Run engine until it has reached operating temperature.
4. Place throttle lever in full idle position.
5. Adjust idle speed screw for proper idle RPM as listed in the "Tune-up Specifications" chart. Turning the screw clockwise will increase the idle RPM and counterclockwise will reduce the idle RPM.
6. If a proper idle adjustment could not be made, the jet needle might need to be adjusted. Refer to "Fuel System" for more information.
7. Stop the engine.
8. Reinstall the engine cover.
9. Remove the flush device, if used.

Engines with a Single Carburetor and Tiller Handle

IDLE SPEED AND MIXTURE ADJUSTMENT

▸ See Figures 53, 54, 55, 56 and 57

➡There is no timing adjustment on these model engines. Early production units may have points and condenser under the flywheel that would alter the timing. Point and condenser replacement will be covered in the "Ignition and Electrical" chapter.

1. Remove engine cover to gain access to carburetors and link rods.
2. Set idle mixture screw to proper base setting as listed in the "Carburetor Specifications" chart.
3. Adjust the throttle arm or throttle plate to make contact at the mark.
4. Install flush device or submerse lower unit in water.
5. Start engine and allow it to reach operating temperature.
6. Idle engine and adjust the idle mixture screw as follows:
   a. Turn screw clockwise ⅛ at a time until the engines RPM increases then starts to sneeze or run lean.

Fig. 52 Location of the idle speed screw—engines with face plate mounted controls

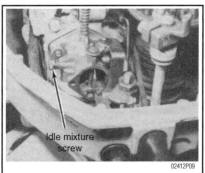

Fig. 53 Idle mixture screw location—engines with a single carburetor and tiller handle

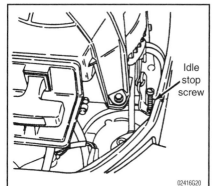

Fig. 54 Idle stop screw location—engines with a single carburetor and tiller handle

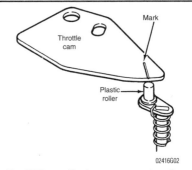

Fig. 55 On smaller horsepower and earlier 9.9 horsepower units the throttle cam is mounted below the magneto plate. It needs to be moved to make the pick up adjustment. Loosen the two screws that would be in the slots shown, adjust the cam and secure the screws after the adjustment

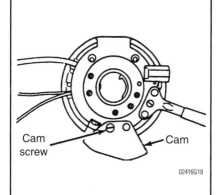

Fig. 56 An underside view of a magneto plate with adjustment screws on the throttle cam

Fig. 57 Idle mixture screw location on later models—engines with a tiller handle having a single carburetor

MAINTENANCE AND TUNE-UP   3-19

   b. Then turn the mixture screw counter clockwise ⅛ at a time until the engine RPM starts to decrease or run too rich.

   c. Calculate the total number of turns between the rich and lean settings and set the mixture screw half way or in the middle of the range.

   7. Adjust the idle screw to the proper idle RPM as listed in the "Tune-up Specifications Chart".

   8. Stop the engine.
   9. Reinstall the engine cover.
   10. Remove the flush device, if used.

1984–89 Engines with a Single Carburetor and Remote Control

➡ Engines of this type that are below 50hp do not have a timing adjustment.

CHOKE ROD ADJUSTMENT

♦ See Figures 58 and 59

   1. Remove engine cover to gain access to carburetors and link rods.
   2. Loosen, but do not remove, the screw securing the choke solenoid rod inside the swivel.

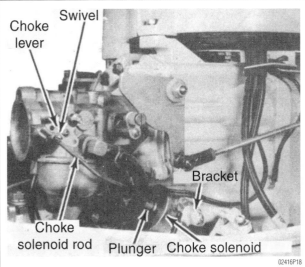

Fig. 58 Loosen, but do not remove, the screw securing the choke solenoid rod inside the swivel

   3. Place a slip of paper approximately ½" wide into the carburetor throat.
   4. Hold the choke lever down in the closed position trapping the paper under the choke shutter plate.
   5. Push the choke plunger into the solenoid as far as it will go. The choke solenoid rod will side down inside the swivel to a new position.
   6. Hold the solenoid rod in this position and tighten the screw on the swivel.
   7. Slowly pull the paper from under the closed choke shutter plate. A slight amount of drag should be felt as the paper is removed. If the drag is not felt, adjust the choke solenoid rod position at the swivel.

THROTTLE ROD ADJUSTMENT

♦ See Figure 60

   1. Remove the throttle rod from throttle cam and manually rotate the cam to contact the throttle pickup roller.
   2. Adjust the throttle cam pick up point by loosening the nut on the eccentric cam on the throttle pick up roller and rotating it until it contacts the cam at the pickup mark. Retighten eccentric cam lock nut.
   3. Reattach the throttle rod and move the throttle to the wide-open throttle position.
   4. Look down the carburetor bore and make certain the throttle plate is open all the way.
   5. If the throttle plate is not in the proper position adjust the length of the throttle rod until the correct result is achieved.

IDLE SPEED & MIXTURE ADJUSTMENT

♦ See Figures 61 and 62

   1. Return the throttle to idle position.
   2. Set idle mixture screw to proper base setting as listed in the "Carburetor Specifications" chart.
   3. Install a flushing device or submerse lower unit in water.
   4. Start the engine and allow it to reach operating temperature.
   5. Idle engine and adjust the idle mixture screw as follows:

   a. Turn screw clockwise ⅛ at a time until the engines RPM increases then starts to sneeze or run lean.

   b. Then turn the mixture screw counter clockwise ⅛ at a time until the engine RPM starts to decrease or run too rich.

   c. Calculate the total number of turns between the rich and lean settings and set the mixture screw half way or in the middle of the range.

   6. Adjust the idle screw to the proper idle RPM as listed in the "Tune-up Specifications" chart.
   7. Stop the engine.

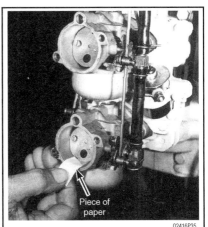

Fig. 59 Hold the choke lever down in the closed position trapping the paper under the choke shutter plate

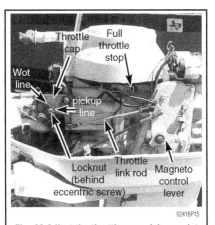

Fig. 60 Adjust the throttle cam pick up point by loosening the nut on the eccentric cam on the throttle pick up roller and rotating it until it contacts the cam at the pickup mark

Fig. 61 Idle mixture screw location—1984–1989 engines with a single carburetor and remote control

3-20 MAINTENANCE AND TUNE-UP

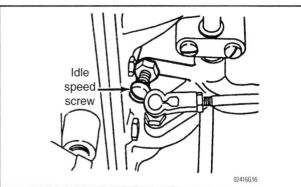

Fig. 62 Idle speed screw location—1984–89 engines with a single carburetor and remote control

IGNITION TIMING ADJUSTMENT

➥Only 50hp or larger engines of this type have adjustable max timing.

✲✲ CAUTION

Do not make any timing adjustments with the engine running. Be sure to keep all loose clothing, hair and your hands away from the flywheel. Serious injury can occur if this warning were to be ignored.

1. Install the proper size test propeller.
2. The boat needs to be launched or the engine secured in a shop test tank.
3. Secure the engine for a wide-open throttle test.

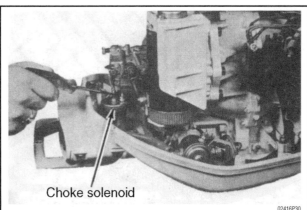

Fig. 63 Loosen, but do not remove, the screw securing the choke solenoid rod inside the swivel

4. Connect a timing light to the number one cylinder (the cylinder closest to the flywheel).
5. Start the engine and allow it to reach operating temperature.
6. Shift the engine into **FORWARD** gear and run engine at wide-open throttle.
7. Using a timing light, check max ignition timing.
8. Return engine to idle let it stabilize then shut it down.
9. Compare the timing to the specifications as listed in the "Tune-up Specifications" chart.
10. Make any necessary timing changes by adjusting the timing rod connected to the trigger housing. If necessary repeat the max timing procedure until the settings are correct.
11. Install engine cover.

1990–99 Engines with a Single Carburetor and Remote Control

CHOKE ROD ADJUSTMENT

♦ See Figure 63

➥Late model engines are equipped with a primer system as a starting aid therefore they have no choke plates to adjust.

1. Remove engine cover to gain access to carburetors and link rods.
2. Loosen, but do not remove, the screw securing the choke solenoid rod inside the swivel. Place a slip of paper approximately ½ wide into the carburetor throat.
3. Hold the choke lever down in the closed position trapping the paper under the choke shutter plate. Push the choke plunger into the solenoid as far as it will go. The choke solenoid rod will side down inside the swivel to a new position.
4. Hold the solenoid rod in this position and tighten the screw on the swivel.
5. Slowly pull the paper from under the closed choke shutter plate. A slight amount of drag should be felt as the paper is removed. If the drag is not felt, adjust the choke solenoid rod position at the swivel.

THROTTLE ROD ADJUSTMENT

♦ See Figures 64, 65 and 66

1. Remove the throttle rod from throttle cam and manually rotate the cam to contact the throttle pickup roller.
2. Adjust the throttle cam pick up point by loosening the nut on the eccentric cam on the throttle pick up roller and rotating it until it contacts the cam at the pickup mark. Retighten eccentric cam lock nut.
3. Reattach the throttle rod and move the throttle to the full open position or wide-open throttle
4. Look down the carburetor bore and make certain the throttle plate is open all the way. If the throttle plate is not in the proper position adjust the length of the throttle rod until the correct result is achieved.
5. Return the throttle to idle position.

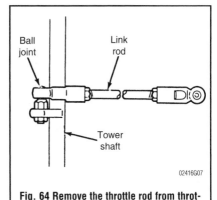

Fig. 64 Remove the throttle rod from throttle cam

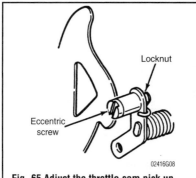

Fig. 65 Adjust the throttle cam pick up point by loosening the locknut on the eccentric cam on the throttle pick up roller

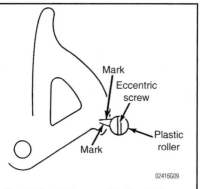

Fig. 66 Rotate the eccentric it until it contacts the cam at the pickup mark

MAINTENANCE AND TUNE-UP 3-21

IDLE SPEED ADJUSTMENT

♦ See Figures 67, 68 and 69

1. Set idle mixture screw to proper base setting as listed in the "Carburetor Specifications" chart.
2. Install a flushing device or submerse lower unit in water.
3. Run engine until it has reached operating temperature.
4. Idle engine and adjust the idle mixture screw as follows:
    a. Turn screw clockwise ⅛ at a time until the engines RPM increases then starts to sneeze or run lean.
    b. Then turn the mixture screw counter clockwise ⅛ at a time until the engine RPM starts to decrease or run too rich.
    c. Calculate the total number of turns between the rich and lean settings and set the mixture screw half way or in the middle of the range.
5. Adjust the idle screw to the proper idle RPM, as listed in the "Tune-Up Specifications" chart.
6. Stop the engine.

IGNITION TIMING ADJUSTMENT

✷✷ CAUTION

Do not make any timing adjustments with the engine running. Be sure to keep all loose clothing, hair and your hands away from the flywheel. Serious injury can occur if this warning were to be ignored.

1. Install the proper size test propeller.
2. The boat needs to be launched or the engine secured in a shop test tank.
3. Secure the engine for a wide-open throttle test.
4. Connect a timing light to the number one cylinder.
5. Start the engine and allow it to reach operating temperature.
6. Shift the engine into forward gear and run engine up to wide-open throttle.
7. Using a timing light, check max ignition timing.
8. Return engine to idle let it stabilize then shut it down.
9. Compare the timing to the specifications as listed in the "Tune-up Specifications" chart.
10. Make any necessary timing changes by adjusting the timing rod connected to the trigger housing. If necessary repeat the max timing procedure until the settings are correct.

Multiple Carburetor Engines with Motorola® Distributor Ignition

CHOKE ROD ADJUSTMENT

♦ See accompanying illustrations

1. Remove engine cover to gain access to carburetors and link rods.
2. Loosen, but do not remove, the choke rod securing screws at the swivel joints. Push the rod up slightly until ¹⁄₁₆" to ⅛" (1.59 to 3.18mm), of the rod shows above the top carburetor swivel joint. Tighten the securing screw on the job joint.

➥ On two carburetor installations, reference to the center carburetor applies to the bottom carburetor.

3. Hold the top and center carburetor shutter plates fully closed and tighten the securing screw on the center swivel joint.
4. Hold the center and bottom carburetor shutter plates fully closed, and tighten the securing screw on the bottom swivel joint.

Fig. 67 Idle mixture screw location on later model carburetor—1990–99 engines with a single carburetor and remote control

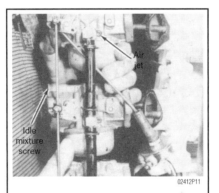

Fig. 68 Idle mixture screw location on earlier model carburetor—1990–99 engines with a single carburetor and remote control

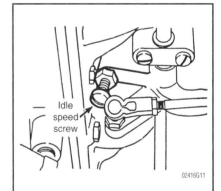

Fig. 69 Idle screw location—1990–99 engines with a single carburetor and remote control

Step 2

Step 3

Step 4

3-22 MAINTENANCE AND TUNE-UP

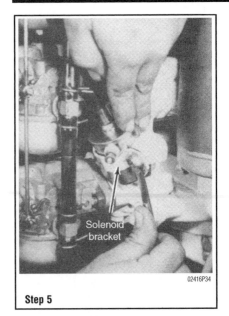
Step 5

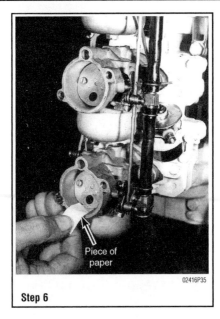
Step 6

Step 7

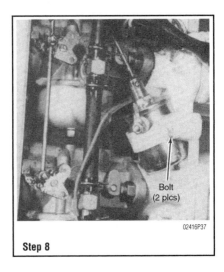
Step 8

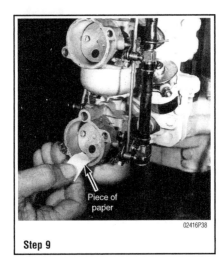
Step 9

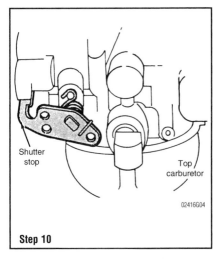
Step 10

5. Loosen the two bolts securing the choke solenoid to the powerhead just enough to permit the solenoid to slide up and down inside the retaining bracket. Do not remove these bolts.

6. Place a slip of paper approximately ½" wide into each of the carburetor throats.

7. Hold the choke plunger down in the closed position trapping the papers under the choke shutter plates. Push the choke solenoid up against the plunger.

8. Tighten the two bolts on the solenoid retaining bracket to hold the solenoid in place. Release pressure on the plunger allowing it to remain inside the solenoid.

9. Slowly pull the paper from under the choke shutter plate. A slight amount of drag should be felt as the paper is removed. If the drag is not felt, adjust the choke solenoid position in the retaining bracket.

10. Open the choke shutter valves and check to be sure all the shutters are horizontal and move together. IF necessary, adjust the position of the shutters by bending the shutter stop at the top carburetor.

THROTTLE ROD ADJUSTMENT

♦ See accompanying illustrations

1. Remove the throttle rod from throttle cam and move the cam away from the throttle pick up roller.

2. Loosen the screws on the carburetor link rod so that the throttle plates will all move separately.

3. Check to be certain the top carburetor throttle plate is completely closed.

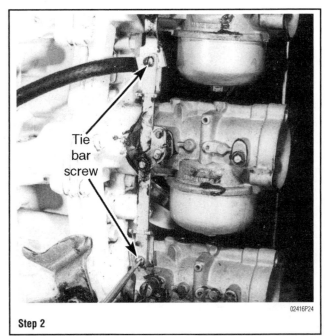
Step 2

MAINTENANCE AND TUNE-UP   3-23

4. Then check the second and third carburetor as equipped for completely closed throttle plates.

5. At this time tighten the carburetor link rod retaining screws.

6. Manually open and close the throttle and check that the throttle plates are all fully closed at rest.

7. Remove the throttle rod from throttle cam and manually rotate the cam to contact the throttle pickup roller.

8. Adjust the throttle cam pick up point by loosening the nut on the eccentric cam on the throttle pick up roller and rotating it until it contacts the cam at the pickup mark. Retighten eccentric cam lock nut.

9. Reattach the throttle rod and move the throttle to the full open position or wide-open throttle.

10. Look down the carburetor bore and make certain the throttle plate is open all the way. If the throttle plate is not in the proper position adjust the length of the throttle rod until the correct result is achieved.

11. Return the throttle to idle position.

TIMING BELT ADJUSTMENT

▶ See Figures 70 and 72

➡If the belt tension is too strong, the bushings inside the distributor and the tower shaft bearings will wear quickly. Incorrect belt tension may cause the tower shaft linkage to bind. A too loose belt condition will cause vibration in the distributor shaft. Such vibration could very likely result in timing change. There is also a danger the timing belt may slip free of the pulley.

1. Rotate the flywheel clockwise until the TDC mark on the flywheel is aligned with the "I" mark on the timing pointer.

➡Do not rotate the flywheel counterclockwise. Such action could cause damage to the water pump vanes.

2. The outer rim of the flywheel should align with the embossed curved line on the distributor pulley.

3. If the flywheel and distributor pulley do not align, begin by sliding the timing belt up and free of the pulley.

4. Rotate the pulley until the curved line follows the rim of the flywheel, and then carefully install the timing belt over the distributor pulley without disturbing the alignment.

5. Check the timing belt tension by attempting to deflect the belt at a point just under the flywheel. The belt should deflect 3/16 to 1/4" for correct tension.

6. To tighten the belt, loosen the two bolts on the distributor bracket. Position the distributor until the tension on the belt is correct. Tighten the bolts.

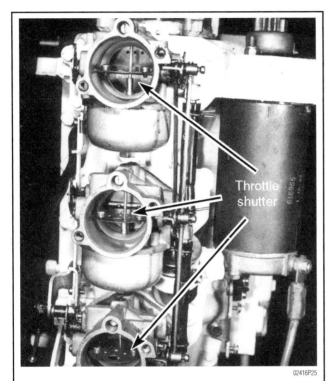

Step 6

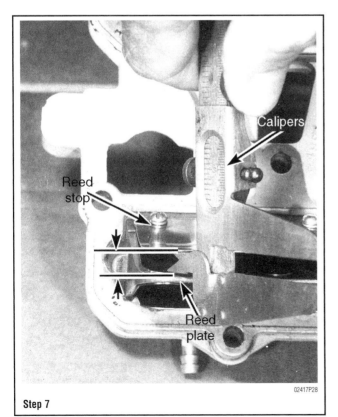

Step 7

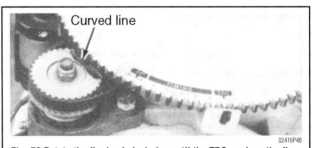

Fig. 70 Rotate the flywheel clockwise until the TDC mark on the flywheel is aligned with the "I" mark on the timing pointer

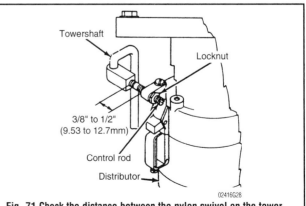

Fig. 71 Check the distance between the nylon swivel on the tower shaft and the nylon swivel on the distributor

3-24 MAINTENANCE AND TUNE-UP

7. Check the distance between the nylon swivel on the tower shaft and the nylon swivel on the distributor. The measurement should be between ⅜–½ inch.
8. If an adjustment is required, loosen the lock nut on the threaded control rod and rotate the slotted end until the distance is within specifications. Tighten the lock nut.
9. Align the outer rim of the flywheel with the embossed curved line on the distributor pulley.
10. If the flywheel and distributor pulley do not align, begin by sliding the timing belt up and free of the pulley.
11. Rotate the pulley until the curved line follows the rim of the flywheel, and then carefully install the timing belt over the distributor pulley without disturbing the alignment.

IDLE SPEED ADJUSTMENT

1. Set idle mixture screw to proper base setting.
2. Install flush device or submerse lower unit in water.
3. Run engine until it has reached operating temperature.
4. Idle engine and adjust the idle mixture screw as follows:
    a. Turn screw clockwise ⅛ at a time until the engines RPM increases then starts to sneeze or run lean.
    b. Then turn the mixture screw counter clockwise ⅛ at a time until the engine RPM starts to decrease or run too rich.
    c. Calculate the total number of turns between the rich and lean settings and set the mixture screw half way or in the middle of the range.
5. Repeat this procedure for all carburetors.
6. Adjust the idle screw to the proper idle RPM as specified in the "Tune-up Specifications" chart.
7. Stop the engine.

IGNITION TIMING ADJUSTMENT

Static Timing

♦ See Figure 72

✶✶ WARNING

These tests can be performed using either a voltmeter or ohmmeter. If the test is performed using an ohmmeter, do not apply battery voltage or damage will result to the meter.

BREAKERLESS IGNITION

1. The following adjustments are made with the powerhead not operating.
2. Disconnect both cables from the battery.
3. Disconnect the White/Black wire from the left terminal on the distributor housing (distributors with breaker points will not have this terminal).
4. Rotate the tower shaft to the wide-open throttle position. Align the specified degrees of spark advance with the timing pointer.

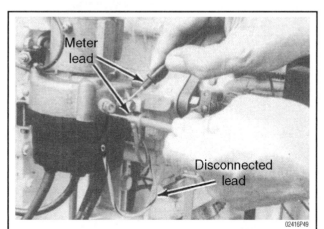

Fig. 72 Obtain a multimeter and connect it to the distributor as described in the text

5. Make contact with the Red meter lead to the right terminal on the distributor. Make contact with the Black meter lead to the left terminal (where the White/Black lead was attached).
6. Connect both battery cables to the battery.
7. Turn the ignition switch to the **ON** position.
8. Loosen the lock nut on the distributor control rod.
9. Rotate the slotted end of the control rod until the voltmeter needle swings from zero to approximately 12 volts.

➡ **If using an ohmmeter, the needle will swing from no continuity to continuity.**

10. Use a screwdriver to prevent the slotted end of the control rod from rotating, and at the same time tighten the lock nut to secure the rod in place.
11. Lightly deflect the timing belt to rock the distributor pulley back and forth.
12. If the rod length is correct, the meter needle should swing between zero and 12 volts.

DISTRIBUTOR WITH BREAKER POINTS

1. Rotate the tower shaft to the wide-open throttle position. Align the specified degrees of spark advance with the timing pointer.
2. Make contact with the Red meter lead to the single terminal on the distributor housing. Make contact with the Black meter lead to a suitable powerhead ground.
3. Connect both battery cables to the battery.
4. Turn the ignition switch to the **ON** position.
5. Loosen the lock nut on the distributor control rod.
6. Rotate the slotted end of the control rod until the voltmeter needle swings from zero to approximately 12 volts.

➡ **If using an ohmmeter, the needle will swing from no continuity to continuity.**

7. Use a screwdriver to prevent the slotted end of the control rod from rotating, and at the same time tighten the lock nut to secure the rod in place.
8. Lightly deflect the timing belt to rock the distributor pulley back and forth.
9. If the rod length is correct, the meter needle should swing between zero and 12 volts.

Dynamic Timing

♦ See Figure 73

✶✶ CAUTION

Do not make any timing adjustments with the engine running. Be sure to keep all loose clothing, hair and your hands away from the flywheel. Serious injury can occur if this warning were to be ignored.

➡ **Dynamic timing is generally more accurate that static timing.**

1. Install the proper size test propeller.
2. Launch the boat or the place the engine into shop test tank.
3. Secure the engine for a wide-open throttle test.
4. Connect a timing light to the number one cylinder.
5. Start the engine and allow it to reach operating temperature.
6. Shift the engine into forward gear and run engine up to wide-open throttle.
7. Check max ignition timing and compare it to the specifications in the "Tune-Up Specifications" chart.
8. Return engine to idle, let it stabilize, then shut it down.
9. If a correction is required adjust the length of the distributor control rod. To advance the timing, rotate the slotted screw to increase the distance between the two nylon stops on the rod. To retard the timing, rotate the slotted screw to decrease the distance between the two stops.
10. Start the engine and recheck the timing.
11. Repeat the maximum timing test until the setting is correct.

Multiple Carburetor Engines with No Distributor

➡ **Idle speed and timing settings can be found at the end of this chapter in the "Tune-Up Specifications".**

MAINTENANCE AND TUNE-UP 3-25

Fig. 73 If a timing correction is required, adjust the length of the distributor control rod using the slotted screw

CHOKE ROD ADJUSTMENT

♦ See accompanying illustrations

➡ Late model engines are equipped with a primer system as a starting aid therefore they have no choke plates to adjust.

1. Remove engine cover to gain access to carburetors and link rods.
2. Loosen, but do not remove, the choke rod securing screws at the swivel joints. Push the rod up slightly until 1/16–1/8 in. (1.59 to 3.18mm) of the rod shows above the top carburetor swivel joint. Tighten the securing screw on the job joint.

➡ On two carburetor installations, references to the center carburetor apply to the bottom carburetor.

3. Hold the top and center carburetor shutter plates fully closed and tighten the securing screw on the center swivel joint.
4. Hold the center and bottom carburetor shutter plates fully closed, and tighten the securing screw on the bottom swivel joint.
5. Loosen the two bolts securing the choke solenoid to the powerhead just enough to permit the solenoid to slide up and down inside the retaining bracket. Do not remove these bolts.
6. Place a slip of paper approximately 1/2" wide into each of the carburetor throats.
7. Hold the choke plunger down in the closed position trapping the papers under the choke shutter plates. Push the choke solenoid up against the plunger.

Step 2

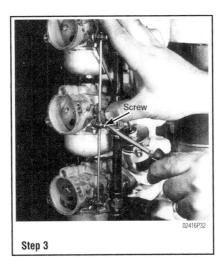

Step 3

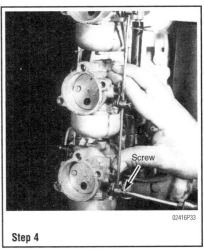

Step 4

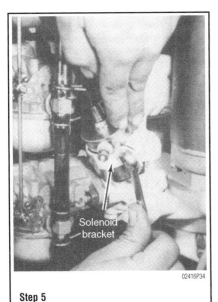

Step 5

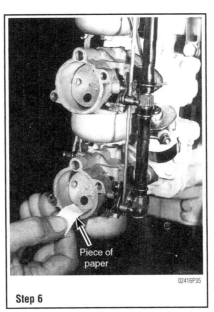

Step 6

Step 7

3-26 MAINTENANCE AND TUNE-UP

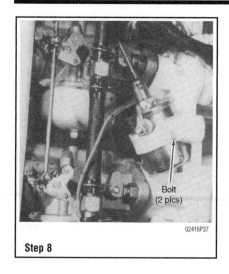
Step 8

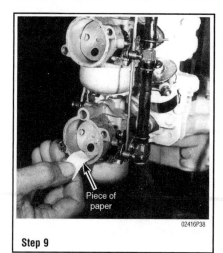
Step 9

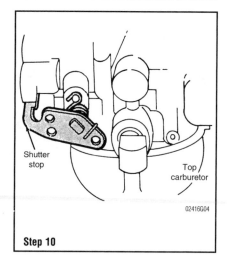
Step 10

8. Tighten the two bolts on the solenoid retaining bracket to hold the solenoid in place. Release pressure on the plunger allowing it to remain inside the solenoid.

9. Slowly pull the paper from under the choke shutter plate. A slight amount of drag should be felt as the paper is removed. If the drag is not felt, adjust the choke solenoid position in the retaining bracket.

10. Open the choke shutter valves and check to be sure all the shutters are horizontal and move together. If necessary, adjust the position of the shutters by bending the shutter stop at the top carburetor.

THROTTLE ROD ADJUSTMENT

♦ See accompanying illustrations

1. Remove engine cover to gain access to carburetors and link rods.
2. Remove the throttle rod from throttle cam and move the cam away from the throttle pick up roller.
3. Loosen the screws on the carburetor link rod so that the throttle plates will all move separately.
4. Check to be certain the top carburetor throttle plate is completely closed.
5. Then check the second and third carburetor as equipped for completely closed throttle plates.
6. At this time tighten the carburetor link rod retaining screws.
7. Manually open and close the throttle and check that the throttle plates are all fully closed at rest.
8. Rotate the throttle cam to contact the throttle pickup roller.
9. Adjust the throttle cam pick up point by loosening the nut on the eccentric cam on the throttle pick up roller and rotating it until it contacts the cam at the pickup mark. Retighten eccentric cam lock nut.
10. Reattach the throttle rod and move the throttle to the full open position or wide-open throttle.

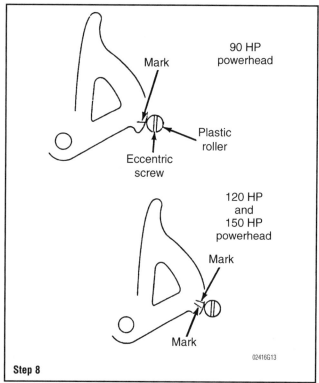
Step 8

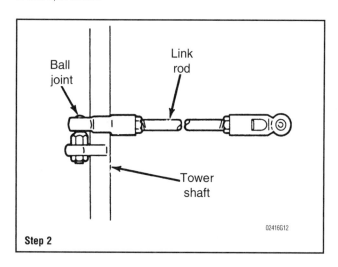
Step 2

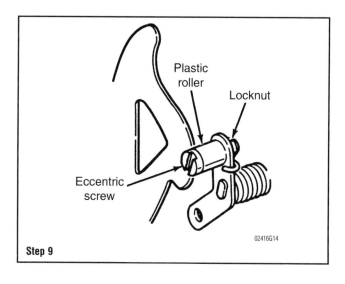
Step 9

MAINTENANCE AND TUNE-UP 3-27

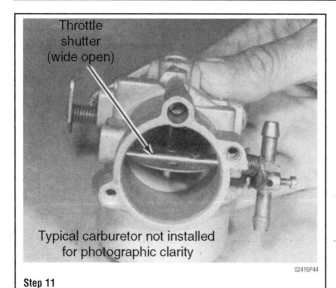

Step 11

11. Look down the carburetor bore and make certain the throttle plate is open all the way. If the throttle plate is not in the proper position adjust the length of the throttle rod until the correct result is achieved.
12. Return the throttle to idle position.

IDLE SPEED ADJUSTMENT

▶ See Figures 74 and 75

1. Remove engine cover to gain access to carburetors and link rods.
2. Set idle mixture screw to proper base setting as listed in the "Carburetor Specifications" chart.
3. Install a flushing device or submerse lower unit in water.
4. Run engine until it has reached operating temperature.
5. Idle engine and adjust the idle mixture screw as follows:
   a. Turn screw clockwise 1/8 turn at a time until the engines RPM increases then starts to sneeze or run lean.
   b. Then turn the mixture screw counter clockwise 1/8 turn at a time until the engine RPM starts to decrease or run too rich.
   c. Calculate the total number of turns between the rich and lean settings and set the mixture screw half way or in the middle of the range. Repeat this procedure for all carburetors.
6. Adjust the idle screw to the proper idle RPM as listed in the "Tune-Up Specifications" chart.
7. Stop the engine.

IGNITION TIMING

▶ See Figures 76, 77 and 78

Static Engine Timing

1. Remove spark plugs from the engine.
2. Connect a spark tester or spark board to all spark plug wires. This will prevent the outboard from starting during the test.

Fig. 74 Idle mixture screw location—multiple carburetor engines with no distributor

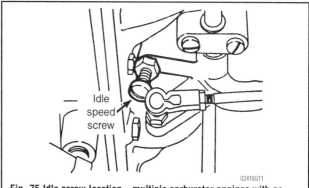

Fig. 75 Idle screw location—multiple carburetor engines with no distributor

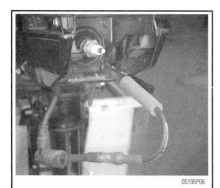

Fig. 76 Connect a spark tester or spark board to all spark plug wires

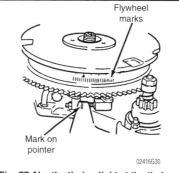

Fig. 77 Aim the timing light at the timing marks on the flywheel and the timing pointer

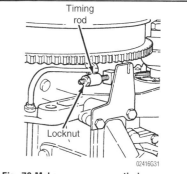

Fig. 78 Make any necessary timing changes by adjusting the timing rod connected to the trigger housing

3-28 MAINTENANCE AND TUNE-UP

3. Remove the propeller.
4. Attach a timing light to the number one cylinder.
5. Shift engine into forward gear then to the wide-open throttle position.
6. Crank engine over while aiming the timing light at the timing marks and note the reading.
7. The timing should be within the specifications given in the "Tune-Up Specifications" chart.
8. Make any necessary timing changes by adjusting the timing rod connected to the trigger housing.
9. If necessary repeat the max timing procedure until the settings are correct.

Dynamic Engine Timing

✳✳ CAUTION

Do not make any timing adjustments with the engine running. . Be sure to keep all loose clothing, hair and your hands away from the flywheel. Serious injury can occur if this warning were to be ignored.

1. Install the proper size test propeller.
2. The boat needs to be launched or the engine bolted into shop test tank.
3. Secure the engine for a wide-open throttle test.
4. Connect a timing light to the number one cylinder.
5. Start the engine and allow it to reach operating temperature.
6. Shift the engine into forward gear and run engine up to wide-open throttle.
7. Check max ignition timing and compare it to the specifications listed in the "Tune-Up Specifications" chart.
8. Return engine to idle, let it stabilize, then shut it down.
9. Make any necessary timing changes by adjusting the timing rod connected to the trigger housing.
10. If necessary repeat the max timing procedure until the settings are correct.

WINTER STORAGE CHECKLIST

Taking extra time to store the boat properly at the end of each season will increase the chances of satisfactory service at the next season. Remember that storage is the greatest enemy of an outboard motor. The unit should be run on a monthly basis. The boat steering and shifting mechanism should also be worked through complete cycles several times each month. If a small amount of time is spent in such maintenance, the reward will be satisfactory performance, increased longevity and greatly reduced maintenance expenses.

For many years there has been the widespread belief simply shutting off the fuel at the tank and then running the powerhead until it stops is the proper procedure before storing the engine for any length of time. This method is absolutely wrong!

First, it is not possible to remove all fuel in the carburetor by operating the powerhead until it stops. Considerable fuel is trapped in the float chamber and other passages and in the line leading to the carburetor. The only guaranteed method of removing all fuel is to take the time to remove the carburetors and drain the fuel.

Proper storage involves adequate protection of the unit from physical damage, rust, corrosion and dirt. The following steps provide an adequate maintenance program for storing the unit at the end of a season.

1. Stabilize your fuel with an appropriate amount of stabilizer chemical.
2. Install an engine flushing attachment or submerge lower unit in a fresh water drum or container.

✳✳ WARNING

Never run the engine without a water supply to prevent water pump damage and engine overheating

3. Start engine in neutral at an idle only to get the stabilized fuel into the engine and flush out the powerhead.
4. Once the engine is up to operating temperature slowing spray an engine storage solution into the carburetors until the engine starts to smoke heavily.
5. Remove the spark plugs then squirt a small quantity of storage solution into each spark plug hole and crank the engine over to distribute the oil around the engine internals. Reinstall the old spark plugs (you will install new spark plugs in the spring).
6. Drains all fuel from the carburetor float bowls.
7. Store the fuel tank in a cool dry area with the vent OPEN to allow air to circulate through the tank. Do not store the fuel tank in an in a closed area or on bare concrete. Position the tank to allow air to circulate around it.
8. Change the fuel filter.
9. Drain and then fill the lower unit with new lower unit gear oil.
10. Lubricate the throttle and shift linkage and the steering pivot shaft.
11. Clean the outboard unit thoroughly then coat the powerhead with a corrosion and rust preventative spray.
12. Remove the propeller. Apply Perfect Seal®or a waterproof sealer to the propeller shaft splines and then install the propeller back in position.
13. Be sure all drain holes in the gear housing are open and free of obstructions. Check to be sure that the flush plug has been removed to allow all the water to drain. Trapped water could freeze, expand and cause expensive castings to crack.
14. Be sure to consult your owners' manual for any particular storage procedures applicable to your specific model.

SPRING COMMISSIONING CHECKLIST

A spring tune-up is essential to getting the most out of your engine. If the engine has been properly winterized, it is usually no problem to get it in top running condition again in the springtime. If the engine has just been put in the garage and forgotten for the winter, then it is doubly important to do a complete tune up before putting the engine back into service. If you have ever been stranded out on the water because your engine has died and you had to suffer the embarrassment of having to be towed back to the marina, now is the time to prevent that from occurring.

Satisfactory performance and maximum enjoyment can be realized if a little time is spent in preparing the outboard unit for service at the beginning of the season. Assuming the unit has been properly stored, a minimum amount of work is required to prepare the unit for use. The following steps outline an adequate and logical sequence of tasks to be performed before using the outboard the first time in a new season.

1. Lubricate the outboard according to the manufacturer's recommendations.
2. Perform a tune-up on the engine. This should include replacing the spark plugs and making a thorough check of the ignition system. The ignition system check should include the ignition coils, stator assembly, and condition of the wiring and the battery.

3. If a built-in fuel tank is installed, take time to check the gasoline tank and all fuel lines, fittings, couplings, valves, including the flexible tank fill and vent. Turn on the fuel supply valve at the tank. If the fuel was not drained at the end of the previous season, make a careful inspection for gum formation. If a six-gallon fuel tank is used, take the same action. When gasoline is allowed to stand for long periods of time, particularly in the presence of copper, gummy deposits form. This gum can clog the filters, lines and passageways in the carburetor.
4. Replace the oil in the lower unit.
5. Replace the fuel filter.
6. The engine can be run with the lower unit in water to flush it. If this is not practical, a flush attachment may be used. This unit is attached to the water pick-up in the lower unit. Attach a garden hose, turn on the water, and allow the water to flow into the engine for awhile and then run the engine.

✳✳ CAUTION

Water must circulate through the lower unit to the powerhead anytime the powerhead is operating to prevent the engine from overheating and damage to the water pump in the lower unit. Just five seconds without water will damage the water pump impeller.

MAINTENANCE AND TUNE-UP 3-29

7. Check the exhaust outlet for water discharge. Check for leaks. Check operation of the thermostat.

8. Check the electrolyte level in the battery and the voltage for a full charge. Clean and inspect the battery terminals and cable connections. Take time to check the polarity, if a new battery is being installed. Cover the cable connections with grease or special protective compound as prevention to corrosion formation. Check all electrical wiring and grounding circuits.

9. Check all electrical parts on the engine and lower portions of the hull. Rubber boots help keep electrical connections clean and reduce the possibility of arcing.

10. If a water separating filter is installed between the fuel tank and the powerhead fuel filter, replace the element at least once each season. This filter removes water and fuel system contaminants such as dirt, rust and other solids, thus reducing potential problems.

11. As a last step in spring commissioning, perform a full engine tune-up.

✻✻ CAUTION

Before putting the boat in the water, take time to verify the drain plugs are installed. Countless numbers of boating excursions have had a very sad beginning because the boat was eased into the water only to have the boat begin to fill with the water.

Maintenance Interval Chart

Component	First 1mth/10hrs	Every 3mths/50hrs	Every 6mths/100hrs	Off Season
Lower Unit Lubricant	R		R	R
Carburetor	I		I	I
Cylindar Head Bolts	T		T	T
Exhaust Cover Bolts	T		T	T
Spark Plugs	I		I	C & A
Fuel Line	I	I	I	I
Fuel Filter/Screen	I	I	I	C & I
Lubrication fittings		I	L	C & L
Prop Shaft		I	L	L
Propeller	I		I	I
Emergency Switch	I	I	I	I
Starter Bendix	I	I	L	C & L
Zinc Annodes	I	I	I	I & R
Distributor & Drive Belt	I	I	I	I
Water Pump Impeller				R
Drive Shaft Splines				L
Cylinder Head Gasket				I ①
Thermostat			I	I ①
Power Trim Fluid	I	I	I	I

T-Tighten

A-Adjust

C-Clean

I-Inspect and Clean, Adjust, Lubricate or Replace

L-Lubricate

R-Replace

① Replace on a semi-annual basis when running in salt or corrosive waters

General Engine Specifications

Year	Horsepower	Cyl.	Displacement cu.in. (cc)	Bore and Stroke	Ignition System ①	Operating Range RPM	Cooling System
1992-93	3	1	4.6 (75.4)	1.850x1.69	Thunderbolt/pointless	4500-5500	Impeller pump
1984-87	4	1	5 (82.0)	2.0x1.5937	Magneto	4750-5750	Impeller pump
1987-89	5	1	5 (82.0)	2.0x1.5624	B.I.M.	4750-5750	Impeller pump
1990-98	5	1	5.6 (91.8)	2.116x1.59	S.E.M.	5500-6500	Impeller pump
1985	7.5	2	10 (163.9)	2.0x1.5937	Magneto	4250-5250	Impeller pump
1984	9.9	2	13.15 (215.5)	2.1875x1.75	Magneto	4250-5250	Impeller pump
1985-89	9.9	2	15.41 252.6	2.25x1.9375	Magneto	4250-5250	Impeller pump
1990-93	9.9	2	15.41 252.6	2.25x1.9375	Magneto	4500-5500	Impeller pump
1994-98	9.9	2	15.41 252.6	2.25x1.9375	B.I.M.	4000-5000	Impeller pump
1984-89	15	2	15.41 252.6	2.25x1.9375	Magneto	4600-5600	Impeller pump
1990-98	15	2	15.41 252.6	2.25x1.9375	B.I.M.	5500-6500	Impeller pump
1994-98	25	3	25.75 (442.0)	2.375x1.94	S.E.M.	5000-6000	Impeller pump
1986-91	35	2	34.1 (558.9)	3.0x2.414	Prestolite	4500-5500	Impeller pump/thermostat control
1992-96	40	2	50 (820)	3.375x2.8	Thunderbolt	4500-5500	Impeller pump/thermostat control
1996-99	40	2	50 (820)	3.375x2.8	C.D.M.	4500-5500	Impeller pump/thermostat control
1984-89	50	2	44.7 (732.6)	3.1875x2.8	Prestolite	4500-5500	Impeller pump/thermostat control
1990-91	50	2	48.3 (791.6)	3.313x2.8	Thunderbolt	5000-5500	Impeller pump/thermostat control
1992-99	50	2	50 (820)	3.375x2.8	C.D.M.	5000-5500	Impeller pump/thermostat control
1985	60	2	49.9 (817.9)	3.375x2.79	Prestolite	5000-5500	Impeller pump/thermostat control
1991-93	70	3	75.2 (1232.5)	3.375x2.87	Thunderbolt	4500-5500	Impeller pump/thermostat control
1994-96	75	3	75.1(1230.1)	3.375x2.8	Thunderbolt	4750-5250	Impeller pump/thermostat control
1997-99	75	3	75.1(1230.1)	3.375x2.8	C.D.M.	4750-5250	Impeller pump/thermostat control
1984-86	85	3	72.4 (1186.6)	3.312x2.8	Motorola	4500-5500	Impeller pump/thermostat control
1984-91	85	3	72.4 (1186.6)	3.312x2.8	Prestolite	4500-5500	Impeller pump/thermostat control
1990A	90	3	72.4 (1186.6)	3.312x2.8	Prestolite	4500-5500	Impeller pump/thermostat control
1990-96	90	3	75.2 (1232.5)	3.375x2.8	Thunderbolt	5000-5500	Impeller pump/thermostat control
1997-99	90	3	75.2 (1232.5)	3.375x2.8	C.D.M.	4750-5250	Impeller pump/thermostat control
1990	120	4	99.2 (1625.9)	3.312x2.8	Prestolite	4500-5500	Impeller pump/thermostat control
1990-96	120	4	102.9 (1686.5)	3.375x2.8	Thunderbolt	5000-5500	Impeller pump/thermostat control
1997-99	120	4	102.9 (1686.5)	3.375x2.8	C.D.M.	4750-5250	Impeller pump/thermostat control
1984-86	125	4	99.2 (1625.9)	3.3125x2.87	Motorola	4500-5500	Impeller pump/thermostat control
1984-89	125	4	99.2 (1625.9)	3.3125x2.87	Prestolite	4500-5500	Impeller pump/thermostat control
1990	150	5	124 (2032.4)	3.312x2.8	Prestolite	5000-5500	Impeller pump/thermostat control
1990-94	150	5	128.6 (2107.8)	3.375x2.8	Thunderbolt	5000-5500	Impeller pump/thermostat control

① Some ignition systems overlap model years. Refer to the "Ignition and Electrical" section to properly identify systems.

MAINTENANCE AND TUNE-UP 3-31

Tuneup Specifications Chart

Horsepower	Years	Spark Plug NGK	Spark Plug Champion	Spark Plug Gap Inch(mm)	Maximum Ignition Timing Static	Maximum Ignition Timing Dynamic	Point Gap Inch(mm)	Max RPM	Idle RPM (Neutral)	Firing Order
3	1990-94	BP6HS	L87YC	.040(1.0)	Not Adjustable	Not Adjustable	—	4500-5500	900-1000	One Cyl
4	1984-87	B6HS	L86	.030(.76)	Not Adjustable	Not Adjustable	.020(.50)	4750-5750	800-1000	One Cyl
5	1987-91	B6HS	L86	.030(.76)	Not Adjustable	Not Adjustable	.020(.50)	4750-5750	800-1000	One Cyl
5	1992-93	BP6HS	L87YC	.030(.76)	Not Adjustable	Not Adjustable	—	5500-6500	800-1000	One Cyl
5	1994-98	BP6HS	L87YC	.035(.89)	Not Adjustable	Not Adjustable	—	5500-6500	950-1050	One Cyl
7.5	1985	B7HS	L82C	.030(.76)	Not Adjustable	Not Adjustable	.020(.50)	4250-5250	600-750	1-2
9.9	1984-89	B7HS	L82C	.030(.76)	Not Adjustable	Not Adjustable	.020(.50)	4250-5250	600-750	1-2
9.9	1990-91	B7HS	L82C	.030(.76)	Not Adjustable	Not Adjustable	—	4500-5500	700-800	1-2
9.9	1992-98	BP6HS	L87YC	.030(.76)	Not Adjustable	Not Adjustable	—	4000-5000	700-800	1-2
15	1984-89	B7HS	L82C	.030(.76)	Not Adjustable	Not Adjustable	.020(.50)	4600-5600	600-750	1-2
15	1990-98	BP7HS	L82YC	.040(1.0)	Not Adjustable	Not Adjustable	—	5500-6500	700-800	1-2
25	1994-99	B7HS	L82C	.035(.89)	Not Adjustable	Not Adjustable	—	5000-6000	700-800	1-2
35	1986-91	B6HS	L86C	.040(1.0)	Not Adjustable	Not Adjustable	—	4500-5500	600-750	1-2
40	1992-99	BUHW	L76V	②	32 BTDC	30 BTDC	—	4500-5500	700-800	1-2
50	1984-89	BUHX	UL18V	②	30 BTDC	28 BTDC	—	4500-5500	600-750	1-2
50	1990-91	BUHX	UL18V	②	32 BTDC	30 BTDC	—	5000-5500	700-800	1-2
50	1992-99	BUHW	L76V	②	34 BTDC	32 BTDC	—	5000-5500	700-800	1-2
60	1985	BUHX	UL18V	②	32 BTDC	30 BTDC	—	5000-5500	600-750	1-2
70	1991-93	BUHW	L76V	②	32 BTDC	30 BTDC	—	4500-5500	700-800	1-2-3
75	1994-99	BUHW	L76V	②	32 BTDC	30 BTDC	—	4750-5250	700-800	1-2-3
85	1984-86	BUHX	UL18V	②	34 BTDC	32 BTDC	.014 ①	4500-5500	700-750	1-2-3
85	1987-89	BUHX	UL18V	②	34 BTDC	32 BTDC	—	4500-5500	700-750	1-2-3
85	1990-91	BUHX	UL18V	②	32 BTDC	30 BTDC	—	4500-5500	700-800	1-2-3
90	1990-91	BUHX	UL18V	②	32 BTDC	30 BTDC	—	4500-5500	700-800	1-2-3
90	1992-99	BUHW	L76V	②	32 BTDC	30 BTDC	—	4500-5500	700-800	1-3-2
120	1990-91	BUHX	UL18V	②	32 BTDC	30 BTDC	—	4750-5250	700-800	1-3-2-4
120	1992-99	BUHW	L76V	②	32 BTDC	30 BTDC	—	4750-5250	700-800	1-3-2-4
125	1984-89	BUHX	UL18V	②	34 BTDC	32 BTDC	.010 ①	4500-5500	700-800	1-3-2-4
150	1990-94	BUHW	L76V	②	32 BTDC	30 BTDC	—	5000-5500	700-800	1-5-2-3-4

① Early models with distributors still had points
② Some models may use surface gap spark plugs. Standard spark plug gap is .030(.76)

Lower Unit Specifications

Horsepower	Year	Gear Ratio	Capacity Ounces	Capacity Litres
3	1990-94	2.18:1	3	0.09
4	1987-87	2.01:1	4	0.12
5	1987-98	2.01:1	4	0.12
7.5	1985	2.01:1	5	0.15
9.9	1984-98	1.57:1	5	0.15
15	1984-98	1.57:1	5	0.15
25	1994-98	2.25:1	7.6	0.23
35	1986-91	2.01:1	12	0.36
40	1992-94	1.62:1	12	0.36
40	1995-99	2.0:1	14.9	0.44
50	1984-94	1.62:1	12	0.36
50	1995-99	2.0:1	14.9	0.44
60	1985	1.62:1	6.5	0.19
70	1991-99	1.64:1	11.5	0.34
85	1984-91	2.01:1	26	0.77
90	1990-92	1.93:1	26	0.77
90	1993-99	1.93:1	35	1.04
90	1992-94	1.93:1	22.5	0.67
120	1995-99	2.3:1	26	0.77
120	1990-92	1.93:1	35	1.04
120	1993-94	1.93:1	22.5	0.67
125	1995-99	2.07:1	26	0.77
125	1984-89	1.73:1	26	0.77
150	1989	1.93:1	26	0.77
150	1989-92	1.93:1	26	0.77
150	1993-94	1.93:1	35	1.04

Carburetor Specifications

Model	Year	Initial Idle Screw Setting	Float Height①
3	1990-94	1 turn out from light seat	.090" Carb body to top of hinge
4	1984-87	1 turn out from light seat	1/8" Carb body to float bottom
5	1987-89②	1 turn out from light seat	1/8" Carb body to float bottom
5	1989-98③	1 turn out from light seat	Level & parallel with carb body
7.5	1985	1 turn out from light seat	1/8" Carb body to float bottom
9.9	1984-89②	1 turn out from light seat	1/8" Carb body to float bottom
9.9	1989-98②	1 turn out from light seat	Level & parallel with carb body
15	1984-89②	1 turn out from light seat	1/8" Carb body to float bottom
15	1989-98②	1 turn out from light seat	Level & parallel with carb body
35	1984-91	1 turn out from light seat	3/16" Carb body to float bottom
40	1992-99	1 turn out from light seat	Level & parallel with carb body
50	1984-95	1 turn out from light seat	3/16" Carb body to float bottom
50	1996-99	1 turn out from light seat	Level & parallel with carb body
60	1985	1 turn out from light seat	3/16" Carb body to float bottom
70	1991-93	1 turn out from light seat	Level & parallel with carb body
75	1994-99	1 turn out from light seat	Level & parallel with carb body
85	1984-89	1 turn out from light seat	13/32" Carb body to float top
90	1990-92	1 turn out from light seat	3/16" Carb body to float bottom
90	1993-99	1 turn out from light seat	Level & parallel with carb body
120	1990-92	1 turn out from light seat	3/16" Carb body to float bottom
120	1993-99	1 turn out from light seat	Level & parallel with carb body
125	1984-89	1 turn out from light seat	13/32" Carb body to float top
150	1989-94	1 turn out from light seat	Level & parallel with carb body

① All float heights are measured with carburetor inverted
② "A" Model units
③ "B" Model units & newer

FUEL AND COMBUSTION 4-2
FUEL 4-2
    OCTANE RATING 4-2
    VAPOR PRESSURE 4-2
    ALCOHOL-BLENDED FUELS 4-2
    HIGH-ALTITUDE OPERATION 4-2
    RECOMMENDATIONS 4-2
COMBUSTION 4-2
CARBURETED FUEL SYSTEM 4-3
DESCRIPTION AND OPERATION 4-3
    BASIC FUNCTIONS 4-3
    FUEL & AIR METERING 4-3
    CARBURETOR CIRCUITS 4-4
    FUEL PUMP 4-5
TROUBLESHOOTING THE FUEL
  SYSTEM 4-5
    COMMON PROBLEMS 4-6
    COMBUSTION RELATED PISTON
      FAILURES 4-7
CARBURETOR 4-7
    REMOVAL & INSTALLATION 4-7
    OVERHAUL 4-10
FUEL PRIMER 4-22
    TESTING 4-23
FUEL PUMP 4-23
    TESTING 4-23
    REMOVAL & INSTALLATION 4-23
    OVERHAUL 4-24
FUEL LINES 4-26
RECIRCULATING SYSTEM 4-26

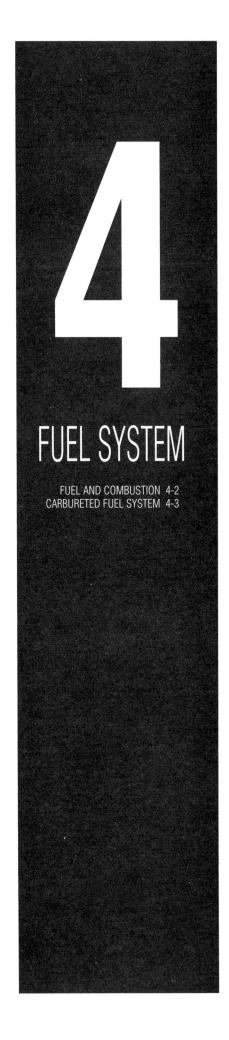

FUEL SYSTEM

FUEL AND COMBUSTION 4-2
CARBURETED FUEL SYSTEM 4-3

4-2 FUEL SYSTEM

FUEL AND COMBUSTION

Fuel

Fuel recommendations have become more complex as the chemistry of modern gasoline changes. The major driving force behind the changes in gasoline chemistry is the search for additives to replace lead as an octane booster and lubricant. These new additives are governed by the types of emissions they produce in the combustion process. Also, the replacement additives do not always provide the same level of combustion stability, making a fuel's octane rating less meaningful.

In the 1960's and 1970's, leaded fuel was common. The lead served two functions. First, it served as an octane booster (combustion stabilizer) and second, in 4-Stroke engines, it served as a valve seat lubricant. For 2-stroke engines, the primary benefit of lead was to serve as a combustion stabilizer. Lead served very well for this purpose, even in high heat applications.

Today, all lead has been removed from the refining process. This means that the benefit of lead as an octane booster has been eliminated. Several substitute octane boosters have been introduced in the place of lead. While many are adequate in an automobile engines, most do not perform nearly as well as lead did, even though the octane rating of the fuel is the same.

OCTANE RATING

A fuel's octane rating is a measurement of how stable the fuel is when heat is introduced. Octane rating is a major consideration when deciding whether a fuel is suitable for a particular application. For example, in an engine, we want the fuel to ignite when the spark plug fires and not before, even under high pressure and temperatures. Once the fuel is ignited, it must burn slowly and smoothly, even though heat and pressure are building up while the burn occurs. The unburned fuel should be ignited by the traveling flame front, not by some other source of ignition, such as carbon deposits or the heat from the expanding gasses. A fuel's octane rating is known as a measurement of the fuel's anti-knock properties (ability to burn without exploding).

Usually a fuel with a higher octane rating can be subjected to a more severe combustion environment before spontaneous or abnormal combustion occurs. To understand how two gasoline samples can be different, even though they have the same octane rating, we need to know how octane rating is determined.

The American Society of Testing and Materials (ASTM) has developed a universal method of determining the octane rating of a fuel sample. The octane rating you see on the pump at a gasoline station is known as the pump octane number. Look at the small print on the pump. The rating has a formula. The rating is determined by the R+M/2 method. This number is the average of the research octane reading and the motor octane rating.

• The Research Octane Rating is a measure of a fuel's anti-knock properties under a light load or part throttle conditions. During this test, combustion heat is easily dissipated.

• The Motor Octane Rating is a measure of a fuel's anti-knock properties under a heavy load, or full throttle conditions, when heat buildup is at maximum.

Because a 2-stroke engine has a power stroke every revolution, with heat buildup every revolution, it tends to respond more to the motor octane rating of the fuel than the research octane rating. Therefore, in an outboard motor, the motor octane rating of the fuel is the best indication of how it will perform.

VAPOR PRESSURE

Fuel vapor pressure is a measure of how easily a fuel sample evaporates. Many additives used in gasoline contain aromatics. Aromatics are light hydrocarbons distilled off the top of a crude oil sample. They are effective at increasing the research octane of a fuel sample, but can cause vapor lock (bubbles in the fuel line) on a very hot day. If you have an inconsistent running engine and you suspect vapor lock, use a piece of clear fuel line to look for bubbles, indicating that the fuel is vaporizing.

One negative side effect of aromatics is that they create additional combustion products such as carbon and varnish. If your engine requires high octane fuel to prevent detonation, de-carbon the engine more frequently with an internal engine cleaner to prevent ring sticking due to excessive varnish buildup.

ALCOHOL-BLENDED FUELS

When the Environmental Protection Agency mandated a phase-out of the leaded fuels in January of 1986, fuel suppliers needed an additive to improve the octane rating of their fuels. Although there are multiple methods currently employed, the addition of alcohol to gasoline seems to be favored because of its favorable results and low cost. Two types of alcohol are used in fuel today as octane boosters, methanol (wood alcohol) or ethanol (grain alcohol).

When used as a fuel additive, alcohol tends to raise the research octane of the fuel, so these additives will have limited benefit in an outboard motor. There are, however, some special considerations due to the effects of alcohol in fuel.

• Since alcohol contains oxygen, it replaces gasoline without oxygen content and tends to cause the air/fuel mixture to become leaner.

• On older outboards, the leaching affect of alcohol will, in time, cause fuel lines and plastic components to become brittle to the point of cracking. Unless replaced, these cracked lines could leak fuel, increasing the potential for hazardous situations.

• When alcohol blended fuels become contaminated with water, the water combines with the alcohol then settles to the bottom of the tank. This leaves the gasoline (and the oil for models using premix) on a top layer.

➡**Modern outboard fuel lines and plastic fuel system components have been specially formulated to resist alcohol leaching effects.**

HIGH ALTITUDE OPERATION

At elevated altitudes there is less oxygen in the atmosphere than at sea level. Less oxygen means lower combustion efficiency and less power output. As a general rule, power output is reduced three percent for every thousand feet above sea level.

On carbureted engines, re-jetting for high altitude does not restore lost power, it simply corrects the air-fuel ratio for the reduced air density and makes the most of the remaining available power. The most important thing to remember when re-jetting for high altitude is to reverse the jetting when return to sea level. If the jetting is left lean when you return to sea level conditions, the correct air/fuel ratio will not be achieved and possible powerhead damage may occur.

RECOMMENDATIONS

According to the fuel recommendations that come with your outboard, the minimum recommended fuel is 87 octane. An 87 octane rating generally means regular grade unleaded. Premium unleaded is more stable under severe conditions, but also produces more combustion products. Therefore, when using premium unleaded, more frequent de-carboning may be necessary.

Combustion

♦ See Figure 1

Unlike a 4-stroke engine, a 2-stroke engine has a power stroke every revolution of the crankshaft. Therefore, the 2-stroke engine has twice as many power strokes for any given RPM. If the displacement of the two types of engines is identical, then the 2-stroke engine has to dissipate twice as much heat as the 4-stroke engine.

In such a high heat environment, the fuel must be very stable to avoid detonation. If any parameters affecting combustion change suddenly (the engine runs lean for example), uncontrolled heat buildup will occurs very rapidly.

The combustion process is affected by several interrelated factors. This means that when one factor is changed, the other factors also must be changed to maintain the same controlled burn and level of combustion stability.

• Compression—determines the level of heat buildup in the cylinder when the air-fuel mixture is compressed. As compression increases, so does the potential for heat buildup

FUEL SYSTEM 4-3

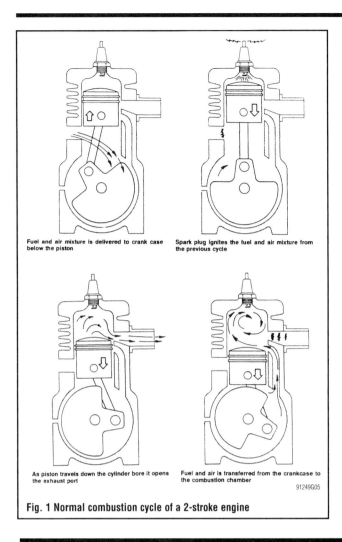

Fig. 1 Normal combustion cycle of a 2-stroke engine

- Ignition Timing—determines when the gasses will start to expand in relation to the motion of the piston. If the ignition timing is too advanced, gasses will be ignited and begin to expand too soon, such as they would during preignition. The motion of the piston opposes the expansion of the gasses, resulting in extremely high combustion chamber pressures and heat. If the ignition timing is retarded, the gases are ignited later in relation to piston position. This means that the piston has already traveled back down the bore toward the bottom of the cylinder, resulting in less usable power.
- Fuel Mixture—determines how efficient the burn will be. A rich mixture burns slower than a lean one. If the mixture is too lean, it can't become explosive. The slower the burn, the cooler the combustion chamber, because pressure buildup is gradual.
- Fuel Quality (Octane Rating)—determines how much heat is necessary to ignite the mixture. Once the burn is in progress, heat is on the rise. The unburned poor quality fuel is ignited all at once by the rising heat instead of burning gradually as a flame front of the burn passing by. This action results in detonation (pinging).

There are two types of abnormal combustion—preignition and detonation.

- Preignition—occurs when the air-fuel mixture is ignited by some other incandescent source other than the correctly timed spark from the spark plug.
- Detonation—occurs when excessive heat and or pressure ignites the air/fuel mixture rather than the spark plug. The burn becomes explosive.

In general, anything that can cause abnormal heat buildup can be enough to push an engine over the edge to abnormal combustion, if any of the four basic factors previously discussed are already near the danger point, for example, excessive carbon buildup raises the compression and retains heat as glowing embers.

CARBURETED FUEL SYSTEM

Description and Operation

BASIC FUNCTIONS

♦ See Figure 2

Traditional carburetor theory often involves a number of laws and principles. The diagram illustrates several carburetor basics. If you blow air across a straw inserted into a container of liquid, a pressure drop is created in the straw column. As the liquid in the column is expelled, an atomized mixture (air and fuel droplets) is created. In a carburetor this is mostly air and a little fuel.

The actual ratio of air to fuel differs with engine conditions but is usually from 15 parts air to one part fuel at optimum cruise to as little as 7 parts air to one part fuel at full choke.

Using our example, what if the top of the container is covered and sealed around the straw, what will happen? There would be no flow. This is typical of a clogged carburetor bowl vent. If the base of the straw is clogged or restricted what will happen? There would be a very low flow rate or none at all. This represents a clogged main jet. If the liquid in the glass is lowered and you blow through the straw with the same force what will happen? Not as much fuel will flow. A lean condition occurs. If the fuel level is raised and you blow again at the same velocity what happens? The result is a richer mixture.

FUEL & AIR METERING

♦ See Figures 3 and 4

The carburetor is merely a metering device for mixing fuel and air in the proper proportions for efficient engine operation. At idle speed, an outboard engine requires a mixture of about 8 parts air to 1 part fuel. At high speed or under heavy duty service, the mixture may change to as much as 12 parts air to 1 part fuel.

In order to start the engine, the fuel must be moved from the tank to the carburetor by a squeeze bulb installed in the fuel line. This action is necessary because the fuel pump does not have sufficient pressure to draw fuel from the tank during cranking before the engine starts.

The fuel for some small horsepower units is gravity fed from a tank mounted at the rear of the powerhead. Even with the gravity feed method, a small fuel pump may be an integral part of the carburetor.

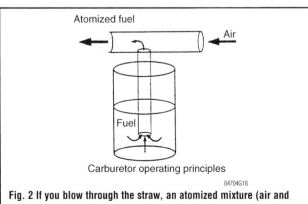

Fig. 2 If you blow through the straw, an atomized mixture (air and fuel droplets) comes out

4-4 FUEL SYSTEM

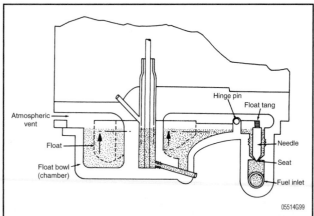

Fig. 3 Fuel passes through the inlet passage to the needle and seat and then into the float chamber. A float in the chamber rides up and down on the surface of the fuel. As the fuel level rises to a predetermined point, a tang on the float closes the inlet needle and the flow entering the chamber is cut off

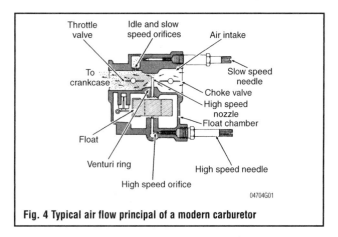

Fig. 4 Typical air flow principal of a modern carburetor

After the engine starts, the fuel passes through the pump to the carburetor. All systems have some type of filter installed somewhere in the line between the tank and the carburetor. Many units have a filter as an integral part of the carburetor.

At the carburetor, the fuel passes through the inlet passage to the needle and seat and then into the float chamber (reservoir). A float in the chamber rides up and down on the surface of the fuel. After fuel enters the chamber and the level rises to a predetermined point, a tang on the float closes the inlet needle and the flow entering the chamber is cut off. When fuel leaves the chamber as the engine operates, the fuel level drops and the float tang allows the inlet needle to move off its seat and fuel once again enters the chamber. In this manner, a constant reservoir of fuel is maintained in the chamber to satisfy the demands of the engine at all speeds.

If the carburetor has more than one reservoir, the fuel level in each reservoir (chamber) is controlled by identical float systems. Fuel level in each chamber is extremely critical and must be maintained accurately. Accuracy is obtained through proper adjustment of the floats. This adjustment will provide a balanced metering of fuel to each cylinder at all speeds.

A fuel chamber vent hole is located near the top of the carburetor body to permit atmospheric pressure to act against the fuel in each chamber. This pressure assures an adequate fuel supply to the various operating systems of the powerhead.

After the outboard is started, a suction effect is created each time the piston moves upward in the cylinder. This suction draws air through the throat of the carburetor. A restriction in the throat, called a venturi, controls air velocity and has the effect of reducing air pressure at this point.

The difference in air pressures at the throat and in the fuel chamber, causes fuel to be pushed out of metering jets extending down into the fuel chamber. When the fuel leaves the jets, it mixes with the air passing through the venturi.

This fuel/air mixture should then be in the proper proportion for burning in the cylinders for maximum engine performance.

In order to obtain the proper air/fuel mixture for all engine speeds, some models have high and low speed jets. These jets have adjustable needle valves that are used to compensate for changing atmospheric conditions. In almost all cases, the high-speed circuit has fixed high-speed jets and is not adjustable.

A throttle valve controls the flow of air/fuel mixture drawn into the combustion chambers. A cold powerhead requires a richer fuel mixture to start and during the brief period it is warming to normal operating temperature. A choke valve is placed ahead of the metering jets and venturi. As this valve begins to close, the volume of air intake is reduced, thus enriching the mixture entering the cylinders.

When this choke valve is fully closed, a very rich fuel mixture is drawn into the cylinders.

The throat of the carburetor is usually referred to as the barrel. Force Outboard® carburetors are single barrel are fed by a single float and chamber and have individual metering jets, needle valves, throttle and choke plates for each barrel.

➥Some later model carburetors no longer have choke plates. The engines have a solenoid operated valve that allows fuel to be fed into the intake and reed valve area.

CARBURETOR CIRCUITS

◆ See Figures 5, 6 and 7

When the starting circuit is operating, the choke plate is closed, creating a partial vacuum in the venturi. As the piston rises, negative pressure in the crankcase draws the rich air-fuel mixture from the float bowl into the venturi and on into the engine

The low speed circuit functions with the throttle plates at the zero to one-eighth position. When the pressure in the crankcase is lowered, the air-fuel mixture is discharged into the venturi through the pilot outlet because the throttle plate is closed. No other outlets are exposed to low venturi pressure. A fixed idle jet meters the fuel. The pilot air jet meters the air. The idle mixture screw regulates the combined air-fuel mixture.

The mid-range circuit functions with the throttle plates at the one-eighth to three-eighths throttle position. As the throttle plate continues to open, the air-fuel mixture is discharged into the venturi through the bypass holes. As the throttle plate uncovers more bypass holes, increased fuel flow results because of the low pressure in the venturi. Depending on the model, there could be two, three or four bypass holes

The high speed circuit functions with the throttle plates at the three-eighths to wide-open throttle. As the throttle plates move toward wide open, maximum airflow occurs with very low pressure. The fuel is metered through the main jet and is drawn into the main discharge nozzle. Air is metered by the main air jet

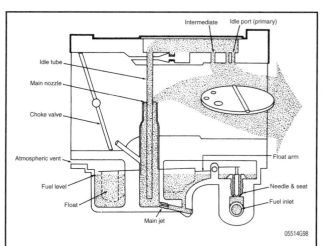

Fig. 5 When the starting circuit is operating, the choke plate is closed, creating a partial vacuum in the venturi. As the piston rises, negative pressure in the crankcase draws the rich air-fuel mixture from the float bowl into the venturi and on into the engine

FUEL SYSTEM 4-5

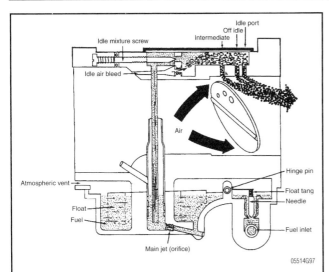

Fig. 6 The low speed circuit functions with the throttle plates at the zero to one-eighth position. When the pressure in the crankcase is lowered, the air-fuel mixture is discharged into the venturi through the pilot outlet because the throttle plate is closed. No other outlets are exposed to low venturi pressure. A fixed fuel jet meters the fuel. A fixed air jet meters the air. The idle mixture screw regulates the combined air-fuel mixture.

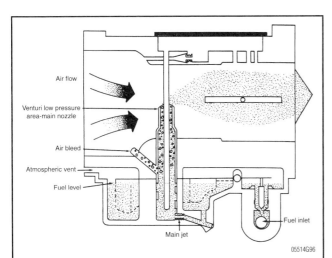

Fig. 7 The high speed circuit functions with the throttle plates at the three-eighths to wide-open throttle. As the throttle plates move toward wide open, maximum airflow occurs with very low pressure. The fuel is metered through the main jet and is drawn into the main discharge nozzle. Air is metered by the main air jet and enters the discharge nozzle, where it combines with fuel. The mixture atomizes, enters the venturi and is drawn into the engine

and enters the discharge nozzle, where it combines with fuel. The mixture atomizes, enters the venturi and is drawn into the engine

FUEL PUMP

▶ See Figure 8

A fuel pump is a basic mechanical device that utilizes crankcase positive and negative pressures to pump fuel from the fuel tank to the carburetors. This device contains a flexible diaphragm and two check valves (flappers or fingers) that control flow.

As the piston goes up, crankcase pressure drops (negative pressure) and the inlet valve opens, pulling fuel from the tank. As the piston nears TDC, pressure

Fig. 8 Typical fuel pump location

in the pump area is neutral (atmospheric pressure). At this point both valves are closed. As the piston comes down, pressure goes up (positive pressure) and the fuel is pushed toward the carburetor bowl by the diaphragm through the now open outlet valve.

This is a reliable method to move fuel but can have several problems. Sometimes an engine backfire can rupture the diaphragm. The diaphragm and valves are moving parts subject to wear. The flexibility of the diaphragm material can go away, reducing or stopping flow. Rust or dirt can hang one of the valves open and reduce or stop fuel flow.

The fuel pumps on these engines could be single or double chambered but work on the same principles.

Troubleshooting the Fuel System

Troubleshooting fuel systems requires the same techniques used in troubleshooting any other area of the outboard. A thorough, systematic approach will pay big rewards. Build your troubleshooting checklist, with the most likely offenders at the top. Use your experience, or the advice of a professional to adjust your list for local conditions. Everyone has been tempted to jump into the carburetor on a vague hunch. Pause a moment and review the facts when this urge occurs.

In order to accurately troubleshoot a carburetor or fuel system problem, you must first verify that the problem is fuel related. Many symptoms can have several different possible causes. Be sure to eliminate mechanical and electrical systems as the potential fault. Carburetion is the number one cause of most engine problems, but there are other possibilities.

When testing the outboard, observe the throttle position when the problem occurs. This will help you pinpoint the circuit that is malfunctioning. Carburetor troubleshooting and repair is very demanding. You must pay close attention to the location, position and sometimes the numbering on each part removed. The ability to identify a circuit by the operating RPM it affects is important. Often your best troubleshooting tool is a can of cleaner. This can be used to trace

4-6 FUEL SYSTEM

those mystery circuits and find that last speck of dirt. Be careful and wear safety glasses when using this method.

One of the toughest tasks with a fuel system is the actual troubleshooting. The following steps provide an orderly sequence of tests to pinpoint problems in the fuel system.

1. Gather as much information as you can.
2. Duplicate the condition. Take the boat out and verify the complaint. If the problem cannot be duplicated, you cannot fix it. Once the problem has been duplicated, you can begin troubleshooting.
3. Give the entire outboard a careful visual inspection. What's the condition of the propeller and the lower unit? Remove the hood and look for any visible signs of failure. Are there any signs of head gasket leakage. Is the outboard paint discolored from high temperature or are there any holes or cracks in the outboard block?
4. Spark plug tip appearance is a good indication of combustion efficiency. The tip should be a light tan. A white insulator or small beads on the insulator indicate too much heat. A dark or oil fouled insulator indicates incomplete combustion. To properly read spark plug tip appearance, run the engine at the RPM you are testing for about 15 second and then immediately turn the engine OFF without changing the throttle position.
5. Perform compression and leak down tests. While cranking the outboard during the compression test, listen for any abnormal sounds. If the outboard passes these simple tests we can assume that the mechanical condition of the outboard is good. All other outboard mechanical inspection would be too time consuming at this point.
6. Your next step is to isolate the fuel system into two sub-systems. Separate the fuel delivery components from the carburetors. To do this, substitute the boat's fuel supply with a known good supply. Use a 6 gallon portable tank and fuel line. Connect the portable fuel supply directly to the outboard fuel pump, bypassing the boat fuel delivery system. Now test the outboard. If the problem is no longer present, you know where to look. If the problem is still present, further troubleshooting is required.
7. Pressure testing fuel pump output can determine whether the fuel spray is adequate and if the fuel pump diaphragms are functioning correctly. A pressure gauge placed between the fuel pumps and the carburetors will test the entire fuel delivery system. Normally a fuel system problem will show up at high speed where the fuel demand is the greatest. A common symptom of a fuel pump output problem is surging at wide open throttle, but normal operation at slower speeds. To check the fuel pumps output, install the pressure gauge and accelerate the engine to wide open throttle. Observe the pressure gauge needle. It should always swing up to the specified value and remain steady. This would indicate the system that is functioning properly.

If the needle gradually swings down toward zero, fuel demand is greater than the fuel system can supply. This reading isolates the problem to the fuel delivery system (fuel tank or line). To confirm this, an auxiliary tank should be installed and the engine re-tested. Be aware that a bad anti-siphon valve on a built-in tank can create enough restriction to cause a lean condition and serious engine damage.

If the needle movement becomes erratic, suspect a ruptured diaphragm in the fuel pump. A quick way to check for a ruptured fuel pump diaphragm is while the engine is at idle speed, to squeeze the primer bulb and hold steady firm pressure on it. If the diaphragm is ruptured, this will cause a rough running condition because of the extra fuel passing through the diaphragm into the crankcase. After performing this test you should check the spark plugs for cylinders that the fuel pump supplies. If the spark plugs are OK, but the fuel pumps are still suspected you should remove the fuel pumps and completely disassemble them. Rebuild or replace the pumps as needed.

8. To check the boat's fuel system for a restriction, install a vacuum gauge in the line before the fuel pump. Run the engine under load at wide open throttle to get a reading. Vacuum should read no more than 4.5 in. Hg (15.2 kPa).
9. To check for air entering the fuel system, install a clear fuel hose between the fuel screen and fuel pump. Air should be visible as bubbles in the clear line. If air is in the line, check all fittings back to the boat's fuel tank.
10. Use a timing light to observe carburetor spray patterns. Look for the proper amount of fuel and for proper atomization in the two fuel outlet areas (main nozzle and bypass holes). The strobe effect of the lights helps you see in detail the fuel being drawn through the throat of the carburetor. On multiple carburetor engines, always attach the timing light to the cylinder you are observing so the strobe doesn't change the appearance of the patterns. If you need to compare two cylinders, change the timing light hookup each time you observe a different cylinder.

✳✳ WARNING

This procedure is to be done with the engine at idle in neutral. If this test is done at higher RPM or with the engine running at speed the reaction could be severe and mechanical damage or personal injury could result.

An effective troubleshooting technique on multiple carburetor engines is to remove the air silencer and or covers. Run the engine with an appropriate water supply, use a clean shop rag and individually choke the carburetors off one at a time and note the results. This will help isolate which one is causing the problem or help diagnose which carburetor circuit the problem might be in. If this method is used and there is no change in the running of the engine, either a no fuel or flooding (too rich) condition exists the carburetor would need to be serviced.

COMMON PROBLEMS

The major causes of carburetor trouble are dirty fuel and improper storage.

To prevent contaminated fuel from entering the engine it is highly recommended that a fuel water separator filter be installed in the boat fuel system to remove all of the dirt and moisture possible before the fuel enters the engine.

Put a stabilizing chemical into the fuel before long periods of storage or inactivity in order to avoid having a stale or varnished fuel system. If possible, in addition to treating the fuel, manually drain each carburetor.

Fuel Delivery

Many times fuel system troubles are caused by a plugged fuel filter, a defective fuel pump or by a leak in the line from the fuel tank to the fuel pump. A defective choke may also cause problems. Would you believe, majorities of starting troubles that are traced to the fuel system are the result of an empty fuel tank or aged sour fuel.

Sour Fuel

Under average conditions (temperate climates), fuel will begin to break down in about four months. A gummy substance forms in the bottom of the fuel tank and in other areas. The filter screen between the tank and the carburetor and small passages in the carburetor will become clogged. The gasoline will begin to give off an odor similar to rotten eggs. Such a condition can cause the owner much frustration, time in cleaning components, and the expense of replacement or overhaul parts for the carburetor.

Even with the high price of fuel, removing gasoline that has been standing unused over a long period of time is still the easiest and least expensive preventative maintenance possible. In most cases, this old gas can be used without harmful effects in an automobile using regular gasoline.

A gasoline preservative for 2 cycle engines will keep the fuel fresh for up to twelve months. These products are available in most areas under various trade names.

Choke Problems

When the engine is hot, the fuel system can cause starting problems. After a hot engine is shut down, the temperature inside the fuel bowl may rise to 200 degrees F and cause the fuel to actually boil. All carburetors are vented to allow this pressure to escape to the atmosphere. However, some of the fuel may percolate over the high-speed nozzle.

If the choke should stick in the open position, the engine will be hard to start. If the choke should stick in the closed position, the engine will flood, making it very difficult to start.

In order for this raw fuel to vaporize enough to burn, considerable air must be added to lean out the mixture. Therefore, the only remedy is to remove the spark plugs, ground the leads, crank the powerhead through about ten revolutions, clean the plugs, reinstall the plugs, and start the engine.

If the needle valve and seat assembly is leaking, an excessive amount of fuel may enter the reed housing in the following manner. After the powerhead is shut down, the pressure left in the fuel line will force fuel past the leaking needle valve. This extra fuel will raise the level in the fuel bowl and cause fuel to overflow into the reed housing.

A continuous overflow of fuel into the reed housing may be due to a sticking inlet needle or to a defective float, which would cause an extra high level of fuel in the bowl and overflow into the reed housing.

FUEL SYSTEM 4-7

Rough Engine Idle

If an engine does not idle smoothly, the most reasonable approach to the problem is to perform a tune-up to eliminate such areas as:
- Defective points
- Faulty spark plugs
- Timing out of adjustment

Other problems that can prevent an engine from running smoothly include:
- An air leak in the intake manifold
- Uneven compression between the cylinders
- Sticky or broken reeds

Of course any problem in the carburetor affecting the air/fuel mixture will also prevent the engine from operating smoothly at idle speed. These problems usually include:
- Too high a fuel level in the bowl
- A heavy float
- Leaking needle valve and seat
- Defective automatic choke
- Improper adjustments for idle mixture or idle speed

Excessive Fuel Consumption

Excessive fuel consumption can be the result of any one of four conditions, or a combination of all.
- Inefficient engine operation.
- Faulty condition of the hull, including excessive marine growth.
- Poor boating habits of the operator.
- Leaking or out of tune carburetor.

If the fuel consumption suddenly increases over what could be considered normal, then the cause can probably be attributed to the engine or boat and not the operator.

Marine growth on the hull can have a very marked effect on boat performance. This is why sailboats always try to have a haul-out as close to race time as possible.

While you are checking the bottom, take note of the propeller condition. A bent blade or other damage will definitely cause poor boat performance.

If the hull and propeller are in good shape, then check the fuel system for possible leaks. Check the line between the fuel pump and the carburetor while the engine is running and the line between the fuel tank and the pump when the engine is not running. A leak between the tank and the pump many times will not appear when the engine is operating, because the suction created by the pump drawing fuel will not allow the fuel to leak. Once the engine is turned off and the suction no longer exists, fuel may begin to leak.

If a minor tune-up has been performed and the spark plugs, points, and timing are properly adjusted, then the problem most likely is in the carburetor and an overhaul is in order.

Check the needle valve and seat for leaking. Use extra care when making any adjustments affecting the fuel consumption, such as the float level or automatic choke.

Engine Surge

If the engine operates as if the load on the boat is being constantly increased and decreased, even though an attempt is being made to hold a constant engine speed. The problem can most likely be attributed to the fuel pump or a restriction in the fuel line between the tank and the carburetor.

COMBUSTION RELATED PISTON FAILURES

♦ See Figure 9

When an engine has a piston failure due to abnormal combustion, fixing the mechanical portion of the engine is the easiest part of the equation. The hard part is determining what caused the problem in order to prevent a repeat failure. Think back to the four basic areas that affect combustion to find the cause of the failure.

Since you probably removed the cylinder head. Inspect the failed piston, look for excessive deposit buildup that could raise compression, or retain heat in the combustion chamber. Statically check the wide open throttle timing. Be sure that the timing is not over advanced. It is a good idea to seal these adjustments with paint to detect tampering.

Look for a fuel restriction that could cause the engine to run lean. Don't forget to check the fuel pump, fuel tank and lines, especially if a built in tank is used. Be sure to check the anti-siphon valve on built in tanks. If everything else looks good, the final possibility is poor quality fuel.

Carburetor

The following carburetor service procedures are listed by application. They can be identified by the specific exploded views with in this chapter and their settings found in the "Carburetor Specifications" chart at the end of "Maintenance and Tune-Up".

REMOVAL & INSTALLATION

Carburetor with Faceplate Mounted Control

♦ See Figures 10, 11, 12 and 13

1. Remove engine cover.
2. Close the fuel valve between the fuel tank and the carburetor.

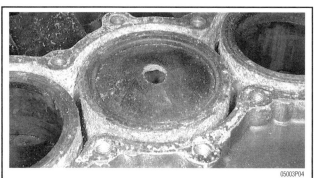

Fig. 9 This burned piston is typical of a combustion relate failure. The combustion chamber temperature got so hot that it melted the top of the piston (hole in the top of the piston)

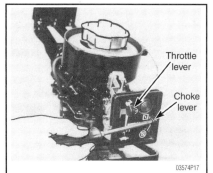

Fig. 10 Remove the screws securing the knobs at the end of the choke and throttle levers

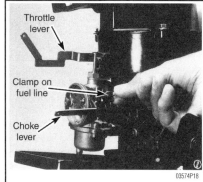

Fig. 11 Close the fuel valve between the fuel tank and the carburetor

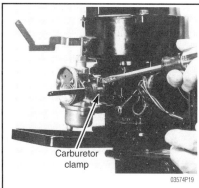

Fig. 12 Loosen the clamp screw and remove the carburetor from the powerhead

4-8 FUEL SYSTEM

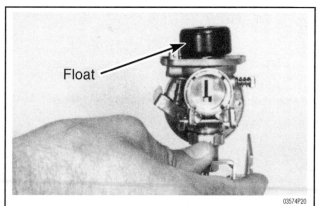

Fig. 13 Remove the carburetor-to-powerhead seal from the recess in the carburetor throat

3. Squeeze the wire-type hose clamp on the fuel line and remove the hose from the inlet fuel fitting.
4. Remove the screws securing the knobs at the end of the choke and throttle levers.
5. Remove the two screws securing the rectangular intake cover to the carburetor body.
6. Lift the cover free with the stop switch wires attached.
7. Loosen the clamp screw and remove the carburetor from the powerhead.
8. Remove the carburetor-to-powerhead seal from the recess in the carburetor throat.

To install:
9. Position a new carburetor-to-power-head seal in the recess of the carburetor throat.
10. Slide the carburetor onto the crankcase cover and secure it in place with the clamp screw making sure that it is square and level.
11. Connect the fuel line to the fuel inlet fitting.
12. Open the fuel valve between the fuel tank and the carburetor.
13. Install the front cover and the knobs onto the choke and throttle levers.
14. Reinstall engine cover.

1984–1988 Single Carburetor with Tiller Handle

▶ See Figure 14

→Some early model smaller horsepower engines incorporate an integral fuel tank with the manual starter built on to it. On these units the fuel tank is the engine cover and needs to be removed and the fuel line properly secured before any other carburetor work can be accomplished.

1. Remove engine cover or integral fuel tank assembly.
2. Remove the three mounting bolts securing the hand rewind starter to the powerhead.
3. Lift the rewind starter clear of the powerhead.
4. Disconnect the high-tension leads from the spark plugs.
5. Disconnect the carburetor linkage.
6. Remove the intake cover, if equipped.
7. Remove the two carburetor retaining nuts.
8. Lift the carburetor free of the powerhead.
9. Move the carburetor to a suitable clean work surface.

To install:
10. Clean the carburetor mounting gasket surfaces.
11. Install using the two retaining nuts.
12. Connect throttle linkage.
13. Install intake cover if one was removed.
14. Install fuel line and/or fuel tank.
15. Install rewind starter.
16. Connect the high-tension leads to the spark plugs.
17. Place cover on engine

1989–1999 Single Carburetor with Tiller Handle

▶ See Figures 15 and 16

1. Remove the powerhead cover.
2. Disconnect the high-tension lead/s from the spark plug/s.
3. Remove the three mounting bolts securing the hand rewind starter to the powerhead.
4. Lift the rewind starter clear of the powerhead.
5. Disconnect the primer hose from the carburetor.
6. Remove the "E" clip and washer retaining the throttle linkage shaft.
7. Slide the throttle linkage shaft out.
8. Remove the two nuts securing the carburetor to the powerhead studs.
9. Move the carburetor away from the studs.
10. Disconnect the fuel line, choke rod and the throttle link from the throttle shaft.
11. Place the carburetor on a clean work surface.
12. Remove the mounting gasket from the studs.

To install:
13. Clean the carburetor mounting gasket surfaces.
14. Install the gasket on the carburetor mounting studs.
15. Place the carburetor in the general vicinity to permit making fuel line connections.
16. Connect the fuel line and the throttle link to the throttle shaft.
17. Move the carburetor onto the two studs and secure it in place with the nuts.
18. Connect the throttle linkage shaft and then secure it with a washer and an "E" clip.

Fig. 14 On engines with an integral fuel tank, the tank needs to be removed to gain access to the carburetor

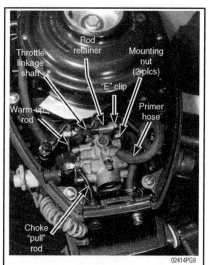

Fig. 15 Remove these components prior to removing the carburetor from the powerhead

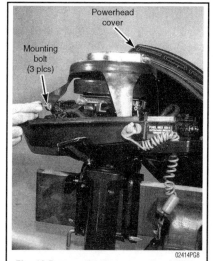

Fig. 16 Remove the three mounting bolts securing the hand rewind starter to the powerhead

FUEL SYSTEM   4-9

19. Connect the primer hose to the carburetor.
20. Slide the choke rod into place and push down on the rod retainer to secure the rod in place.
21. Mount the hand rewind starter and secure it in place with the attaching leg hardware.
22. Connect the high-tension lead/s to the spark plug/s.
23. Install the powerhead cover

Carburetors with Remote Control

♦ See Figures 17 thru 24

1. Remove the full length intake cover if equipped.
2. Disconnect the fuel hose and the fuel primer hoses.
3. Remove the retaining screws at each carburetor, and then remove the tie bar from the throttle levers, if equipped.
4. Remove the two retaining nuts on each carburetor.
5. Remove carburetion system from the powerhead.
6. Loosen the choke swivel screws allowing the choke link to slide out of each swivel. Straighten the cotter pin securing both the choke rod and top swivel to the lever removing the pin, washer, choke rod, two O-rings and the swivel, if equipped
7. Move the carburetor to a suitable work surface.
8. Remove the mounting gasket material from intake area.

To install:

9. Install the swivel to the top choke lever inserting the pin, washer, choke rod and two O-rings. Slide the choke rod through the carburetor choke swivels.
10. Slide a new gasket onto the mounting studs.
11. Mount the carburetion system in place and secure it with the attaching nuts.
12. Position the tie bar and secure it in place with the retaining rings.

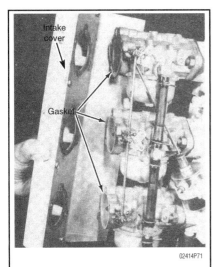

Fig. 17 Remove the full length intake cover if equipped

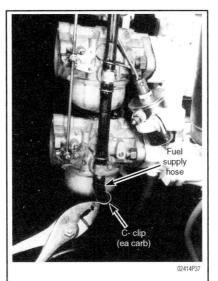

Fig. 18 Remove the fuel inlet hose . . .

Fig. 19 . . . and the fuel primer hose prior to carburetor removal

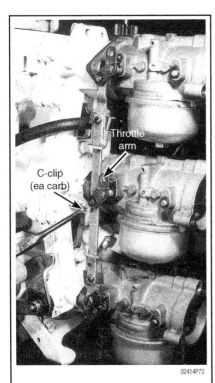

Fig. 20 Remove the retaining screws at each carburetor . . .

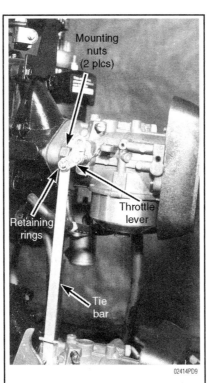

Fig. 21 . . . then remove the tie bar from the throttle levers

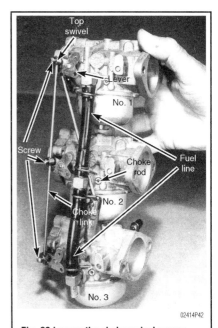

Fig. 22 Loosen the choke swivel screws allowing the choke link to slide out of each swivel. Straighten the cotter pin securing both the choke rod and top swivel to the lever removing the pin, washer, choke rod, two O-rings and the swivel, if equipped

4-10 FUEL SYSTEM

13. Connect all fuel hoses and secure them with a hose clamp.
14. Connect the primer system hoses and secure them in place with tie wraps, if equipped.
15. Install the one piece intake cover and secure with fasteners if applicable.

OVERHAUL

The following carburetor overhaul procedures are listed by application. They can be identified by the specific exploded views in this chapter and their settings found in the "Carburetor Specifications" chart at the end of "Maintenance and Tune-Up".

➡ It is recommended that the carburetor be carefully drained into a safe container prior to disassembly. This can be accomplished by either removing a drain or bowl screw at the bottom of the unit or inverting the carburetor allowing it to drain through it's atmospheric vent at the top of the body.

Engines With Faceplate Mounted Control

DISASSEMBLY

♦ See accompanying illustrations

1. Back out the two screws securing the float bowl to the carburetor.
2. Remove the float from the carburetor.
3. Slide the hinge pin free and remove the float hinge and inlet needle.
4. Lift off and remove the float bowl gasket.

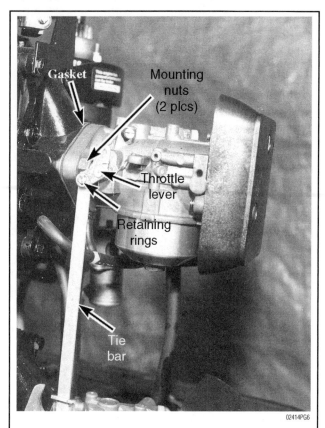

Fig. 23 Slide a new gasket onto the mounting studs

Fig. 24 Connect all fuel hoses and secure them with a hose clamp

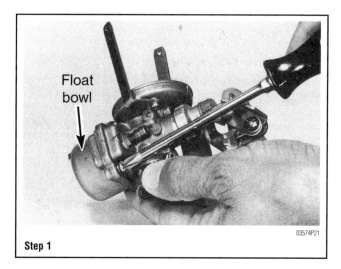

Step 1

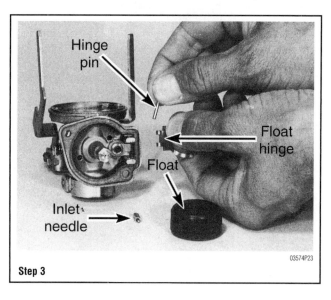

Step 3

FUEL SYSTEM 4-11

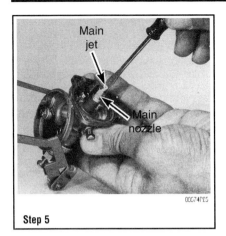
Step 5

Step 7

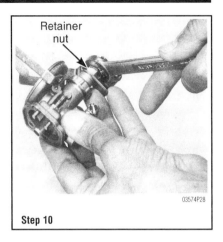
Step 10

5. Unscrew the main jet from the center of the main nozzle.
6. Use the proper size wrench and remove the main nozzle.
7. Rotate the idle speed screw clockwise and count the number of turns necessary to lightly seat it.
8. Record the number of turns necessary to seat the idle speed screw.
9. Remove the idle speed screw and tension spring.
10. Loosen the retainer nut several turns counterclockwise but do not remove it.
11. Unscrew the mixing chamber cover with a pair of pliers.
12. Lift off the throttle valve assembly disconnecting the throttle lever from the bracket.
13. Compress the spring in the throttle valve assembly to allow the throttle cable end to clear the recess in the base of the throttle valve and to slide down the slot.

➡It is not necessary to remove the E-clip from the jet needle unless replacement is required or adjustment seems necessary.

14. Disassemble the throttle valve assembly consisting of the throttle valve, spring, jet needle jet retainer and throttle cable end.

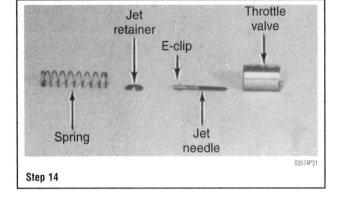
Step 14

CLEANING & INSPECTION

♦ See Figure 25

✳✳ CAUTION
Always wear safety glasses when using compressed air and aerosol solvents.

✳✳ WARNING
Use compressed air only at a low pressure! High pressure air could cause damage to the carburetor by loosening pressed in jets or cause lodged objects to shoot from the unit.

✳✳ WARNING
Never soak components that are or contain diaphragms, rubber or plastic parts in carburetor cleaner. They tend to absorb liquid and expand rendering them unusable. These parts should be cleaned only in solvent and then blown dry with compressed air immediately.

✳✳ WARNING
If manual jet or passage cleaning is necessary always use a tool softer that the component as not to damage or alter the part.

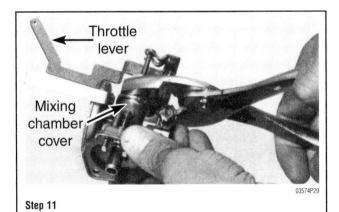
Step 11

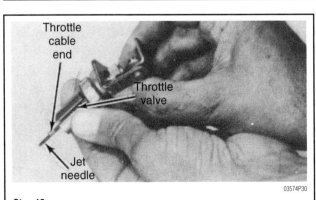
Step 13

1. Place all of the metal parts in a screen type tray and dip them in carburetor cleaner until they appear completely clean.
2. Blow the parts dry with compressed air.
3. Blow out all of the carburetor passageways to be sure they are clear and not blocked or clogged.
4. Carefully inspect the carburetor body for stripped threads or cracks.
5. Check any lead shot or core plugs for any sign of leakage.

4-12 FUEL SYSTEM

6. Inspect the float hinge and hinge pin area for wear and the float for any sign of deterioration.
7. Examine the inlet needle and seat for wear

ASSEMBLY

♦ See accompanying illustrations

Carburetor service kits are available from the local marine dealer. These kits contain all necessary parts to perform the usual carburetor overhaul work. Check the kits part list to be sure you have all necessary parts including any damaged parts possibly not in the standard rebuild kit.

All worn and damaged parts along with any gaskets and seals that need to be replaced should be in hand at this time.

➡ The E-clip must be installed into the same groove from which it was removed.

1. Assemble the throttle valve components in the following order.
2. Insert the E-clip end of the jet needle into the throttle valve. The standard setting is the second groove.
3. Place the needle retainer into the throttle valve over the E-clip and align the retainer slot with the slot in the throttle valve.
4. Thread the spring over the end of the throttle cable and insert the cable into the retainer end of the throttle valve.
5. Compress the spring and at the same time guide the cable end through the slot until the end locks into place in the recess.
6. Position the assembled throttle valve in such a manner to permit the slot to slide over the alignment pin while the throttle valve is lowered into the carburetor.
7. Attach the throttle lever to the bracket on top of the throttle valve assembly.
8. Carefully tighten the mixing chamber cover with a pair of pliers.
9. Position the bracket to allow the throttle to just clear the front of the carburetor, and then tighten the retainer nut.
10. Slide the spring onto the idle speed screw and thread the screw into the carburetor body until it seats lightly.
11. From this position back the screw out the same number of complete turns as noted during removal.
12. Thread the main nozzle into the carburetor body being careful not to over tighten.
13. Install the main jet into the main nozzle being careful not to over tighten.
14. Position the float bowl gasket in place and install the inlet needle.

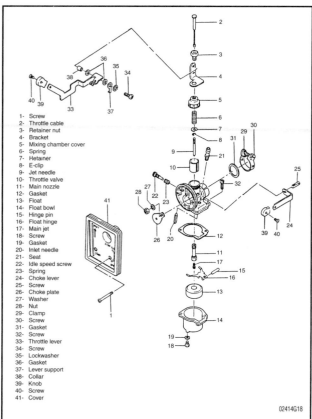

Fig. 25 An exploded view of a faceplate mounted control carburetor

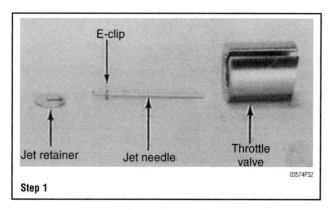

Step 1

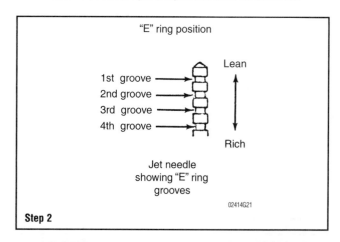

Step 2

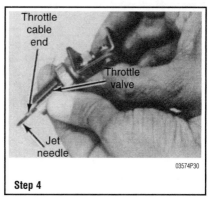

Step 4

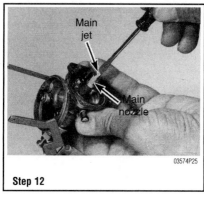

Step 12

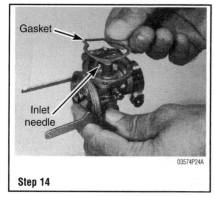

Step 14

FUEL SYSTEM 4-13

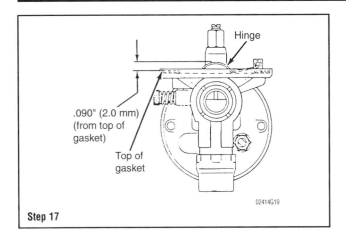

Step 17

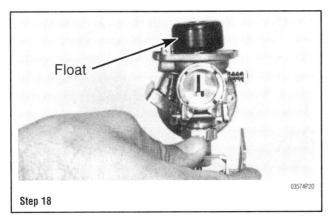

Step 18

3. Remove the fuel bowl.
4. Remove the fuel bowl gasket.

➡Some models of this carburetor have a spring support. If equipped observe the position of the float spring as an aid during assembling.

5. Remove the hinge pin by simply pushing it out.
6. Lift the float straight up and the needle valve will come up with the float.
7. Remove the spring, if equipped.
8. Using a small screwdriver remove the main nozzle from the carburetor body.
9. Remove the needle valve seat and gasket with an appropriate screwdriver.

CLEANING & INSPECTION

▸ See Figures 26, 27 and 28

※ CAUTION

Always wear safety glasses when using compressed air and aerosol solvents.

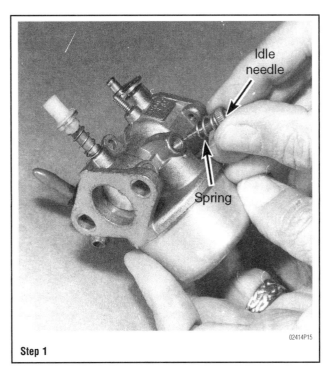

Step 1

15. Hold the carburetor body in a perfect upright position on a firm surface.
16. Set the hinge in position then slide the hinge pin into place through the hinge.
17. Check the float hinge adjustment. Carefully bend the hinge as necessary to achieve the required measurement.
18. Install the float.
19. Place the float bowl in position on the carburetor body and secure it with the two screws.

1984–88 Single Carburetor with Tiller Handle

DISASSEMBLY

▸ See accompanying illustrations

1. Remove the idle needle and spring.
2. Remove the main jet and gasket from the underneath side of the fuel bowl.

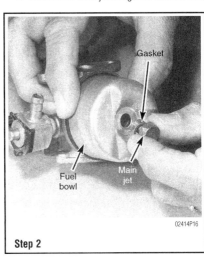

Step 2

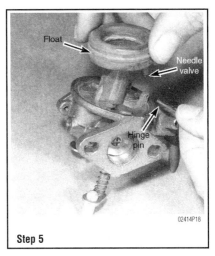

Step 5

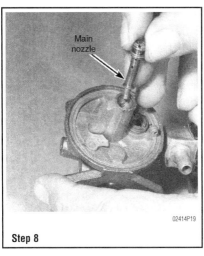

Step 8

4-14 FUEL SYSTEM

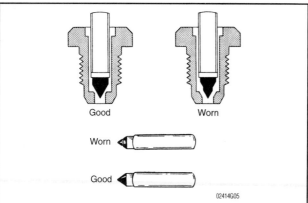

Fig. 26 Needle and seat arrangement on the carburetor covered in this section showing a worn and new needle for comparison

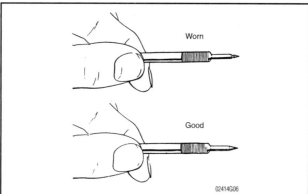

Fig. 27 Carburetor idle mixture adjustment needles. The top needle is worn and unfit for further service. The bottom needle is new

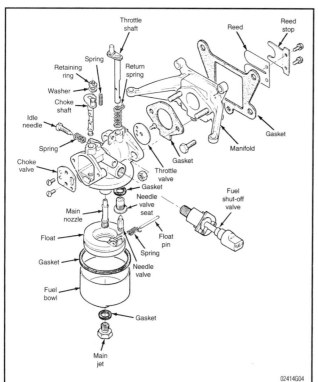

Fig. 28 Exploded view of an early style tiller handle model carburetor

✷✷ WARNING

Use compressed air only at a low pressure! High pressure air could cause damage to the carburetor by loosening pressed in jets or cause lodged objects to shoot from the unit.

✷✷ WARNING

If manual jet or passage cleaning is necessary always use a tool softer that the component as not to damage or alter the part.

➡If the carburetor body is worn or damaged in any way it will need to be replaced as an assembly.

1. Place all of the metal parts in a screen type tray and dip them in carburetor cleaner until they appear completely clean.
2. Blow the parts dry with compressed air.
3. Blow out all of the carburetor passageways to be sure they are clear and not blocked or clogged.
4. Carefully inspect the carburetor body for stripped threads or cracks.
5. Check any lead shot or core plugs for any sign of leakage.
6. Inspect the float hinge and hinge pin area for wear and the float for any sign of deterioration.
7. Inspect the float bowl & bowl gasket area for damage.
8. Examine the inlet needle and seat for wear
9. Inspect mixture screws for wear.
10. Inspect the throttle shaft for wear and excessive play.

ASSEMBLY

♦ See accompanying illustrations

Carburetor service kits are available from the local marine dealer. These kits contain all necessary parts to perform the usual carburetor overhaul work. Check the kits part list to be sure you have all necessary parts including any damaged parts possibly not in the standard rebuild kit.

All worn and damaged parts along with any gaskets and seals that need to be replaced should be in hand at this time.

1. Install the seat gasket and thread into the carburetor body.
2. Install the main nozzle into the carburetor body be careful not to over tighten.

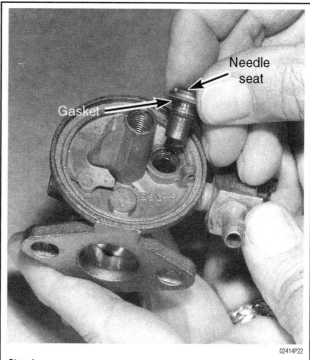

Step 1

FUEL SYSTEM 4-15

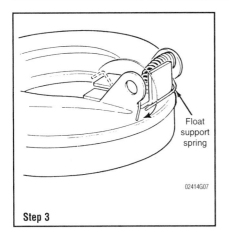
Step 3

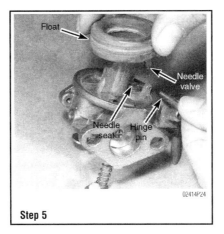
Step 5

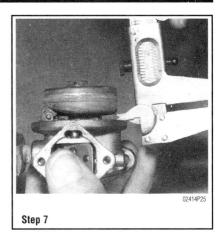
Step 7

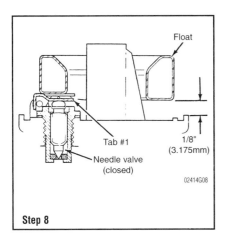
Step 8

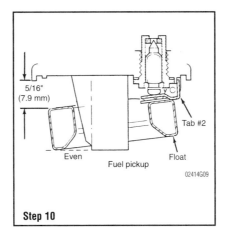
Step 10

Step 12

3. If the carburetor is equipped with a float support spring, temporarily install the spring, as shown in the accompanying illustration.
4. Place the retainer clip onto the needle valve and then attach it to the float.
5. Carfully lower the needle valve and float into place.
6. Slide the hinge pin into place with the pin entering the center of the spring, if equipped.
7. Invert the carburetor and check the float level. As illustrated.
8. Using a needle nose pliers carefully adjust Tab "1" to obtain the correct float level measurement.
9. Hold the carburetor in its normal upright position and measure the float drop.
10. Using a needle nose pliers carefully adjust Tab "2" to obtain the correct float drop measurement.
11. After making the float adjustments recheck them before continuing.
12. Install the float bowl gasket onto the gasket surface of the carburetor.
13. Install the float bowl with the sealing surface squarely on the gasket. Be sure the indented side of the float bowl is over the inlet needle & seat for proper float clearance, if equipped.
14. Install the gasket onto the main jet.
15. Thread the jet into place securing the float bowl.
16. Slide the spring onto the idle needle and thread it into the carburetor body until it just seats.
17. Back the needle out one full turn as a preliminary adjustment.

1989–99 Single Carburetor with a Tiller Handle Having

DISASSEMBLY

▶ See accompanying illustrations

1. Remove the bolt and gasket securing the bowl to the carburetor.
2. Remove the float bowl.

✳✳ WARNING

The float pin has a knurled end and a smooth end. The float pin must be removed by driving the knurled end out. If the knurled end is driven in the hinge pin boss could be damaged rendering the carburetor unusable.

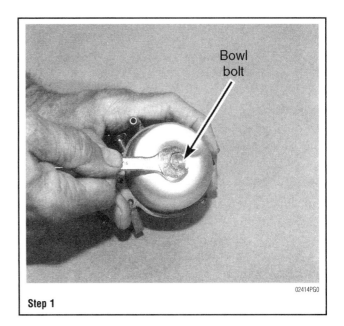
Step 1

4-16 FUEL SYSTEM

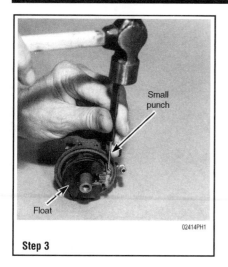
Step 3

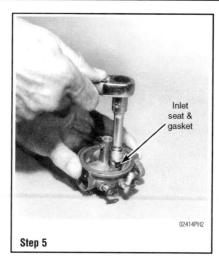
Step 5

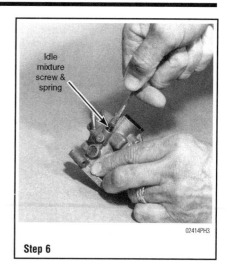
Step 6

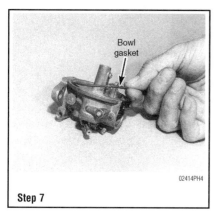
Step 7

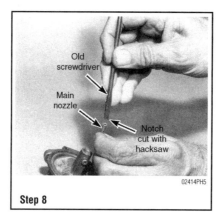
Step 8

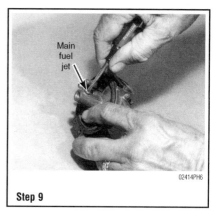
Step 9

3. Using a small punch and hammer drive the float pin free with the knurled end coming out first.
4. Remove the float and inlet needle.
5. Using the proper size socket remove the inlet seat and gasket.
6. Remove the idle screw and spring.
7. Remove the fuel bowl gasket from the recess in the carburetor body.

➡ Cutting a notch in the end of an old screwdriver makes an excellent tool to remove the main nozzle. The edges of the screwdriver will bear against the perimeter of the nozzle and permit the nozzle to be backed out without damaging the emulsification tube that runs through the center of the nozzle.

8. Remove the main nozzle.
9. Remove the main fuel jet from the carburetor body.
At this point in the disassembly process the carburetor has been torn down enough to do a professional level rebuild and clean up job. If your engine has an off idle running problem and that circuit can not be cleaned thoroughly by spraying carburetor cleaner through it the next step would be necessary.

10. If the intermediate circuit in the carburetor in restricted or plugged or the welch plug is damaged it will need to be removed. To remove the welch plug drill a hole small through it and pry it out with a small screwdriver. Be extremely careful not to drill into the carburetor body.

CLEANING & INSPECTION

♦ See Figures 29 and 30

✲✲ CAUTION

Always wear safety glasses when using compressed air and aerosol solvents.

✲✲ WARNING

Use compressed air only at a low pressure! High pressure air could cause damage to the carburetor by loosening pressed in jets or cause lodged objects to shoot from the unit.

✲✲ WARNING

Never soak components that are or contain diaphragms, rubber or plastic parts in carburetor cleaner. They tend to absorb liquid and expand rendering them unusable. These parts should be cleaned only in solvent and then blown dry with compressed air immediately.

➡ If the carburetor body is worn or damaged in any way it will need to be replaced as an assembly.

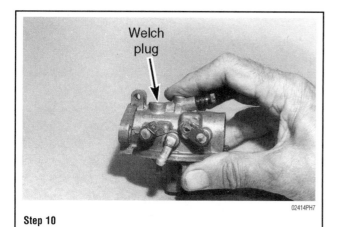
Step 10

FUEL SYSTEM 4-17

1. Place all of the metal parts in a screen type tray and dip them in carburetor cleaner until they appear completely clean.
2. Blow the parts dry with compressed air.
3. Blow out all of the carburetor passageways to be sure they are clear and not blocked or clogged.
4. Carefully inspect the carburetor body for stripped threads or cracks.
5. Check any lead shot or core plugs for any sign of leakage.
6. Inspect the float hinge and hinge pin area for wear and the float for any sign of deterioration.
7. Inspect the float bowl & bowl gasket area for damage.
8. Examine the inlet needle and seat for wear
9. Inspect mixture screws for wear.
10. Inspect the throttle shaft for wear and excessive play.

ASSEMBLY

♦ See accompanying illustrations

Carburetor service kits are available from the local marine dealer. These kits contain all necessary parts to perform the usual carburetor overhaul work. Check the kits part list to be sure you have all necessary parts including any damaged parts possibly not in the standard rebuild kit.

All worn and damaged parts along with any gaskets and seals that need to be replaced should be in hand at this time.

1. Install a new welch plug if it had been removed.
2. Tap the plug into place by striking the plug in the center with a small blunt punch.
3. Once the plug has expanded coat with fingernail polish to seal.
4. Install the main fuel jet into the carburetor body channel.
5. Install the main nozzle into the carburetor body channel.
6. Insert a new fuel bowl gasket into the recess in the carburetor body.
7. Slide the spring onto the idle needle and thread it into the carburetor body until it just seats.
8. Back the needle out one full turn as a preliminary adjustment.
9. Install the inlet seat and gasket.
10. Tighten the seat using the proper size socket.
11. Slide the inlet needle into the inlet seat.

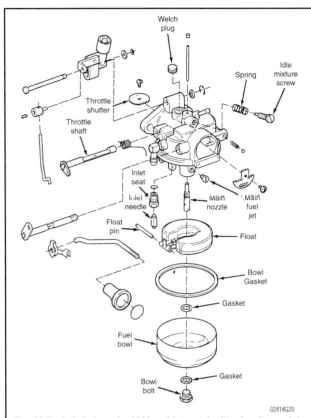

Fig. 29 Exploded view of a 1989 and later style tiller handle model carburetor

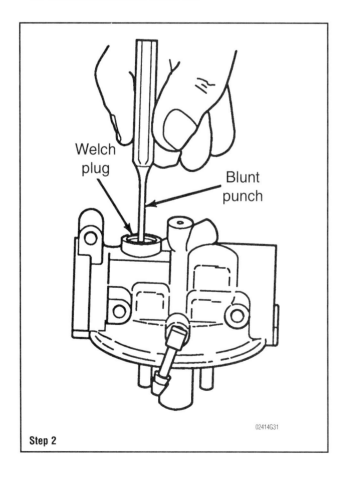

Step 2

Fig. 30 Exploded view of the most current style tiller handle model carburetor

4-18 FUEL SYSTEM

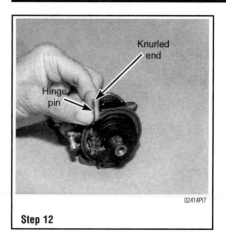
Step 12

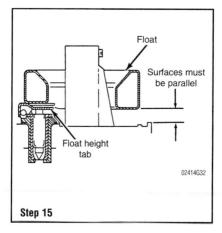
Step 15

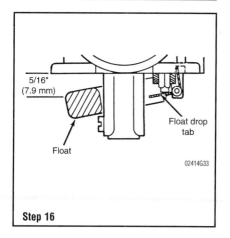
Step 16

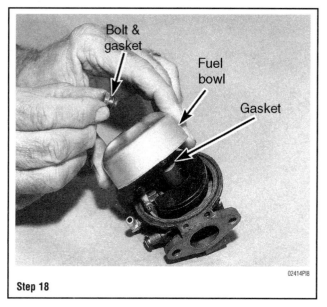
Step 18

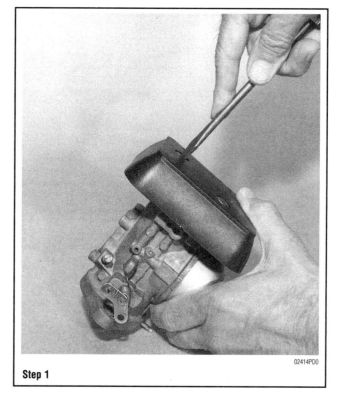
Step 1

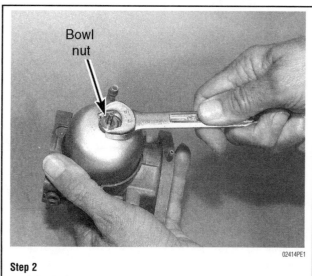
Step 2

12. Place the float in position with the holes in the hinge aligned with the holes in the bracket.

13. Slide the hinge pin through with the smooth end going in first.

14. Tap the knurled end of the hinge pin carefully with a small punch to secure the pin in place.

15. Adjust the float height by bending the inlet needle tab on the float ever so slightly until the float is parallel to the surface of the carburetor body.

16. Check the float drop by holding the carburetor in the upright position and measuring the distance from the upper surface of the float to the surface of the carburetor flange.

17. Make the adjustments as necessary.

18. Set one small gasket onto the center post. Slide another gasket onto the bowl bolt. Lower the bowl down over the float and seat it on the gasket. Secure the bowl in place with the bolt and gasket.

Single or Multiple Carburetors with Remote Control

DISASSEMBLY

▶ See accompanying illustrations

1. Back out the retaining screws, and then remove the carburetor cover, if not already removed.

2. Remove the screw securing the bowl and remove the bowl

➡ The screw has a gasket under the head and another screw gasket inside the bowl.

➡ Models without choke plates will have a fuel nipple incorporated into the bowl screw or the carburetor bowl on the top carburetor to supply fuel to the primer system.

FUEL SYSTEM 4-19

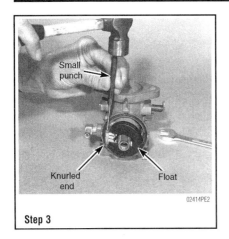
Step 3

Step 5

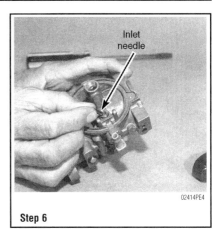
Step 6

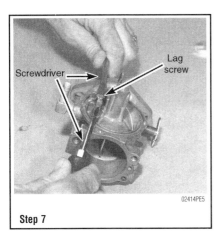
Step 7

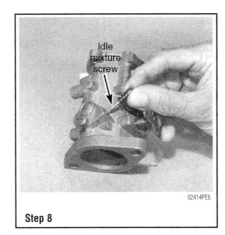
Step 8

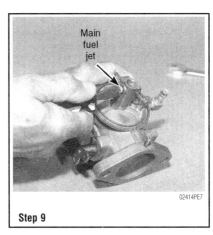
Step 9

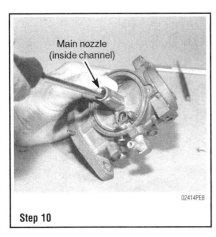
Step 10

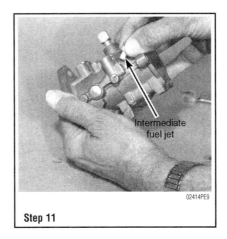
Step 11

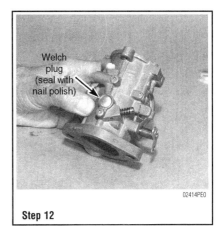
Step 12

3. Using a small punch and hammer drive the float pin free with the knurled end coming out first.
4. Remove the float.
5. Remove the bowl gasket from the recess in the carburetor body.
6. Turn the carburetor over and allow the inlet needle to fall free.

➡ Not all carburetor models have a removable inlet seat. Check the rebuild kit to be certain a replacement was supplied before removing.

7. The inlet needle seat may be removed by threading a screw into the seat and prying up on it using two screwdrivers on each side of the screw head.
8. Back out and remove the idle mixture screw.
9. Back out and remove the main fuel jet. Each carburetor is equipped with a different fuel jet. Therefore, keep the jet with the carburetor from which it was removed to ensure proper installation.

10. Back out the main nozzle and remove it from the carburetor body channel.
11. Remove the sealing screw, and then the intermediate fuel jet. Be sure to clearly identify the carburetor when purchasing the jet to ensure the proper part number jet is obtained and installed.

At this point in the disassembly process the carburetor has been torn down enough to do a professional level rebuild and clean up job. If your engine has an off idle running problem and that circuit can not be cleaned thoroughly by spraying carburetor cleaner through it the next step would be necessary.

12. If the intermediate circuit in the carburetor in restricted or plugged or the welch plug is damaged it will need to be removed. To remove the welch plug drill a hole small through it and pry it out with a small screwdriver. Be extremely careful not to drill into the carburetor body.

4-20 FUEL SYSTEM

CLEANING & INSPECTION

▶ See Figures 31, 32, 33 and 34

> **CAUTION**
> Always wear safety glasses when using compressed air and aerosol solvents.

> **WARNING**
> Use compressed air only at a low pressure! High pressure air could cause damage to the carburetor by loosening pressed in jets or cause lodged objects to shoot from the unit.

> **WARNING**
> Never soak components that are or contain diaphragms, rubber or plastic parts in carburetor cleaner. They tend to absorb liquid and expand rendering them unusable. These parts should be cleaned only in solvent and then blown dry with compressed air immediately.

➥ If the carburetor body is worn or damaged in any way it will need to be replaced as an assembly.

1. Place all of the metal parts in a screen type tray and dip them in carburetor cleaner until they appear completely clean.
2. Blow the parts dry with compressed air.
3. Blow out all of the carburetor passageways to be sure they are clear and not blocked or clogged.
4. Carefully inspect the carburetor body for stripped threads or cracks.
5. Check any lead shot or core plugs for any sign of leakage.
6. Inspect the float hinge and hinge pin area for wear and the float for any sign of deterioration.
7. Inspect the float bowl & bowl gasket area for damage.
8. Examine the inlet needle and seat for wear
9. Inspect mixture screws for wear.
10. Inspect the throttle shaft for wear and excessive play.

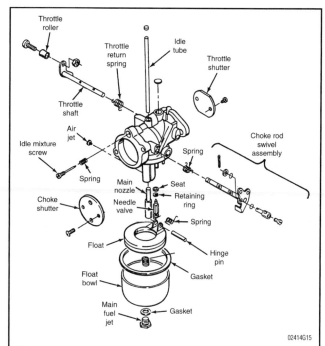

Fig. 32 Exploded drawing of a carburetor used on earlier single applications and on some multiple applications up to 1992

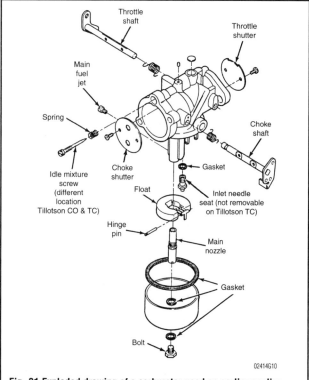

Fig. 31 Exploded drawing of a carburetor used on earlier applications 85 & 125HP

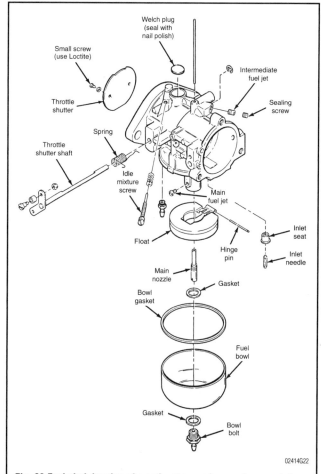

Fig. 33 Exploded drawing of a carburetor used on engines model year 1989 and newer, note the location of the idle mixture screw and the absence of the choke plate on later versions

FUEL SYSTEM 4-21

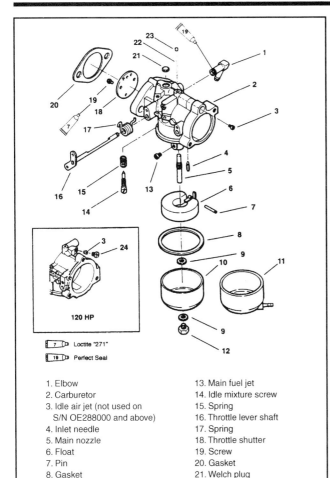

Fig. 34 Exploded drawing of a carburetor used on the latest engines up to 1999

1. Elbow
2. Carburetor
3. Idle air jet (not used on S/N OE288000 and above)
4. Inlet needle
5. Main nozzle
6. Float
7. Pin
8. Gasket
9. Gasket
10. Fuel bowl
11. Fuel bowl, top carburetor
12. Screw
13. Main fuel jet
14. Idle mixture screw
15. Spring
16. Throttle lever shaft
17. Spring
18. Throttle shutter
19. Screw
20. Gasket
21. Welch plug
22. Idle tube
23. Plug
24. Sealing screw (Not used on S/N OE288000 and above)

7 — Loctite "271"
19 — Perfect Seal

ASSEMBLY

♦ See accompanying illustrations

Carburetor service kits are available from the local marine dealer. These kits contain all necessary parts to perform the usual carburetor overhaul work. Check the kits part list to be sure you have all necessary parts including any damaged parts possibly not in the standard rebuild kit.

All worn and damaged parts along with any gaskets and seals that need to be replaced should be in hand at this time.

1. If the welch plug was removed install a new plug.
2. Install the plug by striking it in the center with a pin punch; seal the plug with some fingernail polish.
3. Install the main fuel jet into the same carburetor from which it was removed. The main jet is slightly different for each carburetor, especially the top carburetor. Tighten the jet in place with a screwdriver.
4. Install the main nozzle into the carburetor body channel, and tighten it with a screwdriver.
5. Install the intermediate fuel jet, and the sealing screw on top of the jet, as equipped.
6. Insert the inlet needle seat into the opening in the carburetor body. Press the seat down until it is firmly seated.

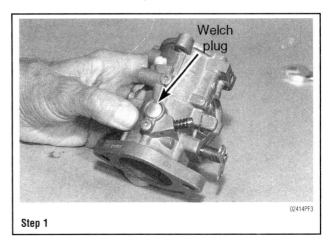

Step 1

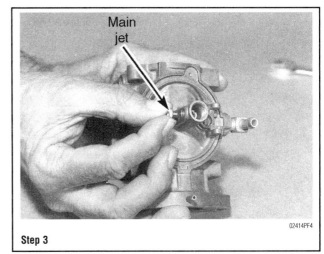

Step 3

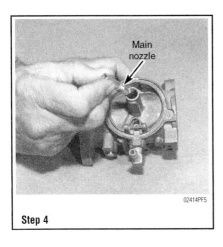

Step 4

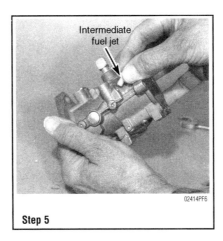

Step 5

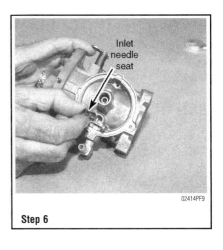

Step 6

4-22 FUEL SYSTEM

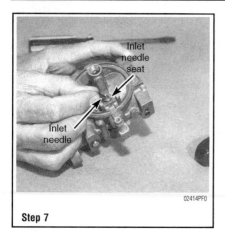
Step 7

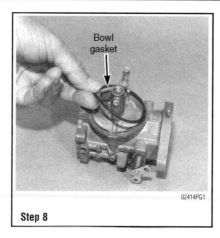
Step 8

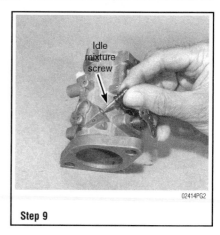
Step 9

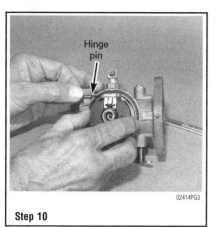
Step 10

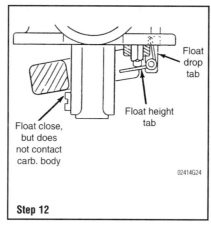
Step 12

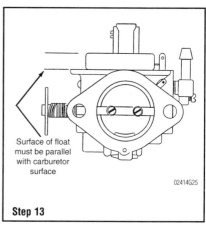
Step 13

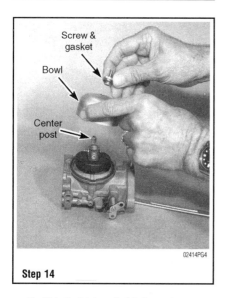
Step 14

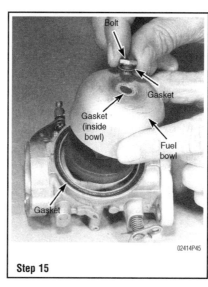
Step 15

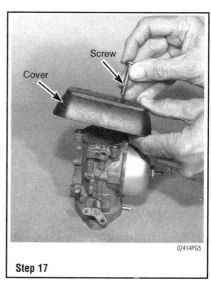
Step 17

7. Slide the inlet needle into the seat.
8. Position a new bowl gasket into the recess in the carburetor body.
9. Thread the idle mixture screw and spring into the carburetor body until the screw barely seats, and then back it out one full turn
10. Set the float into position with the holes in the hinge and bracket aligned.
11. Slide the hinge pin through, with the plain end going in first.
12. After the hinge pin is in place hold the carburetor and allow the float to hang free. The inner edge of the float should come close but not make contact with the carburetor body, adjust as necessary
13. Turn the carburetor with the float on top it should be parallel to the carburetor body, adjust as necessary
14. Set one small gasket onto the center post.
15. Place another gasket onto the bowl screw.
16. Place the bowl on the carburetor and secure it in position with the screw.
17. Install the carburetor cover and secure it in place with the attaching screws, if individual cover system is used.

Fuel Primer

▶ See Figure 35

The primer valve system is a starting aid only. It is fed fuel under pressure from the bowl area of the top or only carburetor. Pressing in the key switch while cranking the engine activates it.

FUEL SYSTEM 4-23

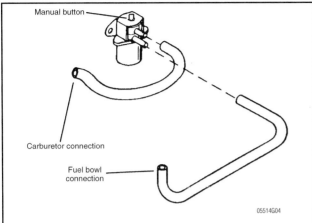

Fig. 35 The primer valve is located on the port side of the intake mounted between the carburetors

TESTING

Mechanical

1. Remove the hoses form the carburetor and intake manifold.
2. Safely drain the fuel from the primer vale and the removed hoses.
3. While pushing down on the button at the top of the primer valve, blow through the one hose and check that the air is passing through the valve.
4. This is the only mechanical test for this system. Replace the valve if this test failed to pass air through the valve.

Electrical

The electrical portion of the test involves testing the valve to ensure it operates properly when activated by the key switch. Repeat the previous test by pushing in the key switch or activating the valve by putting 12 volts to the Black/Yellow wire on the valve.

Fuel Pump

TESTING

Fuel pump testing is an excellent way to pinpoint air leaks, restricted fuel lines and fittings or other fuel supply related performance problems. When a fuel starvation problem is suspected such as engine hesitation or engine stopping, perform the following fuel system test:
1. Connect the piece of clear fuel hose to a side barb of the "T" fitting.
2. Connect one end of the long piece of fuel hose to the vacuum gauge and the other end to the center barb of the "T" fitting.

➡ Use a long enough piece of fuel hose so the vacuum gauge may be read at the helm.

3. Remove the existing fuel hose from the fuel tank side of the fuel pump and connect the remaining barb of the "T" fitting to the fuel hose.
4. Connect the short piece of clear fuel hose to the fuel check valve leading from the fuel filter. If a check valve does not exist, connect the clear fuel hose directly to the fuel filter.
5. Check the vacuum gauges reading after running the engine long enough to stabilize at full power. The vacuum is to not exceed 4.5 in. Hg (15.2 kPa).
6. An anti-siphon valve (required if the fuel system drops below the top of the fuel tank) will cause a 1.5 to 2.5 in. Hg (8.4 kPa) increase in vacuum.
7. If high vacuum is noted, move the T-fitting to the fuel filter outlet and retest.
8. Continue to the fuel filter inlet and along the remaining fuel system until a large drop in vacuum locates the problem.
9. A good clean water separator fuel filter will increase vacuum about 0.5 in. Hg (1.7 kPa).
10. Small internal passages inside a fuel selector valve, fuel tank pickup, or fuel line fittings may cause excessive fuel restriction and high vacuum.
11. Unstable and slowly rising vacuum readings, especially with a full tank of fuel, usually indicates a restricted vent line.
12. Bubbles in the clear fuel line section indicate an air leak, making for an inaccurate vacuum test. Check all fittings for tightened clamps and a tight fuel filter.

➡ Vacuum gauges are not calibrated and some may read as much as 2 in. Hg (6.8 kPa) lower than the actual vacuum. It is recommended to perform a fuel system test while no problems exist to determine vacuum gauge accuracy.

REMOVAL & INSTALLATION

Single Stage Fuel Pump

SMALL HORSEPOWER MODELS

♦ See Figure 36

1. Disconnect the engine fuel supply.
2. Remove the fuel inlet hose from the filter fitting on the pump cover.
3. Remove the fuel outlet hose located on the upper part of the pump cover.
4. Remove the three screws securing the cover and the reed plate to the powerhead.

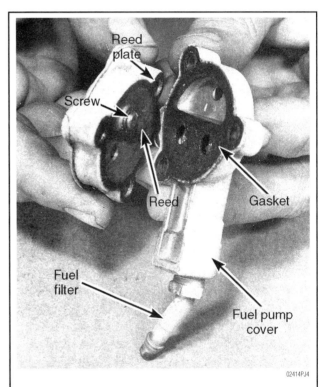

Fig. 36 Single stage, small horsepower fuel pump components

LARGE HORSEPOWER MODELS

1. Disconnect the engine fuel supply.
2. Remove the fuel inlet hose.
3. Remove the fuel pump outlet hose.
4. Remove the two screws fastening the pump to the engine block.
5. Remove the gasket from between the pump and engine block.

To install:
6. Install the gasket between the pump and engine block.
7. Install and securely tighten the two screws fastening the pump to the engine block.
8. Install the fuel pump outlet hose.
9. Install the fuel inlet hose.
10. Connect the engine fuel supply.

4-24 FUEL SYSTEM

Dual Stage Fuel Pumps

♦ See Figures 37, 38, 39 and 40

1. Disconnect the engine fuel supply.
2. Remove the inlet fuel hose from the nipple on the plastic sediment bowl.
3. Remove the fuel outlet hose from the discharge side of the fuel pump.
4. Back off the knurled screw and remove the sediment bowl and screen from the pump body.
5. Remove the sediment bowl gasket.
6. If one of the fuel pump mounting screws is covered by the ignition coil mounting bracket removed to gain access to the mounting screw.
7. Remove the fuel pump body and mounting gasket from the powerhead.

➥It is not necessary to remove the rear cover of the pump from the block to overhaul.

To install:

8. Install the fuel pump body and mounting gasket on the powerhead.
9. Install and tighten the screws securely.
10. Install the sediment bowl gasket, sediment bowl and screen.
11. Install the fuel outlet hose on the discharge side of the fuel pump.
12. Install the inlet fuel hose on the nipple at the plastic sediment bowl.
13. Connect the engine fuel supply.

OVERHAUL

Small Horsepower Single Stage Pump

♦ See Figures 41 and 42

1. Scribe a line on the pump body to aid in proper alignment when reassembled.
2. Separate the pump cover from the reed plate.
3. Remove the gasket from the cover.
4. Remove the reed by loosening the attaching screw.
5. Remove the fuel filter from the pump cover.
6. Remove the powerhead mounting gasket and the diaphragm from the reed.

To assemble:

7. Throughly clean the fuel pump parts in a mild solvent.
8. Inspect the housing parts for cracks.
9. Inspect valves for any deformities or broken pieces.
10. Inspect the gaskets & diaphragms for tears break or signs of being brittle.
11. Install the reed assemblies replace as necessary.
12. Install gaskets and diaphragms replace as necessary.
13. Install fuel filter into pump cover.
14. Connect inlet and discharge hoses with appropriate clamps.
15. Install pump on powerhead with the mounting gasket in place.
16. Connect to fuel supply and prime the system to check for leaks.

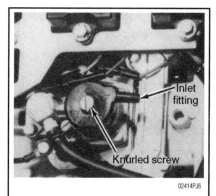

Fig. 37 Remove the inlet fuel hose from the nipple on the plastic sediment bowl

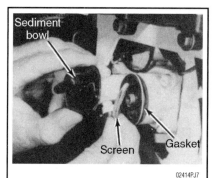

Fig. 38 Back off the knurled screw and remove the sediment bowl and screen from the pump body

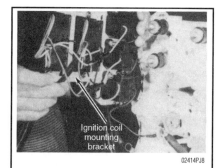

Fig. 39 If one of the fuel pump mounting screws is covered by the ignition coil mounting bracket removed to gain access to the mounting screw

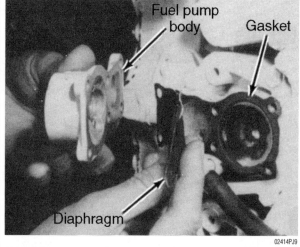

Fig. 40 Remove the fuel pump body and mounting gasket from the powerhead

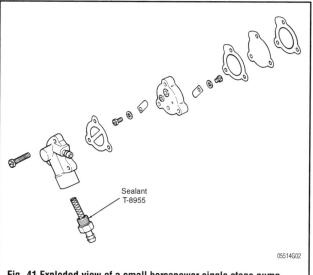

Fig. 41 Exploded view of a small horsepower single stage pump

FUEL SYSTEM 4-25

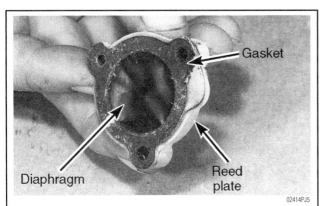

Fig. 42 Remove the powerhead mounting gasket and the diaphragm from the reed

Large Horsepower Single Stage Pump

♦ See Figure 43

These fuel pumps have plastic center bodies. The reed valves are installed with non-reusable rivet type fasteners. Do not remove the reed valves unless they are to be replaced.

➡There are diaphragm support discs and springs that are under tension inside the pump assembly be careful not to lose them.

1. Scribe a line on the pump body to aid in proper alignment when reassembled.
2. Remove the two screws holding the assembly together.
3. Separate the housing parts.

To assemble:

4. Throughly clean the fuel pump parts in a mild solvent.
5. Inspect the housing parts for cracks.
6. Inspect valves for any deformities or broken pieces.
7. Inspect the gaskets & diaphragms for tears break or signs of being brittle.
8. Install the reed assemblies replace as necessary.
9. Install gaskets and diaphragms replace as necessary.
10. Connect inlet and discharge hoses with appropriate clamps.
11. Install pump on powerhead with the mounting gasket in place.
12. Connect to fuel supply and prime the system to check for leaks.

Dual Stage Pump

♦ See Figures 44 thru 50

➡The two outside valves are pressed into the pump body making it necessary to replace them if they need to be removed.

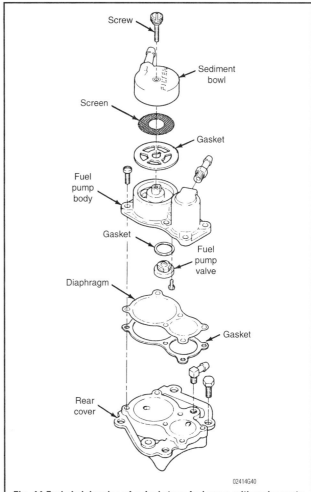

Fig. 44 Exploded drawing of a dual stage fuel pump with major parts identified

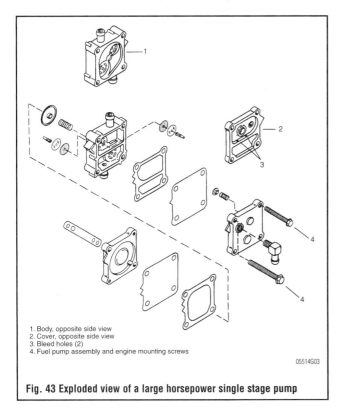

1. Body, opposite side view
2. Cover, opposite side view
3. Bleed holes (2)
4. Fuel pump assembly and engine mounting screws

Fig. 43 Exploded view of a large horsepower single stage pump

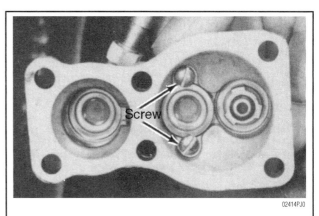

Fig. 45 Remove the center valve through the attaching screws

4-26 FUEL SYSTEM

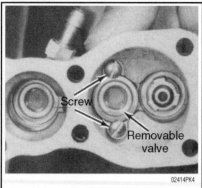

Fig. 46 Insert a gasket into the check valve recess of the pump body

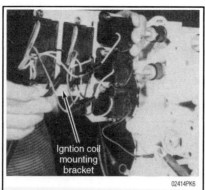

Fig. 47 Secure the pump in place with the attaching screws

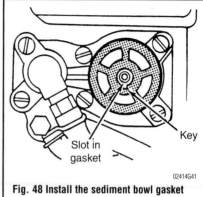

Fig. 48 Install the sediment bowl gasket with the slot in the gasket over the key

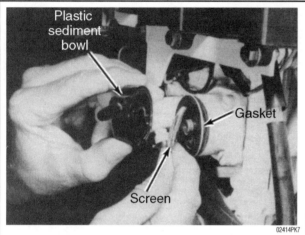

Fig. 49 Insert the screen onto the pump body with the turned edge facing the block

Fig. 50 Connect the outlet fuel line to the fitting on the aft side of the pump

1. Remove the center valve through the attaching screws.
2. The two outside valves are pressed into the pump body.

➥If these two valves are pressed out, they will be rendered unfit for further service and would have to be replaced. Be sure they must be replaced and new valves have been purchased and are on hand, before removing the valves.

To assemble:
3. Throughly clean the fuel pump parts in a mild solvent.
4. Inspect the housing parts for cracks.
5. Inspect valves for any deformities or broken pieces.
6. Inspect the gaskets & diaphragms for tears break or signs of being brittle.
7. Install the reed assemblies replace as necessary.
8. Install gaskets and diaphragms replace as necessary.
9. Insert a gasket into the check valve recess of the pump body.
10. Install the check valve with the spring side going into the recess first.
11. Secure the valve in place with the two attaching screws.
12. Place the mounting screws into the holes in the fuel pump body for ease of alignment.
13. Gently put the diaphragm & mounting gasket in place on the pump body.
14. Secure the pump in place with the attaching screws.
15. If the ignition coil mounting bracket was removed install the bracket.
16. Install the sediment bowl gasket with the slot in the gasket over the key.
17. Insert the screen onto the pump body with the turned edge facing the block.
18. Install the sediment bowl with the inlet nipple in the four o'clock position.

➥Do not over tighten

19. Connect the outlet fuel line to the fitting on the aft side of the pump.
20. Connect the inlet fuel line to the plastic sediment bowl fitting.

Fuel Lines

▶ See Figures 51, 52 and 53

On most installations, the fuel line is provided with quick-disconnect fittings at the tank and at the engine. If there is reason to believe the problem is at the quick-disconnects, the hose ends should be replaced as an assembly. For a small additional expense, the entire fuel line can be replaced and thus eliminate this entire area as a problem source for many future seasons.

The primer squeeze bulb can be replaced in a short time. First, cut the hose line as close to the old bulb as possible. Slide a small clamp over the end of the fuel line from the tank. Next, install the small end of the check valve assembly into this side of the fuel line. The check valve always goes towards the fuel tank. Place a large clamp over the end of the check valve assembly. Use Primer Bulb Adhesive when the connections are made. Tighten the clamps. Repeat the procedure with the other side of the bulb assembly and the line leading to the engine.

Recirculating System

▶ See Figures 54, 55 and 56

This system is designed to recalculate unburned fuel and oil back to the intake where it can reenter the combustion process and be properly burned off. It is primarily an idle speed circuit that prevents the lower cylinders from flooding and fouling out the spark plugs.

Over time the valves in this system can carbon up and not function properly. The most common symptom is a plug fouling on a lower cylinder but be certain to check all other systems first.

Early model recirculation systems have the valves and passages built into cover plates bolted onto the side of the engine block and generally incorporate non serviceable valves.

This system is not considered a regular maintenance area but is a mandatory inspection during a rebuild.

FUEL SYSTEM 4-27

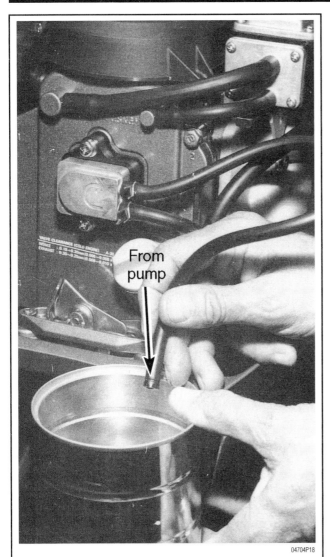

Fig. 51 To test the fuel pickup in the fuel tank, operate the squeeze bulb and observe fuel flowing from the disconnected line at the fuel pump. Discharge fuel into an approved container.

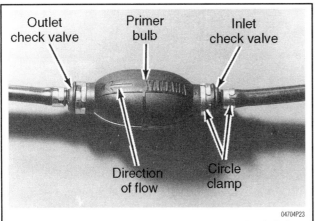

Fig. 53 Major parts of a typical fuel line squeeze bulb. The bulb is used to prime the fuel system until the powerhead is operating and the pump can deliver the required amount of fuel to run the engine

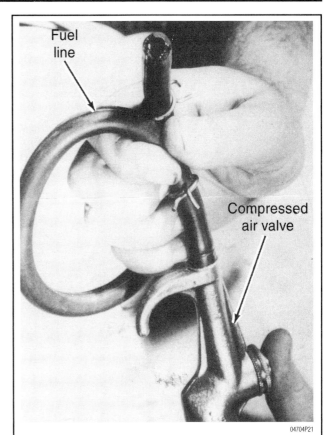

Fig. 52 Many times restrictions such as foreign material may be cleared from the fuel lines using compressed air. Ensure the open end of the hose is pointing in a clear direction to avoid personal injury

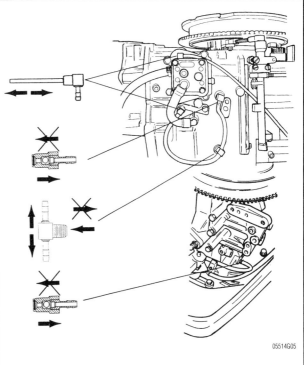

Fig. 54 Hose routing and valve placement late model 40–50hp

4-28 FUEL SYSTEM

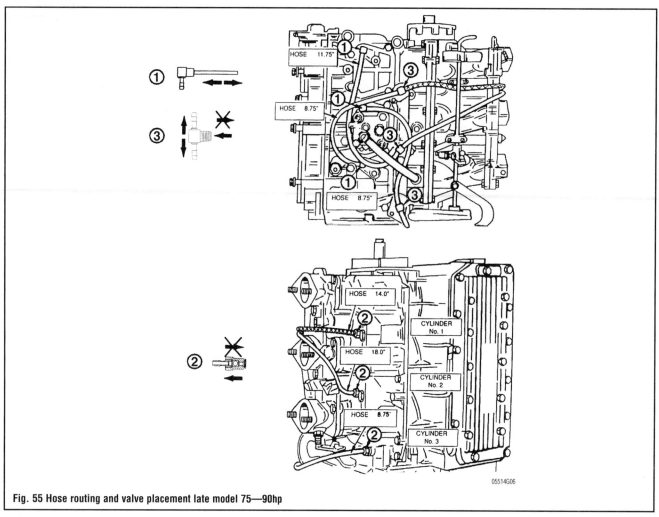

Fig. 55 Hose routing and valve placement late model 75—90hp

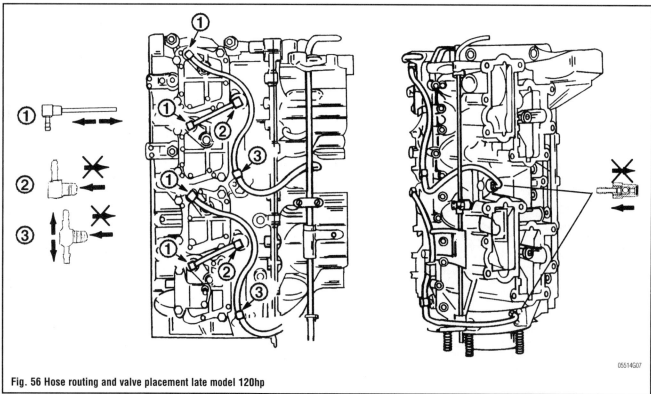

Fig. 56 Hose routing and valve placement late model 120hp

UNDERSTANDING AND TROUBLESHOOTING
ELECTRICAL SYSTEMS 5-2
BASIC ELECTRICAL THEORY 5-2
    HOW ELECTRICITY WORKS: THE WATER
      ANALOGY 5-2
    OHM'S LAW 5-2
ELECTRICAL COMPONENTS 5-2
    POWER SOURCE 5-2
    GROUND 5-2
PROTECTIVE DEVICES 5-3
    SWITCHES & RELAYS 5-3
    LOAD 5-4
    WIRING & HARNESSES 5-4
    CONNECTORS 5-4
TEST EQUIPMENT 5-5
    JUMPER WIRES 5-5
    TEST LIGHTS 5-5
    MULTIMETERS 5-6
TROUBLESHOOTING THE ELECTRICAL SYSTEM 5-6
    VOLTAGE 5-6
    VOLTAGE DROP 5-7
    RESISTANCE 5-7
    OPEN CIRCUITS 5-7
    SHORT CIRCUITS 5-8
WIRE AND CONNECTOR REPAIR 5-8
BREAKER POINTS (MAGNETO) IGNITION 5-8
TROUBLESHOOTING THE BREAKER POINTS (MAGNETO)
    IGNITION 5-9
FLYWHEEL 5-10
    REMOVAL & INSTALLATION 5-10
    INSPECTION 5-11
BREAKER POINTS 5-11
    POINT GAP ADJUSTMENT 5-11
    TESTING 5-12
    REMOVAL & INSTALLATION 5-12
CONDENSER 5-12
    TESTING 5-12
    REMOVAL & INSTALLATION 5-12
IGNITION COIL 5-13
    TESTING 5-13
    REMOVAL & INSTALLATION 5-13
MOTOROLA® DISTRIBUTOR IGNITION 5-13
DESCRIPTION AND OPERATION 5-13
TROUBLESHOOTING THE MOTOROLA® DISTRIBUTOR
    IGNITION 5-14
FLYWHEEL 5-14
    REMOVAL & INSTALLATION 5-14
    INSPECTION 5-15
DISTRIBUTOR 5-15
    TESTING 5-15
    REMOVAL & INSTALLATION 5-16
    DISASSEMBLY 5-17
    CLEANING & INSPECTION 5-19
    ASSEMBLY 5-19
IGNITION COILS 5-21
    TESTING 5-21
    REMOVAL & INSTALLATION 5-21
IGNITION MODULE 5-21
    TESTING 5-21
    REMOVAL & INSTALLATION 5-22
**ELECTRONIC IGNITION SYSTEMS UP TO
25 HP 5-22**
DESCRIPTION AND OPERATION 5-22
TROUBLESHOOTING ELECTRONIC IGNITION SYSTEMS UP
    TO 25 HP 5-22
FLYWHEEL 5-22
    REMOVAL & INSTALLATION 5-22
    INSPECTION 5-24
IGNITION COILS 5-24
    TESTING 5-24
    REMOVAL & INSTALLATION 5-24
PRESTOLITE® ELECTRONIC IGNITION 5-24
DESCRIPTION AND OPERATION 5-24
TROUBLESHOOTING THE PRESTOLITE® ELECTRONIC
    IGNITION 5-25
FLYWHEEL 5-25
    REMOVAL & INSTALLATION 5-25
    INSPECTION 5-27
CHARGING COILS 5-27
    TESTING 5-27
    REMOVAL & INSTALLATION 5-28
TRIGGER COILS 5-29
    TESTING 5-29
    REMOVAL & INSTALLATION 5-30
IGNITION COILS 5-31
    TESTING 5-31
    REMOVAL & INSTALLATION 5-31
IGNITION MODULE 5-32
    TESTING 5-32
**THUNDERBOLT® ELECTRONIC IGNITION
SYSTEM 5-32**
DESCRIPTION AND OPERATION 5-32
TROUBLESHOOTING THE THUNDERBOLT® ELECTRONIC
    IGNITION SYSTEM 5-32
FLYWHEEL 5-34
    REMOVAL & INSTALLATION 5-34
    INSPECTION 5-35
CHARGING COILS 5-35
    TESTING 5-35
    REMOVAL & INSTALLATION 5-36
TRIGGER COILS 5-36
    TESTING 5-36
    REMOVAL & INSTALLATION 5-36
IGNITION COILS 5-37
    TESTING 5-37
    REMOVAL & INSTALLATION 5-38
SWITCH BOX 5-38
    TESTING 5-38
    REMOVAL & INSTALLATION 5-38
**CAPACITOR DISCHARGE MODULE (CDM)
SYSTEM 5-38**
DESCRIPTION AND OPERATION 5-38
TROUBLESHOOTING THE CDM SYSTEM 5-39
FLYWHEEL 5-40
    REMOVAL & INSTALLATION 5-40
    INSPECTION 5-41
CHARGING COILS 5-41
    TESTING 5-41
    REMOVAL & INSTALLATION 5-42
TRIGGER COILS 5-42
    TESTING 5-42
    REMOVAL 5-42
CDM IGNITION COIL MODULE 5-43
    TESTING 5-43
    REMOVAL & INSTALLATION 5-43
CHARGING CIRCUIT 5-45
DESCRIPTION AND OPERATION 5-45
TROUBLESHOOTING THE CHARGING SYSTEM 5-45
STATOR 5-45
    TESTING 5-45
    REMOVAL & INSTALLATION 5-46
RECTIFIER 5-46
    TESTING 5-46
    REMOVAL & INSTALLATION 5-47
RECTIFIER/REGULATOR 5-47
    TESTING 5-47
    REMOVAL & INSTALLATION 5-48
BATTERY 5-48
    BATTERY CONSTRUCTION 5-48
    MARINE BATTERIES 5-48
    BATTERY RATINGS 5-49
    BATTERY LOCATION 5-49
    BATTERY CHARGERS 5-49
    BATTERY CABLES 5-50
STARTING CIRCUIT 5-50
DESCRIPTION AND OPERATION 5-50
TROUBLESHOOTING THE STARTING SYSTEM 5-51
STARTER MOTOR 5-51
    REMOVAL & INSTALLATION 5-52
    DISASSEMBLY 5-53
    CLEANING & INSPECTION 5-54
    TESTING 5-55
    ASSEMBLY 5-57
STARTER RELAY 5-58
    TESTING 5-58
    REMOVAL & INSTALLATION 5-58
**IGNITION AND ELECTRICAL WIRING
DIAGRAMS 5-59**
SPECIFICATIONS CHART
    IGNITION SYSTEM SPECIFICATIONS 5-44

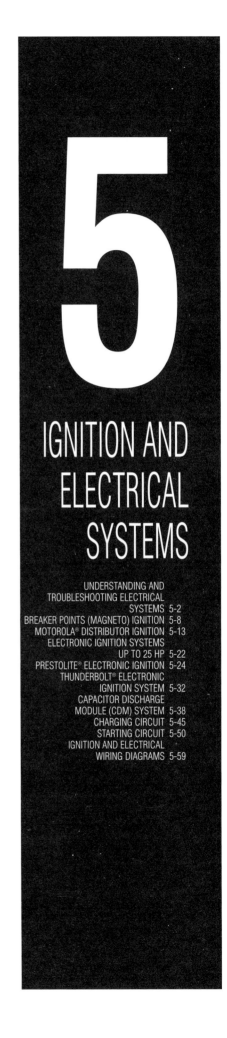

5

IGNITION AND ELECTRICAL SYSTEMS

UNDERSTANDING AND
TROUBLESHOOTING ELECTRICAL
SYSTEMS 5-2
BREAKER POINTS (MAGNETO) IGNITION 5-8
MOTOROLA® DISTRIBUTOR IGNITION 5-13
ELECTRONIC IGNITION SYSTEMS
UP TO 25 HP 5-22
PRESTOLITE® ELECTRONIC IGNITION 5-24
THUNDERBOLT® ELECTRONIC
IGNITION SYSTEM 5-32
CAPACITOR DISCHARGE
MODULE (CDM) SYSTEM 5-38
CHARGING CIRCUIT 5-45
STARTING CIRCUIT 5-50
IGNITION AND ELECTRICAL
WIRING DIAGRAMS 5-59

IGNITION AND ELECTRICAL SYSTEMS

UNDERSTANDING AND TROUBLESHOOTING ELECTRICAL SYSTEMS

Basic Electrical Theory

♦ See Figure 1

For any 12 volt, negative ground, electrical system to operate, the electricity must travel in a complete circuit. This simply means that current (power) from the positive terminal (+) of the battery must eventually return to the negative terminal (-) of the battery. Along the way, this current will travel through wires, fuses, switches and components. If for any reason the flow of current through the circuit is interrupted, the component(s) fed by that circuit will cease to function properly.

Perhaps the easiest way to visualize a circuit is to think of connecting a light bulb (with two wires attached to it) to the battery—one wire attached to the negative (-) terminal of the battery and the other wire to the positive (+) terminal. With the two wires touching the battery terminals, the circuit would be complete and the light bulb would illuminate. Electricity would follow a path from the battery to the bulb and back to the battery. It's easy to see that with longer wires on our light bulb, it could be mounted anywhere. Further, one wire could be fitted with a switch so that the light could be turned on and off.

The normal marine circuit differs from this simple example in two ways. First, instead of having a return wire from each bulb to the battery, the current travels through a single ground wire which handles all the grounds for a specific circuit. Secondly, most marine circuits contain multiple components which receive power from a single circuit. This lessens the amount of wire needed to power components.

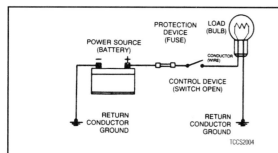

Fig. 1 This example illustrates a simple circuit. When the switch is closed, power from the positive (+) battery terminal flows through the fuse and the switch and then to the light bulb. The light illuminates and the circuit is completed through the ground wire back to the negative (-) battery terminal.

HOW ELECTRICITY WORKS: THE WATER ANALOGY

Electricity is the flow of electrons—the sub-atomic particles that constitute the outer shell of an atom. Electrons spin in an orbit around the center core of an atom. The center core is comprised of protons (positive charge) and neutrons (neutral charge). Electrons have a negative charge and balance out the positive charge of the protons. When an outside force causes the number of electrons to unbalance the charge of the protons, the electrons will split off the atom and look for another atom to balance out. If this imbalance is kept up, electrons will continue to move and an electrical flow will exist.

Many people have been taught electrical theory using an analogy with water. In a comparison with water flowing through a pipe, the electrons would be the water and the wire is the pipe.

The flow of electricity can be measured much like the flow of water through a pipe. The unit of measurement used is amperes, frequently abbreviated as amps (a). You can compare amperage to the volume of water flowing through a pipe. When connected to a circuit, an ammeter will measure the actual amount of current flowing through the circuit. When relatively few electrons flow through a circuit, the amperage is low. When many electrons flow, the amperage is high.

Water pressure is measured in units such as pounds per square inch (psi); electrical pressure is measured in units called volts (v). When a voltmeter is connected to a circuit, it is measuring the electrical pressure. When electrical pressure is low, then voltage is considered to be low. When electrical pressure is high, then voltage is considered to be high.

The actual flow of electricity depends not only on voltage and amperage but also on the resistance of the circuit. The higher the resistance, the higher the force necessary to push the current through the circuit. The standard unit for measuring resistance is an ohm (Ω). Resistance in a circuit varies depending on the amount and type of components used in the circuit and the overall condition of the components and wires. If we assume that everything in our circuit is new, then the main factors which determine resistance are:

- **Material**—some materials have more resistance than others. Those with high resistance are said to be insulators. Rubber materials (or rubber-like plastics) are some of the most common insulators used, as they have a very high resistance to electricity. Very low resistance materials are said to be conductors. Copper wire is among the best conductors. Silver is actually a superior conductor to copper and is used in some relay contacts but its high cost prohibits its use as common wiring. Most marine wiring is made of copper.
- **Size**—the larger the wire size being used, the less resistance the wire will have. This is why components which use large amounts of electricity usually have large wires supplying current to them.
- **Length**—for a given thickness of wire, the longer the wire, the greater the resistance. The shorter the wire, the less the resistance. When determining the proper wire for a circuit, both size and length must be considered to design a circuit that can handle the current needs of the component.
- **Temperature**—with many materials, the higher the temperature, the greater the resistance (positive temperature coefficient). Some materials exhibit the opposite trait of lower resistance with higher temperatures (negative temperature coefficient). These principles are used in many of the sensors on the engine.

OHM'S LAW

There is a direct relationship between current, voltage and resistance which can be summed up by a statement known as Ohm's law.
- Voltage (E) is equal to amperage (I) times resistance (R): $E = I \times R$
- Other forms of the formula are $R = E/I$ and $I = E/R$

In each of these formulas, E is the voltage in volts, I is the current in amps and R is the resistance in ohms. The basic point to remember is that as the resistance of a circuit goes up, the amount of current that flows in the circuit will go down, if voltage remains the same.

The amount of work that the electricity can perform is expressed as power. A unit of power is known as a watt (w).

There is a direct relationship between power, voltage and current which can be summed up by the following formula:
- Power (W) is equal to amperage (I) times voltage (E): $W = I \times E$

➡ This formula is only true for direct current (DC) circuits; The alternating current formula is a tad different but since the electrical circuits in most boats are DC type, we need not get into AC circuit theory.

Electrical Components

POWER SOURCE

♦ See Figure 2

Power is supplied to the boat by two devices: The battery and the stator. The battery supplies electrical power during starting or during periods when the current demand of the boat's electrical system exceeds the output capacity of the alternator. The stator supplies electrical current when the engine is running. The stator does not just supply the current needs of the boat but it also recharges the battery.

In most modern boats, the battery is a lead/acid electrochemical device consisting of six 2 volt subsections (cells) connected in series, so that the unit is capable of producing approximately 12 volts of electrical pressure. Each subsection consists of a series of positive and negative plates held a short distance apart in a solution of sulfuric acid and water.

The two types of plates are of dissimilar metals, which sets up a chemical reaction inside the battery case. It is this reaction which produces current flow from the battery when its positive and negative terminals are connected to an electrical load. The power removed from the battery is replaced by the alternator, restoring the battery to its original chemical state.

IGNITION AND ELECTRICAL SYSTEMS 5-3

Fig. 2 The stator is the component part that produces the electrical power for the ignition and charging systems

The following is a brief description of how the system works:

The battery stores electricity and acts as a "sponge" for the whole system. It mops up generated current until it's fully charged. and it release s energy on demand.

The flywheel holds the permanent magnets that create the moving magnetic field- If your engine has good spark, you can take it for granted that the magnets are in working order because the ignition and charging systems share the same magnets.

The stator windings are the stationary coils of wire the flywheel magnets rotate around. They produce the electrical charge. Simply put. the more windings in your stator, the greater the potential output in amps your charging system you'll have.

The rectifier consists of a series of diodes or electrical one-way valves. The rectifier overcomes one of the disadvantages of a current-generating system using permanent magnets and stator windings, which is that the current produced within the windings is alternating current (AC). You can't use AC to charge batteries. They accept only direct current (DC). So the rectifier is designed to convert AC current to a usable form of DC current simply called "rectified AC."

On larger outboards, there may be a voltage regulator. either combined with the rectifier or standing alone. The regulator automatically reduces the output of generated current as the battery becomes fully charged.

GROUND

All boats use some sort of a ground return circuit. Direct ground components are grounded to an electrically conductive metal component through their mounting points. These electrically conductive metal components are then grounded to the battery.

All other components use some sort of ground wire which leads directly back to the battery. The electrical current runs through the ground wire and returns to the battery through the ground (-) cable. If you look, you'll see that the battery ground cable connects between the battery and a heavy gauge ground wire.

➡It should be noted that a good percentage of electrical problems can be traced to bad grounds.

PROTECTIVE DEVICES

▶ See Figure 3

It is possible for large surges of current to pass through the electrical system of your boat. If this surge of current were to reach components in the circuit, the surge could burn them out or severely damage them. Surges can also overload the wiring, causing the harness to get hot and melt the insulation. To prevent this, fuses, circuit breakers and/or fusible links are connected into the supply wires of the electrical system. These items are nothing more than a built-in weak spot in the system. When an abnormal amount of current flows through the system, these protective devices work as follows to protect the circuit:

- Fuse—when an excessive electrical current passes through a fuse, the fuse "blows" (the conductor melts) and opens the circuit, preventing the passage of current.
- Circuit Breaker—a circuit breaker is basically a self-repairing fuse. It will open the circuit in the same fashion as a fuse but when the surge subsides, the circuit breaker can be reset and does not need replacement.
- Fusible Link—a fusible link (fuse link or main link) is a short length of special, high temperature insulated wire that acts as a fuse. When an excessive electrical current passes through a fusible link, the thin gauge wire inside the link melts, creating an intentional open to protect the circuit.

To repair the circuit, the link must be replaced. Some newer type fusible links are housed in plug-in modules, which are simply replaced like a fuse, while older type fusible links must be cut and spliced if they melt. Since this link is very early in the electrical path, it's the first place to look if nothing on the boat works, yet the battery seems to be charged and is properly connected.

✲✲ CAUTION

Always replace fuses, circuit breakers and fusible links with identically rated components. Under no circumstances should a component of higher or lower amperage rating be substituted.

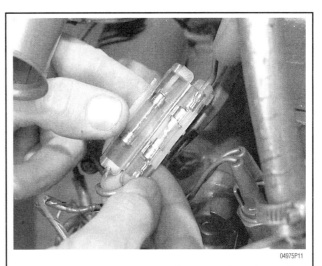

Fig. 3 Fuses protect the vessel's electrical system from abnormally high amounts of current flow

SWITCHES & RELAYS

▶ See Figure 4

Switches are used in electrical circuits to control the passage of current. The most common use is to open and close circuits between the battery and the various electric devices in the system. Switches are rated according to the amount of amperage they can handle. If a sufficient amperage rated switch is not used in a circuit, the switch could overload and cause damage.

Some electrical components which require a large amount of current to oper-

5-4 IGNITION AND ELECTRICAL SYSTEMS

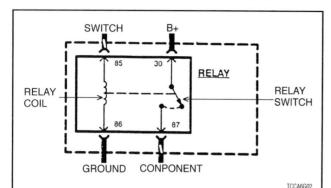

Fig. 4 Relays are composed of a coil and a switch. These two components are linked together so that when one operates, the other operates at the same time. The large wires in the circuit are connected from the battery to one side of the relay switch (B+) and from the opposite side of the relay switch to the load (component). Smaller wires are connected from the relay coil to the control switch for the circuit and from the opposite side of the relay coil to ground

ate use a special switch called a relay. Since these circuits carry a large amount of current, the thickness of the wire in the circuit is also greater. If this large wire were connected from the load to the control switch, the switch would have to carry the high amperage load and the space needed for wiring in the boat would be twice as big to accommodate the increased size of the wiring harness. To prevent these problems, a relay is used.

Relays are composed of a coil and a set of contacts. When the coil has a current passed though it, a magnetic field is formed and this field causes the contacts to move together, completing the circuit. Most relays are normally open, preventing current from passing through the circuit but they can take any electrical form depending on the job they are intended to do. Relays can be considered "remote control switches." They allow a smaller current to operate devices that require higher amperages. When a small current operates the coil, a larger current is allowed to pass by the contacts. Some common circuits which may use relays are horns, lights, starter, electric fuel pumps and other high draw circuits.

LOAD

Every electrical circuit must include a "load" (something to use the electricity coming from the source). Without this load, the battery would attempt to deliver its entire power supply from one pole to another. This would result in a "short circuit" of the battery. All this electricity would take a short cut to ground and cause a great amount of damage to other components in the circuit by developing a tremendous amount of heat. This condition could develop sufficient heat to melt the insulation on all the surrounding wires and reduce a multiple wire cable to a lump of plastic and copper.

WIRING & HARNESSES

The average boat contains miles of wiring, with hundreds of individual connections. To protect the many wires from damage and to keep them from becoming a confusing tangle, they are organized into bundles, enclosed in plastic or taped together and called wiring harnesses. Different harnesses serve different parts of the boat. Individual wires are color coded to help trace them through a harness where sections are hidden from view.

Marine wiring can be either single strand wire, multi-strand wire or printed circuitry. Single strand wire has a solid metal core and is usually used inside such components as alternators, motors, relays and other devices. Multi-strand wire has a core made of many small strands of wire twisted together into a single conductor. Most of the wiring in a marine electrical system is made up of multi-strand wire, either as a single conductor or grouped together in a harness. All wiring is color coded on the insulator, either as a solid color or as a colored wire with an identification stripe. A printed circuit is a thin film of copper or other conductor that is printed on an insulator backing. Occasionally, a printed circuit is sandwiched between two sheets of plastic for more protection and flexibility. A complete printed circuit, consisting of conductors, insulating material and connectors is called a printed circuit board. Printed circuitry is used in place of individual wires or harnesses in places where space is limited, such as behind instrument panels.

Since marine electrical systems are very sensitive to changes in resistance, the selection of properly sized wires is critical when systems are repaired. A loose or corroded connection or a replacement wire that is too small for the circuit will add extra resistance and an additional voltage drop to the circuit.

The wire gauge number is an expression of the cross-section area of the conductor. Boats from countries that use the metric system will typically describe the wire size as its cross-sectional area in square millimeters. In this method, the larger the wire, the greater the number. Another common system for expressing wire size is the American Wire Gauge (AWG) system. As gauge number increases, area decreases and the wire becomes smaller. An 18 gauge wire is smaller than a 4 gauge wire. A wire with a higher gauge number will carry less current than a wire with a lower gauge number. Gauge wire size refers to the size of the strands of the conductor, not the size of the complete wire with insulator. It is possible, therefore, to have two wires of the same gauge with different diameters because one may have thicker insulation than the other.

It is essential to understand how a circuit works before trying to figure out why it doesn't. An electrical schematic shows the electrical current paths when a circuit is operating properly. Schematics break the entire electrical system down into individual circuits. In a schematic, usually no attempt is made to represent wiring and components as they physically appear on the boat; switches and other components are shown as simply as possible. Face views of harness connectors show the cavity or terminal locations in all multi-pin connectors to help locate test points.

CONNECTORS

♦ See Figures 5, 6, 7 and 8

Weatherproof connectors are most commonly used where the connector is exposed to the elements. Terminals are protected against moisture and dirt by

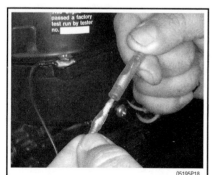

Fig. 5 Bullet connectors are some of the more common electrical connectors found on an outboard engine

Fig. 6 A typical weatherproof electrical connector

Fig. 7 Hard shell (left) and weatherproof (right) connectors have replaceable terminals

IGNITION AND ELECTRICAL SYSTEMS

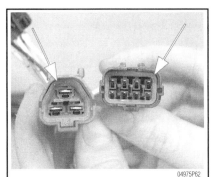

Fig. 8 The seals on weatherproof connectors must be kept in good condition to prevent the terminals from corroding

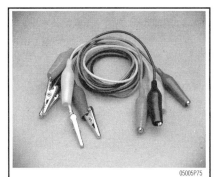

Fig. 9 Jumper wires are simple, yet extremely valuable, pieces of test equipment

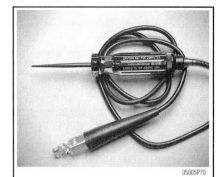

Fig. 10 A test light is used to detect the presence of voltage in a circuit

sealing rings which provide a weather tight seal. All repairs require the use of a special terminal and the tool required to service it.

Unlike standard blade type terminals, these weatherproof terminals cannot be straightened once they are bent. Make certain that the connectors are properly seated and all of the sealing rings are in place when connecting leads.

Test Equipment

Pinpointing the exact cause of trouble in an electrical circuit is most times accomplished by the use of special test equipment. The following sections describe different types of commonly used test equipment and briefly explain how to use them in diagnosis. In addition to the information covered below, the tool manufacturer's instruction manual (provided with most tools) should be read and clearly understood before attempting any test procedures.

JUMPER WIRES

♦ See Figure 9

✳✳ CAUTION

Never use jumper wires made from a thinner gauge wire than the circuit being tested. If the jumper wire is of too small a gauge, it may overheat and possibly melt. Never use jumpers to bypass high resistance loads in a circuit. Bypassing resistances, in effect, creates a short circuit. This may, in turn, cause damage and fire. Jumper wires should only be used to bypass lengths of wire or to simulate switches.

Jumper wires are simple, yet extremely valuable, pieces of test equipment. They are basically test wires which are used to bypass sections of a circuit. Although jumper wires can be purchased, they are usually fabricated from lengths of standard marine wire and whatever type of connector (alligator clip, spade connector or pin connector) that is required for the particular application being tested. In cramped, hard-to-reach areas, it is advisable to have insulated boots over the jumper wire terminals in order to prevent accidental grounding.

It is also advisable to include a standard marine fuse in any jumper wire. This is commonly referred to as a "fused jumper". By inserting an in-line fuse holder between a set of test leads, a fused jumper wire can be used for bypassing open circuits. Use a 5 amp fuse to provide protection against voltage spikes.

Jumper wires are used primarily to locate open electrical circuits. If an electrical component fails to operate, connect the jumper wire between the component and a good ground. If the component operates only with the jumper installed, the ground circuit is open.

If the ground circuit is good but the component does not operate, the circuit between the power feed and component may be open. By moving the jumper wire successively back from the component toward the power source, you can isolate the area of the circuit where the open is located. When the component stops functioning or the power is cut off, the open is in the segment of wire between the jumper and the point previously tested.

You can sometimes connect the jumper wire directly from the battery to the "hot" terminal of the component but first make sure the component uses 12 volts in operation. Some electrical components, such as fuel injectors or sensors are designed to operate on about 4 to 5 volts and running 12 volts directly to these components will cause damage.

TEST LIGHTS

♦ See Figure 10

The test light is used to check circuits and components while electrical current is flowing through them. It is used for voltage and ground tests. To use a 12 volt test light, connect the ground clip to a good ground and probe wherever necessary with the pick. The test light will illuminate when voltage is detected. This does not necessarily mean that 12 volts (or any particular amount of voltage) is present; it only means that some voltage is present.

➡ It is advisable before using the test light to touch its ground clip and probe across the battery posts or terminals to make sure the light is operating properly.

✳✳ WARNING

Do not use a test light to probe electronic ignition, spark plug or coil wires. Never use a pick-type test light to probe wiring on electronically controlled systems unless specifically instructed to do so. Any wire insulation that is pierced by the test light probe should be taped and sealed with silicone after testing.

Like the jumper wire, the 12 volt test light is used to isolate opens in circuits. But, whereas the jumper wire is used to bypass the open to operate the load, the 12 volt test light is used to locate the presence of voltage in a circuit. If the test light illuminates, there is power up to that point in the circuit; if the test light does not illuminate, there is an open circuit (no power). Move the test light in successive steps back toward the power source until the light in the handle illuminates. The open is between the probe and a point which was previously probed.

The self-powered test light is similar in design to the 12 volt test light but contains a 1.5 volt penlight battery in the handle. It is most often used in place of a multimeter to check for open or short circuits when power is isolated from the circuit (continuity test).

The battery in a self-powered test light does not provide much current. A weak battery may not provide enough power to illuminate the test light even when a complete circuit is made (especially if there is high resistance in the circuit). Always make sure that the test battery is strong. To check the battery, briefly touch the ground clip to the probe; if the light glows brightly, the battery is strong enough for testing.

✳✳ WARNING

A self-powered test light should not be used on any electronically controlled system or component. The small amount of electricity transmitted by the test light is enough to damage many electronic marine components.

5-6 IGNITION AND ELECTRICAL SYSTEMS

MULTIMETERS

▶ See Figure 11

Multimeters are an extremely useful tool for troubleshooting electrical problems. They can be purchased in either analog or digital form and have a price range to suit any budget. A multimeter is a voltmeter, ammeter and ohmmeter (along with other features) combined into one instrument. It is often used when testing solid state circuits because of its high input impedance (usually 10 megaohms or more). A brief description of the multimeter main test functions follows:

• **Voltmeter**—the voltmeter is used to measure voltage at any point in a circuit or to measure the voltage drop across any part of a circuit. Voltmeters usually have various scales and a selector switch to allow the reading of different voltage ranges.

The voltmeter has a positive and a negative lead. To avoid damage to the meter, always connect the negative lead to the negative (-) side of the circuit (to ground or nearest the ground side of the circuit) and connect the positive lead to the positive (+) side of the circuit (to the power source or the nearest power source).

➡ The negative voltmeter lead will always be Black and that the positive voltmeter will always be some color other than Black (usually red).

• **Ohmmeter**—the ohmmeter is designed to read resistance (measured in ohms) in a circuit or component. Most ohmmeters will have a selector switch which permits the measurement of different ranges of resistance (usually the selector switch allows the multiplication of the meter reading by 10, 100, 1,000 and 10,000). Some ohmmeters are "auto-ranging" which means the meter itself will determine which scale to use.

Since the meters are powered by an internal battery, the ohmmeter can be used like a self-powered test light. When the ohmmeter is connected, current from the ohmmeter flows through the circuit or component being tested. Since the ohmmeter's internal resistance and voltage are known values, the amount of current flow through the meter depends on the resistance of the circuit or component being tested.

The ohmmeter can also be used to perform a continuity test for suspected open circuits. In using the meter for making continuity checks, do not be concerned with the actual resistance readings. Zero resistance (or any ohm reading), indicates continuity in the circuit. Infinite resistance indicates an opening in the circuit. A high resistance reading where there should be none indicates a problem in the circuit.

Checks for short circuits are made in the same manner as checks for open circuits, except that the circuit must be isolated from both power and normal ground. Infinite resistance indicates no continuity, while zero resistance indicates a dead short.

✳✳ WARNING

Never use an ohmmeter to check the resistance of a component or wire while there is voltage applied to the circuit.

• **Ammeter**—an ammeter measures the amount of current flowing through a circuit in units called amps (amperes). At normal operating voltage, most circuits have a characteristic amount of amps, called "current draw" which can be measured using an ammeter. By referring to a specified current draw rating, then measuring the amps and comparing the two values, one can determine what is happening within the circuit to aid in diagnosis.

For example, an open circuit will not allow any current to flow, so the ammeter reading will be zero. A damaged component or circuit will have an increased current draw, so the reading will be high.

The ammeter is always connected in series with the circuit being tested. All of the current that normally flows through the circuit must also flow through the ammeter; if there is any other path for the current to follow, the ammeter reading will not be accurate. The ammeter itself has very little resistance to current flow and, therefore, will not affect the circuit but it will measure current draw only when the circuit is closed and electricity is flowing. Excessive current draw can blow fuses and drain the battery, while a reduced current draw can cause motors to run slowly, lights to dim and other components to not operate properly.

Troubleshooting the Electrical System

When diagnosing any electrical problem, organized troubleshooting is a must. The complexity of electrical systems on modern boats and their powerplants demands that you approach any problem in a logical organized manner. There are certain troubleshooting techniques, which are standard:

• **Establish when the problem occurs**—Does the problem appear only under certain conditions? Were there any noises, odors or other unusual symptoms?

• **Check for obvious problems**—Problems such as broken wires and loose or dirty connections can cause major problems. Always check the obvious before assuming something complicated (or expensive) is the cause.

➡ Experience has shown that most problems tend to be the result of a fairly simple and obvious cause, such as loose or corroded connectors, bad grounds or damaged wire insulation which causes a short. This makes careful visual inspection of components during testing essential to quick and accurate troubleshooting.

• **Isolate the problem area**—Make some simple tests and observations, then eliminate the systems that are working properly. Test for problems systematically to determine the cause once the problem area is isolated. Are all the components functioning properly? Is there power going to electrical switches and motors. Performing careful, systematic checks will often turn up most causes on the first inspection, without wasting time checking components that have little or no relationship to the problem.

• **Verify all systems after repairs are completed**—Some causes can be traced to more than one component, so a careful verification of repair work is important in order to pick up additional malfunctions that may cause a problem to reappear or a different problem to arise. A blown fuse, for example, is a simple problem that may require more than another fuse to repair. If you don't look for a problem that caused a fuse to blow, a shorted wire (for example) may go undetected.

VOLTAGE

▶ See Figure 12

This test determines voltage available from the battery and should be the first step in any electrical troubleshooting procedure after visual inspection. Many electrical problems, especially on electronically controlled systems, can be caused by a low state of charge in the battery. Excessive corrosion at the battery cable terminals can cause poor contact that will prevent proper charging and full battery current flow.

1. Set the voltmeter selector switch to the 10 volt position.
2. Connect the multimeter negative lead to the battery's negative (-) terminal and the positive lead to the battery's positive (+) terminal.
3. Turn the battery (ignition) switch **ON** to provide a load.

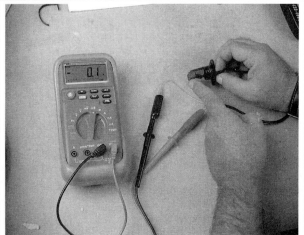

Fig. 11 Multimeters, such as this one from UEI, are essential for diagnosing faulty wires, switches and other electrical components

IGNITION AND ELECTRICAL SYSTEMS 5-7

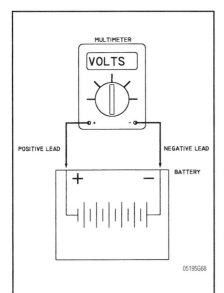

Fig. 12 The voltage test determines voltage available from the battery and should be the first step in any electrical troubleshooting procedure after visual inspection

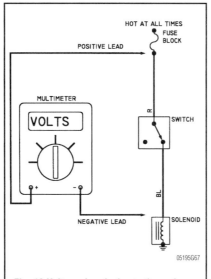

Fig. 13 Voltage drop is due to the resistance created by the load and also by small resistances created by corrosion at the connectors and damaged insulation on the wires

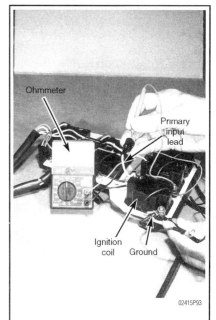

Fig. 14 Checking the primary resistance of an ignition coil using a multimeter

4. A well charged battery should register over 12 volts. If the meter reads below 11.5 volts, the battery power may be insufficient to operate the electrical system properly.
5. Charge the battery and retest.

VOLTAGE DROP

♦ See Figure 13

When current flows through a load, the voltage beyond the load drops. This voltage drop is due to the resistance created by the load and also by small resistances created by corrosion at the connectors and damaged insulation on the wires. The maximum allowable voltage drop under load is critical, especially if there is more than one load in the circuit, since all voltage drops are cumulative.

1. Set the voltmeter selector switch to the 10 volt position.
2. Connect the multimeter negative lead to the battery negative (-) terminal or another good ground.
3. Touch the multimeter positive lead to the battery's positive (+) terminal to determine battery voltage.
4. Operate the circuit and check the voltage prior to the first component (load).
5. There should be little or no voltage drop (from battery voltage) in the circuit prior to the first component. If a voltage drop exists, the wire or connectors in the circuit are suspect.
6. While operating the first component in the circuit, probe the ground side of the component with the positive meter lead and observe the voltage readings. A small voltage drop should be noticed. This voltage drop is caused by the resistance of the component.
7. Repeat the test for each component (load) down the circuit.
8. If a large voltage drop is noticed, the preceding component, wire or connector is suspect.

RESISTANCE

♦ See Figures 14 and 15

※※ WARNING

Never use an ohmmeter with power applied to the circuit. The ohmmeter is designed to operate on its own power supply. The normal 12 volt electrical system voltage can damage the meter!

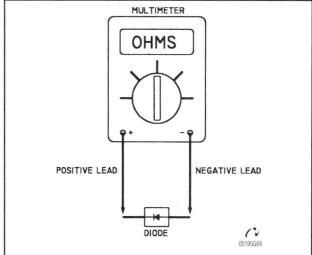

Fig. 15 When performing resistance tests, always isolate the circuit from power

1. Isolate the circuit from the boat's power source.
2. Ensure that the battery (ignition) switch is **OFF**.
3. Isolate at least one side of the circuit to be checked, in order to avoid reading parallel resistances. Parallel circuit resistances will always give a lower reading than the actual resistance of either of the branches.
4. Connect the meter leads to both sides of the circuit (wire or component) and read the actual resistance measured in ohms on the meter scale. Make sure the selector switch is set to the proper ohm scale for the circuit being tested, to avoid misreading the ohmmeter test value.
5. Compare this reading to the resistance specification for the component or formulate the theoretical resistance using Ohms Law.

OPEN CIRCUITS

♦ See Figures 16 and 17

This test already assumes the existence of an open in the circuit and it is used to help locate the open portion.

1. Isolate the circuit from power and ground.

5-8 IGNITION AND ELECTRICAL SYSTEMS

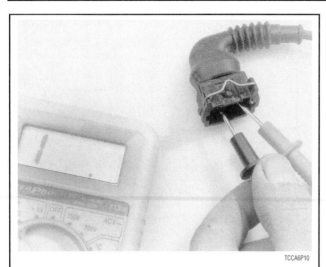

Fig. 16 The infinite display on this multimeter (1.) indicates that the circuit is open

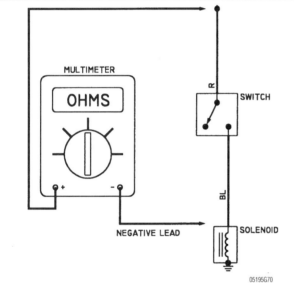

Fig. 17 The easiest way to illustrate an open circuit is to consider an example circuit with a switch. When the switch is turned OFF, power does not flow through the circuit to the load. Thus, the circuit is open

2. Connect the self-powered test light or ohmmeter ground clip to the ground side of the circuit and probe sections of the circuit sequentially.
3. If the light is out or there is infinite resistance, the open is between the probe and the circuit ground.
4. If the light is on or the meter shows continuity, the open is between the probe and the end of the circuit toward the power source.

BREAKER POINTS (MAGNETO) IGNITION

♦ See Figures 19 and 20

The breaker point ignition consists of the rotor assembly, contact point assembly, ignition coil, condenser spark plug, spark plug cap and the engine stop switch. The breaker points ignition system uses a mechanically switched, collapsing field to induce spark at the plug. A magnet moving by a coil produces current in the primary coil winding. The current in the primary winding creates a magnetic field.

At the proper time, the breaker points are separated by action of a cam designed into the collar of the crankshaft and the primary circuit is broken.

SHORT CIRCUITS

♦ See Figure 18

➡ Never use a self-powered test light to perform checks for opens or shorts when power is applied to the circuit under test. The test light can be damaged by outside power.

1. Isolate the circuit from power and ground.
2. Connect the self-powered test light or ohmmeter ground clip to a good ground and probe any easy-to-reach point in the circuit.
3. If the light comes on or there is continuity, there is a short somewhere in the circuit.
4. To isolate the short, probe a test point at either end of the isolated circuit (the light should be on or the meter should indicate continuity).
5. Leave the test light probe engaged and sequentially open connectors or switches, remove parts, etc. until the light goes out or continuity is broken.
6. When the light goes out, the short is between the last two circuit components which were opened.

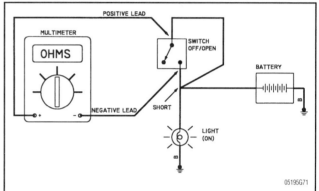

Fig. 18 In this illustration, the circuit between the battery and light should be open because the switch is turned OFF. However, battery voltage is reaching the light at the point of the short. This could possibly be caused by chaffed wires

Wire and Connector Repair

Almost anyone can replace damaged wires, as long as the proper tools and parts are available. Wire and terminals are available to fit almost any need. Even the specialized weatherproof, molded and hard shell connectors are now available from aftermarket suppliers.

Be sure the ends of all the wires are fitted with the proper terminal hardware and connectors. Wrapping a wire around a stud is never a permanent solution and will only cause trouble later. Replace wires one at a time to avoid confusion. Always route wires in the same manner of the manufacturer.

When replacing connections, make absolutely certain that the connectors are certified for marine use. Automotive wire connectors may not meet United States Coast Guard (USCG) specifications.

➡ If connector repair is necessary, only attempt it if you have the proper tools. Weatherproof connectors require special tools to release the pins inside the connector. Attempting to repair these connectors with conventional hand tools will damage them.

When the circuit is broken, the flow of primary current stops and causes the magnetic field about the coil to break down instantly. At this precise moment, an electrical current of extremely high voltage is induced in the fine secondary windings of the coil. This high voltage is conducted to the spark plug where it jumps the gap between the points of the plug to ignite the compressed charge of air-fuel mixture in the cylinder.

The breaker points must be aligned accurately to provide the best contact surface. This is the only way to assure maximum contact area between the point surfaces; accurate setting of the point gap; proper synchronization; and satisfactory point life. If the points are not aligned properly, the result will be premature

IGNITION AND ELECTRICAL SYSTEMS

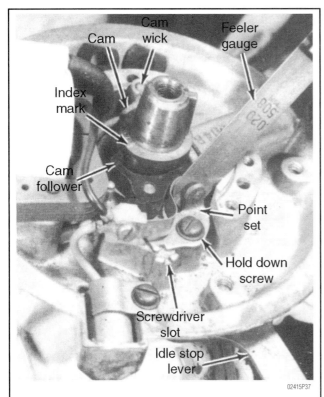

Fig. 19 A typical Force outboard single cylinder engine with breaker point ignition

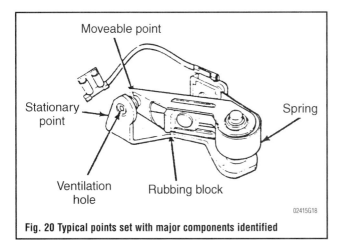

Fig. 20 Typical points set with major components identified

wear or pitting. This type of damage may change the cam angle, although the actual distance will remain the same.

The breaker point ignition system contains a condenser that works like a sponge in the circuit. Current that is flowing through the primary circuit tries to keep going. When the breaker point switch opens the current will arc over the widening gap. The condenser is wired in parallel with the points. The condenser absorbs some of the current flow as the points open. This reduces arc over and extends the life of the points.

In simple terms, a condenser is composed of two sheets of tin or aluminum foil laid one on top of the other but separated by a sheet of insulating material such as waxed paper, etc. The sheets are rolled into a cylinder to conserve space and then inserted into a metal case for protection and to permit easy assembling.

The purpose of the condenser is to prevent excessive arcing across the points and to extend their useful life. When the flow of primary current is brought to a sudden stop by the opening of the points, the magnetic field in the primary windings collapses instantly and is not allowed to fade away, which would happen if the points were allowed to arc.

The condenser stores the electricity that would have arced across the points and discharges that electricity when the points close again. This discharge is in the opposite direction to the original flow and tends to smooth out the current. The more quickly the primary field collapses, the higher the voltage produced in the secondary windings and delivered to the spark plugs. In this way, the condenser (in the primary circuit) affects the voltage (in the secondary circuit) at the spark plugs.

Modern condensers seldom cause problems, therefore, it is not necessary to install a new one each time the points are replaced. However, if the points show evidence of arcing, the condenser may be at fault and should be replaced. A faulty condenser may not be detected without the use of special test equipment. Testing will reveal any defects in the condenser but will not predict the useful life left in the unit.

The modest cost of a new condenser justifies its purchase and installation to eliminate this item as a source of trouble.

The coil is the heart of the ignition system. Essentially, it is nothing more than a transformer which takes the relatively low voltage (12 volts) available from the primary coil and increases it to a point where it will fire the spark plug as much as 20,000 volts.

Once the voltage is discharged from the ignition coil the secondary circuit begins and only stretches from the ignition coil to the spark plugs via extremely large high tension leads. At the spark plug end, the voltage arcs in the form of a spark, across from the center electrode to the outer electrode and then to ground via the spark plug threads. This completes the ignition circuit.

Breaker point ignition systems installed on outboard engines will usually operate over extremely long periods of time without requiring adjustment or repair. However, if ignition system problems are encountered and the usual corrective actions such as replacement of spark plugs does not correct the problem, the magneto output should be checked to determine if the unit is functioning properly.

Troubleshooting the Breaker Points (Magneto) Ignition

▶ See Figure 21

As with any electrical problem, the first step in troubleshooting a breaker points ignition system is to look for the obvious. Check all wiring connections for loose or broken wires. Look for signs of corrosion on the terminals and at the connectors.

The next step in checking out your ignition system is to verify that you're getting spark to each cylinder.

✱✱ WARNING

When checking the spark, make sure there is no fuel on either the engine or the spark plug. Also keep your hands away from high voltage electrical components.

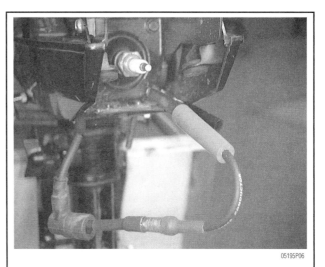

Fig. 21 Spark tester installed on a single cylinder outboard with breaker points ignition system

5-10 IGNITION AND ELECTRICAL SYSTEMS

➡ If you are working in bright daylight, it's a good idea to create some shade near the plug wire you're checking. It's very difficult to see a spark in sunlight.

The best tool for checking spark is an inexpensive spark tester. This very useful unit is marketed by several tool manufacturers and can be had for a reasonable cost. The air gap between the two contacts is adjustable. This is important because some manufacturers give a air-gap specification in their manuals. The wider the air gap a spark will jump, the higher the total ignition system output. So, by comparing the maximum gap jump for each ignition coil, you can isolate a weak or faulty coil.

1. Carefully remove the plug wire from the spark plug you wish to test using a twisting, pulling motion on the boot only. Never pull on the wire itself, as you may damage the connection inside the boot where the wire attaches to its end connector.
2. Insert the heavy brass terminal end of the tool into the plug wire end, making sure it fits snugly into the boot. Now take the wire lead from the spark tester and attach the clip to a good ground point on the engine. Make sure the ground connection is good.

➡ A bare metal connection is best as paint acts as an insulator.

3. Adjust the knurled knob on the tool to a gap equal to or larger than that of the spark plug gap.
4. Hold the tool so you can see inside the cylinder and crank the engine over. You should see a bright Blue spark jumping between the two contact points. If you do, then total ignition output is satisfactory. If you don't, or if the spark is Yellow, further investigation will be needed.
5. In the event that there is no spark before continuing be sure to check or disconnect the engine stop button. A shorted switch in this area will cause a no spark condition.
6. If the engine is skipping or misfiring, check all of the plug wires to be certain the each coil is sending spark through its respective plug wire to the plug.

Flywheel

REMOVAL & INSTALLATION

✴✴ CAUTION

Always wear safety glasses when removing flywheels. Extreme force might be necessary to remove the unit and metal slivers or rust could fly up into your face.

✴✴ WARNING

Flywheels that have never been removed or have sheared keys are typically very difficult to remove. The use of heat from a torch or soaking with penetrating oil is recommended to ease removal.

Flywheels Without Puller Bolt Holes

▶ See Figures 22, 23 and 24

1. Remove the engine cover and manual starter as necessary.
2. Hold the flywheel with a flywheel holding tool or strap wrench.
3. Remove the fastener holding the flywheel in place.
4. Pull up on the outer edge of the flywheel with a prybar.
5. Tap the crankshaft with a brass hammer to loosen the flywheel.
6. Carefully lift the flywheel from the powerhead.

To install:

7. Install the flywheel on the powerhead making certain the flywheel key is properly installed in the keyway and not damaged.
8. Install the fastener holding the flywheel in place.
9. Hold the flywheel with a flywheel holding tool or strap wrench and tighten to the specified torque.
10. Install the manual starter and engine cover.

Flywheels With Puller Bolt Holes

▶ See Figures 25, 26 and 27

1. Hold the Flywheel with a strap wrench or a flywheel holder.
2. Remove the flywheel nut.
3. Install a puller on to the flywheel using the proper thread size and length bolt.
4. Thread the puller bolt in putting pressure on the crankshaft and removing the flywheel.

➡ It may be necessary to hit the puller bolt with a hammer to shock the crankshaft to release the flywheel.

5. Carefully lift the flywheel from the powerhead.

To install:

6. Install the flywheel on the powerhead making certain the flywheel key is properly installed in the keyway and not damaged.
7. Install the fastener holding the flywheel in place.
8. Hold the flywheel with a flywheel holding tool or strap wrench and tighten to the specified torque.
9. Install the manual starter and engine cover.

Fig. 22 Remove the engine cover and manual starter as necessary

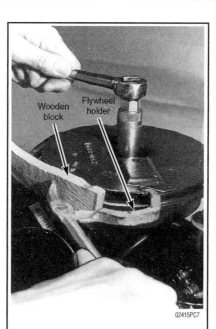

Fig. 23 Hold the flywheel with a strap wrench

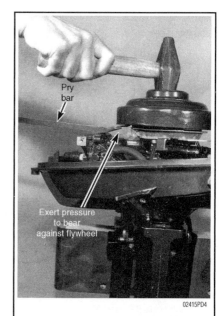

Fig. 24 Pull up on the outer edge of the flywheel with a prybar

IGNITION AND ELECTRICAL SYSTEMS

Fig. 25 Hold the Flywheel with a strap wrench or a flywheel holder

Fig. 26 Install a puller on to the flywheel using the proper thread size and length bolt

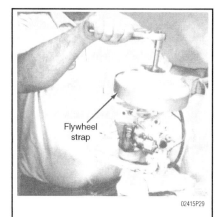
Fig. 27 Reverse this procedure to install the flywheel making certain the flywheel key in properly installed in the keyway and not damaged

Flywheels With Threaded Center

♦ See Figures 28 and 29

1. While holding the flywheel with the flywheel holder or strap wrench, remove the flywheel nut and washer.
2. Install a crankshaft protector cap (91–24161), then install a flywheel puller on the flywheel.
3. Hold the flywheel tooth with the wrench while tightening the bold down on the protector cap. Tighten the bolt until the flywheel comes free.
4. Carefully lift the flywheel from the powerhead.

To install:

5. Install the flywheel on the powerhead making certain the flywheel key is properly installed in the keyway and not damaged.
6. Install the fastener holding the flywheel in place.
7. Hold the flywheel with a flywheel holding tool or strap wrench and tighten to the specified torque.
8. Install the manual starter and engine cover.

INSPECTION

Visually inspect the flywheel for cracks, broken teeth and deformities.
Inspect the magnets around the inside of the flywheel for cracks or damage.
Test the strength of the magnets by placing a medium size flat blade screwdriver against it. The magnet should be able to hold the screwdriver. All outer magnets should be the same strength.
Replace the flywheel as necessary.

Breaker Points

POINT GAP ADJUSTMENT

♦ See Figures 30 and 31

1. Remove the flywheel.
2. Remove the spark plugs.

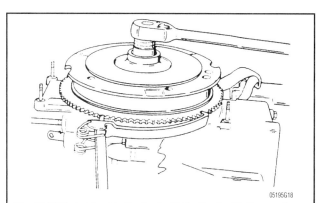
Fig. 28 While holding the flywheel with the flywheel holder or strap wrench, remove the flywheel nut and washer

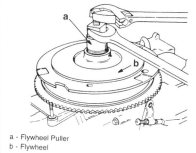
Fig. 29 Hold the flywheel tooth with the wrench while tightening the bold down on the protector cap. Tighten the bolt until the flywheel comes free

Fig. 30 Turn the crankshaft clockwise until the cam follower lines up with the index mark on the cam and check the point gap with a feeler gauge

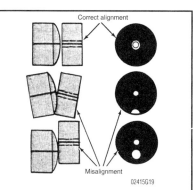
Fig. 31 Points should be properly aligned prior to adjustment. Always bend the stationary point, never the breaker lever

5-12 IGNITION AND ELECTRICAL SYSTEMS

➡ Before performing any type of test or adjustment, be sure that the contact point faces are in good condition. Breaker points should be replaced when they appear worn burned or pitted.

3. Turn the crankshaft clockwise until the cam follower lines up with the index mark on the cam.
4. Check the point gap with a feeler gauge and compare to the specifications listed in the "Tune-Up Specifications" chart.
5. Adjust as necessary by loosening the hold down screw on the breaker point base plate.
6. Use a flat blade screwdriver to rotate the breaker base with the slot provided until the correct gap is achieved.
7. Repeat these procedures for each set of breaker points.
8. Install spark plugs.
9. Install flywheel.

TESTING

♦ See Figure 32

➡ Breaker points are a regular service item and should be replaced if any of the test results are questionable.

1. Perform a visual inspection of the overall condition of the breaker points.
2. Check the contact points for wear and that they are flat and can mate correctly.
3. Remove the screw and wire connections from the condenser, ignition coil and stop switch.
4. Using a multimeter on the high ohms scale check that there is continuity across the points.
5. Open the points and check that there is no continuity.
6. Check that the points are not shorted to ground by placing one meter lead on the base plate and the other on the tension spring.

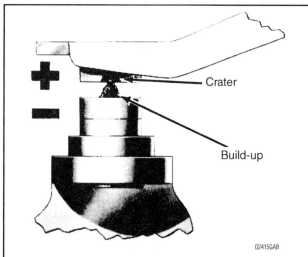

Fig. 32 An enlarged view of a visual inspection shows this set of breaker points unusable due to transfer deposits

REMOVAL & INSTALLATION

♦ See Figure 33

1. Remove the flywheel.
2. Remove the spark plugs.
3. Remove the screw and wire connections from the stop switch, condenser and ignition coil.
4. Remove the screw holding the breaker point base.
5. Slide the breaker point set up and off the pivot stud.
6. Repeat procedure for each set of breaker points.

To install:

7. Slide the breaker point set onto the pivot stud.
8. Install the screw holding the breaker point base.

Fig. 33 Breaker point and condenser installation

➡ Lay the connected wires toward the ignition coil before securing the screw to avoid them from contacting the flywheel after assembly.

9. Install the screw and wire connections from the stop switch, condenser and ignition coil.
10. Set the points as described in gap adjustment.
11. Install the spark plugs.
12. Install the flywheel.

Condenser

The condenser is a regular service item and should be replaced if any of the test results are questionable. If during the breaker point inspection the contacts appeared burned or had a deposit build up the condenser should be replaced.

TESTING

1. The condenser does not have to be removed to perform this test.
2. Place a multimeter test lead on the condenser output wire and another on the condenser body. The meter should read 0.22–0.28 microfarads.
3. If the condenser measurement does not meet specifications, it may be faulty and will need to be replaced.

REMOVAL & INSTALLATION

♦ See Figure 33

1. Remove the flywheel
2. Remove the screw and wire connections at the breaker points for the condenser.
3. Remove the screw holding the condenser clamp.
4. Remover the condenser.
5. Repeat procedure for each condenser.
6. Reverse these procedures to install.

IGNITION AND ELECTRICAL SYSTEMS 5-13

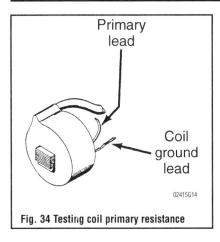

Fig. 34 Testing coil primary resistance

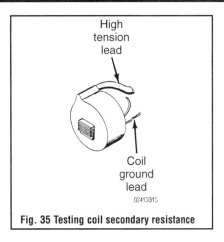

Fig. 35 Testing coil secondary resistance

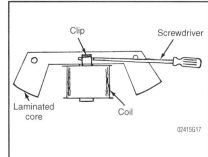

Fig. 36 Pull back the spring clip holding the coil on the lamination base. Carefully remove the high tension lead with the coil assembly

Ignition Coil

TESTING

Resistance

♦ See Figures 34 and 35

1. Visually inspect the coil for cracks electrical arching or burn marks.
2. Remove the wire connection at the breaker points.
3. Connect the Black multimeter lead to ground and the Red lead to the primary coil wire.
4. Compare this reading of the coil primary resistance to the specifications listed in the "Ignition System Specifications" chart.
5. Connect the Black multimeter lead to the primary coil wire and the Red lead to the spark plug high tension lead.
6. Compare this reading of the coil secondary resistance to the specifications listed in the "Ignition System Specifications" chart.
7. Repeat these procedures for each ignition coil.
8. Replace any coil that has failed any of the tests.

REMOVAL & INSTALLATION

Early Models

♦ See Figure 36

1. Remove the flywheel.
2. Remove the screw and wire connections at the breaker points for the ignition coil.
3. Remove the spark plug boot and spring.
4. Pull back the spring clip holding the coil on the lamination base.
5. Carefully remove the high tension lead with the coil assembly and note the wire routing for reinstallation.

To install:

6. Install coil by feeding the high tension lead back into place.
7. Place the coil on the lamination base and secure the spring clip.
8. Connect wiring on to the breaker point base.
9. Install the spark plug spring and boot.
10. Repeat procedure for each ignition coil.
11. Install flywheel.

Late Models

1. Remove the flywheel.
2. Remove the screw and wire connections at the breaker points for the ignition coil.
3. Remove the spark plug boot and spring.
4. Remove the ignition coil as an assembly by removing the attaching hardware holding the lamination plate.
5. Carefully remove the high tension lead with the coil assembly and note the wire routing for reinstallation.

To install:

6. Install coil by feeding the high tension lead back into place.
7. Locate coil laminations onto magneto plate and secure with attaching hardware.
8. Connect wiring on to the breaker point base.
9. Install the spark plug spring and boot.
10. Repeat procedure for each ignition coil.
11. Install flywheel.

MOTOROLA® DISTRIBUTOR IGNITION

Description and Operation

This system is known as battery CD ignition which needs an external 12 volt source to enable it to operate.

A properly timed distributor is belt driven off the underside of the flywheel. The rotating of the distributor activates the sensor or breaker points signaling the ignition coil to fire.

The internal construction of a typical ignition coil includes primary and secondary windings. It also uses the principle of magnetic induction with the magnetism generated by the primary (lower -voltage) winding creating a magnetic field around the secondary winding, which has many more windings than the primary coil. The rapid movement of this magnetic field past the secondary windings induces electrical current flow. The are greater the number of turns of wire in the secondary winding, the higher the voltage produced.

As this secondary voltage leaves the center tower of the ignition coil, it travels along the spark-plug wire, which is heavily insulated and designed to carry this high voltage. This current travels to the distributor and directed to the proper cylinder in the firing order.

If all is well, the high voltage will jump the gap on the spark plug between the center electrode and the ground electrode. On larger engines with surface-gap plugs, the high voltage current will jump from the center electrode to the side of the plug assembly itself, completing a circuit to ground via the engine block.

Last, but certainly not least, is the stop control. You need a means to shut your engine off and a good way is to stop the spark plugs from working. Depending on your engine, this may be accomplished by a simple stop button or a key switch on larger engines. This disables the whole ignition system. On newer engines, an emergency stop button with an overboard clip and lanyard attachment is standard. This system is wired directly into the ignition coils. It functions by creating a momentary short circuit inside the coil unit, grounding the current intended for the high tension coils and thus shutting off the ignition long enough to stop the engine. Faulty stop circuits are frequently the cause of a no spark condition.

5-14 IGNITION AND ELECTRICAL SYSTEMS

Troubleshooting the Motorola® Distributor Ignition

→The following conditions apply to the Motorola® Distributor Ignition System. Troubleshooting procedure are courtesy of our friends at CDI Electronics (256-772-3829).

→Maintenance-free batteries are not recommended for use with these systems.

Check for broken wires and terminals, especially inside the plastic plug-in connectors. Removal of the pins from the connectors using the Rapair (R554-9706) Pin Removal Tool and will allow you to visually inspect the pins for damage.

Check the flywheel for broken or loose magnets.

Disconnect the kill wires from the CD and connect a DC voltmeter between the kill wires and engine ground, turn the ignition switch on and off several times. If at any time you see voltage appearing on the meter, there is a problem in the harness or ignition switch.

→At no time should you see battery voltage on a kill circuit.

Visually inspect stator for burned or discolored areas. If found, replace the stator. If the areas are on the battery charge windings, it indicates a possible problem with the rectifier.

IF NO FIRE ON ANY CYLINDER: Disconnect kill wire at the pack. Check for broken or bare wires on the unit, stator and trigger. Using a multimeter with a peak reading adapter, or and piercing probes, measure DVA voltage of the stator between the output wire sets. With everything connected, readings should be approximately 180 volts or, more. The OEM stator resistance between the blue and yellow wires is 700-900 ohms. CDI Electronics stators should read 300-500 ohms. Disconnect the rectifier. If the engine fires, replace the rectifier.

NO FIRE OR INTERMITTENT FIRE ON ONE CYLINDER: Check stator and trigger resistance. Trigger wire sets read approximately 50 ohms between the wire sets (DVA 5V or more), OEM stators read 700-900 ohms (DVA 180V or more) from Blue to Yellow, while CDI Electronics stators read 300-500 ohms. If readings are good, disconnect kill wire from one pack. If the dead cylinder starts firing, the problem is likely the blocking diode in the other pack.

NO FIRE ON TWO CYLINDERS: If two cylinders from the same CD unit will not fire, the problem is usually in the stator. Test per above.

ENGINE WILL NOT KILL: Check kill circuit in the pack by using a jumper wire connected to the kill wire coming out of the pack and shorting it to ground. If this kills the pack, the kill circuit in the harness or on the boat is bad, possibly the ignition switch.

COILS ONLY FIRE WITH THE SPARK PLUGS OUT: Check for dragging starter or low battery causing slow cranking speed. DVA test stator and trigger.

HIGH SPEED MISS: Using a multimeter with a peak reading adapter and piercing probes, DVA check stator voltage to each pack at high speed. If it exceeds 400 volts, replace the pack. Disconnect the rectifier and retest. If the miss is gone replace the rectifier.

Flywheel

REMOVAL & INSTALLATION

※※ **CAUTION**

Always wear safety glasses when removing flywheels. Extreme force might be necessary to remove the unit and metal slivers or rust could fly up into your face.

※※ **WARNING**

Flywheels that have never been removed or have sheared keys are typically very difficult to remove. The use of heat from a torch or soaking with penetrating oil is recommended to ease removal.

Flywheels Without Puller Bolt Holes

♦ See Figures 37, 38 and 39

1. Remove the engine cover and manual starter as necessary.
2. Hold the flywheel with a flywheel holding tool or strap wrench.
3. Remove the fastener holding the flywheel in place.
4. Pull up on the outer edge of the flywheel with a prybar.
5. Tap the crankshaft with a brass hammer to loosen the flywheel.
6. Carefully lift the flywheel from the powerhead.

To install:

7. Install the flywheel on the powerhead making certain the flywheel key is properly installed in the keyway and not damaged.
8. Install the fastener holding the flywheel in place.
9. Hold the flywheel with a flywheel holding tool or strap wrench and tighten to the specified torque.
10. Install the manual starter and engine cover.

Fig. 37 Remove the engine cover and manual starter as necessary

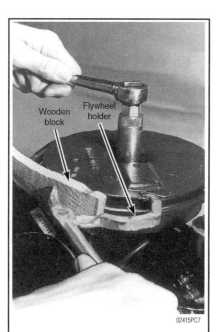

Fig. 38 Hold the flywheel with a strap wrench

Fig. 39 Pull up on the outer edge of the flywheel with a prybar

IGNITION AND ELECTRICAL SYSTEMS   5-15

Fig. 40 Hold the Flywheel with a strap wrench or a flywheel holder

Fig. 41 Install a puller on to the flywheel using the proper thread size and length bolt

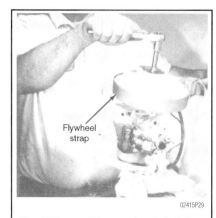
Fig. 42 Reverse this procedure to install the flywheel making certain the flywheel key in properly installed in the keyway and not damaged

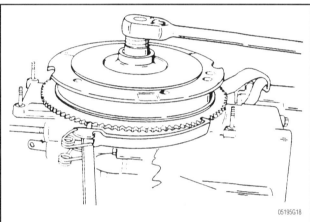
Fig. 43 While holding the flywheel with the flywheel holder or strap wrench, remove the flywheel nut and washer

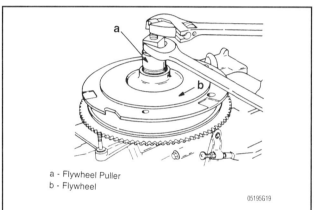
a - Flywheel Puller
b - Flywheel

Fig. 44 Hold the flywheel tooth with the wrench while tightening the bold down on the protector cap. Tighten the bolt until the flywheel comes free

Flywheels With Puller Bolt Holes

▶ See Figures 40, 41 and 42

1. Hold the Flywheel with a strap wrench or a flywheel holder.
2. Remove the flywheel nut.
3. Install a puller on to the flywheel using the proper thread size and length bolt.
4. Thread the puller bolt in putting pressure on the crankshaft and removing the flywheel.

➡ It may be necessary to hit the puller bolt with a hammer to shock the crankshaft to release the flywheel.

5. Carefully lift the flywheel from the powerhead.

To install:
6. Install the flywheel on the powerhead making certain the flywheel key is properly installed in the keyway and not damaged.
7. Install the fastener holding the flywheel in place.
8. Hold the flywheel with a flywheel holding tool or strap wrench and tighten to the specified torque.
9. Install the manual starter and engine cover.

Flywheels With Threaded Center

▶ See Figures 43 and 44

1. While holding the flywheel with the flywheel holder or strap wrench, remove the flywheel nut and washer.
2. Install a crankshaft protector cap (91–24161), then install a flywheel puller on the flywheel.
3. Hold the flywheel tooth with the wrench while tightening the bold down on the protector cap. Tighten the bolt until the flywheel comes free.
4. Carefully lift the flywheel from the powerhead.

To install:
5. Install the flywheel on the powerhead making certain the flywheel key is properly installed in the keyway and not damaged.
6. Install the fastener holding the flywheel in place.
7. Hold the flywheel with a flywheel holding tool or strap wrench and tighten to the specified torque.
8. Install the manual starter and engine cover.

INSPECTION

Visually inspect the flywheel for cracks, broken teeth and deformities.
Inspect the magnets around the inside of the flywheel for cracks or damage.
Test the strength of the magnets by placing a medium size flat blade screwdriver against it. The magnet should be able to hold the screwdriver. All outer magnets should be the same strength.
Replace the flywheel as necessary.

Distributor

TESTING

1. To perform the following tests the distributor drive belt needs to be removed.

5-16 IGNITION AND ELECTRICAL SYSTEMS

2. A visual inspection is necessary to check for any discoloration due to heat and or a shaft bearing failure.
3. Rotate the distributor shaft to inspect for freedom of movement and smooth operation.
4. A distributor with breaker points can be tested for function by performing a simple continuity test Points closed there is continuity, points open there is no continuity.

➡ A breakerless distributor uses a preamplifier to activate the spark.

5. Disconnect the White/Black lead from the distributor.
6. Connect the Red lead from a volt meter to the terminal with the Blue lead.
7. Connect the Black meter lead to the terminal that had the White/Black lead attached to it.
8. Turn the key switch to the **ON** position.
9. Rotate the distributor shaft, the meter should read 12 volts then no voltage.
10. There should be an on/off pattern one for each cylinder during one rotation of the shaft.

REMOVAL & INSTALLATION

▶ See Figures 45 thru 54

1. Identify each high tension lead with a cylinder number and remove the leads from the spark plug.
2. If servicing a breaker less distributor, disconnect the White/Black and the Blue leads from the terminals on the distributor housing. If servicing a breaker point distributor, disconnect the White/Blue lead.

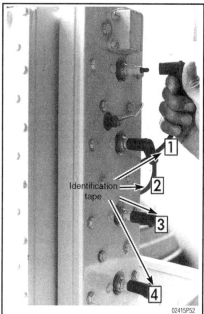

Fig. 45 Identify each high tension lead with a cylinder number and remove the leads from the spark plug

Fig. 46 If servicing a breakerless distributor, disconnect the White/Black and the Blue leads from the terminals on the distributor housing. If servicing a breaker point distributor, disconnect the White/Blue lead

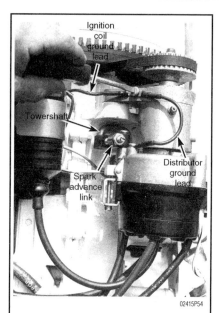

Fig. 47 Disconnect the ignition coil ground lead and connect the distributor ground lead back onto the terminal with the securing screw

Fig. 48 Remove the two bolts under the flywheel securing the distributor to the powerhead

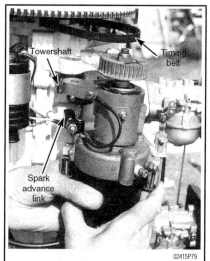

Fig. 49 Slide the end of the tower shaft into the spark advance link rod. Install the distributor onto the powerhead. Install but do not tighten the two securing bolts, at this time

Fig. 50 Connect the White/Black lead to the left terminal. Breakerless distributor: Connect the Blue lead to the right terminal

IGNITION AND ELECTRICAL SYSTEMS

Fig. 51 Connect the coil grounding lead to the ground terminal on the distributor

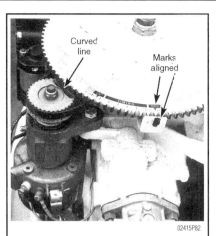

Fig. 52 Rotate the flywheel clockwise until the TDC mark on the flywheel is aligned with the "I" mark on the timing pointer. Do not rotate the flywheel counterclockwise. Such action could cause damage to the water pump vanes

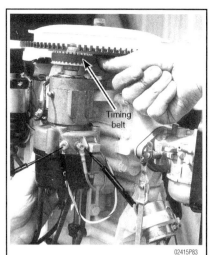

Fig. 53 Check the timing belt tension by attempting to deflect the belt at a point just under the flywheel. The belt should deflect 3/16 to 1/4" for correct tension

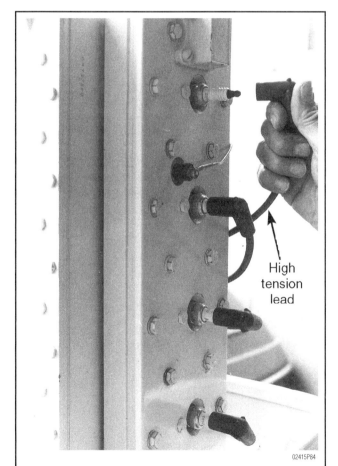

Fig. 54 Connect the high tension leads to the correct spark plugs, as identified before removal

3. If an interlock switch is mounted on the exterior of the distributor housing, disconnect the Yellow and the Yellow/Black leads.
4. Disconnect the ignition coil ground lead and connect the distributor ground lead back onto the terminal with the securing screw.
5. Remove the two bolts under the flywheel securing the distributor to the powerhead.

6. Slip the timing belt free of the distributor pulley and lift the distributor from the powerhead, allowing the end of the towershaft to slide out of the spark advance link rod.

To install:
7. Slide the end of the towershaft into the spark advance link rod. Install the distributor onto the powerhead. Install but do not tighten the two securing bolts, at this time. Do not install the timing belt at this time.
8. Connect the White/Black lead to the left terminal. Breakerless distributor: Connect the Blue lead to the right terminal.
9. If the distributor being serviced is equipped with an interlock switch, connect the Yellow/Black lead to the outermost (left) terminal. Connect the Yellow lead to the inner (right) terminal.
10. Connect the coil grounding lead to the ground terminal on the distributor.
11. Rotate the flywheel clockwise until the TDC mark on the flywheel is aligned with the "I" mark on the timing pointer. Do not rotate the flywheel counterclockwise. Such action could cause damage to the water pump vanes.
12. Rotate the pulley until the curved line follows the rim of the flywheel, and then Carefully install the timing belt over the distributor pulley without disturbing the alignment.
13. Check the timing belt tension by attempting to deflect the belt at a point just under the flywheel. The belt should deflect 3/16 to 1/4" for correct tension.
14. Tighten the two bolts on the distributor bracket. If the belt tension is too strong, the bushings inside the distributor and the towershaft bearings will wear quickly. Incorrect belt tension may cause the towershaft linkage to bind.
15. A belt that is too loose will cause vibration in the distributor shaft. Such vibration could likely result in a timing change. There is also a danger the timing belt may slip free of the pulley.
16. Connect the high tension leads to the correct spark plugs, as identified before removal

DISASSEMBLY

▶ See Figures 55 thru 63

1. Remove the two screws securing the distributor cover to the cap.
2. Separate the cover from the cap with the high tension leads attached.
3. Pull the high tension leads free of the cover one by one.
4. Remove the cover seal.
5. Check to be sure that a flash proofing screen is in place in the distributor cover.

✱✱ WARNING

Never operate a powerhead unless the flash proofing is properly in place.

5-18 IGNITION AND ELECTRICAL SYSTEMS

Fig. 55 Remove the two screws securing the distributor cover to the cap

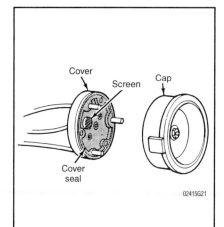
Fig. 56 Check to be sure that a flash proofing screen is in place in the distributor cover

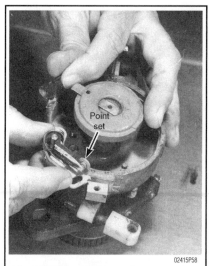
Fig. 57 Lifting a breaker point set out of the distributor housing

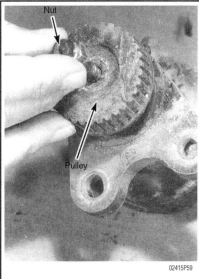
Fig. 58 Remove the nut securing the distributor pulley to the shaft

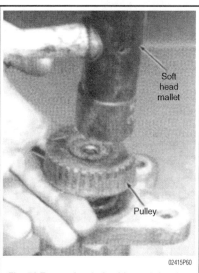
Fig. 59 Tap on the shaft with a soft head mallet and drive the shaft down and through to release the pulley

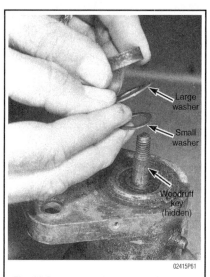
Fig. 60 Remove the thick spacer, large washer, and the small washer from the distributor shaft.

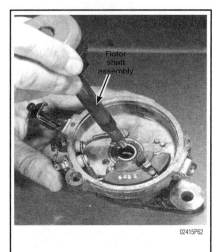
Fig. 61 Pull the rotor and shaft assembly from the housing

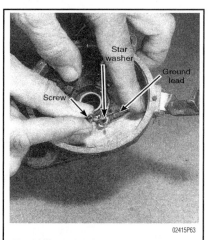
Fig. 62 Removing the screw and star washer securing the preamplifier ground lead to the distributor housing

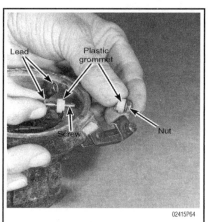
Fig. 63 Disconnect the two remaining preamplifier leads by removing the nuts, terminal screws, star washers, and plastic grommets from the distributor housing

IGNITION AND ELECTRICAL SYSTEMS 5-19

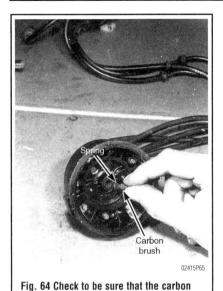

Fig. 64 Check to be sure that the carbon brush is attached firmly to the tension spring

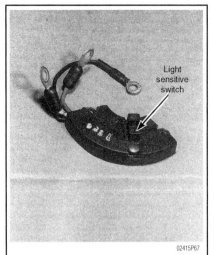

Fig. 65 Take time to clean the groove in the light sensitive switch. An oily film on the switch surface will impair performance of the ignition system

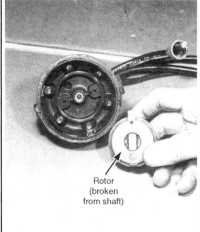

Fig. 66 The rotor is secured to the distributor shaft with epoxy. If the rotor is the least bit loose, the shaft and rotor assembly must be replaced

6. Loosen and remove the two screws, lift the breaker point set out of the distributor housing, if equipped
7. Remove the screw securing the Black lead to the breaker point set, if equipped
8. Remove the nut securing the distributor pulley to the shaft.
9. Remove the pulley from the shaft. To release the pulley tap on the shaft with a soft head mallet and drive the shaft down and through the pulley.
10. Pry the Woodruff key free of the shaft.
11. Remove the thick spacer, large washer, and the small washer from the distributor shaft.
12. Pull the rotor and shaft assembly from the housing.
13. Slide the second spacer free of the shaft.

✱✱ WARNING

Do not attempt to remove the rotor from the shaft. The rotor is secured to the shaft with epoxy at the factory. If the rotor is no longer fit for service, the shaft is purchased with the rotor attached.

14. Remove the screw and star washer securing the preamplifier ground lead to the distributor housing, if equipped
15. Disconnect the two remaining preamplifier leads by removing the nuts, terminal screws, star washers, and plastic grommets from the distributor housing.
16. Remove the attaching screws, and then lift out the preamplifier.

CLEANING & INSPECTION

▶ See Figures 64, 65 and 66

Check the distributor housing for dirt, rust, or corrosion. Remove rust or corrosion with No. 320 carborundum paper, and then wipe the housing clean.

Inspect the rotor and shaft assembly for rust and corrosion. Clean these two items thoroughly with No. 320 carborundum paper, and then wipe the housing clean. Never use a wire brush to clean distributor parts.

If the end of the rotor contact requires attention, use a fine file and remove only enough material to obtain a bright clean finish. If excessive material is removed from the end, the contact will arc to the cap contacts.

Check the rotor on the shaft. If there is any evidence the rotor is loose, do not attempt to secure it in place with epoxy. If the rotor were to come free during high speed powerhead operation, considerable damage would be caused to the distributor, the ignition would be "shorted", and the powerhead would shut down.

Check the condition of the high tension leads.

Inspect and replace the distributor shaft bearings, if there is any indication of wear or roughness. Never use solvent to clean bearings.

Check the distributor cap for leakage paths caused by broken leads, poor connections, moisture entering the distributor, dirt, carbon, or corrosion.

Inspect the carbon brush in the center of the distributor cap for wear and to be sure the brush is attached to a tension spring, but free to move.

Verify the flash proofing screen is in place.

Replace any questionable parts with identical items, as a guarantee of satisfactory powerhead performance after the work is complete.

Inspect the condition of the windows on the sides of the rotor. Check the rotor carefully. A crack could mean an additional window for the light to pass through. If the rotor is cracked or chipped, replace the rotor shaft assembly.

Gently clean the groove between the light emitting diode and the light sensitive switch on the preamplifier.

ASSEMBLY

▶ See Figures 67 thru 74

1. Lay out the cleaned and replacement parts on a clean work surface.
2. Place the preamplifier in position inside the distributor housing, secure it with the two retaining screws, if equipped.

Fig. 67 Slide that screw through a star washer a plastic grommet and through the hole in the distributor housing closest to the distributor cap clip

5-20 IGNITION AND ELECTRICAL SYSTEMS

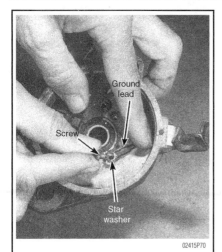

Fig. 68 Slide the preamplifier ground lead retaining screw through a star washer, and then secure the lead to the distributor housing

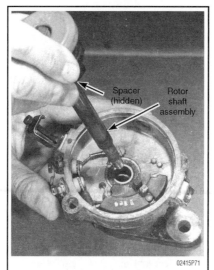

Fig. 69 Slide the spacer onto the distributor shaft, and then insert the shaft into the distributor housing

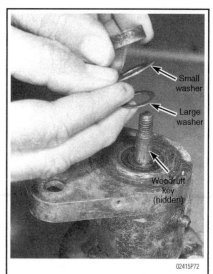

Fig. 70 Slide the small washer, the large washer, and the second spacer onto the shaft

Fig. 71 Slide the distributor pulley onto the shaft with the slot in the pulley indexed over the woodruff key

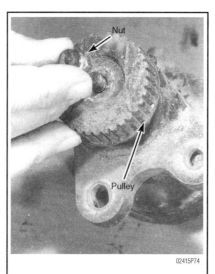

Fig. 72 Install and tighten the pulley nut

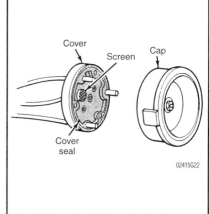

Fig. 73 Insert the high tension leads through the cap cover until each lead extends inside the cover at least 1 in. (2.5cm)

Fig. 74 Position the distributor cap over the housing with the post on the housing indexed into the hole in the cap

   3. Place the mid length preamplifier lead onto a terminal screw.
   4. Slide that screw through a star washer a plastic grommet and through the hole in the distributor housing closest to the distributor cap clip.
   5. On the outside of the housing install a plastic grommet, star washer and locknut.
   6. Tighten to secure in place.
   7. Install the short preamplifier lead in the same manner just described into the remaining hole in the distributor housing.
   8. Slide the preamplifier ground lead retaining screw through a star washer, and then secure the lead to the distributor housing.
   9. Slide the spacer onto the distributor shaft, and then insert the shaft into the distributor housing.
   10. Slide the small washer, the large washer, and the second spacer onto the shaft.
   11. Install the woodruff key into the shaft slot. Slide the distributor pulley onto the shaft with the slot in the pulley indexed over the woodruff key.
   12. Install and tighten the pulley nut securely.
   13. Install the breaker points with the attaching screws, if equipped.
   14. Insert the high tension leads through the cap cover until each lead extends inside the cover at least 1 in. (2.5cm).
   15. Install the cap seal.

IGNITION AND ELECTRICAL SYSTEMS 5-21

16. Align the screw holes in the cap and cover to position the vent hole in the cover with the vent flash proofing screen in the cap seal.
17. Seat the high tension lead ends onto the pointed contacts on the cover. Bring the cap and cover together, and then secure them with the two attaching screws.
18. Position the distributor cap over the housing with the post on the housing indexed into the hole in the cap.

Slide the clips over the two ridges in the cap and tighten the retaining screws just snug.

Ignition Coils

TESTING

Resistance

1. Visually inspect the coil for cracks electrical arching or burn marks.
2. Remove the wire connection at the breaker points.
3. Connect the Black multimeter lead to ground and the Red lead to the primary coil wire.
4. Compare this reading of the coil primary resistance to the specifications listed in the "Ignition System Specifications" chart.
5. Connect the Black multimeter lead to the primary coil wire and the Red lead to the spark plug high tension lead.
6. Compare this reading of the coil secondary resistance to the specifications listed in the "Ignition System Specifications" chart.
7. Repeat these procedures for each ignition coil.
8. Replace any coil that has failed any of the tests.

Voltage Output

One of the most accurate ways of determining coil condition is to do a dynamic coil output voltage test. This test will show the actual coil output under varying conditions and will not only determine if the coil defective, but will also alert the user to when coil is about to fail. Coil Voltage Output Testers are available from CDI Electronics (256-772-3829).

1. Remove the cowling.
2. Disconnect the spark plug high tension lead and connect the voltage output tester between the spark plug and the high tension lead.
3. Start (crank) the outboard and read the voltage from the tester.
4. Voltage should be 18–20 Kv.
5. If voltage is not as specified, the coil may be defective.
6. Disconnect the tester and connect the spark plug high tension lead.
7. Install the cowling.

REMOVAL & INSTALLATION

1. Remove the wires on the coil and mark them for location.
2. Remove the high tension lead between the coil and distributor.
3. Remove the fasteners holding the coil strap or bracket.
4. Remove ignition coil.

To install:

5. Install the ignition coil.
6. Install the fasteners holding the coil strap or bracket.
7. Install the high tension lead between the coil and distributor.
8. Install the wires on the coil.

Ignition Module

TESTING

▶ See Figures 75, 76, 77 and 78

1. Set the voltmeter to the 12 volts DC scale.
2. Make contact with the Black meter lead to a suitable powerhead ground. Make contact with the Red meter lead to the Blue lead terminal from the CD module at the terminal block.
3. If the meter registers 12 volts sufficient voltage is being supplied to the ignition module.

If the meter registers less than 12 volts another component in the system is at fault. This condition needs to be repaired before continuing the ignition module test.

4. Rotate the flywheel clockwise until the TDC mark on the flywheel is aligned with the "I" mark on the timing pointer
5. The outer rim of the flywheel should align with the embossed curved line on the distributor pulley. If the flywheel and distributor pulley do not align slide the belt off the distributor pulley, turn the pulley to properly align the marks and install the belt.
6. Disconnect the White/Black wire from the left terminal on the distributor housing.
7. Rotate the shaft to the wide open throttle position. Align the specified degrees of spark advance with the timing pointer.
8. Make contact with the Red meter lead to the right terminal on the distributor.
9. Make contact with the Black meter lead to the left terminal.

➡Breaker point systems have one terminal, contact the Red meter lead to the single terminal on the distributor housing. Make contact with the Black meter lead to a suitable powerhead ground.

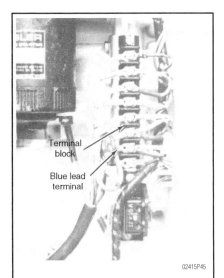

Fig. 75 Place the Red meter lead on the Blue lead terminal from the CD module at the terminal block

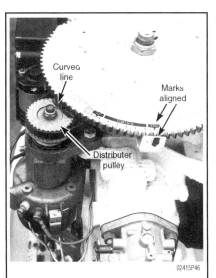

Fig. 76 Rotate the flywheel clockwise until the TDC mark on the flywheel is aligned with the "I" mark on the timing pointer

Fig. 77 To test for voltage being applied to the breaker point set or through the ignition module, disconnect the White/Black wire from the left terminal on the distributor housing and measure voltage between the two terminals on the distributor with the ignition ON

5-22 IGNITION AND ELECTRICAL SYSTEMS

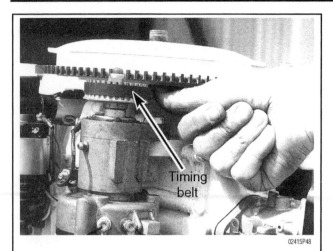

Fig. 78 Lightly deflect the timing belt to rock the distributor pulley back and forth. The meter needle should swing between zero and 12 volts

10. Turn the ignition switch to the ON position.
11. Lightly deflect the timing belt to rock the distributor pulley back and forth. The meter needle should swing between zero and 12 volts.
12. The meter needle failing to register between zero and 12 volts indicates no voltage is being applied to the breaker point set or through the ignition module.
13. With test voltage into the ignition module and no voltage output to the distributor and a no spark condition the module is bad.

REMOVAL & INSTALLATION

1. Remove the wires connecting the ignition coil, distributor, the module ground and terminal block mark them for location.
2. Remove the fasteners holding the module in place.
3. Remove the ignition module.

To install:
4. Install the ignition module.
5. Install the fasteners holding the module in place.
6. Install the wires connecting the ignition coil, distributor, the module ground and terminal block.

ELECTRONIC IGNITION SYSTEMS UP TO 25 HP

Description and Operation

The engine's flywheel contains magnets carefully positioned to create an electric current as they rotate past specially designed coils of wire. Current is created by magnetic induction. Simply put, that means that a magnet moving rapidly near a conductor will induce electrical flow within the conductor.

Conversely, a wire that moves rapidly through a magnetic field will also generate electrical flow. This principle governs the working of electric motors, alternators and generators.

Ignition coils are mounted under the flywheel with no other external components

➥The 3 hp engine model falls into this category and does have an external mounted ignition coil and CDI module. Therefore the troubleshooting and repair procedures will best coincide with the instructions in the Prestolite® electronic ignition section.

The internal construction of a typical ignition coil includes primary and secondary windings. It also uses the principle of magnetic induction with the magnetism generated by the primary (lower-voltage) winding creating a magnetic field around the secondary winding, which has many more windings than the primary coil. The rapid movement of this magnetic field past the secondary windings induces electrical current flow. The are greater the number of turns of wire in the secondary winding, the higher the voltage produced.

As this secondary voltage leaves the center tower of the ignition coil, it travels along the spark-plug wire, which is heavily insulated and designed to carry this high voltage.

If all is well, the high voltage will jump the gap on the spark plug between the center electrode and the ground electrode. On larger engines with surface-gap plugs, the high voltage current will jump from the center electrode to the side of the plug assembly itself, completing a circuit to ground via the engine block.

Last, but certainly not least, is the stop control. You need a means to shut your engine off and a good way is to stop the spark plugs from working. Depending on your engine, this may be accomplished by a simple stop button or a key switch on larger engines. This disables the whole ignition system. On newer engines, an emergency stop button with an overboard clip and lanyard attachment is standard. This system is wired directly into the ignition coils. It functions by creating a momentary short circuit inside the coil unit, grounding the current intended for the high tension coils and thus shutting off the ignition long enough to stop the engine. Faulty stop circuits are frequently the cause of a no spark condition.

Troubleshooting Electronic Ignition Systems Up To 25 HP

These systems are basic and have very few component parts. The parts are a flywheel, an ignition coil and spark plug wire per cylinder and a stop switch.

Conduct a spark test to check to see if there is any spark on any cylinder. With a no spark condition before continuing disconnect the stop and or emergency switch and retest. If there is now spark the trouble lies in the disconnected switch or switches.

In the event the spark did not return the problem is with a coil or a bad wire that can not be seen without removing the flywheel.

➥Remember that the magnets in the flywheel generate the electrical field and even though it is very rare bad or weak flywheel magnets can be the cause of the no spark condition.

Flywheel

REMOVAL & INSTALLATION

✱✱ CAUTION

Always wear safety glasses when removing flywheels. Extreme force might be necessary to remove the unit and metal slivers or rust could fly up into your face.

✱✱ WARNING

Flywheels that have never been removed or have sheared keys are typically very difficult to remove. The use of heat from a torch or soaking with penetrating oil is recommended to ease removal.

Flywheels Without Puller Bolt Holes

♦ See Figures 79, 80 and 81

1. Remove the engine cover and manual starter as necessary.
2. Hold the flywheel with a flywheel holding tool or strap wrench.
3. Remove the fastener holding the flywheel in place.
4. Pull up on the outer edge of the flywheel with a prybar.
5. Tap the crankshaft with a brass hammer to loosen the flywheel.
6. Carefully lift the flywheel from the powerhead.

IGNITION AND ELECTRICAL SYSTEMS 5-23

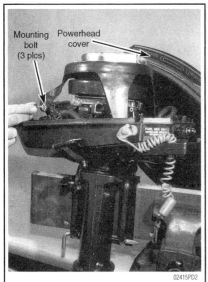

Fig. 79 Remove the engine cover and manual starter as necessary

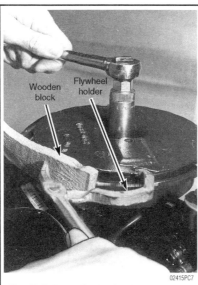

Fig. 80 Hold the flywheel with a strap wrench

Fig. 81 Pull up on the outer edge of the flywheel with a prybar

To install:

7. Install the flywheel on the powerhead making certain the flywheel key is properly installed in the keyway and not damaged.
8. Install the fastener holding the flywheel in place.
9. Hold the flywheel with a flywheel holding tool or strap wrench and tighten to the specified torque.
10. Install the manual starter and engine cover.

Flywheels With Puller Bolt Holes

▶ See Figures 82, 83 and 84

1. Hold the Flywheel with a strap wrench or a flywheel holder.
2. Remove the flywheel nut.
3. Install a puller on to the flywheel using the proper thread size and length bolt.
4. Thread the puller bolt in putting pressure on the crankshaft and removing the flywheel.

➡ It may be necessary to hit the puller bolt with a hammer to shock the crankshaft to release the flywheel.

5. Carefully lift the flywheel from the powerhead.

To install:

6. Install the flywheel on the powerhead making certain the flywheel key is properly installed in the keyway and not damaged.

7. Install the fastener holding the flywheel in place.
8. Hold the flywheel with a flywheel holding tool or strap wrench and tighten to the specified torque.
9. Install the manual starter and engine cover.

Flywheels With Threaded Center

▶ See Figures 85 and 86

1. While holding the flywheel with the flywheel holder or strap wrench, remove the flywheel nut and washer.
2. Install a crankshaft protector cap (91-24161), then install a flywheel puller on the flywheel.
3. Hold the flywheel tooth with the wrench while tightening the bold down on the protector cap. Tighten the bolt until the flywheel comes free.
4. Carefully lift the flywheel from the powerhead.

To install:

5. Install the flywheel on the powerhead making certain the flywheel key is properly installed in the keyway and not damaged.
6. Install the fastener holding the flywheel in place.
7. Hold the flywheel with a flywheel holding tool or strap wrench and tighten to the specified torque.
8. Install the manual starter and engine cover.

Fig. 82 Hold the Flywheel with a strap wrench or a flywheel holder

Fig. 83 Install a puller on to the flywheel using the proper thread size and length bolt

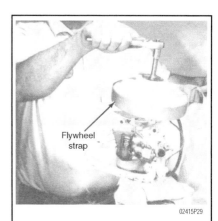

Fig. 84 Reverse this procedure to install the flywheel making certain the flywheel key in properly installed in the keyway and not damaged

5-24 IGNITION AND ELECTRICAL SYSTEMS

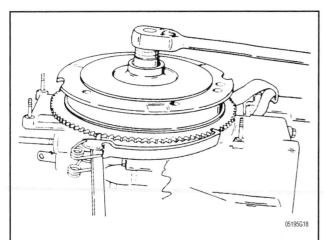

Fig. 85 While holding the flywheel with the flywheel holder or strap wrench, remove the flywheel nut and washer

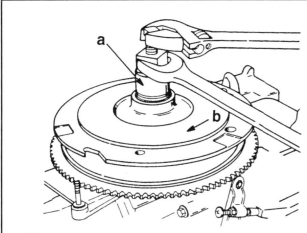

a - Flywheel Puller
b - Flywheel

Fig. 86 Hold the flywheel tooth with the wrench while tightening the bold down on the protector cap. Tighten the bolt until the flywheel comes free

INSPECTION

Visually inspect the flywheel for cracks, broken teeth and deformities.
Inspect the magnets around the inside of the flywheel for cracks or damage.

Test the strength of the magnets by placing a medium size flat blade screwdriver against it. The magnet should be able to hold the screwdriver. All outer magnets should be the same strength.
Replace the flywheel as necessary.

Ignition Coils

TESTING

Resistance

1. Visually inspect the coil for cracks electrical arching or burn marks.
2. Remove the wire connection at the breaker points.
3. Connect the Black multimeter lead to ground and the Red lead to the primary coil wire.
4. Compare this reading of the coil primary resistance to the specifications listed in the "Ignition System Specifications" chart.
5. Connect the Black multimeter lead to the primary coil wire and the Red lead to the spark plug high tension lead.
6. Compare this reading of the coil secondary resistance to the specifications listed in the "Ignition System Specifications" chart.
7. Repeat these procedures for each ignition coil.
8. Replace any coil that has failed any of the tests.

Voltage Output

One of the most accurate ways of determining coil condition is to do a dynamic coil output voltage test. This test will show the actual coil output under varying conditions and will not only determine if the coil defective, but will also alert the user to when coil is about to fail. Coil Voltage Output Testers are available from CDI Electronics (256-772-3829).

1. Remove the cowling.
2. Disconnect the spark plug high tension lead and connect the voltage output tester between the spark plug and the high tension lead.
3. Start (crank) the outboard and read the voltage from the tester.
4. Voltage should be 18–20 Kv.
5. If voltage is not as specified, the coil may be defective.
6. Disconnect the tester and connect the spark plug high tension lead.
7. Install the cowling.

REMOVAL & INSTALLATION

1. Remove the flywheel.
2. Remove any wires attached to the coils such as grounds or stop switch connections.
3. Remove the attaching hardware screws (some coils are held by spring clips).
4. Remove the coil from the magneto plate.

To install:

5. Install the coil on the magneto plate.
6. Install the attaching hardware screws or spring clips.
7. Install wires attached to the coils and tighten fasteners securely.
8. Install the flywheel.

PRESTOLITE® ELECTRONIC IGNITION

Description and Operation

There are a number of variations to this system that are used on the Force outboard engines. The basic system uses separate components and mostly terminal board connections. The two cylinder variation used on some models has a coil and ignition module built into one unit. Another system uses plug connections rather than a terminal strip. The systems all work in the same manner and are tested in the same steps. The Prestolite® system operates as follows.

The engine's flywheel contains magnets carefully positioned to create an electric current as they rotate past specially designed coils of wire. Current is created by magnetic induction. Simply put, that means that a magnet moving rapidly near a conductor will induce electrical flow within the conductor.

Conversely, a wire that moves rapidly through a magnetic field will also generate electrical flow. This principle governs the working of electric motors, alternators and generators.

One of the coils under your flywheel is called a charge coil. As the flywheel magnets spin past this coil, they generate in it a fairly high-voltage alternating current that travels to your system's ignition module. This voltage is often in the region of 200 volts AC.

The other ignition system coil(s) found under the flywheel are called the sensor coils, pulsar coils or trigger coils. They send an electrical signal to the ignition module to tell it which cylinder to work with at the correct time.

The ignition module is the brains of the system and serves several functions. First, it converts the alternating current (AC) from the charge coil into usable direct current (DC). Next it stores the current in a built-in capacitor. The module also interprets the timing signal from the trigger coil. This changes constantly

IGNITION AND ELECTRICAL SYSTEMS

with engine speed, and moving the trigger coil's position relative to the flywheel magnets brings about the change. A device called a timing plate where the trigger coils are mounted controls the coil's movement. The timing plate moves in response to changes in throttle opening, to which it is mechanically linked. The ignition module also controls the discharge of the capacitor and sends this voltage to the primary winding of the ignition coil for the correct cylinder.

Also (depending on the system) the module may incorporate electronic circuits that limit engine speed and prevent over-revving. Some modules even have a circuit that reduces engine speed if the engine begins to run too hot for any reason. Larger engines often have an automatic ignition advance for initial start-up and for when the engine is running at temperatures less than approximately 100°F.

The voltage from the module goes to the primary winding of the ignition coil or high-tension coil. You may know this type of coil as a step-up transformer. Here, the voltage is stepped up to between 15,000 and 40,000 volts. That's the 'kind of voltage needed to jump the air gap on the spark plug and ignite the air/fuel mixture in the cylinder. Your high-tension ignition coil has two sides, the primary side and the secondary side. It is really two coil assemblies combined into one neat, compact case.

The internal construction of a typical ignition coil includes primary and secondary windings. It also uses the principle of magnetic induction with the magnetism generated by the primary (lower -voltage) winding creating a magnetic field around the secondary winding, which has many more windings than the primary coil. The ignition module controls the rapid turning on and off of electrical flow in the primary winding, thereby turning this magnetic field on and off. The rapid movement of this magnetic field past the secondary windings induces electrical current flow. The are greater the number of turns of wire in the secondary winding, the higher the voltage produced.

As this secondary voltage leaves the center tower of the ignition coil, it travels along the spark-plug wire, which is heavily insulated and designed to carry this high voltage.

If all is well, the high voltage will jump the gap on the spark plug between the center electrode and the

ground electrode. On larger engines with surface-gap plugs, the high voltage current will jump from the center electrode to the side of the plug assembly itself, completing a circuit to ground via the engine block.

Last, but certainly not least, is the stop control. You need a means to shut your engine off and a good way is to stop the spark plugs from working. Depending on your engine, this may be accomplished by a simple stop button or a key switch on larger engines. This disables the whole ignition system. On newer engines, an emergency stop button with an overboard clip and lanyard attachment is standard. This system is wired directly into the ignition module. It functions by creating a momentary short circuit inside the module unit, grounding the current intended for the high tension coils and thus shutting off the ignition long enough to stop the engine. Faulty stop circuits are frequently the cause of a no spark condition.

Troubleshooting the Prestolite® Electronic Ignition

➡ The following conditions apply to the Prestolite® Electronic Ignition System. Troubleshooting procedure are courtesy of our friends at CDI Electronics (256-772-3829)

➡ Maintenance-free batteries are not recommended for use with these systems.

Check for broken wires and terminals, especially inside the plastic plug-in connectors. Removal of the pins from the connectors using the Rapair (R554-9706) Pin Removal Tool and will allow you to visually inspect the pins for damage.

Check the flywheel for broken or loose magnets.

Disconnect the kill wires from the CD and connect a DC voltmeter between the kill wires and engine ground, turn the ignition switch on and off several times. If at any time you see voltage appearing on the meter, there is a problem in the harness or ignition switch.

➡ At no time should you see battery voltage on a kill circuit.

Visually inspect stator for burned or discolored areas. If found, replace the stator. If the areas are on the battery charge windings, it indicates a possible problem with the rectifier.

IF NO FIRE ON ANY CYLINDER: Disconnect kill wire at the pack. Check for broken or bare wires on the unit, stator and trigger. Using a multimeter with a peak reading adapter, or and piercing probes, measure DVA voltage of the stator between the output wire sets. With everything connected, readings should be approximately 180 volts or, more. The OEM stator resistance between the blue and yellow wires is 700-900 ohms. CDI Electronics stators should read 300-500 ohms. Disconnect the rectifier. If the engine fires, replace the rectifier.

NO FIRE OR INTERMITTENT FIRE ON ONE CYLINDER: Check stator and trigger resistance. Trigger wire sets read approximately 50 ohms between the wire sets (DVA 5V or more), OEM stators read 700-900 ohms (DVA 180V or more) from Blue to Yellow, while CDI Electronics stators read 300-500 ohms. If readings are good, disconnect kill wire from one pack. If the dead cylinder starts firing, the problem is likely the blocking diode in the other pack.

NO FIRE ON TWO CYLINDERS: If two cylinders from the same CD unit will not fire, the problem is usually in the stator. Test per above.

ENGINE WILL NOT KILL: Check kill circuit in the pack by using a jumper wire connected to the kill wire coming out of the pack and shorting it to ground. If this kills the pack, the kill circuit in the harness or on the boat is bad, possibly the ignition switch.

COILS ONLY FIRE WITH THE SPARK PLUGS OUT: Check for dragging starter or low battery causing slow cranking speed. DVA test stator and trigger.

HIGH SPEED MISS: Using a multimeter with a peak reading adapter and piercing probes, DVA check stator voltage to each pack at high speed. If it exceeds 400 volts, replace the pack. Disconnect the rectifier and retest. If the miss is gone replace the rectifier.

Flywheel

REMOVAL & INSTALLATION

✷✷ CAUTION

Always wear safety glasses when removing flywheels. Extreme force might be necessary to remove the unit and metal slivers or rust could fly up into your face.

✷✷ WARNING

Flywheels that have never been removed or have sheared keys are typically very difficult to remove. The use of heat from a torch or soaking with penetrating oil is recommended to ease removal.

Flywheels Without Puller Bolt Holes

▶ See Figures 87, 88 and 89

1. Remove the engine cover and manual starter as necessary.
2. Hold the flywheel with a flywheel holding tool or strap wrench.

Fig. 87 Remove the engine cover and manual starter as necessary

5-26 IGNITION AND ELECTRICAL SYSTEMS

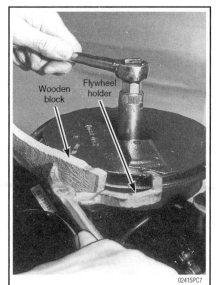

Fig. 88 Hold the flywheel with a strap wrench

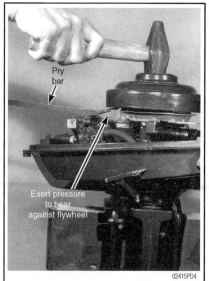

Fig. 89 Pull up on the outer edge of the flywheel with a prybar

Fig. 90 Hold the Flywheel with a strap wrench or a flywheel holder

Fig. 91 Install a puller on to the flywheel using the proper thread size and length bolt

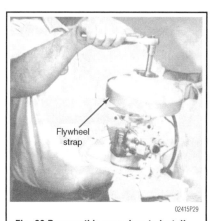

Fig. 92 Reverse this procedure to install the flywheel making certain the flywheel key in properly installed in the keyway and not damaged

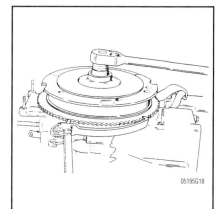

Fig. 93 While holding the flywheel with the flywheel holder or strap wrench, remove the flywheel nut and washer

  3. Remove the fastener holding the flywheel in place.
  4. Pull up on the outer edge of the flywheel with a prybar.
  5. Tap the crankshaft with a brass hammer to loosen the flywheel.
  6. Carefully lift the flywheel from the powerhead.

To install:
  7. Install the flywheel on the powerhead making certain the flywheel key is properly installed in the keyway and not damaged.
  8. Install the fastener holding the flywheel in place.
  9. Hold the flywheel with a flywheel holding tool or strap wrench and tighten to the specified torque.
  10. Install the manual starter and engine cover.

Flywheels With Puller Bolt Holes

♦ See Figures 90, 91 and 92

  1. Hold the Flywheel with a strap wrench or a flywheel holder.
  2. Remove the flywheel nut.
  3. Install a puller on to the flywheel using the proper thread size and length bolt.
  4. Thread the puller bolt in putting pressure on the crankshaft and removing the flywheel.

➡It may be necessary to hit the puller bolt with a hammer to shock the crankshaft to release the flywheel.

  5. Carefully lift the flywheel from the powerhead.
To install:
  6. Install the flywheel on the powerhead making certain the flywheel key is properly installed in the keyway and not damaged.
  7. Install the fastener holding the flywheel in place.
  8. Hold the flywheel with a flywheel holding tool or strap wrench and tighten to the specified torque.
  9. Install the manual starter and engine cover.

Flywheels With Threaded Center

♦ See Figures 93 and 94

  1. While holding the flywheel with the flywheel holder or strap wrench, remove the flywheel nut and washer.
  2. Install a crankshaft protector cap (91–24161), then install a flywheel puller on the flywheel.
  3. Hold the flywheel tooth with the wrench while tightening the bold down on the protector cap. Tighten the bolt until the flywheel comes free.
  4. Carefully lift the flywheel from the powerhead.
To install:
  5. Install the flywheel on the powerhead making certain the flywheel key is properly installed in the keyway and not damaged.
  6. Install the fastener holding the flywheel in place.

IGNITION AND ELECTRICAL SYSTEMS

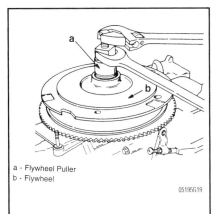

Fig. 94 Hold the flywheel tooth with the wrench while tightening the bold down on the protector cap. Tighten the bolt until the flywheel comes free

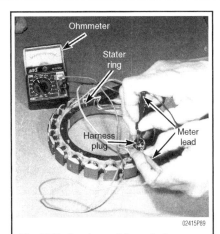

Fig. 95 Performing resistance tests on a stator ring removed from the powerhead, using an ohmmeter

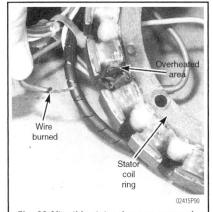

Fig. 96 After this stator ring was removed from the powerhead, a burned wire and an overheated area were discovered. This unit was tested and had no continuity

7. Hold the flywheel with a flywheel holding tool or strap wrench and tighten to the specified torque.
8. Install the manual starter and engine cover.

INSPECTION

Visually inspect the flywheel for cracks, broken teeth and deformities.
Inspect the magnets around the inside of the flywheel for cracks or damage. Test the strength of the magnets by placing a medium size flat blade screwdriver against it. The magnet should be able to hold the screwdriver. All outer magnets should be the same strength.
Replace the flywheel as necessary.

Charging Coils

TESTING

▶ See Figures 95 and 96

The wires for the charge coil are generally Yellow and Blue. Before continuing refer to the ignition wiring diagrams at the end of this chapter to fully identify the system that is being tested. In the event the wire colors are different test the charge coil accordingly.

Units With Plug in Connectors

RESISTANCE TEST

1. Disconnect the stator leads at the harness plug.
2. Using a multimeter on the appropriate scale test the resistance of the charge coil.
3. Connect the Black meter lead to the Yellow wire terminal in the harness plug from the stator.
4. Connect the Red meter lead to the Blue wire terminal in the harness plug from the stator.
5. Record the reading and compare it to the specification listed in the ignition chart at the end of the chapter.
6. In the event that there is no specific value listed in the chart a reading of continuity or a resistance value shows the coil to be intact and most likely ok.
7. If the meter registers infinity, there is an "open" in the charge coil indicating the coil to be bad.
8. The final resistance test is to check the coil for a short to ground. Set the meter to the high ohms scale and connect the Black lead to engine ground and the Red lead to the Yellow lead then the Blue lead. Any reading of continuity would show a shorted coil and therefore need replacement.
9. Repeat these tests if the engine is equipped with more than one charge coil.

OUTPUT TEST

1. Output tests are done using a peak reading voltmeter or a multimeter with a peak adapter.
2. Connect the Red meter lead to the Yellow wire terminal in the harness plug from the stator.
3. Connect the Black meter lead to a clean engine ground.
4. Set the meter to the 400DVA scale or equivalent.
5. Crank the powerhead with the cranking motor for a few seconds and observe the meter needle.
6. Record the meter reading.
7. Repeat these tests if the engine is equipped with more than one charge coil.
8. Replace the charge coil itself or the stator ring as necessary.

Units With Terminal Block Connections

RESISTANCE TEST

The wires for the charge coil are generally Yellow and Blue. Before continuing refer to the ignition wiring diagrams at the end of this chapter to fully identify the system that is being tested.

1. Disconnect the stator leads at the terminal block.
2. Using a multimeter on the appropriate scale test the resistance of the charge coil.
3. Connect the Black meter lead to the Yellow wire.
4. Connect the Red meter lead to the Blue wire.
5. Record the reading and compare it to the specification listed in the ignition chart at the end of the chapter.
6. In the event that there is no specific value listed in the chart a reading of continuity or a resistance value shows the coil to be intact and most likely ok.
7. If the meter registers infinity, there is an "open" in the charge coil indicating the coil to be bad.
8. The final resistance test is to check the coil for a short to ground. Set the meter to the high ohms scale and connect the Black lead to engine ground and the Red lead to the Yellow lead then the Blue lead. Any reading of continuity would show a shorted coil and therefore need replacement.
9. Repeat these tests if the engine is equipped with more than one charge coil.

OUTPUT TEST

1. Output tests are done using a peak reading voltmeter or a multimeter with a peak adapter.
2. Connect the Red meter lead to the Yellow wire terminal.
3. Connect the Black meter lead to a clean engine ground.
4. Set the meter to the 400DVA scale or equivalent.
5. Crank the powerhead with the cranking motor for a few seconds and observe the meter needle.

5-28 IGNITION AND ELECTRICAL SYSTEMS

6. Record the meter reading.
7. Repeat these tests if the engine is equipped with more than one charge coil.
8. Replace the charge coil itself or the stator ring as necessary.

REMOVAL & INSTALLATION

▶ See Figures 97 thru 105

1. Remove the spark plugs.
2. Remove the flywheel as instructed in this Section.
3. Identify and mark the two halves of all connections leading from the stator.
4. Disconnect the stator to rectifier and stator to terminal block, or CD module leads.
5. Loosen and remove the stator attaching screws.

➡ It may be necessary to use an impact wrench to loosen these screws.

6. Remove the stator ring.

If any traces of oil are found on the stator or trigger, inspect the condition of the upper oil seal. This seal is located in the stator ring, or in the upper bearing cage.

Inspect the charge coil and stator coils. Inspect for burnt or discolored areas, these are very often accompanied by wiring which becomes loose at a touch. Do not attempt to solder loose connections unless a replacement part is unavailable.

To install:

7. Install the stator plate onto the powerhead. Check to be sure the harness leads and spark control arm face in the same direction they were facing before being disturbed.
8. Install and tighten the stator attaching hardware. No torque specifications are available for these attaching screws, but they should be tightened securely.
9. Connect the stator harness leads to the same plug connectors identified during disassembly. Remember that these plugs are not necessarily matched color-to-color.
10. Install the flywheel.
11. Install the spark plugs.

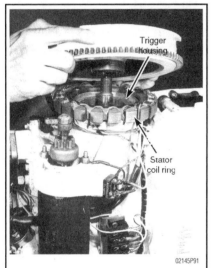

Fig. 97 The flywheel must be pulled to gain access to the trigger assembly or the stator assembly

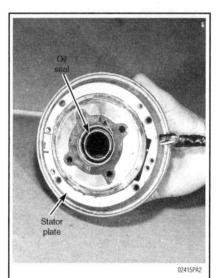

Fig. 98 If any traces of oil are discovered on the stator or trigger assemblies, the oil seal in the stator must be replaced

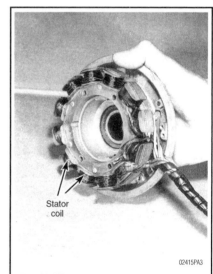

Fig. 99 All stator coils should be bright and shiny a coil that is chocolate colored has been overloaded and could soon fail

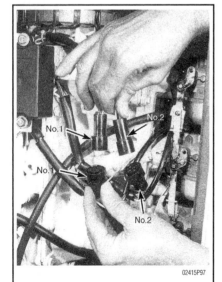

Fig. 100 Identify and mark the two halves of all connections leading from the stator

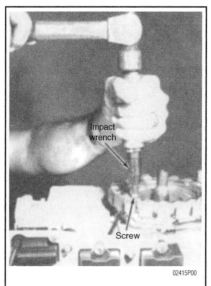

Fig. 101 Loosen and remove the stator attaching screws

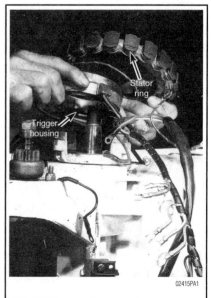

Fig. 102 Remove the stator ring

IGNITION AND ELECTRICAL SYSTEMS 5-29

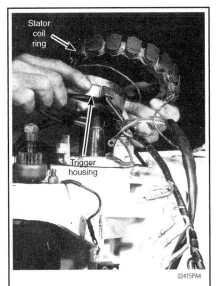

Fig. 103 Installing the trigger housing and stator plate onto the powerhead

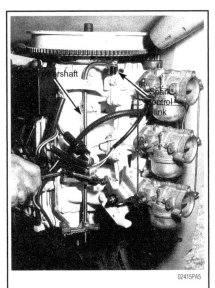

Fig. 104 Install and tighten the stator attaching hardware

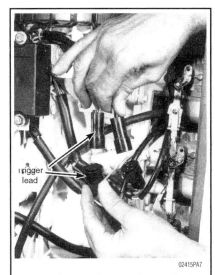

Fig. 105 Connect the stator harness leads to the same plug connectors identified during disassembly

Trigger Coils

♦ See Figure 106

The wires for the trigger coil are generally Green and Orange. Before continuing refer to the ignition wiring diagrams at the end of this chapter to fully identify the system that is being tested. In the event the wire colors are different test the trigger coils accordingly.

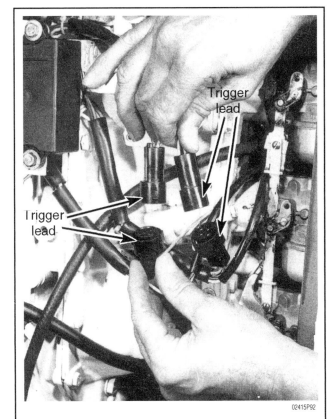

Fig. 106 The trigger harness leads should be identified prior to making a disconnect to ensure proper connection after testing or service work has been completed

TESTING

Units With Plug in Connectors

RESISTANCE TEST

1. Disconnect the trigger leads at the harness plug.
2. Using a multimeter on the appropriate scale test the resistance of the trigger coil.
3. Connect the Black meter lead to the Green wire terminal in the harness plug from the stator.
4. Connect the Red meter lead to the Orange wire terminal in the harness plug from the stator.
5. Record the reading and compare it to the specification listed in the ignition chart at the end of the chapter.
6. In the event that there is no specific value listed in the chart a reading of continuity or a resistance value shows the coil to be intact and most likely ok.
7. If the meter registers infinity, there is an "open" in the trigger coil indicating the coil to be bad.
8. The final resistance test is to check the coil for a short to ground. Set the meter to the high ohms scale and connect the Black lead to engine ground and the Red lead to the Green lead then the Orange lead. Any reading of continuity would show a shorted coil and therefore need replacement.
9. Repeat these tests if the engine is equipped with more than one trigger coil.

OUTPUT TEST

1. Output tests are done using a peak reading voltmeter or a multimeter with a peak adapter.
2. Connect the Red meter lead to the Green wire terminal in the harness plug from the trigger.
3. Connect the Black meter lead to the Orange wire terminal in the harness plug from the trigger.
4. Set the meter to the 50DVA scale or equivalent.
5. Crank the powerhead with the cranking motor for a few seconds and observe the meter needle.
6. Record the meter reading.
7. Repeat these tests if the engine is equipped with more than one trigger coil.
8. Replace the trigger coil itself or the trigger assembly as necessary.

Units With Terminal Block Connections

RESISTANCE TEST

The wires for the trigger coil are generally Green and Orange Before continuing refer to the ignition wiring diagrams at the end of this chapter to fully identify the system that is being tested.

5-30 IGNITION AND ELECTRICAL SYSTEMS

1. Disconnect the trigger leads at the terminal block.
2. Using a multimeter on the appropriate scale test the resistance of the trigger coil.
3. Connect the Black meter lead to the Green wire.
4. Connect the Red meter lead to the Orange wire.
5. Record the reading and compare it to the specification listed in the ignition chart at the end of the chapter.
6. In the event that there is no specific value listed in the chart a reading of continuity or a resistance value shows the trigger coil to be intact and most likely ok.
7. If the meter registers infinity, there is an "open" in the trigger coil indicating the coil to be bad.
8. The final resistance test is to check the coil for a short to ground. Set the meter to the high ohms scale and connect the Black lead to engine ground and the Red lead to the Green lead then the Orange lead. Any reading of continuity would show a shorted coil and therefore need replacement.
9. Repeat these tests if the engine is equipped with more than one trigger coil.

OUTPUT TEST

1. Output tests are done using a peak reading voltmeter or a multimeter with a peak adapter.
2. Connect the Red meter lead to the Green wire terminal.
3. Connect the Black meter lead to the Orange wire terminal.
4. Set the meter to the 50DVA scale or equivalent.
5. Crank the powerhead with the cranking motor for a few seconds and observe the meter needle.
6. Record the meter reading.
7. Repeat these tests if the engine is equipped with more than one trigger coil.
8. Replace the trigger coil itself or the trigger assembly as necessary.

REMOVAL & INSTALLATION

◆ See Figures 107 thru 111

1. Remove the spark plugs.
2. Remove the flywheel as instructed in this Section.
3. Identify and mark the two halves of all connectors leading from the trigger housing, as an assist during installation.
4. Disconnect the stator to rectifier and stator to terminal block, or CD module leads. Disconnect the trigger to CD module leads.
5. Pry the throttle control link rod free of the ball joint on the towershaft. Loosen the two retaining screws on the towershaft support.
6. Separate the towershaft from the spark control link by removing attaching hardware, if any, and lifting the towershaft up about ½" to clear the control link.
7. Loosen and remove the stator attaching screws.

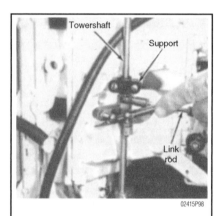

Fig. 107 Pry the throttle control link rod free of the ball joint on the towershaft

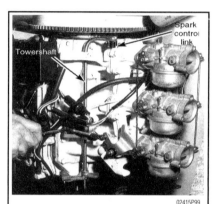

Fig. 108 Separate the towershaft from the spark control link by removing attaching hardware

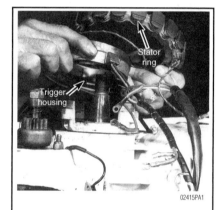

Fig. 109 Remove the stator ring and trigger housing from the powerhead

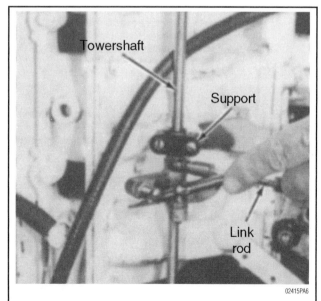

Fig. 110 Tighten the two screws on the towershaft support. Snap the throttle control link rod back onto the ball joint on the towershaft

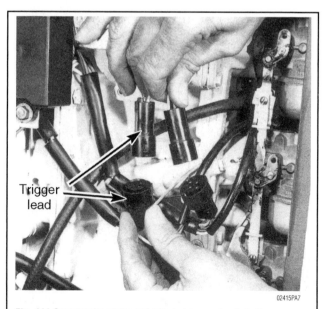

Fig. 111 Connect the stator and trigger harness leads to the same plug connectors identified during disassembly

IGNITION AND ELECTRICAL SYSTEMS 5-31

➙ It may be necessary to use an impact wrench to loosen these screws.

8. Remove the stator ring and trigger housing from the powerhead.
9. If any traces of oil are found on the trigger inspect the condition of the upper oil seal. This seal is located in the stator ring, or in the upper bearing cage.

If any traces of oil are discovered on the trigger assemblies, the oil seal in the stator must be replaced.

10. Inspect the coils of the trigger assembly. Watch for burnt or discolored areas, these are very often accompanied by wiring which becomes loose at a touch. Do not attempt to solder loose connections unless a replacement part is unavailable.

To install:

11. Install the trigger housing and stator plate onto the powerhead. Check to be sure the harness leads and spark control arm face in the same direction they were facing before being disturbed.
12. Install and tighten the stator attaching hardware. No torque specifications are available for these attaching screws, but they should be tightened securely.
13. Attach the spark control arm to the towershaft using the attaching hardware.
14. Tighten the two screws on the towershaft support. Snap the throttle control link rod back onto the ball joint on the towershaft.
15. Connect the stator and trigger harness leads to the same plug connectors identified during disassembly. Remember that these plugs are not necessarily matched color-to-color.
16. Install the flywheel.
17. Install the spark plugs.

Ignition Coils

♦ See Figures 112 and 113

This section covers the testing of ignition coils that are individual units and not part of a coil/ignition module assembly.

TESTING

Resistance

1. Visually inspect the coil for cracks electrical arching or burn marks.
2. Remove the wire connection at the breaker points.
3. Connect the Black multimeter lead to ground and the Red lead to the primary coil wire.
4. Compare this reading of the coil primary resistance to the specifications listed in the "Ignition System Specifications" chart.
5. Connect the Black multimeter lead to the primary coil wire and the Red lead to the spark plug high tension lead.
6. Compare this reading of the coil secondary resistance to the specifications listed in the "Ignition System Specifications" chart.
7. Repeat these procedures for each ignition coil.
8. Replace any coil that has failed any of the tests.

Voltage Output

One of the most accurate ways of determining coil condition is to do a dynamic coil output voltage test. This test will show the actual coil output under varying conditions and will not only determine if the coil defective, but will also alert the user to when coil is about to fail. Coil Voltage Output Testers are available from CDI Electronics (256-772-3829).

1. Remove the cowling.
2. Disconnect the spark plug high tension lead and connect the voltage output tester between the spark plug and the high tension lead.
3. Start (crank) the outboard and read the voltage from the tester.
4. Voltage should be 18–20 Kv.
5. If voltage is not as specified, the coil may be defective.
6. Disconnect the tester and connect the spark plug high tension lead.
7. Install the cowling.

REMOVAL & INSTALLATION

➙ It may be necessary to remove the ignition coil mounting bracket to gain access to the wire connections and fasteners holding the coils in place.

1. Remove the spark plug lead and boot from spark plug.
2. Label and disconnect the primary coil wire from the ignition module.
3. Remove the attaching hardware holding the coil in place.
4. Remove coil.

To install:

5. Install the coil.
6. Install the attaching hardware holding the coil in place.
7. Connect the primary coil wire to the ignition module.
8. Install the spark plug lead and boot on spark plug.

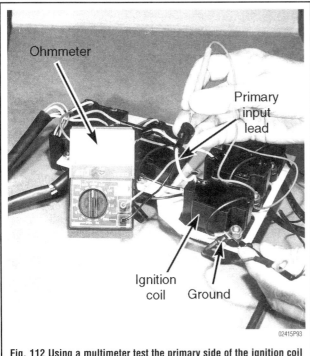

Fig. 112 Using a multimeter test the primary side of the ignition coil

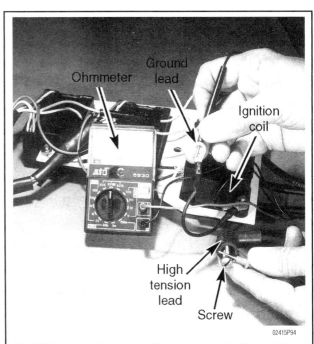

Fig. 113 Using a multimeter test the secondary side of the ignition coil

Ignition Module

There are no resistance tests for the ignition module. It can be tested for output to the ignition coils only. The systems equipped with the ignition module/coil assembly the same applies.

➡ It may be necessary to remove the ignition coil mounting bracket to gain access to the wire connections.

TESTING

1. Output tests are done using a peak reading voltmeter or a multimeter with a peak adapter.
2. Connect the Black meter lead to a clean engine ground.
3. Be certain the ignition module is properly grounded.
4. Set the meter to the 400DVA scale or equivalent.
5. Disconnect the ignition module output wire from the coil to be tested.
6. Connect the Red meter lead to the output wire from the ignition module being tested.
7. Record the meter reading.
8. Repeat this test for the remaining cylinders being tested.
9. All output test readings should be within specification. If not isolate the bad ignition module and replace it.

➡ Remember to test the trigger output and charge coil output before replacing the ignition module.

REMOVAL & INSTALLATION

➡ It may be necessary to remove the ignition coil mounting bracket to gain access to the wire connections and fasteners holding the ignition module in place.

1. Label and disconnect the charge coil output wires or connector.
2. Label and disconnect the trigger output wires or connector.
3. Label and disconnect the ignition module output wires to the ignition coils.
4. Remove the attaching hardware holding the ignition module in place.
5. Remove the ignition module.

To install:

6. Install the ignition module.
7. Install the attaching hardware holding the ignition module in place.
8. Connect the ignition module output wires to the ignition coils.
9. Connect the trigger output wires or connector.
10. Connect the charge coil output wires or connector.

✴✴ WARNING

Be sure to check all the wire connections especially the system grounds before testing or running the engine. Ignition parts are generally not returnable and a loose connection can damage an electronic part immediately.

THUNDERBOLT® ELECTRONIC IGNITION SYSTEM

Description and Operation

The engine's flywheel contains magnets carefully positioned to create an electric current as they rotate past specially designed coils of wire. Current is created by magnetic induction. Simply put, that means that a magnet moving rapidly near a conductor will induce electrical flow within the conductor.

Conversely, a wire that moves rapidly through a magnetic field will also generate electrical flow. This principle governs the working of electric motors, alternators and generators.

One of the coils under your flywheel is called a charge coil. As the flywheel magnets spin past this coil, they generate in it a fairly high-voltage alternating current that travels to your system's switch box. This voltage is often in the region of 200 volts AC.

The other ignition system coil(s) found under the flywheel are called the sensor coils, pulsar coils or trigger coils. They send an electrical signal to the ignition module to tell it which cylinder to work with at the correct time.

The switch box is the brains of the system and serves several functions. First, it converts the alternating current (AC) from the charge coil into usable direct current (DC). Next it stores the current in a built-in capacitor. The module also interprets the timing signal from the trigger coil. This changes constantly with engine speed, and moving the trigger coil's position relative to the flywheel magnets brings about the change. A device called a timing plate where the trigger coils are mounted controls the coil's movement. The timing plate moves in response to changes in throttle opening, to which it is mechanically linked. The switch box also controls the discharge of the capacitor and sends this voltage to the primary winding of the ignition coil for the correct cylinder.

Also (depending on the system) the module may incorporate electronic circuits that limit engine speed and prevent over-revving. Some modules even have a circuit that reduces engine speed if the engine begins to run too hot for any reason

The voltage from the module goes to the primary winding of the ignition coil or high-tension coil. You may know this type of coil as a step-up transformer. Here, the voltage is stepped up to between 15,000 and 40,000 volts. That's the kind of voltage needed to jump the air gap on the spark plug and ignite the air/fuel mixture in the cylinder. Your high-tension ignition coil has two sides, the primary side and the secondary side. It is really two coil assemblies combined into one neat, compact case.

The internal construction of a typical ignition coil includes primary and secondary windings. It also uses the principle of magnetic induction with the magnetism generated by the primary (lower -voltage) winding creating a magnetic field around the secondary winding, which has many more windings than the primary coil. The ignition module controls the rapid turning on and off of electrical flow in the primary winding, thereby turning this magnetic field on and off.

The rapid movement of this magnetic field past the secondary windings induces electrical current flow. The are greater the number of turns of wire in the secondary winding, the higher the voltage produced.

As this secondary voltage leaves the center tower of the ignition coil, it travels along the spark-plug wire, which is heavily insulated and designed to carry this high voltage.

If all is well, the high voltage will jump the gap on the spark plug between the center electrode and the
ground electrode. On larger engines with surface-gap plugs, the high voltage current will jump from the center electrode to the side of the plug assembly itself, completing a circuit to ground via the engine block.

Last, but certainly not least, is the stop control. You need a means to shut your engine off and a good way is to stop the spark plugs from working. Depending on your engine, this may be accomplished by a simple stop button or a key switch on larger engines. This disables the whole ignition system. On newer engines, an emergency stop button with an overboard clip and lanyard attachment is standard. This system is wired directly into the switch box. It functions by creating a momentary short circuit inside the module unit, grounding the current intended for the high tension coils and thus shutting off the ignition long enough to stop the engine. Faulty stop circuits are frequently the cause of a no spark condition.

Troubleshooting the Thunderbolt® Electronic Ignition System

♦ See Figures 114, 115, 116 and 117

➡ The following conditions apply to the Thunderbolt® Electronic Ignition System. Troubleshooting procedure are courtesy of our friends at CDI Electronics (256-772-3829).

➡ Maintenance-free batteries are not recommended for use with these systems.

The trouble shooting chart provided is an easy system to go step by step through the ignition system. The important thing is to work backwards through the system starting with the spark plugs. Remember to keep the troubleshooting as simple as possible. If there is a suspect component that can be swapped with another on the engine such as an ignition coil. Swap the parts and test to see if the no spark condition has moved to swapped component or has it stayed on the same cylinder. This test can also be conducted by switching input wires from the trigger, but be certain to put the wires back in there proper place before attempting to start the engine.

IGNITION AND ELECTRICAL SYSTEMS 5-33

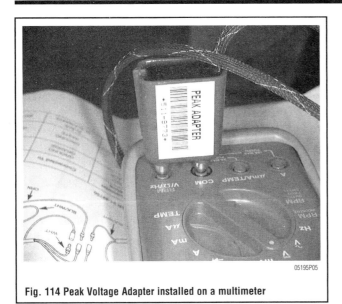

Fig. 114 Peak Voltage Adapter installed on a multimeter

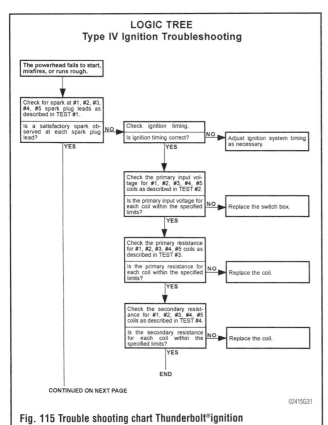

Fig. 115 Trouble shooting chart Thunderbolt® ignition

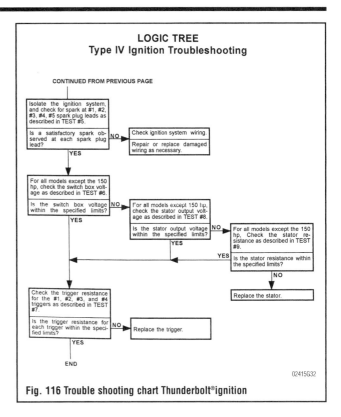

Fig. 116 Trouble shooting chart Thunderbolt® ignition

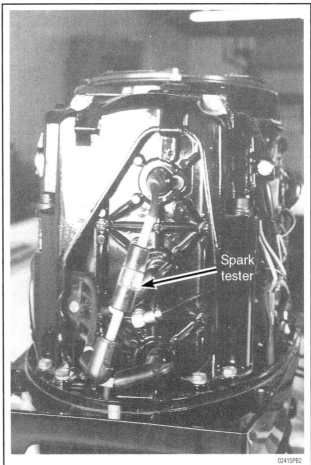

Fig. 117 Typical installation of an inexpensive spark tester hooked up to the No. 1 spark plug.

Check for broken wires and terminals, especially inside the plastic plug-in connectors. Removal of the pins from the connectors using the Rapair (R554-9706) Pin Removal Tool and will allow you to visually inspect the pins for damage.

Check the flywheel for broken or loose magnets.

Disconnect the kill wires from the CD and connect a DC voltmeter between the kill wires and engine ground, turn the ignition switch on and off several times. If at any time you see voltage appearing on the meter, there is a problem in the harness or ignition switch.

➡ At no time should you see battery voltage on a kill circuit.

Visually inspect stator for burned or discolored areas. If found, replace the stator. If the areas are on the battery charge windings, it indicates a possible problem with the rectifier.

5-34 IGNITION AND ELECTRICAL SYSTEMS

IF NO FIRE ON ANY CYLINDER: Disconnect kill wire at the pack. Check for broken or bare wires on the unit, stator and trigger. Using a multimeter with a peak reading adapter, or and piercing probes, measure DVA voltage of the stator between the output wire sets. With everything connected, readings should be approximately 180 volts or, more. The OEM stator resistance between the blue and yellow wires is 700-900 ohms. CDI Electronics stators should read 300-500 ohms. Disconnect the rectifier. If the engine fires, replace the rectifier.

NO FIRE OR INTERMITTENT FIRE ON ONE CYLINDER: Check stator and trigger resistance. Trigger wire sets read approximately 50 ohms between the wire sets (DVA 5V or more), OEM stators read 700-900 ohms (DVA 180V or more) from Blue to Yellow, while CDI Electronics stators read 300-500 ohms. If readings are good, disconnect kill wire from one pack. If the dead cylinder starts firing, the problem is likely the blocking diode in the other pack.

NO FIRE ON TWO CYLINDERS: If two cylinders from the same CD unit will not fire, the problem is usually in the stator. Test per above.

ENGINE WILL NOT KILL: Check kill circuit in the pack by using a jumper wire connected to the kill wire coming out of the pack and shorting it to ground. If this kills the pack, the kill circuit in the harness or on the boat is bad, possibly the ignition switch.

COILS ONLY FIRE WITH THE SPARK PLUGS OUT: Check for dragging starter or low battery causing slow cranking speed. DVA test stator and trigger.

HIGH SPEED MISS: Using a multimeter with a peak reading adapter and piercing probes, DVA check stator voltage to each pack at high speed. If it exceeds 400 volts, replace the pack. Disconnect the rectifier and retest. If the miss is gone replace the rectifier.

Flywheel

REMOVAL & INSTALLATION

❊❊ CAUTION

Always wear safety glasses when removing flywheels. Extreme force might be necessary to remove the unit and metal slivers or rust could fly up into your face.

❊❊ WARNING

Flywheels that have never been removed or have sheared keys are typically very difficult to remove. The use of heat from a torch or soaking with penetrating oil is recommended to ease removal.

Flywheels Without Puller Bolt Holes

♦ See Figures 118, 119 and 120

1. Remove the engine cover and manual starter as necessary.
2. Hold the flywheel with a flywheel holding tool or strap wrench.
3. Remove the fastener holding the flywheel in place.
4. Pull up on the outer edge of the flywheel with a prybar.
5. Tap the crankshaft with a brass hammer to loosen the flywheel.
6. Carefully lift the flywheel from the powerhead.

To install:

7. Install the flywheel on the powerhead making certain the flywheel key is properly installed in the keyway and not damaged.
8. Install the fastener holding the flywheel in place.
9. Hold the flywheel with a flywheel holding tool or strap wrench and tighten to the specified torque.
10. Install the manual starter and engine cover.

Flywheels With Puller Bolt Holes

♦ See Figures 121, 122 and 123

1. Hold the Flywheel with a strap wrench or a flywheel holder.
2. Remove the flywheel nut.
3. Install a puller on to the flywheel using the proper thread size and length bolt.
4. Thread the puller bolt in putting pressure on the crankshaft and removing the flywheel.

➡ It may be necessary to hit the puller bolt with a hammer to shock the crankshaft to release the flywheel.

5. Carefully lift the flywheel from the powerhead.

To install:

6. Install the flywheel on the powerhead making certain the flywheel key is properly installed in the keyway and not damaged.
7. Install the fastener holding the flywheel in place.
8. Hold the flywheel with a flywheel holding tool or strap wrench and tighten to the specified torque.
9. Install the manual starter and engine cover.

Flywheels With Threaded Center

♦ See Figures 124 and 125

1. While holding the flywheel with the flywheel holder or strap wrench, remove the flywheel nut and washer.
2. Install a crankshaft protector cap (91–24161), then install a flywheel puller on the flywheel.

Fig. 118 Remove the engine cover and manual starter as necessary

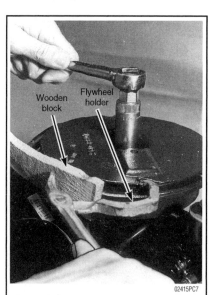

Fig. 119 Hold the flywheel with a strap wrench

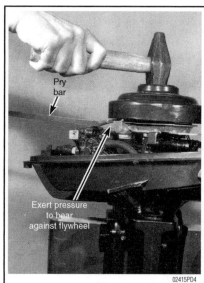

Fig. 120 Pull up on the outer edge of the flywheel with a prybar

IGNITION AND ELECTRICAL SYSTEMS

Fig. 121 Hold the Flywheel with a strap wrench or a flywheel holder

Fig. 122 Install a puller on to the flywheel using the proper thread size and length bolt

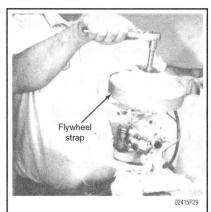

Fig. 123 Reverse this procedure to install the flywheel making certain the flywheel key in properly installed in the keyway and not damaged

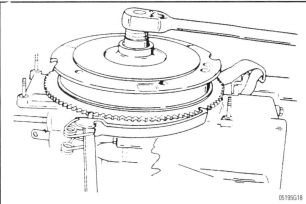

Fig. 124 While holding the flywheel with the flywheel holder or strap wrench, remove the flywheel nut and washer

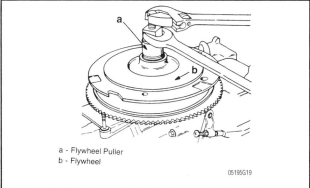

a - Flywheel Puller
b - Flywheel

Fig. 125 Hold the flywheel tooth with the wrench while tightening the bold down on the protector cap. Tighten the bolt until the flywheel comes free

3. Hold the flywheel tooth with the wrench while tightening the bold down on the protector cap. Tighten the bolt until the flywheel comes free.
4. Carefully lift the flywheel from the powerhead.

To install:

5. Install the flywheel on the powerhead making certain the flywheel key is properly installed in the keyway and not damaged.
6. Install the fastener holding the flywheel in place.
7. Hold the flywheel with a flywheel holding tool or strap wrench and tighten to the specified torque.
8. Install the manual starter and engine cover.

INSPECTION

Visually inspect the flywheel for cracks, broken teeth and deformities.
Inspect the magnets around the inside of the flywheel for cracks or damage.
Test the strength of the magnets by placing a medium size flat blade screwdriver against it. The magnet should be able to hold the screwdriver. All outer magnets should be the same strength.
Replace the flywheel as necessary.

Charging Coils

➔ The charge coils in this system are incorporated in the stator assembly.

TESTING

Resistance

A standard multimeter may be used for these tests.

LOW SPEED

1. Set the multimeter to the X1000 ohm scale.
2. Connect the Red meter lead to the Blue wire (low-speed) with the switch box leads disconnected.
3. Connect the Black meter lead to the Red stator lead.
4. The meter should indicate from 5500 to 6200 ohms.
5. If the stator fails this test, replace the stator.

HI SPEED

1. Set the meter to the x1 ohm scale.
2. Connect the Red meter lead to the Red stator lead (hi-speed) and the Black meter to the Black stator lead.
3. The meter should indicate 90 to 140 ohms.
4. If the stator fails this test, replace the stator.

Direct Voltage Adapter

✶✶ CAUTION

Be sure to remove the high tension leads from the spark plugs.

Use a multimeter with peak reading adapter. Set the meter to 400.
Low-Speed—Connect the Red meter lead to the Blue switch box terminal—stator leads are attached to switch box. Connect the Black meter lead to a good ground on the powerhead. The meter should not indicate less than 200 volts.
Hi-Speed—Connect the Red meter lead to the Red switch box terminal and the Black meter lead to a good ground on the powerhead. The meter should not indicate less then 20 volts.
If the stator fails either test conduct a Stator Resistance Test.

5-36 IGNITION AND ELECTRICAL SYSTEMS

REMOVAL & INSTALLATION

▶ See Figure 126

1. Remove the flywheel.
2. Using an Allen wrench remove the screws securing the stator to the powerhead.
3. Lift the stator up and clear of the crankshaft.
4. If the stator is to be replaced, disconnect the wire harness.
5. Label and disconnect the charge coil wires to the switch box.
6. Label and disconnect the stator charge coil wires from the rectifier/regulator.

To install:

7. Connect the stator charge coil wires to the rectifier/regulator.
8. Connect the charge coil wires to the switch box.
9. Connect the wiring harness.
10. Position the stator on the powerhead.
11. Using an Allen wrench install the screws securing the stator to the powerhead.

➡ Be certain that the flywheel key is in place before placing the flywheel over the crankshaft.

12. Install the flywheel.
13. Hold the flywheel and torque the nut to proper specification.

Trigger Coils

TESTING

Resistance

▶ See Figure 127

Use a standard VOA meter set on the x100 ohm scale.
Test each trigger. Connect the Red meter lead to the trigger wire and the black meter lead to the common ground as indicated in the table.

Direct Voltage Adapter

Consult the chart for the appropriate wire colors
1. Remove the trigger leads from the ignition module.
2. Using a DVA meter set on a low voltage scale.
3. Connect the black meter lead to a clean engine ground.
4. Connect the Red meter lead to a trigger output wire.
5. Crank the engine over and read the meter while the engine is cranking.
6. Record the reading.
7. Remove the Red lead from the trigger wire and connect it to the next output wire.
8. Conduct the same test and continue testing through all trigger wires.
9. There will be one trigger output wire for each cylinder.

REMOVAL & INSTALLATION

▶ See Figures 128 thru 135

1. Remove the flywheel.
2. Using an Allen wrench remove the screws securing the stator to the powerhead.
3. Lift the stator up and clear of the crankshaft.
4. Remove the bolt securing the trigger bracket adjacent to the cranking motor.
5. Remove the two top nuts and the one lower bolt securing the ignition plate to the powerhead.
6. Move the plate back a bit to allow the spacers to be removed.
7. Back out the studs by rotating the spacers, then remove the studs and spacers.

Fig. 126 Using an Allen wrench remove the screws securing the stator to the powerhead

Model	Red Meter Lead To Trigger No.	Black Meter Lead to
40 hp	#1 Vio, #2 Wht	Wht Common Ground
50 hp	#1 Vio, #2 Wht	Wht Common Ground
70 hp	#1 Brn, #2 Wht, #3 Vio	Wht Common Ground
90 hp	#1 Brn, #2 Wht, #3 Vio	Wht Common Ground
120 hp	#1 Vio, #2 Wht, #3 Brn #4 Wht/Blk	Wht/Blk Common Ground
150 Hp	#1 Brn, #2 Wht, #3 Vio, #4 Blk, #5 Yel	Wht/Blk Common Ground

Fig. 127 Test each trigger coil individually using the colors indicated

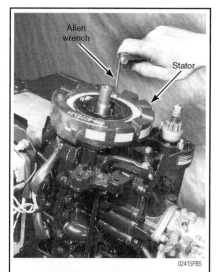

Fig. 128 Using an Allen wrench remove the screws securing the stator to the powerhead

Fig. 129 Remove the bolt securing the trigger bracket adjacent to the cranking motor

IGNITION AND ELECTRICAL SYSTEMS 5-37

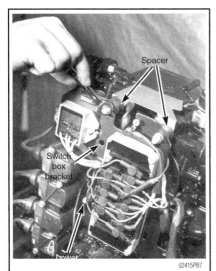

Fig. 130 Remove the two top nuts and the one lower bolt securing the ignition plate to the powerhead

Fig. 131 Lift off the cover plate exposing and releasing the trigger wire harness

Fig. 132 Connect the wire harness to the trigger assembly

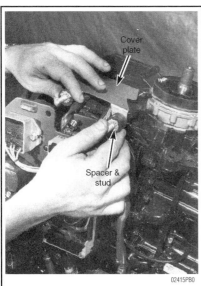

Fig. 133 Position the cover plate in place and secure it with the studs and spacers

Fig. 134 Secure the trigger bracket with the attaching bolt

Fig. 135 Lower the stator assembly into place over the trigger

8. Lift off the cover plate exposing and releasing the trigger wire harness.
9. Disconnect the trigger output leads.
10. The trigger may now be lifted up and clear of the crankshaft.

To install:
11. Connect the wire harness to the trigger assembly.
12. Lower the trigger down over the crankshaft with the hole in the bracket aligned with the mounting hole.
13. Position the cover plate in place and secure it with the studs and spacers.
14. Install the ignition plate and attaching nuts.
15. Install the lower bolt securing the ignition plate.
16. Secure the trigger bracket with the attaching bolt.
17. Lower the stator assembly into place over the trigger.
18. Secure the stator with the attaching Allen head screws.

Install the flywheel being certain that the flywheel key is in place before placing the flywheel over the crankshaft. Hold the flywheel and torque the nut to proper specification.

Ignition Coils

TESTING

Resistance

1. Visually inspect the coil for cracks electrical arching or burn marks.
2. Remove the wire connection at the breaker points.
3. Connect the Black multimeter lead to ground and the Red lead to the primary coil wire.
4. Compare this reading of the coil primary resistance to the specifications listed in the "Ignition System Specifications" chart.
5. Connect the Black multimeter lead to the primary coil wire and the Red lead to the spark plug high tension lead.
6. Compare this reading of the coil secondary resistance to the specifications listed in the "Ignition System Specifications" chart.

5-38 IGNITION AND ELECTRICAL SYSTEMS

7. Repeat these procedures for each ignition coil.
8. Replace any coil that has failed any of the tests.

Voltage Output

One of the most accurate ways of determining coil condition is to do a dynamic coil output voltage test. This test will show the actual coil output under varying conditions and will not only determine if the coil defective, but will also alert the user to when coil is about to fail. Coil Voltage Output Testers are available from CDI Electronics (256-772-3829).
1. Remove the cowling.
2. Disconnect the spark plug high tension lead and connect the voltage output tester between the spark plug and the high tension lead.
3. Start (crank) the outboard and read the voltage from the tester.
4. Voltage should be 18–20 Kv.
5. If voltage is not as specified, the coil may be defective.
6. Disconnect the tester and connect the spark plug high tension lead.
7. Install the cowling.

REMOVAL & INSTALLATION

1. Remove the spark plug lead and boot from spark plug.
2. Remove the attaching hardware holding the ignition coils in place.
3. Disconnect the primary coil wire from the ignition coil positive terminal.
4. Remove the ignition coil ground from the ignition coil negative terminal.
5. Remove coil.

To install:
6. Install coil.
7. Install the ignition coil ground from the ignition coil negative terminal.
8. Disconnect the primary coil wire from the ignition coil positive terminal.
9. Install the attaching hardware holding the ignition coils in place.
10. Install the spark plug lead and boot from spark plug.

Switch Box

The switch box relies on its inputs to function correctly. Be certain that the charge coils are in specification along with the trigger coils before testing the switch box.

TESTING

The switch box can only be tested for output. It has a primary voltage output to each ignition coil and to the engine stop circuit.

※※ CAUTION

Be sure to remove the high tension leads from the spark plugs.

1. Using a DVA meter set at 400V.
2. Remove the Black/Yellow wire from the switch box.
3. Connect the Red meter lead to the switch box terminal.
4. Connect the Black meter lead to the engine block.
5. Crank the engine over and record the meter reading (200 to 360 volts).
6. Connect the Red meter lead to the #1 primary output terminal of the switch box.
7. Crank the engine over and record the meter reading.
8. Repeat this test for the remaining cylinders.
9. In the event the switch box tested bad before continuing disconnect the regulator rectifier and retest the switch box.
10. If output readings are now in specification replace the regulator rectifier.
11. All the outputs to the primary terminals should be the same. Replace the switch box if any cylinder shows no or very low output.

➡In the event that they are close to being the same be certain that the trigger coils and charge coils have been thoroughly tested before replacing the switch box.

12. Place the Black/Yellow lead back on the terminal stud and secure with nut.
13. Retest the primary outputs to be certain the key switch or harness is not shorted causing the loss of output.

REMOVAL & INSTALLATION

1. Label and disconnect all connections to the switch box.
2. Remove the attaching hardware and ground wire, if equipped.
3. Remove the switchbox.

To install:
4. Install the switchbox.
5. Install the attaching hardware and ground wire.
6. Install all connections to the switch box.

CAPACITOR DISCHARGE MODULE (CDM) SYSTEM

Description and Operation

The engine's flywheel contains magnets carefully positioned to create an electric current as they rotate past specially designed coils of wire. Current is created by magnetic induction. Simply put, that means that a magnet moving rapidly near a conductor will induce electrical flow within the conductor.

Conversely, a wire that moves rapidly through a magnetic field will also generate electrical flow. This principle governs the working of electric motors, alternators and generators.

One of the coils under your flywheel is called a charge coil. As the flywheel magnets spin past this coil, they generate in it a fairly high-voltage alternating current that travels to your system's ignition coil modules. This voltage is often in the region of 200 volts AC.

The other ignition system coil(s) found under the flywheel are called the sensor coils, pulsar coils or trigger coils. They send an electrical signal to the ignition coil module to tell it which cylinder to work with at the correct time.

The ignition coil module is the brains of the system and serves several functions. First, it converts the alternating current (AC) from the charge coil into usable direct current (DC). Next it stores the current in a built-in capacitor. The module also interprets the timing signal from the trigger coil. This changes constantly with engine speed, and moving the trigger coil's position relative to the flywheel magnets brings about the change. A device called a timing plate where the trigger coils are mounted controls the coil's movement. The timing plate moves in response to changes in throttle opening, to which it is mechanically linked. The ignition coil module also controls the discharge of the capacitor and sends this voltage to the high tension /spark plug lead for the correct cylinder.

Also (depending on the system) the module may incorporate electronic circuits that limit engine speed and prevent over-revving. Some modules even have a circuit that reduces engine speed if the engine begins to run too hot for any reason. Larger engines often have an automatic ignition advance for initial start-up and for when the engine is running at temperatures less than approximately 100'F.

The internal construction of an ignition coil module includes primary and secondary ignition coil windings a rectifier and capacitor system. Each module in effect is a single switch box coil assembly.

If all is well, the high voltage will jump the gap on the spark plug between the center electrode and the
ground electrode. On larger engines with surface-gap plugs, the high voltage current will jump from the center electrode to the side of the plug assembly itself, competing a circuit to ground via the engine block.

Last, but certainly not least, is the stop control. You need a means to shut your engine off and a good way is to stop the spark plugs from working. Depending on your engine, this may be accomplished by a simple stop button or a key switch on larger engines. This disables the whole ignition system. On newer engines, an emergency stop button with an overboard clip and lanyard attachment is standard. This system is wired directly into the ignition module. It functions by creating a momentary short circuit inside the CDI unit, grounding the current intended for the high tension coils and thus shutting off the ignition long enough to stop the engine. Faulty stop circuits are frequently the cause of a no spark condition.

IGNITION AND ELECTRICAL SYSTEMS   5-39

Troubleshooting the CDM System

♦ See Figure 136

→The following conditions apply to the Motorola® Distributor Ignition System. Troubleshooting procedure are courtesy of our friends at CDI Electronics (256-772-3829)

→Maintenance-free batteries are not recommended for use with these systems.

Check for broken wires and terminals, especially inside the plastic plug-in connectors. Removal of the pins from the connectors using the Rapair (R554-9706) Pin Removal Tool and will allow you to visually inspect the pins for damage.

Check the flywheel for broken or loose magnets.

Disconnect the kill wires from the CD and connect a DC voltmeter between the kill wires and engine ground, turn the ignition switch on and off several times. If at any time you see voltage appearing on the meter, there is a problem in the harness or ignition switch.

→At no time should you see battery voltage on a kill circuit.

Visually inspect stator for burned or discolored areas. If found, replace the stator. If the areas are on the battery charge windings, it indicates a possible problem with the rectifier.

IF NO FIRE ON ANY CYLINDER: Disconnect kill wire at the pack. Check for broken or bare wires on the unit, stator and trigger. Using a multimeter with a peak reading adapter, or and piercing probes, measure DVA voltage of the stator between the output wire sets. With everything connected, readings should be approximately 180 volts or, more. The OEM stator resistance between the blue and yellow wires is 700-900 ohms. CDI Electronics stators should read 300-500 ohms. Disconnect the rectifier. If the engine fires, replace the rectifier.

NO FIRE OR INTERMITTENT FIRE ON ONE CYLINDER: Check stator and trigger resistance. Trigger wire sets read approximately 50 ohms between the wire sets (DVA 5V or more), OEM stators read 700-900 ohms (DVA 180V or more) from Blue to Yellow, while CDI Electronics stators read 300-500 ohms. If readings are good, disconnect kill wire from one pack. If the dead cylinder starts firing, the problem is likely the blocking diode in the other pack.

NO FIRE ON TWO CYLINDERS: If two cylinders from the same CD unit will not fire, the problem is usually in the stator. Test per above.

ENGINE WILL NOT KILL: Check kill circuit in the pack by using a jumper wire connected to the kill wire coming out of the pack and shorting it to ground.

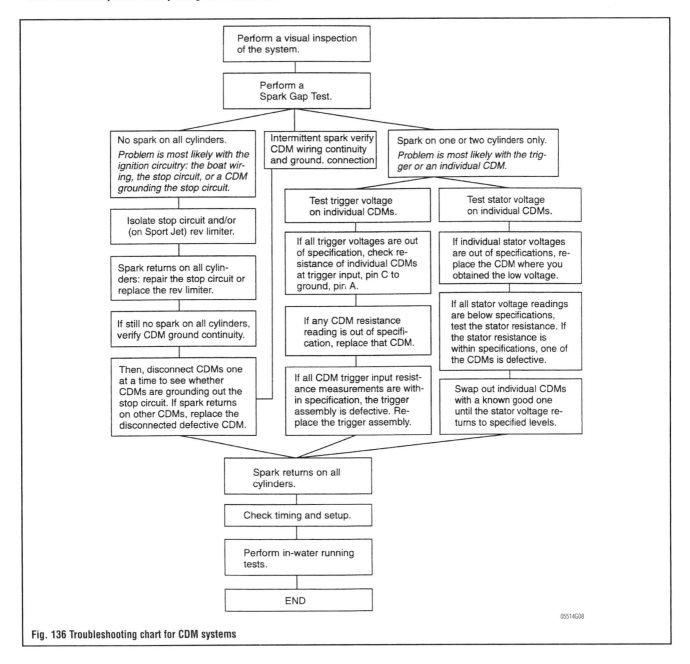

Fig. 136 Troubleshooting chart for CDM systems

5-40 IGNITION AND ELECTRICAL SYSTEMS

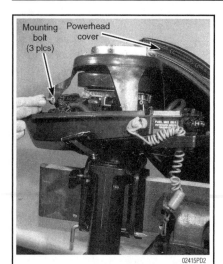

Fig. 137 Remove the engine cover and manual starter as necessary

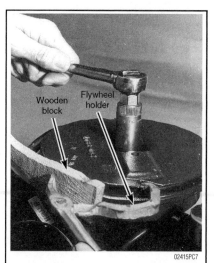

Fig. 138 Hold the flywheel with a strap wrench

Fig. 139 Pull up on the outer edge of the flywheel with a prybar

If this kills the pack, the kill circuit in the harness or on the boat is bad, possibly the ignition switch.

COILS ONLY FIRE WITH THE SPARK PLUGS OUT: Check for dragging starter or low battery causing slow cranking speed. DVA test stator and trigger.

HIGH SPEED MISS: Using a multimeter with a peak reading adapter and piercing probes, DVA check stator voltage to each pack at high speed. If it exceeds 400 volts, replace the pack. Disconnect the rectifier and retest. If the miss is gone replace the rectifier.

Flywheel

REMOVAL & INSTALLATION

✱✱ CAUTION

Always wear safety glasses when removing flywheels. Extreme force might be necessary to remove the unit and metal slivers or rust could fly up into your face.

✱✱ WARNING

Flywheels that have never been removed or have sheared keys are typically very difficult to remove. The use of heat from a torch or soaking with penetrating oil is recommended to ease removal.

Flywheels Without Puller Bolt Holes

♦ See Figures 137, 138 and 139

1. Remove the engine cover and manual starter as necessary.
2. Hold the flywheel with a flywheel holding tool or strap wrench.
3. Remove the fastener holding the flywheel in place.
4. Pull up on the outer edge of the flywheel with a prybar.
5. Tap the crankshaft with a brass hammer to loosen the flywheel.
6. Carefully lift the flywheel from the powerhead.

To install:

7. Install the flywheel on the powerhead making certain the flywheel key is properly installed in the keyway and not damaged.
8. Install the fastener holding the flywheel in place.
9. Hold the flywheel with a flywheel holding tool or strap wrench and tighten to the specified torque.
10. Install the manual starter and engine cover.

Flywheels With Puller Bolt Holes

♦ See Figures 140, 141 and 142

1. Hold the Flywheel with a strap wrench or a flywheel holder.
2. Remove the flywheel nut.
3. Install a puller on to the flywheel using the proper thread size and length bolt.
4. Thread the puller bolt in putting pressure on the crankshaft and removing the flywheel.

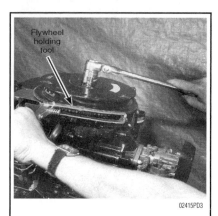

Fig. 140 Hold the Flywheel with a strap wrench or a flywheel holder

Fig. 141 Install a puller on to the flywheel using the proper thread size and length bolt

Fig. 142 Reverse this procedure to install the flywheel making certain the flywheel key in properly installed in the keyway and not damaged

IGNITION AND ELECTRICAL SYSTEMS

➥It may be necessary to hit the puller bolt with a hammer to shock the crankshaft to release the flywheel.

5. Carefully lift the flywheel from the powerhead.

To install:

6. Install the flywheel on the powerhead making certain the flywheel key is properly installed in the keyway and not damaged.
7. Install the fastener holding the flywheel in place.
8. Hold the flywheel with a flywheel holding tool or strap wrench and tighten to the specified torque.
9. Install the manual starter and engine cover.

Flywheels With Threaded Center

♦ See Figures 143 and 144

1. While holding the flywheel with the flywheel holder or strap wrench, remove the flywheel nut and washer.
2. Install a crankshaft protector cap (91–24161), then install a flywheel puller on the flywheel.
3. Hold the flywheel tooth with the wrench while tightening the bold down on the protector cap. Tighten the bolt until the flywheel comes free.
4. Carefully lift the flywheel from the powerhead.

To install:

5. Install the flywheel on the powerhead making certain the flywheel key is properly installed in the keyway and not damaged.
6. Install the fastener holding the flywheel in place.

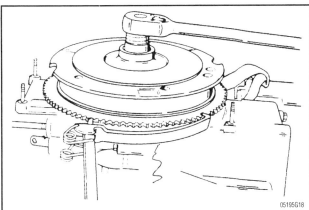

Fig. 143 While holding the flywheel with the flywheel holder or strap wrench, remove the flywheel nut and washer

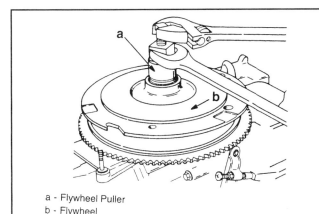

a - Flywheel Puller
b - Flywheel

Fig. 144 Hold the flywheel tooth with the wrench while tightening the bold down on the protector cap. Tighten the bolt until the flywheel comes free

7. Hold the flywheel with a flywheel holding tool or strap wrench and tighten to the specified torque.
8. Install the manual starter and engine cover.

INSPECTION

Visually inspect the flywheel for cracks, broken teeth and deformities.
Inspect the magnets around the inside of the flywheel for cracks or damage.
Test the strength of the magnets by placing a medium size flat blade screwdriver against it. The magnet should be able to hold the screwdriver. All outer magnets should be the same strength.
Replace the flywheel as necessary.

Charging Coils

TESTING

Resistance Tests

➥Resistance testing is generally done at room temperature, therefore if the readings are close to specification and it happens to be very cold or hot the unit could still be good.

1. Using a multimeter on the correct scale, connect the Red meter lead to the White/Green wire and the Black meter lead to the Green/White wire. Record the reading
2. Connect the Red meter lead to the White/Green lead and the Black meter lead there should be no continuity.
3. Connect the Red meter lead to the Green/White lead and the Black meter lead there should be no continuity.
4. Check the resistance readings against the specification in the "Ignition System Specification".
5. Replace when any of the tests proved to be bad or out of specification.

Direct Voltage Adapter Tests

♦ See Figures 145, 146, 147, 148 and 149

1. The output voltage can be tested by checking the voltage between the Green/White wire and engine ground at each ignition coil module input connection with the modules all plugged in.
2. With a spark tester installed crank the engine over and observe the meter reading.
3. Conduct this test for all cylinders.

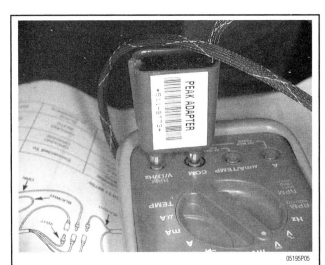

Fig. 145 A CDI Electronics peak reading voltage adapter (#511-9773) plugged into a high quality multimeter

5-42 IGNITION AND ELECTRICAL SYSTEMS

Fig. 146 Using a multimeter and DVA attachment to check the peak voltage on the trigger/charging circuit

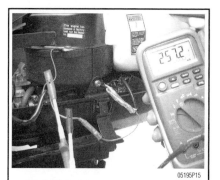

Fig. 147 High quality piercing probes (#511-9770), like those available from CDI Electronics (256-772-3829), allow you to perform dynamic ignition testing without the use of factory wiring harness adapters

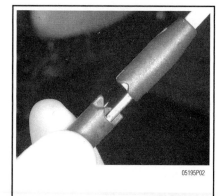

Fig. 148 The tiny pins in the piercing probes contact the wires to read resistance or voltage . . .

Fig. 149 . . . but do not destroy the insulation on the wire

4. All cylinders should read the same voltage.
5. Low voltage on only one module could indicate a bad module.

REMOVAL & INSTALLATION

1. Remove the flywheel.
2. Using an Allen wrench remove the screws securing the stator to the powerhead.
3. Lift the stator up and clear of the crankshaft.
4. If the stator is to be replaced, disconnect the wire harness.
5. Label and disconnect the charge coil wires to the switch box.
6. Label and disconnect the stator charge coil wires from the rectifier/regulator.

To install:

7. Connect the stator charge coil wires to the rectifier/regulator.
8. Connect the charge coil wires to the switch box.
9. Connect the wiring harness.
10. Position the stator on the powerhead.
11. Using an Allen wrench install the screws securing the stator to the powerhead.

➡ Be certain that the flywheel key is in place before placing the flywheel over the crankshaft.

12. Install the flywheel.
13. Hold the flywheel and torque the nut to proper specification.

Trigger Coils

TESTING

1. There is no resistance test for the trigger coils in this system. A low reading on the output would show a bad trigger coil a high reading would probably indicate a bad ignition coil module.
2. Connect a DVA meter with the Red lead to the trigger output wire for that cylinder and the Black lead to the Black lead of the ignition coil module.
3. With a spark tester installed crank the engine over and observe the meter reading.
4. Conduct this test for all cylinders.

Replace the trigger assembly when any of the trigger coils test bad.

REMOVAL

▶ See Figures 150 thru 154

1. Remove the flywheel.
2. Using an Allen wrench remove the screws securing the stator to the powerhead.
3. Lift the stator up and clear of the crankshaft.
4. Remove the bolt securing the trigger bracket adjacent to the cranking motor.
5. Remove the ignition plate for access to connections as necessary.
6. Disconnect the trigger output wires to the harness.
7. Remove the trigger assembly.

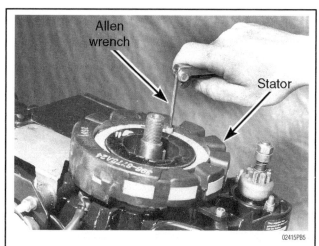

Fig. 150 Using an Allen wrench remove the screws securing the stator to the powerhead

IGNITION AND ELECTRICAL SYSTEMS 5-43

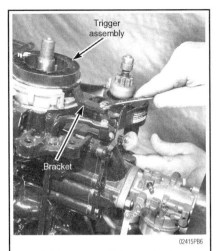

Fig. 151 Remove the bolt securing the trigger bracket adjacent to the cranking motor

Fig. 152 Connect the wire harness to the trigger assembly

Fig. 153 Secure the trigger bracket with the attaching bolt

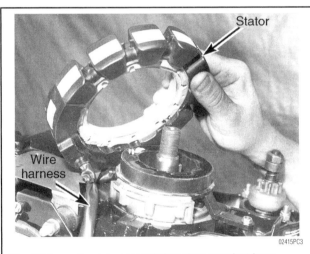

Fig. 154 Lower the stator assembly into place over the trigger

To install:
8. Connect the wire harness to the trigger assembly. Lower the trigger down over the crankshaft with the hole in the bracket aligned with the mounting hole.
9. Install the ignition plate and attaching nuts, if removed.
10. Install the lower bolt securing the ignition plate.
11. Secure the trigger bracket with the attaching bolt.
12. Place the stator assembly over the trigger and secure it with the attaching Allen head screws.

Install the flywheel being certain that the flywheel key is in place before placing the flywheel over the crankshaft. Hold the flywheel and torque the nut to proper specification.

CDM Ignition Coil Module

TESTING

♦ See Figure 155

➡The easiest test to isolate the possibility of a bad ignition coil module is to swap the module in place of another that is functioning properly. The modules are all the same for each cylinder making this test the quickest. Remember that the spark plugs play an important roll in the ability of the unit to function so do not just replace a module without checking or replacing that spark plug.

Resistance

Using a multimeter test the module for primary and secondary coil resistance. The remaining tests are diode tests meaning that the current should only flow in one direction for these components to be in good condition.
1. Remove the connector from the ignition coil module
2. Test primary coil resistance by connecting the Red lead to terminal "A" and the Black lead to terminal "C".
3. Test the secondary coil resistance by connecting the Red meter lead to the spark plug boot and the Black lead to terminal "A".
4. Test for continuity between terminals "D" and "A". There should be continuity with the leads on in one direction but with the leads reversed.
5. Test for continuity between terminals "D" and "B". There should be continuity with the leads on in one direction but with the leads reversed.

REMOVAL & INSTALLATION

1. The ignition coil modules are bolted to the ignition plate therefore making it necessary in most cases to loosed or remove the plate from the engine to gain access to the mounting hardware.
2. Twist and pull the spark plug lead from the spark plug.
3. Remove the connector from the module by releasing the lock tab and gently pulling the connection loose.
4. Remove the fasteners holding the module in place.
5. Remove the ignition module.

To install:
6. Install by mounting the module on the ignition plate.
7. Connect the harness plug and spark plug lead.
8. Install the ignition plate if it had to be loosened or removed.

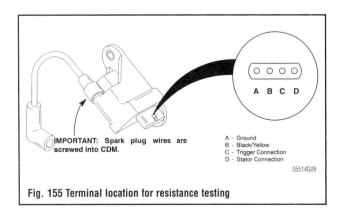

Fig. 155 Terminal location for resistance testing

5-44 IGNITION AND ELECTRICAL SYSTEMS

Ignition System Specifications Chart

Year	Horsepower	Part Number	Trigger Ohms	Ignition Coil Primary Ohms	Ignition Coil Secondary Ohms	Charge Coil High Speed Ohms	Charge Coil Low Speed Ohms	Trigger Cranking DVA	Charge Coil Cranking DVA High Speed	Charge Coil Cranking DVA Low Speed	CD Module (Switch Box) Cranking DVA	Trigger Wire Colors	Charge Coil Wire Colors
1984	9.9-15	510301	48-52	0.7-0.9	800-1100	58-65	1775-1850	0.5V+	25V+	180V+	150V+	Between Wire Pairs	Bl to Ground
1984-91	20-35	529301	48-52	0.7-0.9	800-1100	58-65	1775-1850	0.5V+	25V+	180V+	150V+	Between Wire Pairs	Bl to Y
1988	50 A,B,C	658475	48-52	N/A	120-140	680-800	-	0.5V+	180V+	-	150V+	Between Wire Pairs	Between Bl/Y or Bl/Bl Wire Pairs
1988	50 D	685301	48-52	0.2-0.7	800-1100	680-800	-	0.5V+	180V+	-	150V+	Between Wire Pairs	Between Bl/Y or Bl/Bl Wire Pairs
1989-92	50	685301	48-52	0.2-0.7	800-1100	680-800	-	0.5V+	180V+	-	150V+	Between Wire Pairs	Between Bl/Y or Bl/Bl Wire Pairs
1992-95	50	18495	650-1000	0.2-0.7	850-1200	75-180	2900-3650	4V+	25V+	180V+	150V+	B/W to Br, P	BL & R to Ground
1996-99	50	827509	N/A	N/A	850-1200	500-700	-	0.5V+	180V+	-	N/A	Ground to Br, W, P	G/W to W/G
1985	60	475301	48-52	0.2-0.7	120-160	680-800	-	0.5V+	180V+	-	150V+	Between Wire Pairs	Between Bl/Y or Bl/Bl Wire Pairs
1991-92	70	332-7778	1200-1350	0.2-0.7	850-1200	75-180	2900-3650	4V+	25V+	180V+	150V+	B/W to Br, P	BL & R to Ground
1993-95	70	18495	1200-1350	0.2-0.4	800-1100	75-90	3250-3650	4V+	25V+	180V+	150V+	W/B to All Other Wires	BL & R to Ground
1996-99	75	827509	N/A	N/A	850-1200	500-700	-	0.5V+	180V+	-	N/A	Ground to Br, W, P	G/W to W/G
1984-86	85 A & L	475301	48-52	0.2-0.7	120-160	680-800	-	0.5V+	180V+	-	150V+	Between Wire Pairs	Between Bl/Y or Bl/Bl Wire Pairs
1986	85 B & M	653301	48-52	0.2-0.7	120-160	680-800	-	0.5V+	180V+	-	150V+	Between Wire Pairs	Between Bl/Y or Bl/Bl Wire Pairs
1987-88	85	653301	48-52	0.2-0.7	120-160	680-800	-	0.5V+	180V+	-	150V+	Between Wire Pairs	Between Bl/Y or Bl/Bl Wire Pairs
1989	85	685301	48-52	0.2-0.7	800-1100	680-800	-	0.5V+	180V+	-	150V+	Between Wire Pairs	Between Bl/Y or Bl/Bl Wire Pairs
1989-90	85-90 L Drive	685301	48-52	0.2-0.7	800-1100	680-800	-	0.5V+	180V+	-	150V+	Between Wire Pairs	Between Bl/Y or Bl/Bl Wire Pairs
1991	90 A, C & E	332-7778	1200-1350	0.2-0.4	800-1100	75-90	3250-3650	4V+	25V+	180V+	150V+	W/B to All Other Wires	BL & R to Ground
1991	90 D & H	685301	48-52	0.2-0.7	800-1100	680-800	-	0.5V+	180V+	-	150V+	Between Wire Pairs	Between Bl/Y or Bl/Bl Wire Pairs
1991	90 D & H	332-7778	1200-1350	0.2-0.4	800-1100	75-90	3250-3650	4V+	25V+	180V+	150V+	W/B to All Other Wires	BL & R to Ground
1992	90 C	332-7778	1200-1350	0.2-0.4	800-1100	75-90	3250-3650	4V+	25V+	180V+	150V+	W/B to All Other Wires	BL & R to Ground
1992-95	90	18495	48-52	0.2-0.4	800-1100	75-90	3250-3650	4V+	25V+	180V+	150V+	W/B to All Other Wires	BL & R to Ground
1996-99	90	827509	N/A	N/A	850-1200	500-700	-	0.5V+	180V+	-	N/A	Ground to Br, W, P	G/W to W/G
1990-91	120 (1990A & 91C)	685301	48-52	0.2-0.7	800-1100	680-800	-	0.5V+	180V+	-	150V+	Between Wire Pairs	Between Bl/Y or Bl/Bl Wire Pairs
1991-95	120 (1991D & 92C)	332-6772	1200-1350	0.2-0.4	800-1100	75-90	3250-3650	4V+	25V+	180V+	150V+	W/B to Br, W to P	Bl to B/W & Red to Red/W
1996-99	120	827509	N/A	N/A	850-1200	500-700	-	0.5V+	180V+	-	N/A	Ground to Br, W, P	G/W to W/G
1990	120 L Drive	685301	48-52	0.2-0.7	800-1100	680-800	-	0.5V+	180V+	-	150V+	Between Wire Pairs	Between Bl/Y or Bl/Bl Wire Pairs
1991-92	120 L Drive	332-5772	1200-1350	0.2-0.4	800-1100	75-90	3250-3650	4V+	25V+	180V+	150V+	W/B to Br, W to P	Bl to B/W & Red to Red/W
1985-86	125 A	475301	48-52	0.2-0.7	120-160	680-800	-	0.5V+	180V+	-	150V+	Between Wire Pairs	Between Bl/Y or Bl/Bl Wire Pairs
1986	125 B	653301	48-52	0.2-0.7	120-160	680-800	-	0.5V+	180V+	-	150V+	Between Wire Pairs	Between Bl/Y or Bl/Bl Wire Pairs
1987-88	125 A & B	653301	48-52	0.2-0.7	120-160	680-800	-	0.5V+	180V+	-	150V+	Between Wire Pairs	Between Bl/Y or Bl/Bl Wire Pairs
1988	125 C	685301	48-52	0.2-0.7	800-1100	680-800	-	0.5V+	180V+	-	150V+	Between Wire Pairs	Between Bl/Y or Bl/Bl Wire Pairs
1989	125	685301	48-52	0.2-0.7	800-1100	680-800	-	0.5V+	180V+	-	150V+	Between Wire Pairs	Between Bl/Y or Bl/Bl Wire Pairs
1989-91	150 (1989A-91B)	685301	48-52	0.2-0.7	800-1100	680-800	-	0.5V+	180V+	-	150V+	Between Wire Pairs	Between Bl/Y or Bl/Bl Wire Pairs
1991-92	150 (1991B-92C)	817323A1	1200-1350	0.2-0.4	800-1100	12V Inverter	-	4V+	225V	-	150V+	W/B to All Other Wires	Bl Terminal to Ground

N/A - Not Applicable

- B - Black
- Bl - Blue
- Br - Brown
- G - Green
- P - Purple
- R - Red
- W - White
- Y - Yellow

IGNITION AND ELECTRICAL SYSTEMS

CHARGING CIRCUIT

Description and Operation

The battery stores electricity and acts as a sponge for the whole system. It mops up generated current until it's fully charged, and it releases energy on demand.

The flywheel holds the permanent magnets that create the moving magnetic field. If your engine has good spark, you can take it for granted that the magnets are in working order because the ignition and charging systems share the same magnets.

The stator windings are the stationary coils of wire the flywheel magnets rotate around. They produce the electrical charge. Simply put, the more windings in your stator, the greater the potential output in amps your charging system will have.

The rectifier consists of a series of diodes or electrical one-way valves. The rectifier overcomes one of the disadvantages of a current-generating system using permanent magnets and stator windings, which is that the current produced within the windings is alternating current (AC). You can't use AC to charge batteries. They accept only direct current (DC). So the rectifier is designed to convert AC current to a usable form of DC current simply called rectified AC.

On midsize and large outboard engines, there may be a voltage regulator, either combined with the rectifier or standing alone. The regulator automatically reduces the output of generated current as the battery becomes fully charged.

Troubleshooting the Charging System

The charging system should be inspected if:
- The charging system warning light is illuminated
- The voltmeter on the instrument panel indicates improper charging (either high or low) voltage
- The battery is overcharged (electrolyte level is low and/or boiling out)
- The battery is undercharged (insufficient power to crank the starter)

The starting point for all charging system problems begins with the inspection of the battery, related wiring and charging system components. The battery must be in good condition and fully charged before system testing.

Do a visual check of the battery, wiring and fuses. Are there any new additions to the wiring? An excellent clue might be, everything was working OK until I added that live well pump. With a comment like this you would know where to check first.

➡ **Check battery condition thoroughly. It is the #1 culprit in charging system failures.**

The regulator/rectifier assembly is the brains of the charging system. The regulator controls current flow in the charging system. If battery voltage is below about 14.6 volts the regulator sends the available current to the battery. If the battery is fully charged (about 14.5 to 15 volts) the regulator diverts the current/amps to ground.

Do not expect the regulator to send current to a fully charged battery. Check the battery for a possible draw with the key off. This draw may be the cumulative effect of several radio and/or clock memories. If these accessories are wired to the battery then a complaint of charging system failure may really be excessive draw. Draw in excess of 25 milliamps should arouse your suspicions.

➡ **Do not forget to check through the fuses. It can be embarrassing to overlook a blown fuse.**

You must pull the battery voltage down below 12.5 volts to test charging system output. Running the power trim and tilt will reduce the battery voltage. Once the battery's good condition is verified and it has been reduced to below 12.5 volts you can test further.

Install an ammeter to check actual amperage output. Verify that the system is delivering sufficient amperage. Too much amperage and a battery that goes dry very quickly indicates that the rectifier/regulator should be replaced.

If the system does not put out enough amperage, then test the lighting coil. Isolate the coil and test for correct resistance and short to ground.

During these test procedures the regulator/rectifier has not been bench checked. Usually it is advisable to avoid troubleshooting the regulator/rectifier directly. The procedures listed so far have focused on checking around the rectifier/regulator. If you verify that all other systems stator are good then what is left in the system to cause the verified problem? The process of elimination has declared the rectifier/regulator bad.

Stator

The wires for the battery charging system in the stator are generally either Yellow or Green/Yellow. The function of the stator coil is to produce AC current for battery charging.

TESTING

▶ See Figures 156, 157 and 158

1. Disconnect the battery
2. The stator may be tested without removing the flywheel, by merely disconnecting the output leads and using a multimeter to check the resistance.
3. Check resistance between the two stator wires.
4. Resistance should be as specified in the "Stator Alternator Resistance Specification" chart.
5. If the stator is installed on the powerhead, check continuity between each stator wire and ground.
6. If the stator is not installed on the powerhead, check continuity between each stator wire and the metal stator ring.
7. If resistance is not within specification, or continuity exists, the stator may be faulty.

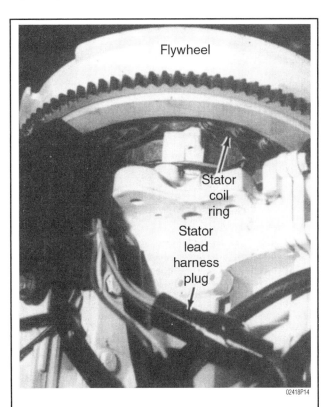

Fig. 156 The stator coils may be tested without removing the flywheel. The leads may be disconnected at the harness plug and tested for continuity.

5-46 IGNITION AND ELECTRICAL SYSTEMS

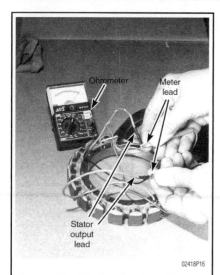

Fig. 157 Bench testing the stator alternator—Prestolite System and earlier

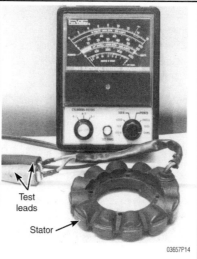

Fig. 158 Bench testing the stator alternator—Thunderbolt System

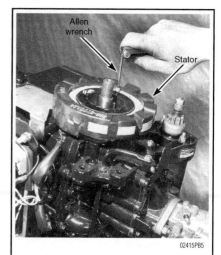

Fig. 159 Using an Allen wrench remove the screws securing the stator to the powerhead

Stator Alternator Resistance

Horsepower	Year	Output	Resistance	Regulated System
3	1990-94	—	—	—
4	1987-87	—	—	—
5	1987-98	—	—	—
7.5	1985	3.5 AMP	—	N
9.9	1984-98	—	—	—
15	1984-98	—	—	—
25	1994-98	—	—	—
35	1986-91	2.0 AMP	1.0-2.0	N
40	1992-94	9.0 AMP	0.8-1.1	N
40	1995-99	9.0 AMP	0.15-0.20	Y
50	1984-87	6.0 AMP	—	N
50	1988-92	7.5 AMP	0.5-1.0	N
50	1995-99	9.0 AMP	0.15-0.20	Y
60	1985	7.0 AMP	1	N
70	1991-99	9.0 AMP	0.6-1.1	Y
70	1991-99	14.0 AMP	0.1-0.5	Y
75	1994-99	14.0 AMP	0.22-0.24	Y
85	1984-91	7.0 AMP	0.5-1.0	N
90	1990-92	7.0 AMP	0.5-1.1	N
90	1992-94	9.0 AMP	0.5-1.1	Y
90	1992-94	14.0 AMP	0.1-0.5	Y
90	1995-99	14.0 AMP	0.15-0.20	Y
120	1990-92	7.0 AMP	0.5-1.0	Y
120	1993-94	9.0 AMP	0.6-1.1	Y
120	1995-99	14.0 AMP	0.15-0.20	Y
125	1984-89	7.0 AMP	0.5-1.0	N
125	1989	7.0 AMP	0.5-1.0	N
150	1989-90C	11.0 AMP	0.1-0.19	Y
150	1989-92	7.0 AMP	0.1-0.19	N
150	1993-94	16.0 AMP	0.17-0.19	Y

REMOVAL & INSTALLATION

♦ See Figure 159

1. Remove the flywheel.
2. Using an Allen wrench remove the screws securing the stator to the powerhead.
3. Lift the stator up and clear of the crankshaft.
4. If the stator is to be replaced, disconnect the wire harness.
5. Label and disconnect the charge coil wires to the switch box.
6. Label and disconnect the stator charge coil wires from the rectifier/regulator.

To install:
7. Connect the stator charge coil wires to the rectifier/regulator.
8. Connect the charge coil wires to the switch box.
9. Connect the wiring harness.
10. Position the stator on the powerhead.
11. Using an Allen wrench install the screws securing the stator to the powerhead.

➡ **Be certain that the flywheel key is in place before placing the flywheel over the crankshaft.**

12. Install the flywheel.
13. Hold the flywheel and torque the nut to proper specification.

Rectifier

♦ See Figures 160 and 161

The rectifier is the component that changes AC current to DC current by removing half of the AC wave. This is accomplished by the use of diodes that are electrical one way switches.

Rectifiers are used on engines with no regulation on the charging system.

TESTING

1. Disconnect the battery.
2. Disconnect the leads from the rectifier. Label for identification when reassembled.
3. Connect the Red meter lead to the positive or output terminal of the rectifier.
4. Connect the Black lead the one of the AC terminals on the rectifier where one of the stator leads was connected.
5. Switch the Red and Black leads.
6. There should have been continuity in one test and none on the other.
7. Repeat this same test using the other AC terminal.
8. Connect the Red meter lead to the negative terminal of the rectifier.
9. Connect the Black lead to one of the AC terminals.
10. Switch the Red and Black leads.
11. There should have been continuity in one test and none on the other.
12. Repeat this same test using the other AC terminal.
13. The four rectifier diodes have now been tested. In any test if there was

IGNITION AND ELECTRICAL SYSTEMS

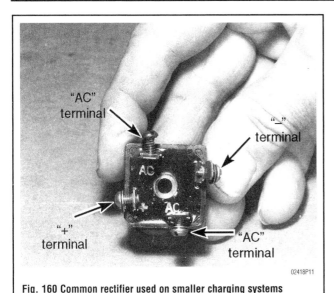

Fig. 160 Common rectifier used on smaller charging systems

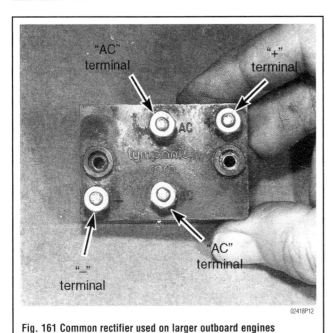

Fig. 161 Common rectifier used on larger outboard engines

an open circuit or closed or shorted circuit in both directions of the test the rectifier is bad and must be replaced.

14. Install the wires to the proper terminals and connect the battery.

REMOVAL & INSTALLATION

1. The rectifier systems only have four terminal connections and are held on by one or two fasteners.
2. Disconnect the battery.
3. Remove and label the wire connections.
4. Remove the fasteners holding the rectifier.

To install:

5. Install the replacement rectifier with the appropriate fasteners.

➡ Do not over tighten the fasteners.

6. Install the wires to their proper terminals.
7. Connect the battery.

Rectifier/Regulator

◆ See Figures 162, 163 and 164

TESTING

Resistance Tests

EARLY REGULATED SYSTEMS

1. Using a multimeter set to the Rx1000 scale check the condition of the rectifier diodes.
2. Disconnect the battery.
3. Disconnect the leads from the rectifier. Label for identification when reassembled.
4. Connect the Red meter lead to the positive or output terminal of the rectifier.
5. Connect the Black lead the one of the AC terminals on the rectifier where one of the stator leads was connected.
6. Switch the Red and Black leads.
7. There should have been continuity in one test and none on the other.
8. Repeat this same test using the other AC terminal.
9. Connect the Red meter lead to the negative terminal of the rectifier.
10. Connect the Black lead to one of the AC terminals.
11. Switch the Red and Black leads.
12. There should have been continuity in one test and none on the other.
13. Repeat this same test using the other AC terminal.
14. The four rectifier diodes have now been tested. If there was an open circuit or closed or shorted circuit in both directions of the test the rectifier is bad and must be replaced.
15. Install the wires to there proper terminals and connect the battery.

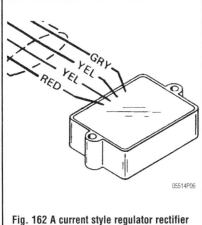

Fig. 162 A current style regulator rectifier used with the thunderbolt and CDM ignition systems

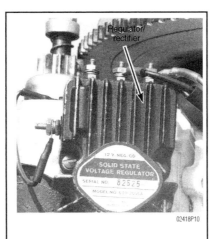

Fig. 163 A regulator/rectifier installed on a late model 4-cylinder powerhead

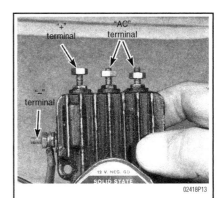

Fig. 164 Identification of the three different design regulator/rectifiers used by the manufacturer. The "+", "-", and two "AC" terminals are identified to assist during testing which begins on the following page

5-48 IGNITION AND ELECTRICAL SYSTEMS

LATE MODEL SYSTEMS

♦ See Figure 165

Running Output Test

1. Connect the voltmeter positive (+) lead to the positive terminal on the battery. Connect the negative (-) meter lead to the negative terminal on the battery.
2. Crank the powerhead with the starter motor while monitoring the voltmeter. If the battery voltage drops below 9.5 volts while cranking, the battery is weak and should be recharged.
3. If the cranking voltage is acceptable, start the powerhead and let it idle. Battery voltage at idle speed should be 12.5 to 14.5 volts. If the voltage is over 14.5 volts, replace the voltage regulator/rectifier.
4. If the voltage at idle speed is acceptable shut the powerhead down and disconnect the battery. Disconnect the Red wire from the center terminal on the voltage regulator/rectifier. Secure the remaining lead on the center terminal with the hex nut previously removed.
5. Connect the positive (+) lead of the ammeter to the center terminal of the voltage regulator/rectifier. Connect the negative (-) meter lead of the ammeter to the Red wire previously removed from the voltage regulator. Secure all loose wires clear of the flywheel.
6. Connect the battery and start the powerhead.
7. Run the engine with an appropriate test propeller or under a load in the water.
8. Slowly advance the throttle and verify the alternator amps output.

REMOVAL & INSTALLATION

1. Disconnect the battery.
2. Remove the leads to the regulator rectifier and mark them as necessary.
3. Remove the fasteners holding the unit in place.

To install:

4. Install the replacement regulator rectifier with the appropriate fasteners.

➡ Do not over tighten the fasteners.

5. Connect the wires to the proper terminals or connections
6. Connect the battery.

Battery

The battery is one of the most important parts of the electrical system. In addition to providing electrical power to start the engine, it also provides power for operation of the running lights, radio and electrical accessories.

Because of its job and the consequences (failure to perform in an emergency), the best advice is to purchase a well-known brand, with an extended warranty period, from a reputable dealer.

The usual warranty covers a pro-rated replacement policy, which means the purchaser is entitled to consideration for the time left on the warranty period if the battery should prove defective before the end of the warranty.

Many manufacturers have specifications on the size and type of battery to use for their engines. If in doubt as to how large the boat requires, make a liberal estimate and then purchase the one with the next higher amp rating.

BATTERY CONSTRUCTION

A battery consists of a number of positive and negative plates immersed in a solution of diluted sulfuric acid. The plates contain dissimilar active materials and are kept apart by separators. The plates are grouped into elements. Plate straps on top of each element connect all of the positive plates and all of the negative plates into groups.

The battery is divided into cells holding a number of the elements apart from the others. The entire arrangement is contained within a hard plastic case. The top is a one-piece cover and contains the filler caps for each cell. The terminal posts protrude through the top where the battery connections for the boat are made. Each of the cells is connected to its neighbor in a positive-to-negative manner with a heavy strap called the cell connector.

MARINE BATTERIES

♦ See Figure 166

Because marine batteries are required to perform under much more rigorous conditions than automotive batteries, they are constructed differently than those used in automobiles or trucks. Therefore, a marine battery should always be the No. 1 unit for the boat and other types of batteries used only in an emergency.

Marine batteries have a much heavier exterior case to withstand the violent pounding and shocks imposed on it as the boat moves through rough water and in extremely tight turns. The plates are thicker and each plate is securely anchored within the battery case to ensure extended life. The caps are spill proof to prevent acid from spilling into the bilge when the boat heels to one side in a tight turn or is moving through rough water. Because of these features, the marine battery will recover from a low charge condition and give satisfactory service over a much longer period of time than any type intended for automotive use.

✳✳ WARNING

Never use a Maintenance-free battery with an outboard engine that is not voltage regulated. The charging system will continue to charge as long as the engine is running and it is possible that the electrolyte could boil out if periodic checks of the cell electrolyte level are not done.

SMALL VOLTAGE REGULATOR TEST (STATIC)

TEST LEADS	OHMS SCALE	METER READING (Ohms)
DIODE CHECK: Connect NEGATIVE (-) test lead to either YELLOW lead. Connect POSITIVE (+) test lead to thick RED lead.	R x 10	100 - 400
DIODE CHECK: Connect NEGATIVE (-) test lead to thick RED lead. Connect POSITIVE (+) test lead to either* YELLOW lead. Note 1	R x 1K	40K to Infinity
SCR CHECKS: Connect NEGATIVE (-) test lead to either YELLOW lead. Connect POSITIVE (+) test lead to case ground.	R x 1K	10K to Infinity
TACHOMETER CIRCUIT CHECK: Connect NEGATIVE(-) test lead to case ground. Connect POSITIVE (+) test lead to GRAY lead.	R x 1K	10K to 30K

Fig. 165 Chart showing the appropriate readings for a resistance test on the newer style thunder bolt or CDM regulated systems

IGNITION AND ELECTRICAL SYSTEMS 5-49

Fig. 166 Cut-away look at a typical marine battery

BATTERY RATINGS

♦ See Figure 167

Three different methods are used to measure and indicate battery electrical capacity:
- Amp/hour rating
- Cold cranking performance
- Reserve capacity

The amp/hour rating of a battery refers to the battery's ability to provide a set amount of amps for a given amount of time under test conditions at a constant temperature. Therefore, if the battery is capable of supplying 4 amps of current for 20 consecutive hours, the battery is rated as an 80 amp/hour battery. The amp/hour rating is useful for some service operations, such as slow charging or battery testing.

Cold cranking performance is measured by cooling a fully charged battery to 0°F (-17°C) and then testing it for 30 seconds to determine the maximum current flow. In this manner the cold cranking amp rating is the number of amps available to be drawn from the battery before the voltage drops below 7.2 volts.

The illustration depicts the amount of power in watts available from a battery at different temperatures and the amount of power in watts required of the engine at the same temperature. It becomes quite obvious—the colder the climate, the more necessary for the battery to be fully charged.

Reserve capacity of a battery is considered the length of time, in minutes, at 80°F (27°C), a 25 amp current can be maintained before the voltage drops below 10.5 volts. This test is intended to provide an approximation of how long the engine, including electrical accessories, could operate satisfactorily if the stator assembly or lighting coil did not produce sufficient current. A typical rating is 100 minutes.

➡If possible, the new battery should have a power rating equal to or higher than the unit it is replacing.

BATTERY LOCATION

♦ See Figure 168

Every battery installed in a boat must be secured in a well protected, ventilated area. If the battery area lacks adequate ventilation, hydrogen gas, which is given off during charging, is very explosive. This is especially true if the gas is concentrated and confined.

Fig. 168 A good example of a secured, well protected and ventilated battery area

BATTERY CHARGERS

♦ See Figure 169

Before using any battery charger, consult the manufacturer's instructions for its use. Battery chargers are electrical devices that change Alternating Current (AC) to a lower voltage of Direct Current (DC) that can be used to charge a marine battery. There are two types of battery chargers—manual and automatic.

A manual battery charger must be physically disconnected when the battery has come to a full charge. If not, the battery can be overcharged and possibly fail. Excess charging current at the end of the charging cycle will heat the electrolyte, resulting in loss of water and active material, substantially reducing battery life.

➡As a rule, on manual chargers, when the ammeter on the charger registers half the rated amperage of the charger, the battery is fully charged. This can vary and it is recommended to use a hydrometer to accurately measure state of charge.

Automatic battery chargers have an important advantage—they can be left connected (for instance, overnight) without the possibility of overcharging the battery. Automatic chargers are equipped with a sensing device to allow the bat-

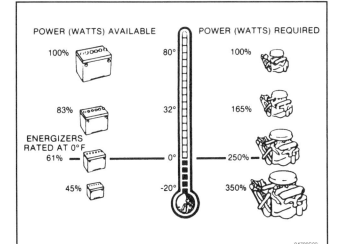

Fig. 167 Comparison of battery efficiency and engine demands at various temperatures

5-50 IGNITION AND ELECTRICAL SYSTEMS

Fig. 169 Automatic chargers, such as the Battery Tender® from Deltran, are equipped with a sensing device to allow the battery charge to taper off to near zero as the battery becomes fully charged

tery charge to taper off to near zero as the battery becomes fully charged. When charging a low or completely discharged battery, the meter will read close to full rated output. If only partially discharged, the initial reading may be less than full rated output, as the charger responds to the condition of the battery. As the battery continues to charge, the sensing device monitors the state of charge and reduces the charging rate. As the rate of charge tapers to zero amps, the charger will continue to supply a few milliamps of current—just enough to maintain a charged condition.

BATTERY CABLES

Battery cables don't go bad very often but like anything else, they can wear out. If the cables on your boat are cracked, frayed or broken, they should be replaced.

When working on any electrical component, it is always a good idea to disconnect the negative (-) battery cable. This will prevent potential damage to many sensitive electrical components

Always replace the battery cables with one of the same length or you will increase resistance and possibly cause hard starting. Smear the battery posts with a light film of dielectric grease or a battery terminal protectant spray once you've installed the new cables. If you replace the cables one at a time, you won't mix them up.

✽✽ WARNING

The battery cables equipped with the engine are of the appropriate size and gauge. In the event the cables need to be longer the wire gauge would most likely need to increase to compensate for the additional resistance in the cable due to the length. Cables that are too small for the application will cause premature starter motor failure and a heavier load on the systems charging system.

➡ Any time you disconnect the battery cables, it is recommended that you disconnect the negative (-) battery cable first. This will prevent you from accidentally grounding the positive (+) terminal when disconnecting it, thereby preventing damage to the electrical system.

Before you disconnect the cable(s), first turn the ignition to the **OFF** position. This will prevent a draw on the battery that could cause arcing. When the battery cable(s) are reconnected (negative cable last), be sure to check all electrical accessories are all working correctly.

STARTING CIRCUIT

Description and Operation

▶ See Figures 170

In the early days, all outboard engines were started by simply pulling on a rope wound around the flywheel. As time passed and owners were reluctant to use muscle power, it was necessary to replace the rope starter with some form of power cranking system. Today, many small engines are still started by pulling on a rope but others have a powered starter motor installed.

The system utilized to replace the rope method was an electric starter motor coupled with a mechanical gear mesh between the starter motor and the powerhead flywheel, similar to the method used to crank an automobile engine.

As the name implies, the sole purpose of the starter motor circuit is to control operation of the starter motor to crank the powerhead until the engine is operating. The circuit includes a relay or magnetic switch to connect or disconnect the motor from the battery. The operator controls the switch with a key switch.

A neutral safety switch is installed into the circuit to permit operation of the starter motor only if the shift control lever is in neutral. This switch is a safety device to prevent accidental engine start when the engine is in gear.

The starter motor is a series wound electric motor which draws a heavy current from the battery. It is designed to be used only for short periods of time to crank the engine for starting. To prevent overheating the motor, cranking should not be continued for more than 30-seconds without allowing the motor to cool for at least three minutes. Actually, this time can be spent in making preliminary checks to determine why the engine fails to start.

Power is transmitted from the starter motor to the powerhead flywheel through a Bendix drive. This drive has a pinion gear mounted on screw threads. When the motor is operated, the pinion gear moves upward and meshes with the teeth on the flywheel ring gear.

When the powerhead starts, the pinion gear is driven faster than the shaft and as a result, it screws out of mesh with the flywheel. A rubber cushion is built into the Bendix drive to absorb the shock when the pinion meshes with the flywheel ring gear. The parts of the drive must be properly assembled for efficient operation. If the drive is removed for cleaning, take care to assemble the parts as illustrated in the accompanying illustrations in this section. If the screw shaft assembly is reversed, it will strike the splines and the rubber cushion will not absorb the shock.

The sound of the motor during cranking is a good indication of whether the starter motor is operating properly or not. Naturally, temperature conditions will

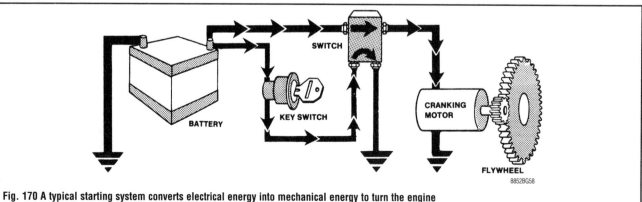

Fig. 170 A typical starting system converts electrical energy into mechanical energy to turn the engine

IGNITION AND ELECTRICAL SYSTEMS 5-51

affect the speed at which the starter motor is able to crank the engine. The speed of cranking a cold engine will be much slower than when cranking a warm engine. An experienced operator will learn to recognize the favorable sounds of the powerhead cranking under various conditions.

The job of the starter motor relay is to complete the circuit between the battery and starter motor. It does this by closing the starter circuit electromagnetically, when activated by the key switch. This is a completely sealed switch, which meets SAE standards for marine applications. DO NOT substitute an automotive-type relay for this application. It is not sealed and gasoline fumes can be ignited upon starting the powerhead. The relay consists of a coil winding, plunger, return spring, contact disc and four externally mounted terminals. The relay is installed in series with the positive battery cables mounted to the two larger terminals. The smaller terminals connect to the neutral switch and ground.

To activate the relay, the shift lever is placed in neutral, closing the neutral switch. Electricity coming through the ignition switch goes into the relay coil winding which creates a magnetic field. The electricity then goes on to ground in the powerhead. The magnetic field surrounds the plunger in the relay, which draws the disc contact into the two larger terminals. Upon contact of the terminals, the heavy amperage circuit to the starter motor is closed and activates the starter motor. When the key switch is released, the magnetic field is no longer supported and the magnetic field collapses. The return spring working on the plunger opens the disc contact, opening the circuit to the starter.

When the armature plate is out of position or the shift lever is moved into forward or reverse gear, the neutral switch is placed in the open position and the starter control circuit cannot be activated. This prevents the powerhead from starting while in gear.

Troubleshooting the Starting System

♦ See Figure 171

If the starter motor spins but fails to crank the engine, the cause is usually a corroded or gummy Bendix drive. The drive should be removed, cleaned and given an inspection.

1. Before wasting too much time troubleshooting the starter motor circuit, the following checks should be made. Many times, the problem will be corrected.
   - Battery fully charged.
   - Shift control lever in neutral.
   - Main 20-amp fuse located at the base of the fuse cover is good (not blown).
   - Check that all electrical connections clean and tight.
   - Wiring in good condition, insulation not worn or frayed.
2. Starter motor cranks slowly or not at all.
   - Faulty wiring connection
   - Short-circuited lead wire
   - Shift control not engaging neutral (not activating neutral start switch)
   - Defective neutral start switch
   - Starter motor not properly grounded
   - Faulty contact point inside ignition switch
   - Bad connections on negative battery cable to ground (at battery side and engine side)
   - Bad connections on positive battery cable to magnetic switch terminal
   - Open circuit in the coil of the magnetic switch (relay)
   - Bad or run-down battery
   - Excessively worn down starting motor brushes
   - Burnt commutator in starting motor
   - Brush spring tension slack
   - Short circuit in starter motor armature
3. Starter motor keeps running.
   - Melted contact plate inside the magnetic switch
   - Poor ignition switch return action
4. Starter motor picks up speed, put pinion will not mesh with ring gear.
   - Worn down teeth on clutch pinion
   - Worn down teeth on flywheel ring gear
5. Two more areas may cause the powerhead to crank slowly even though the starter motor circuit is in excellent condition
   - A tight or frozen powerhead
   - Water in the lower unit.

Starter Motor

♦ See Figure 172

Marine cranking motors are very similar in construction and operation to the units used in the automotive industry.

All marine cranking motors use the inertia type drive assembly. This type assembly is mounted on an armature shaft with external spiral splines which mate with the internal splines of the drive assembly.

Never operate a cranking motor for more than 30 seconds without allowing it to cool for at least three minutes. Continuous operation without the cooling period can cause serious damage to the cranking motor.

One type cranking motor is used on the powerheads covered in this manual. The exterior appearance of the cranking motor may vary. Therefore, the accompanying illustrations may differ slightly from the unit being serviced, but the procedures and maintenance instructions are valid for all cranking motors on all powerheads covered in this manual.

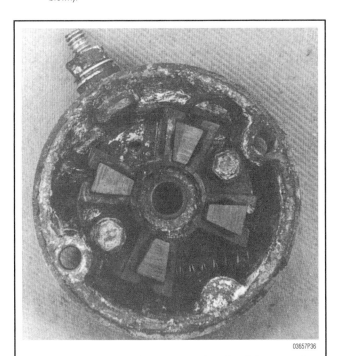

Fig. 171 Corroded or burned brushes greatly reduced the power of the starter

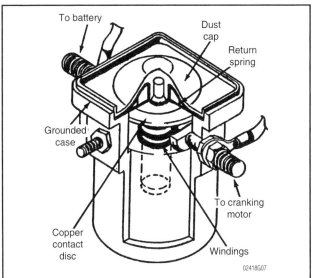

Fig. 172 A cutaway drawing of a cranking motor solenoid with major parts identified.

5-52 IGNITION AND ELECTRICAL SYSTEMS

REMOVAL & INSTALLATION

♦ See Figures 173 thru 178

> ※※ **CAUTION**
>
> Before beginning any work on the cranking motor, disconnect the positive (+) lead from the battery terminal

> ※※ **WARNING**
>
> The mounting bolts that run vertically are the same bolts that hold the starter together. Use care once the starter is loose it comes out as an assembly.

1. Disconnect the large Red cable from the cranking motor relay to the cranking motor.
2. Remove the retainer bracket at the bottom of the starter.
3. Remove the bolts securing the cranking motor to the powerhead. Lift the motor free.

To install:

Once assembled, rotate the armature shaft inside the bushing and check to be sure there is no binding and the brushes sweep across the commutator smoothly.

Slide the free end of the armature shaft into the motor field frame. Take care not to loose the brushes from the commutator. Seat the lower end cap onto the field frame with the marks made prior to disassembly aligned. Install the thrust washer over the armature shaft and then install the upper end cap over the shaft, again with marks made during disassembly aligned.

Hold the assembly together and insert the two thru bolts through the three pieces. If the bolts were not used to mount the cranking motor to the powerhead, install and tighten the retaining nuts.

4. Slide the pinion gear onto the armature shaft, followed by the spring and the spring collar. Start a new nut onto the armature shaft. The nut used is a special locking nut and cannot be used a second time.
5. Place a wrench on the bottom side of the pinion gear and a second wrench on the nut. Hold firm with the first wrench and tighten the nut with the second.
6. Position the cranking motor in place in the mounting bracket. Install and tighten the nuts on the thru-bolts or the bolts securely.

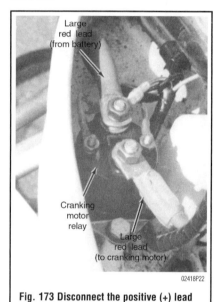

Fig. 173 Disconnect the positive (+) lead from the battery terminal

Fig. 174 Remove the bolts securing the cranking motor to the powerhead

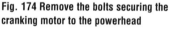

Fig. 175 Slide the pinion gear onto the armature shaft along with the remaining hardware as shown

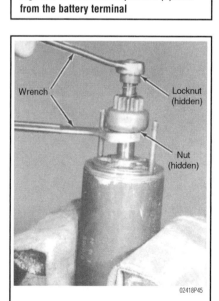

Fig. 176 To tighten the nut place a wrench on the bottom side of the pinion gear and a second wrench on the nut

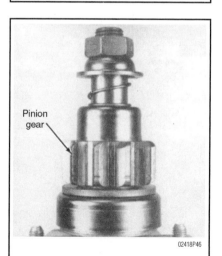

Fig. 177 Typical Bendix spring arrangement on a cranking motor. A small amount of oil on the shaft in the spring area will prolong satisfactory operation

Fig. 178 With the starter in place thread the nuts onto the studs to secure it into place

IGNITION AND ELECTRICAL SYSTEMS 5-53

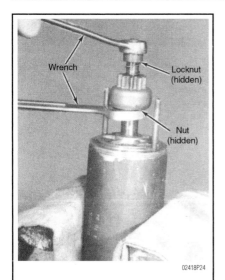

Fig. 179 Move the pinion gear upward and place a wrench on the bottom side of the gear

Fig. 180 Remove the spring collar, spring, and pinion gear

Fig. 181 Observe the caps and find the identifying mark or boss on each indicating their position in relation to the frame

Connect the large Red cable from the battery to the positive terminal on the motor.

7. Mount the outboard unit in a test tank, on the boat in a body of water, or connect a flush attachment and hose to the power unit.

8. Crank the powerhead with the cranking motor and start the unit. Shut the powerhead down and start it several times to check operation of the cranking motor.

DISASSEMBLY

Units With Brushes Within The Frame "Prestolite®"

♦ See Figures 179, 180, 181, 182 and 183

1. Move the pinion gear upward and place a wrench on the bottom side of the gear.
2. Place a second wrench on the top nut while holding both wrenches remove the top nut from the shaft.
3. Remove the spring collar, spring, and pinion gear.
4. Observe the caps and find the identifying mark or boss on each indicating their position in relation to the frame.

➡If the marks are not visible, make a set of identifying marks, one on the cap and a matching mark on the frame, prior to removing the through-bolts. These marks will prove to be an essential aid during assembling.

5. Remove the through-bolts. On some models, the thru-bolts thread into the opposite cap, and on other models, a nut is used.
6. Remove the upper end cap and the thrust washer on the armature shaft.
7. Separate the lower cap from the field frame enough to determine how the brushes are mounted.
8. The brushes are mounted in the motor field frame, remove the lower end cap.
9. Remove the thrust washer and wavy washer.
10. Pull on the armature shaft from the drive gear end and remove the armature from the field frame assembly units with the brushes mounted in the frame.
11. Remove the armature from the frame by holding the frame and pulling the armature out.
12. Remove the old set of ground brushes by cutting off the rivets with a chisel or by drilling them out.
13. The brush spring is removed from the holder by compressing one side of the spring with a small screwdriver until the spring flips out of its seat.
14. After the spring pops out turn the spring clockwise until it is free of the holder.

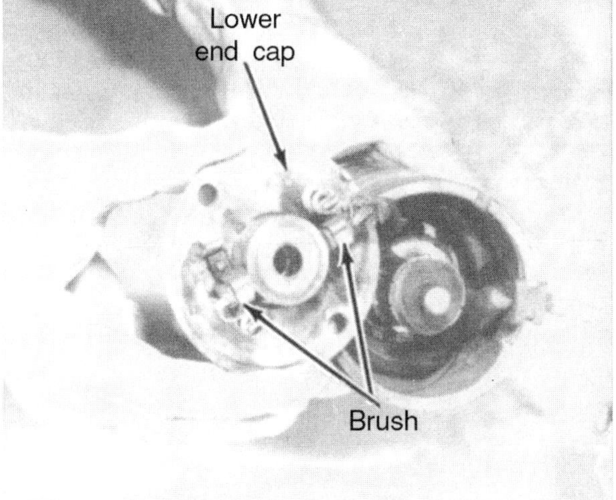

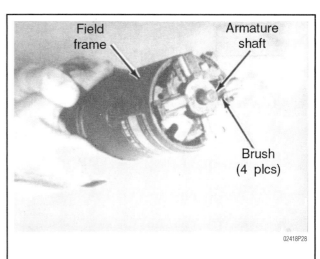

Fig. 182 Separate the lower cap from the field frame enough to determine how the brushes are mounted

Fig. 183 Pull on the armature shaft from the drive gear end and remove the armature from the field frame assembly

5-54 IGNITION AND ELECTRICAL SYSTEMS

Units With Brushes In End Cap "Bosch®"

1. Move the pinion gear upward and place a wrench on the bottom side of the gear.
2. Place a second wrench on the top nut while holding both wrenches remove the top nut from the shaft..
3. Remove the spring collar, spring, and pinion gear.
4. Observe the caps and find the identifying mark or boss on each indicating their position in relation to the frame.

➡️ **If the marks are not visible, make a set of identifying marks, one on the cap and a matching mark on the frame, prior to removing the through-bolts. These marks will prove to be an essential aid during assembling.**

5. Remove the through-bolts. On some models, the thru-bolts thread into the opposite cap, and on other models, a nut is used.
6. Remove the upper end cap and the thrust washer on the armature shaft.
7. Separate the lower cap from the field frame enough to determine how the brushes are mounted.
8. The brushes are mounted in the lower end cap so use caution during removal because the positive brush will be attached to the field frame at the positive terminal.

CLEANING & INSPECTION

▶ See Figures 184, 185, 186 and 187

Inspect the pinion gear teeth for chips, cracks, or a broken tooth. Check the splines inside the pinion gear for burrs and to be sure the gear moves freely on the armature shaft.

Clean the armature shaft and check to be sure that the shaft is free of any burrs. If burrs are discovered, they may be removed with crocus cloth.

Clean the field coils, armature, commutator, armature shaft, brush-end plate and drive-end housing with a brush or compressed air. Wash all other parts in solvent and blow them dry with compressed air.

Inspect the insulation and the unsoldered connections of the armature windings for breaks or burns.

Check the springs in the brush holder to be sure none are broken. Check the spring tension and replace if the tension is not 32–40 ounces (900–1135 gm). Check the insulated brush holders for shorts to ground. If the brushes are worn down to ¼" (6.35mm) or less, they must be replaced.

The lower armature shaft rides in a special bronze bushing, prelubricated at the factory. Some evidence of lubricant must be present when the cranking motor is disassembled.

Normally, it is not necessary to add lubricant to this bushing, but if the bush-

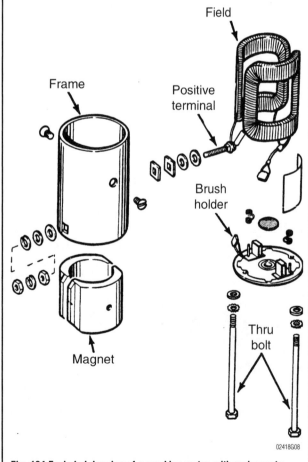

Fig. 184 Exploded drawing of a cranking motor, with major parts identified. The brushes on this motor are mounted inside the end cap

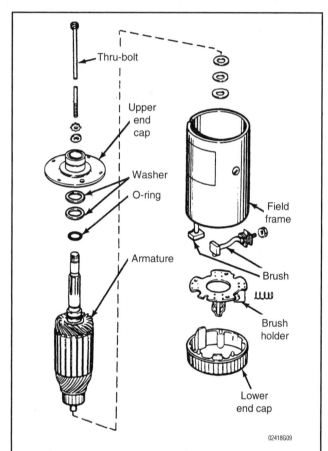

Fig. 185 Exploded diagram of a different cranking motor than the one shown at left. The brushes of this motor are mounted inside the field frame. Other major parts are identified

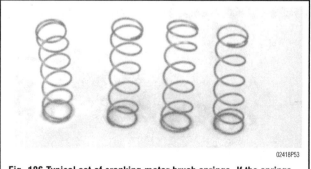

Fig. 186 Typical set of cranking motor brush springs. If the springs have turned blue in color, they must be replaced

IGNITION AND ELECTRICAL SYSTEMS 5-55

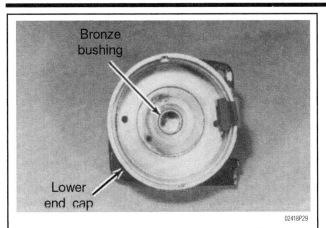

Fig. 187 The lower armature shaft rides in a bronze bushing at the base of the lower end cap

ing is completely dry, fill the base of the bushing with approximately 1/8" (3mm) of Multi Purpose Marine Lubricant. Do not use more lubricant than the 1/8" (3mm) because excessive lubricant would be squeezed out onto the commutator and brushes.

TESTING

♦ See Figures 188, 189, 190 and 191

→Most marine shops and all electrical motor rebuild shops will test an armature for a modest charge. If the armature has a short, it must be replaced.

1. Check the armature for a short circuit by placing it on a growler and holding a hack saw blade over the armature core while the armature is rotated. If the saw blade vibrates, the armature is shorted. Clean between the armature bars, and then check again on the growler. If the saw blade still vibrates, the armature must be replaced. Occasionally carbon dust from the brushes will

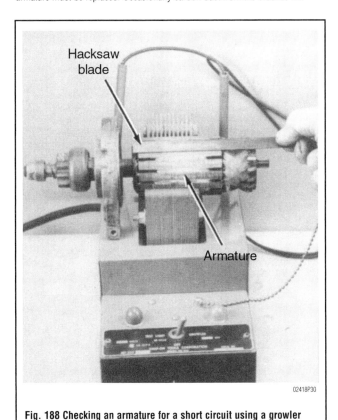

Fig. 188 Checking an armature for a short circuit using a growler

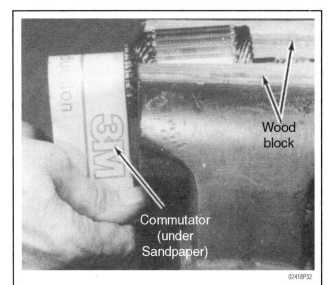

Fig. 189 Testing for continuity between the armature core or shaft and the commutator

Fig. 190 Sanding the commutator to remove burrs following under-cutting

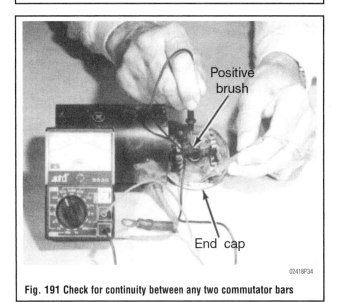

Fig. 191 Check for continuity between any two commutator bars

5-56 IGNITION AND ELECTRICAL SYSTEMS

short the armature. Therefore, blow the slots in the armature clean with compressed air.

2. Make contact with one probe of an ohmmeter on the armature core or shaft. Make contact with the other probe on the commutator. If the meter fails to indicate continuity, the armature is grounded and must be replaced.

3. Check the commutator for run out. Inspect the armature shaft and both bearings for scoring.

4. Turn the commutator in a lathe if it is out-of-round by more than 0.005" (0.13mm).

Never undercut the mica because the brushes are harder than the insulation. Undercut the insulation between the commutator bars 1/32" (0.80mm) to the full width of the insulation and flat at the bottom. A triangular groove is not satisfactory. After the undercutting work is completed, clean out the slots carefully to remove dirt and copper dust. Sand the commutator lightly with No. 600 sandpaper to remove any burrs left from the undercutting.

5. Make contact with the ohmmeter leads to any two commutator bars. The ohm meter should indicate continuity. Check the armature a second time on the growler for possible short circuits.

Brush Testing Bosch®

♦ See Figures 192, 193, 194 and 195

1. Test the brushes by connecting one meter lead to one positive brush and the other meter lead to the second positive brush. The ohmmeter must indicate continuity between the brushes. If the meter indicates any resistance, check the lead to the brush and the lead to the positive terminal solder connection. If the connection cannot be repaired, the brushes must be replaced.

2. Connect a meter lead to one of the positive brushes, and make contact with the other meter lead to the end cap. The ohmmeter must register no continuity.

3. Repeat this test for the other positive brush. If the meter registers continuity in either test, the positive brush is shorted to the end cap. Inspect the connections and repair or replace as necessary.

4. Slide the positive brush/es from the end cap. Make contact with one meter lead to the motor frame on an unpainted surface. Make contact with the other meter lead to the positive brush. The meter should register no continuity. If the meter registers continuity, the positive brush is shorted to the motor frame. Inspect the connections and repair or replace as necessary.

5. Make contact with one meter lead to one of the negative brushes. Make contact with the other meter lead to the end cap. The meter should indicate continuity. If the meter registers no continuity, the negative brush has an "open" in the circuit. Inspect the connections and repair or replace as necessary.

6. Make contact with a meter lead to one of the positive insulated brushes. Make contact with the other meter lead to the positive terminal on the exterior of the motor housing. The meter should register continuity. Repeat the test for the other positive brush. If the meter registers no continuity in either test, the positive brush has an open in the circuit. Inspect the connections and repair or replace as necessary.

7. Make contact with a meter lead to an unpainted portion of the motor frame. Make contact with the other meter lead to the positive terminal on the exterior of the motor housing. The meter should register no continuity. If the meter registers continuity, the terminal stud, or one or both of the positive brushes is grounded to the motor frame. Inspect the connections and repair or replace as necessary.

Brush Testing Prestolite®

♦ See Figure 196

1. Make contact with one meter lead to one of the negative brushes. Make contact with the other meter lead to the end cap. The meter should indicate continuity. If the meter registers no continuity, the negative brush has an open in the circuit. Inspect the connections and repair or replace as necessary.

2. Make contact with a meter lead to one of the negative grounded brushes. Make contact with the other meter lead to the positive terminal on the exterior of the motor frame. The meter should register no continuity. If the meter registers continuity, the terminal stud, or one or both of the positive brushes is grounded to the motor frame. Inspect the connections and repair or replace as necessary.

3. Check the field brush connections and lead insulation. A brush kit and a contact kit are available at your local marine dealer, but all other assemblies must be replaced rather than repaired.

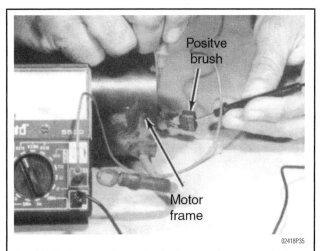

Fig. 192 Testing for a shorted brush, there should be no continuity between the positive brush and the starter case

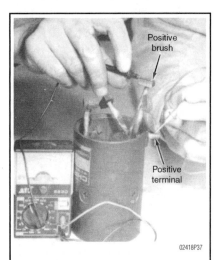

Fig. 193 The positive brushes should have continuity between the brush and the positive terminal post

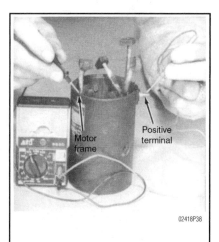

Fig. 194 The positive terminal should not have continuity with the motor housing frame

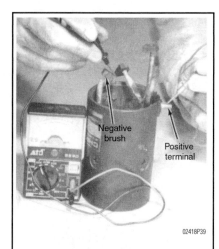

Fig. 195 There should be no continuity between the negative brush and the positive terminal post

IGNITION AND ELECTRICAL SYSTEMS 5-57

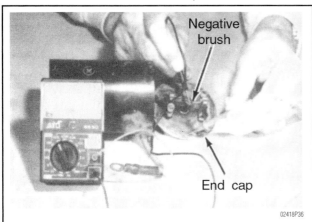

Fig. 196 There should be continuity between the negative brush and the end cap

1. Cut off the old brush leads where they are attached to the field coils.
2. Prepare the ends of the coils for soldering the new brush lead assemblies.
3. Clean the ends of the coils by filing or grinding off the old brush lead connections. Remove the varnish only as far back as necessary to enable a good soldered connection to be made.
4. Use rosin flux and solder the leads to the backsides of the coil to prevent any excess solder from rubbing against the armature. Be sure the leads are in the right position to reach the brush holders. Do not overheat the leads, because the solder will run onto the lead and the lead will lose its flexibility.
5. The positive brush leads are soldered to the positive terminal of the motor.
6. The negative brush leads are soldered to the field.
7. Insert the springs and then the brushes into the brush holders.
8. Place the armature washer centrally between the brushes holding them back inside their holders.
9. Slide the armature shaft assembly into the washer hole. The end of the shaft will displace the washer and the brushes will spring forward against the commutator.
10. Install the wavy washer and then thrust washer onto the end of the armature shaft before final assembly.
11. On units with brushes installed inside the motor frame, insert the small end of the armature into the end cap bushing. Be careful not to disturb the brushes.
12. Install the bottom cap and secure with the screws from bottom to top.

The armature, fields, and brush holders must be checked before assembling the cranking motor.

ASSEMBLY

Brush Installation Prestolite®

♦ See Figures 197 and 198

Replacement brush sets are available and usually contain the following parts:
- Two insulated brushes, with flexible leads attached.
- Two ground brush holders with brushes and leads attached.
- Necessary attaching screws, washers, and nuts.

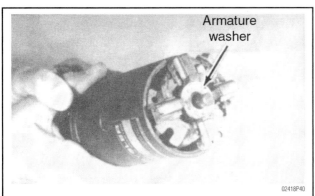

Fig. 197 Insert the springs and then the brushes into the brush holders.

Brush Installation Bosch®

♦ See Figures 199, 200 and 201

The positive brushes are attached to the positive terminal and are sold as an assembled set. The negative brushes are attached to the lower cap with a bolt.

To remove the positive brushes, slip the terminal out of the slot in the cap. The negative brushes are removed by simply removing the two bolts attaching the brush lead to the lower cap. After the brushes are removed, remove the thrust washer and wavy washer from the end cap. These two items are used, as an aid in the installation of the armature shaft, therefore, does not reinstall them at this time.

Installation of the new positive brushes is accomplished by sliding the new positive terminal into the slot of the end cap. Install the negative brushes by positioning them in place in the lower cap, and then securing the leads with the attaching bolts.

1. Insert the springs and then the brushes into the brush holders.
2. Place the armature wavy washer and then thrust washer centrally between the brushes, holding them back inside their holders, as shown.
3. Slide the armature shaft into the washer hole. The end of the shaft will displace the washer and the brushes will spring forward against the commutator.
4. Units with the brushes vertical in the lower housing are held in place with a tool such as the one pictured.
5. With the armature washers correctly installed behind the brushes.
6. Install the bottom cap and secure with the screws from bottom to top.

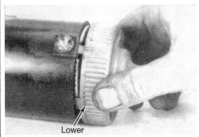

Fig. 198 On units with brushes installed inside the motor frame, insert the small end of the armature into the end cap bushing. Be careful not to disturb the brushes

Fig. 199 Insert the springs and then the brushes into the brush holders

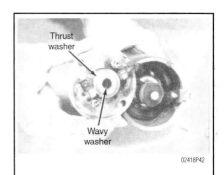

Fig. 200 Place the armature wavy washer and then thrust washer centrally between the brushes, holding them back inside their holders, as shown

5-58 IGNITION AND ELECTRICAL SYSTEMS

Fig. 201 Units with the brushes vertical in the lower housing are held in place with a tool such as the one pictured

Starter Relay

TESTING

Resistance Test

♦ See Figures 202, 203 and 204

➡ The following test MUST be conducted with the solenoid removed from the powerhead.

1. Connect one test lead of a multimeter to each of the large solenoid terminals.
2. Connect the positive (+) lead from a fully charged 12-volt battery to the small solenoid terminal marked "S".
3. Momentarily make contact with the ground lead from the battery to the small solenoid terminal marked I. If a loud "click" sound is heard and the multimeter indicates continuity, the solenoid is in serviceable condition. If, however a "click" sound is not heard and/or the multimeter does not indicate continuity, the solenoid is defective and must be replaced and only with a marine solenoid.

Solenoid Voltage Test

1. Connect the voltmeter between the common powerhead ground and the No. 1 point.

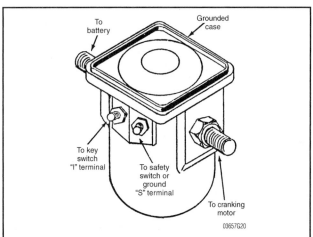

Fig. 202 This solenoid acts as a switch between the battery and starter motor. If the unit is found to be defective, it must be replaced with a marine approved solenoid

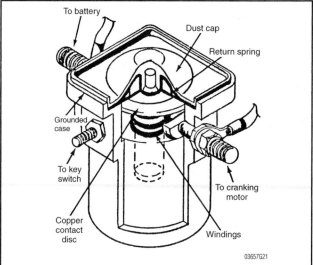

Fig. 203 A cutaway drawing of a starting motor solenoid with major parts identified

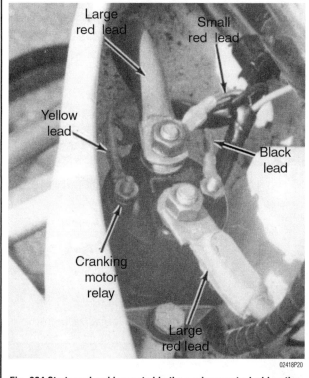

Fig. 204 Starter solenoid mounted in the engine pan typical location

2. Turn the ignition key switch to the start position.
3. Observe the voltmeter. If there is no reading, the starting motor solenoid is defective and must be replaced. If a reading is indicated and a click sound is heard, the solenoid is functioning properly.

REMOVAL & INSTALLATION

1. Disconnect the battery.
2. Remove and label all leads from the starter solenoid.
3. Remove the two fasteners holding the solenoid in place.

To install:

4. Install replacement unit with the fasteners.
5. Install all wires on to the starter solenoid in their proper places.
6. Connect battery and test.

IGNITION AND ELECTRICAL SYSTEMS

IGNITION AND ELECTRICAL WIRING DIAGRAMS

Engine Wiring Diagram—1984–87 3 and 4 HP

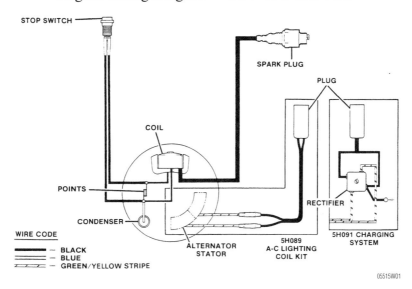

Engine Wiring Diagram—1987–98 5 HP Breaker Point Ignition System

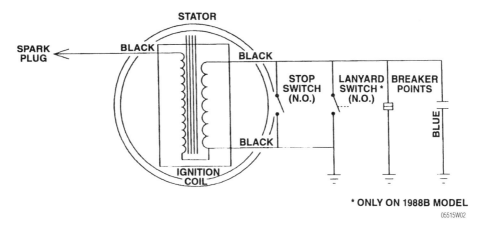

Engine Wiring Diagram—1984—98 9.9, 15 HP Magneto Ignition System

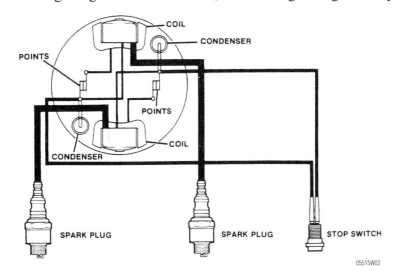

5-60 IGNITION AND ELECTRICAL SYSTEMS

Engine Wiring Diagram—1984–87 9.9, 15 HP Breaker Point Ignition System

Engine Wiring Diagram—1994–99 25 HP

Engine Wiring Diagram—1984–87 250 Sailor Remote

IGNITION AND ELECTRICAL SYSTEMS 5-61

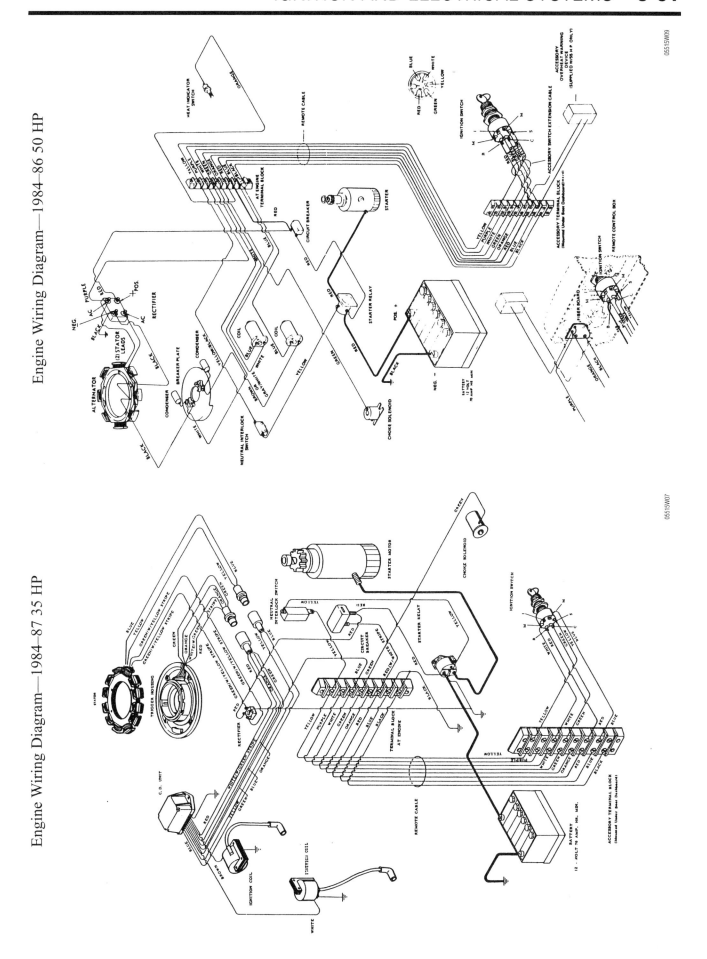

5-62 IGNITION AND ELECTRICAL SYSTEMS

Engine Wiring Diagram—Prior to 1988 "D" 50 HP

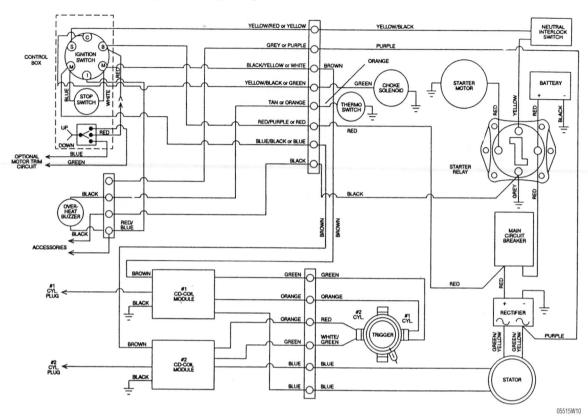

Engine Wiring Diagram—1988 "D"–99 50 HP

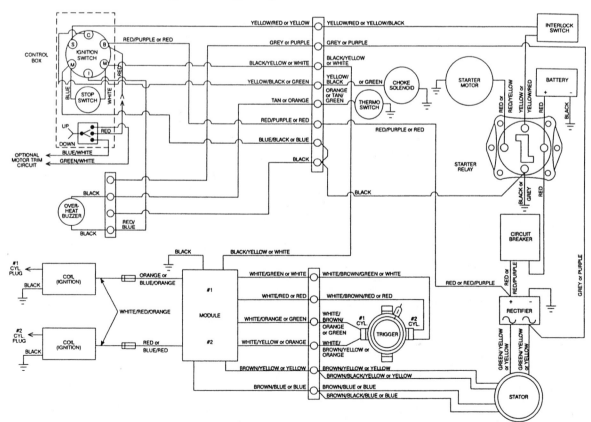

IGNITION AND ELECTRICAL SYSTEMS   5-63

Engine Wiring Diagram—1992–94 40, 50 HP

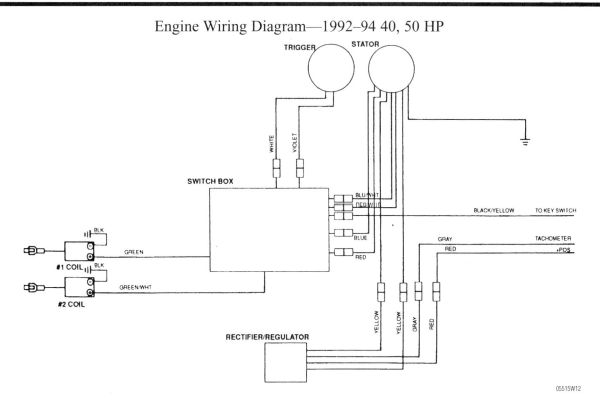

Engine Wiring Diagram—40, 50 HP with starting Serial Numbers OE138600

1 - Flywheel
2 - Stator
3 - Trigger
4 - Voltage Regulator
5 - 20 Ampere Fuse
6 - Cylinder Head Temperature Sender
7 - CDM (Capacitor Discharge Module)
8 - Trip "Up" Solenoid
9 - Trim "Down" Solenoid
10 - Trim Pump Motor
11 - Cowl Mounted Trim Switch
12 - To Remote Control Trim Switch
13 - Engine Harness Connector
14 - To 12V Battery
15 - Starter Solenoid
16 - Starter
17 - Primer Solenoid

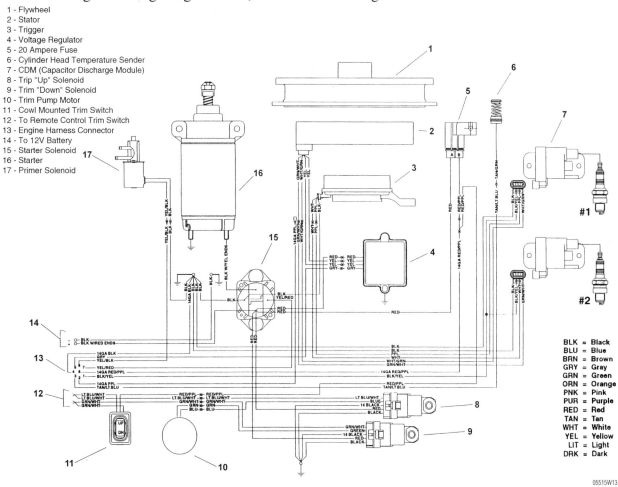

BLK = Black
BLU = Blue
BRN = Brown
GRY = Gray
GRN = Green
ORN = Orange
PNK = Pink
PUR = Purple
RED = Red
TAN = Tan
WHT = White
YEL = Yellow
LIT = Light
DRK = Dark

5-64 IGNITION AND ELECTRICAL SYSTEMS

Engine Wiring Diagram—1991–95 70 HP Type II Ignition System

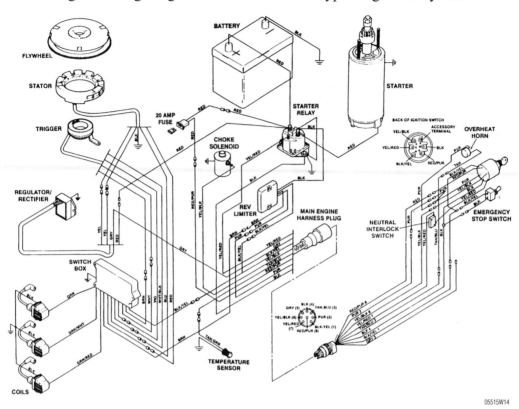

Engine Wiring Diagram—1996–99 75 HP

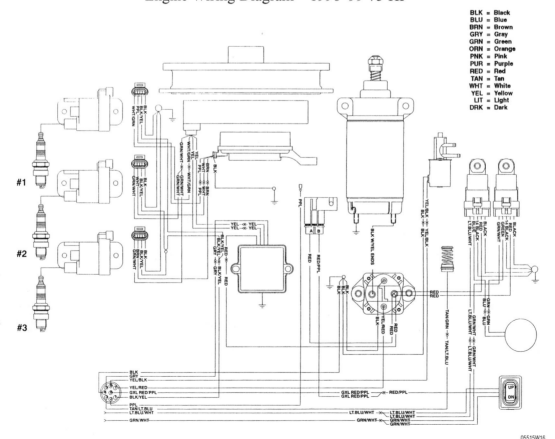

IGNITION AND ELECTRICAL SYSTEMS

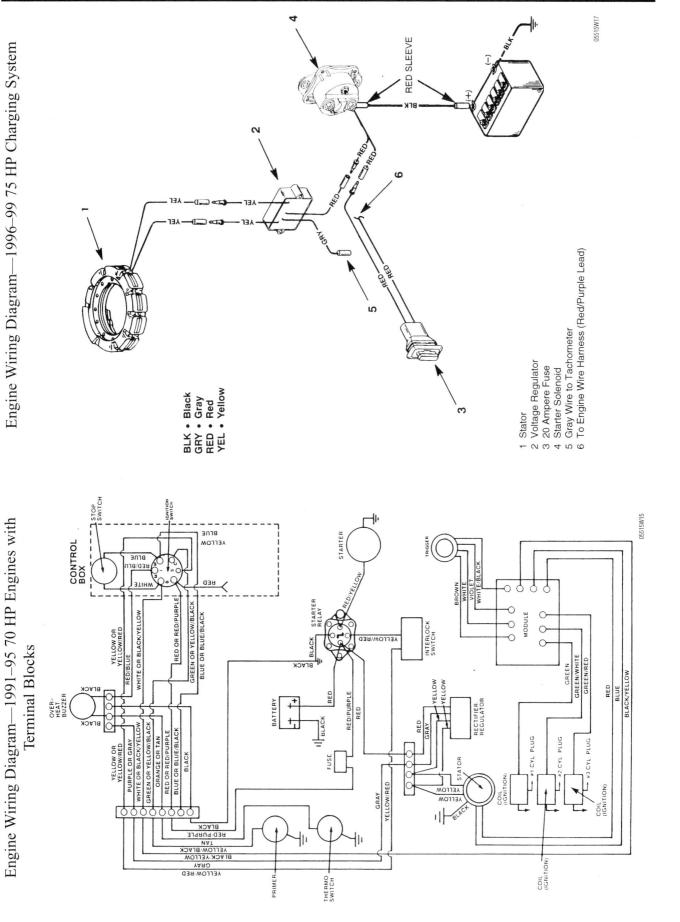

5-66 IGNITION AND ELECTRICAL SYSTEMS

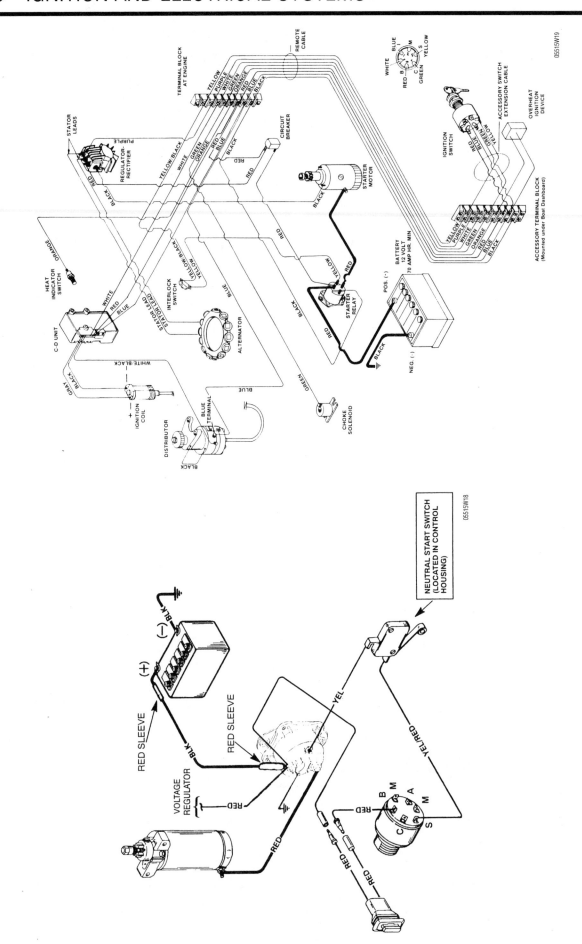

Engine Wiring Diagram—90, 100, 105, 115, 125, 140 HP Motorola System

Engine Wiring Diagram—1996–99 75 HP Starting System

IGNITION AND ELECTRICAL SYSTEMS 5-67

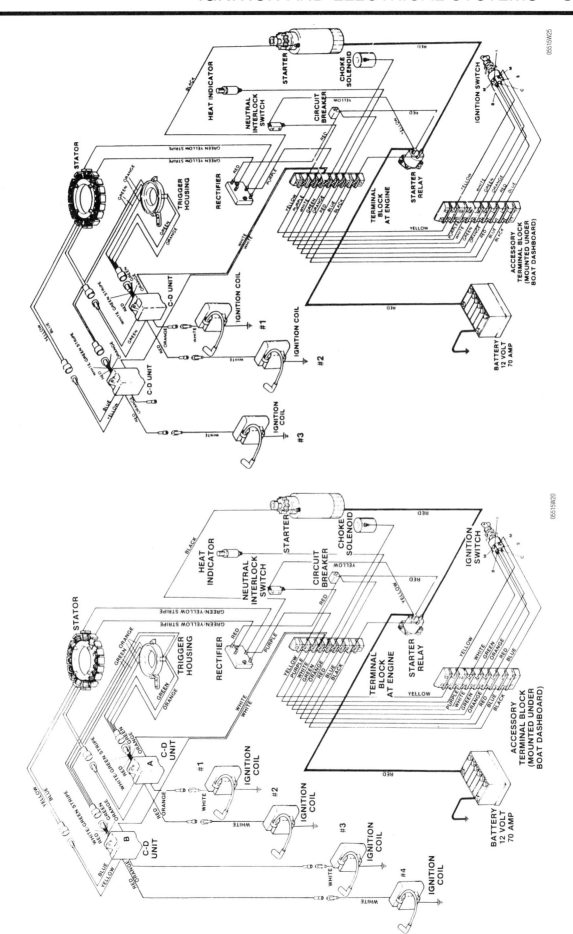

IGNITION AND ELECTRICAL SYSTEMS

Engine Wiring Diagram—75 HP Starting with Serial Numbers OE345000, 90 HP Starting with Serial Numbers OE937000

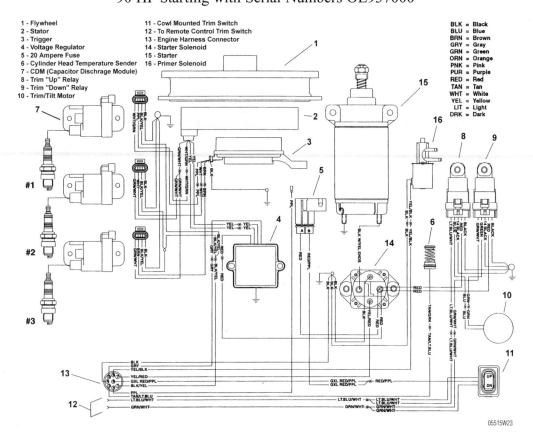

Engine Wiring Diagram—120 HP Starting with Serial Numbers OE093700

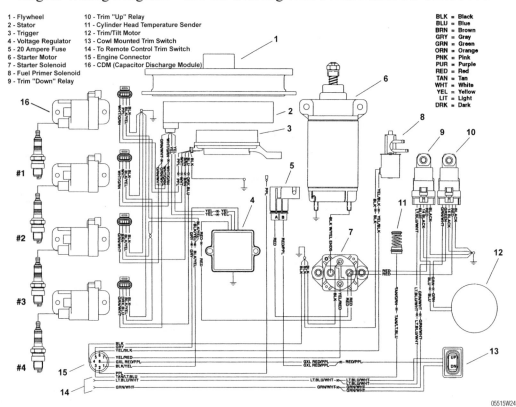

IGNITION AND ELECTRICAL SYSTEMS

Engine Wiring Diagram—90 HP with Force Control Box

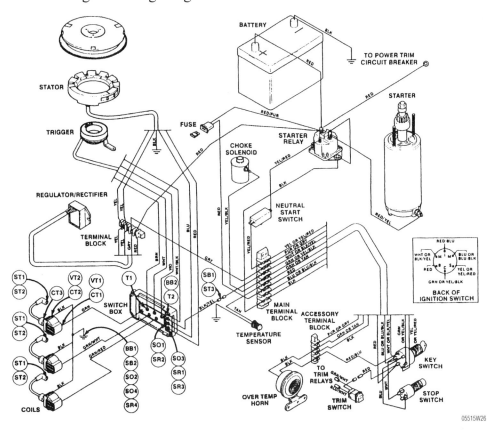

Engine Wiring Diagram—120 HP with Force Control Box

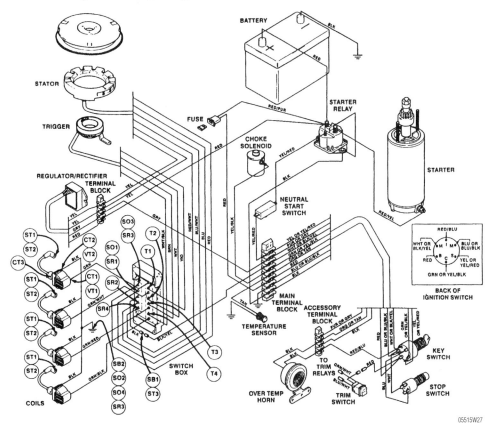

5-70 IGNITION AND ELECTRICAL SYSTEMS

Engine Wiring Diagram—1988–91 85, 90 HP Ignition System

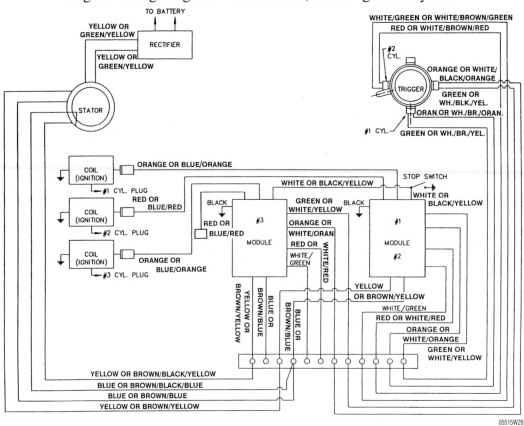

Engine Wiring Diagram—1988–91 120, 125 HP Ignition System

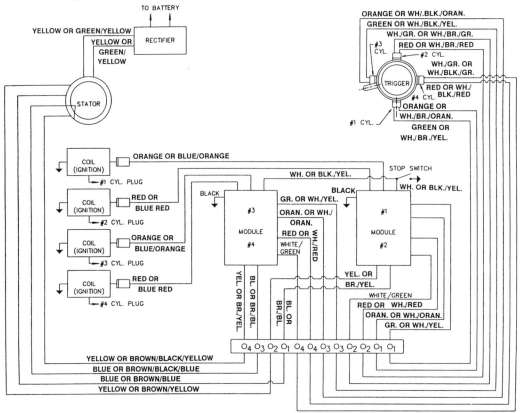

IGNITION AND ELECTRICAL SYSTEMS 5-71

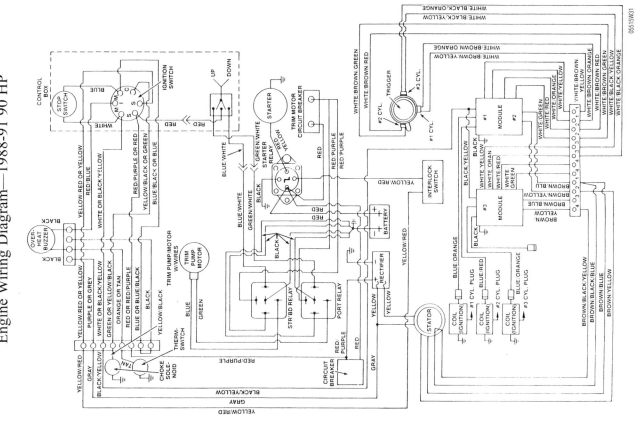

Engine Wiring Diagram—1988-91 90 HP

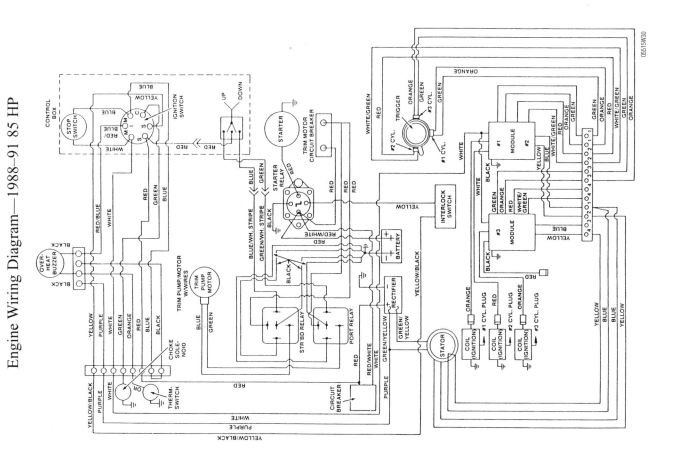

Engine Wiring Diagram—1988-91 85 HP

5-72 IGNITION AND ELECTRICAL SYSTEMS

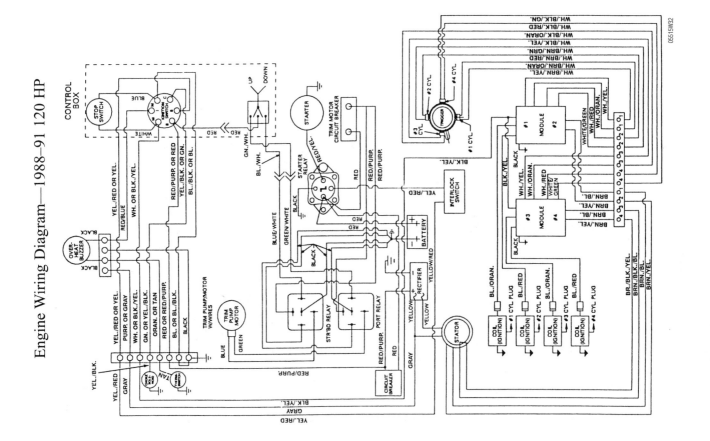

Engine Wiring Diagram—1988–91 125 HP

Engine Wiring Diagram—1988–91 120 HP

IGNITION AND ELECTRICAL SYSTEMS

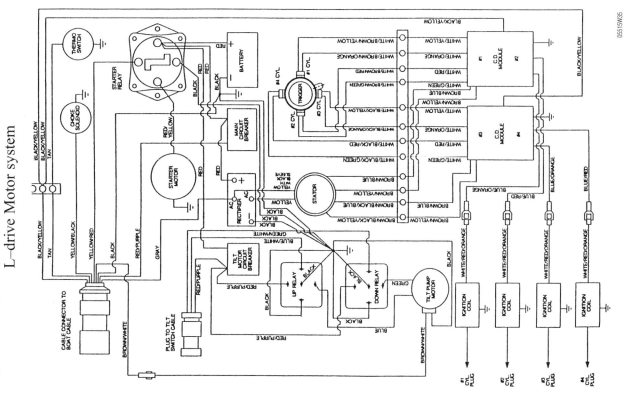

Engine Wiring Diagram—1989-99 120, 125 HP 4-Cylinder L-drive Motor system

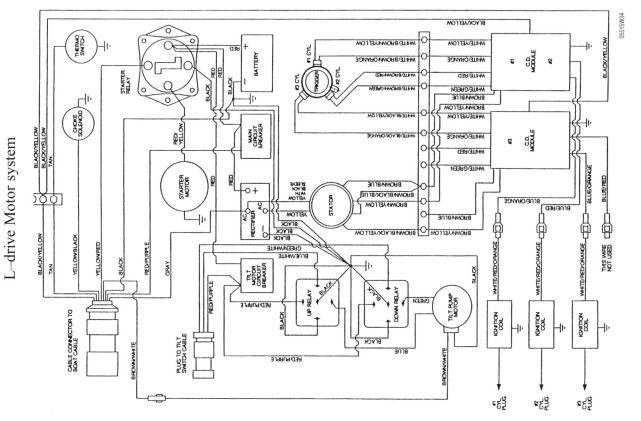

Engine Wiring Diagram—1989-99 85, 90 HP 3-Cylinder L-drive Motor system

5-74 IGNITION AND ELECTRICAL SYSTEMS

Engine Wiring Diagram—1990 "D"–1991 "A" 150 HP

Engine Wiring Diagram—Prior to 1990 "D" 150 HP

IGNITION AND ELECTRICAL SYSTEMS 5-75

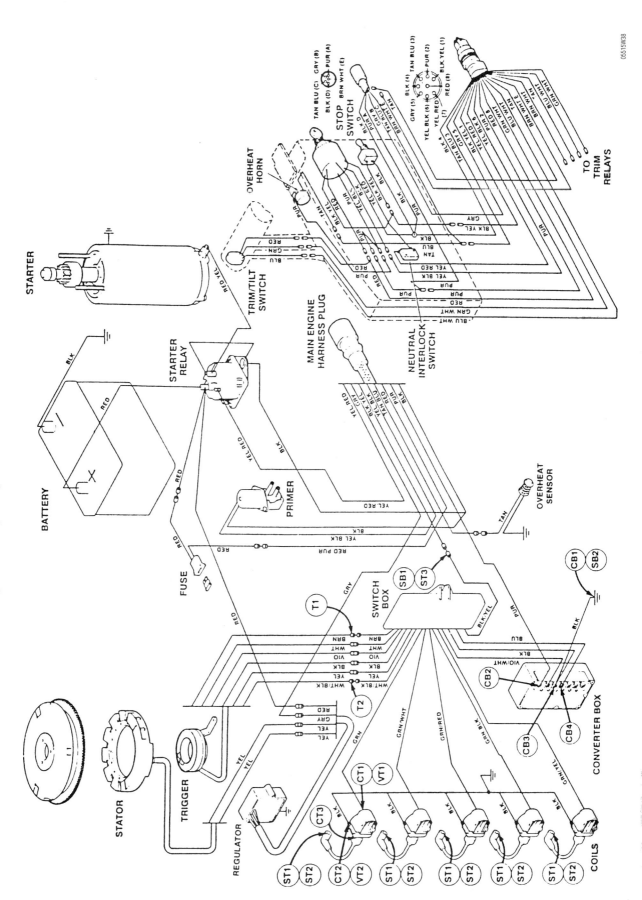

Engine Wiring Diagram—1991 "J"–94 150 HP

5-76 IGNITION AND ELECTRICAL SYSTEMS

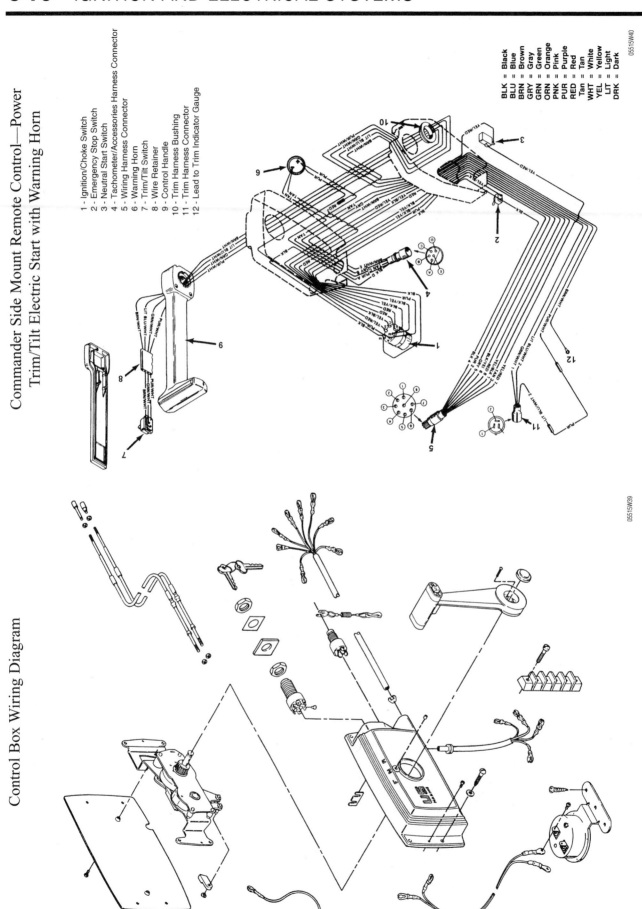

IGNITION AND ELECTRICAL SYSTEMS

Commander 2000 Side Mount Remote Control—
Power Trim/Tilt Electric Start with Warning Horn

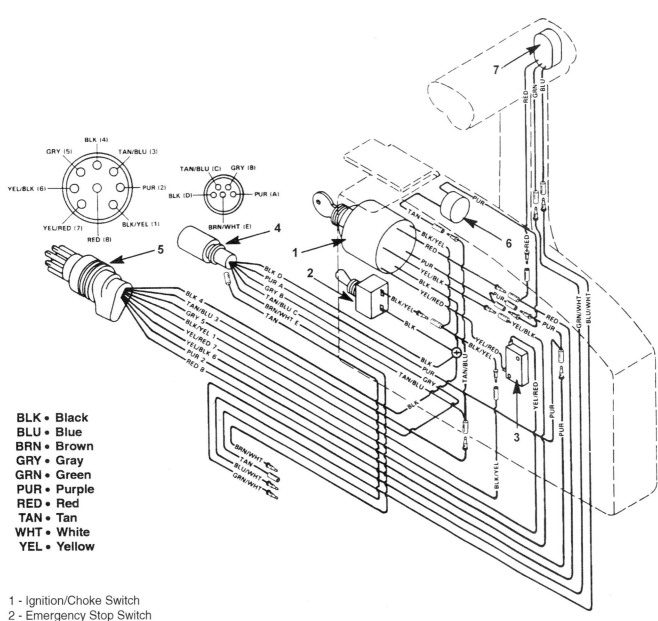

1 - Ignition/Choke Switch
2 - Emergency Stop Switch
3 - Neutral Start Switch
4 - Tachometer/Accessories Harness Connector
5 - Wiring Harness Connector
6 - Warning Horn
7 - Trim/Tilt Switch

5-78 IGNITION AND ELECTRICAL SYSTEMS

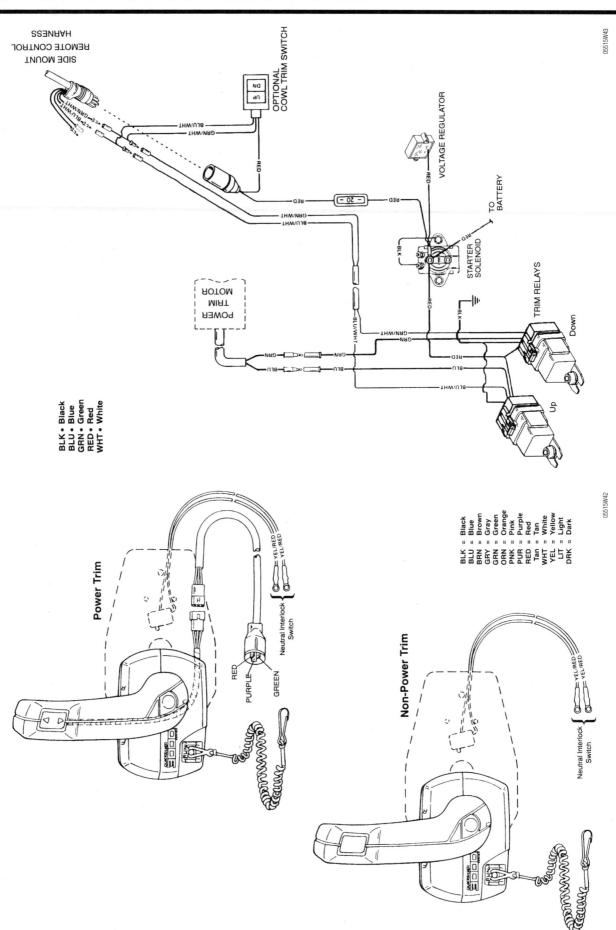

COOLING SYSTEM 6-2
DESCRIPTION AND OPERATION 6-2
TROUBLESHOOTING THE COOLING
 SYSTEM 6-2
WATER PUMP 6-3
  SERVICE CAUTIONS 6-3
  REMOVAL & INSTALLATION 6-3
  CLEANING & INSPECTION 6-9
THERMOSTAT 6-9
  REMOVAL & INSTALLATION 6-9
WARNING SYSTEMS 6-10
DESCRIPTION AND OPERATION 6-10
TROUBLESHOOTING THE WARNING
 SYSTEMS 6-10
CYLINDER HEAD TEMPERATURE
 SENSOR 6-10
  REMOVAL & INSTALLATION 6-10
  TESTING 6-10

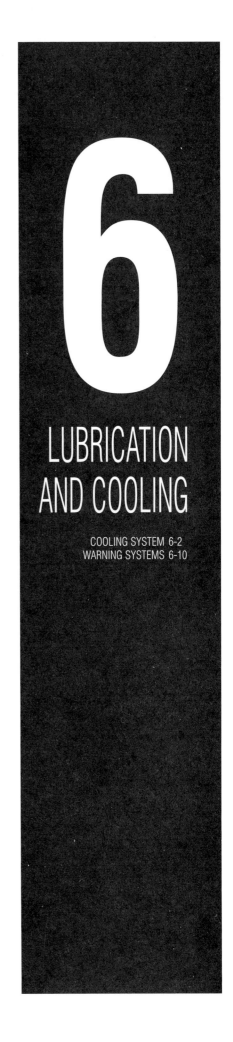

6

LUBRICATION AND COOLING

COOLING SYSTEM 6-2
WARNING SYSTEMS 6-10

6-2 LUBRICATION AND COOLING

COOLING SYSTEM

Description and Operation

♦ See Figure 1

Water cooling is the most popular method in use to cool outboard powerheads. A "raw-water" type pump delivers seawater to the powerhead, circulating it through the cylinder head(s), the thermostat, the exhaust housing, and back down through the outboard. The water runs down the exhaust cavity and away, either through an exhaust tube or through the propeller hub.

Routine maintenance of the cooling system is quite important, as expensive damage can occur if it overheats. The cooling system is so important that many outboards covered in this manual incorporate overheat alarm systems in case the engine's operating temperature exceeds predetermined limits.

Poor operating habits can play havoc with the cooling system. For instance, running the engine with the water pickup out of water can destroy the water pump impeller in a matter of seconds. Running in shallow water, kicking up debris that is drawn through the pump, can not only damage the pump itself, but send the debris throughout the entire system, causing water restrictions that create overheating.

The water pumps used on most outboards are a displacement type water pump. Water pressure is increased by the change in volume between the impeller and the pump case.

On most outboards, the water pump is mounted on top of the lower unit. A driveshaft key engages a flat on the driveshaft and a notch in the impeller hub. As the driveshaft rotates, the impeller rotates with it.

A pellet-type thermostat is used to control the flow of engine water, to provide fast engine warm-up and to regulate water temperatures. A wax pellet element in the thermostat expands when heated and contracts when cooled. The pellet element is connected through a piston to a valve. When the pellet element is heated, pressure is exerted against a rubber diaphragm, which forces the valve to open. As the pellet element is cooled, the contraction allows a spring to close the valve. Thus, the valve remains closed while the water is cold, limiting circulation of water.

As the engine warms, the pellet element expands and the thermostat valve opens, permitting water to flow through the powerhead. This opening and closing of the thermostat permits enough water to enter the powerhead to keep the engine within operating limits.

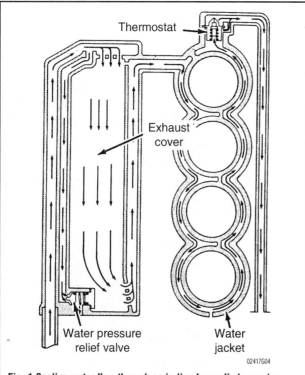

Fig. 1 Cooling water flow through an in-line four cylinder engine

Troubleshooting the Cooling System

♦ See Figure 2

Poor operating habits can play havoc with the cooling system. For instance, running the engine with the water pickup out of water can destroy the water pump impeller in a matter of seconds. Running in shallow water, kicking up

After 60 seconds at 1500 rpm.

After 90 seconds at 1500 rpm.

After 30 seconds at 2000 rpm.

After 45 seconds at 2000 rpm.

After 60 seconds at 2000 rpm.

Fig. 2 Running the engine without a good source of cooling water will cause severe damage to the pump impeller in a very short amount of time

LUBRICATION AND COOLING

debris that is drawn through the pump, can not only damage the pump itself, but send the debris throughout the entire system, causing water restrictions that create overheating.

Symptoms of overheating are numerous and include:
- A "pinging" noise coming from the engine, commonly known as detonation
- Loss of power
- A burning smell coming from the engine
- Paint discoloration on the powerhead in the area of the spark plugs and cylinder heads

➡If these symptoms occur, immediately seek and correct the cause. If the engine has overheated to the point where paint has discolored, it may be too late to save the powerhead. Powerheads in this state usually require at least partial overhaul.

So what are major causes of overheating? Well the most prevalent cause is lack of maintenance. Other causes which are directly attributable to lack of maintenance or poor operating habits are:
- Fuel system problems causing lean mixture
- Incorrect oil mixture in fuel or a problem with the oil injection system
- Spark plugs of incorrect heat range
- Faulty thermostat
- Restricted water flow through the powerhead due to sand or silt buildup
- Faulty water pump impeller
- Sticking thermostat

Water Pump

SERVICE CAUTIONS

1. Since proper water pump operation is critical to outboard operation, all seals and gaskets should be replaced whenever the water pump is removed. Also, installation of a new impeller each time the water pump is disassembled is good insurance against overheating.
2. Never turn a used impeller over and reuse it. The impeller rotates with the driveshaft and the vanes take a set in a clockwise direction. Turning the impeller over will cause the vanes to move in the opposite and result in premature impeller failure.
3. Units that have been used in salt water may need the use of heat to assist in removing fasteners without breaking them. It is also recommended that the drive shaft be polished with a wire brush or sandpaper. This will assist in the removal and installation of the pump parts when they are slid up and down the driveshaft.
4. Prior to any assembly it is recommended that all fasteners be clean and greased to ease installation and help in the removal of the fasteners during future repairs or maintenance.

REMOVAL & INSTALLATION

Up To 15 HP

♦ See Figures 3 thru 10

1. Disconnect the spark plug leads.
2. Place an identifying mark on the shift rod to ensure proper location for assembly.
3. Disconnect the shift rod.
   a. On units with forward and neutral only, remove the cover plug to expose shift rod connection. Loosen to disconnect.
   b. On units with forward, neutral and reverse, remove coupling bolt to disconnect shift rod.
4. Remove the fasteners holding the lower unit in place.
5. Carefully pull the lower unit down to remove.
6. Support the unit in a lower unit holder or a vise.
7. Remove the fasteners holding the water pump housing and/or cover in place.
8. Remove the housing and/or cover and impeller up and slide off the driveshaft.
9. Remove water pump body by removing the stud and lifting the body up and off the shaft.
10. Remove the impeller wear plate and gaskets.

To install:

➡A brush on gasket sealer would be recommended in salt water applications or when the housings are in marginal condition.

11. Install the water pump lower housing.
12. Secure it into place with the stud removed earlier.
13. Install the water pump plate and gasket(s).
14. Slide the water pump impeller down the driveshaft.
15. Place the impeller key into the slot on the driveshaft.
16. Align the slot in the water pump impeller with the key and slide down into place.
17. Lower the water pump housing down the driveshaft.
18. With a clockwise twisting motion of the driveshaft push the impeller into the pump housing.
19. Lubricate and install the fasteners that hold the pump housing in place.

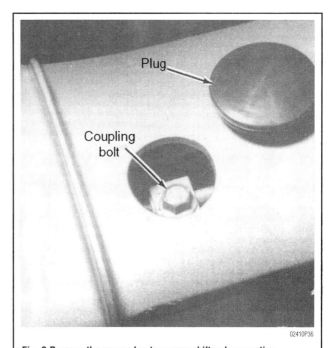

Fig. 3 Remove the cover plug to expose shift rod connection. Loosen to disconnect

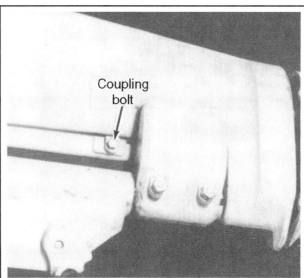

Fig. 4 Remove coupling bolt to disconnect shift rod

6-4 LUBRICATION AND COOLING

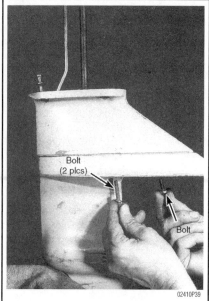

Fig. 5 Remove the fasteners holding the lower unit in place

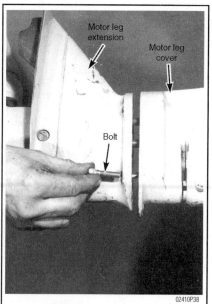

Fig. 6 Carefully pull the lower unit down to remove

Fig. 7 Support the unit in a lower unit holder or a vise

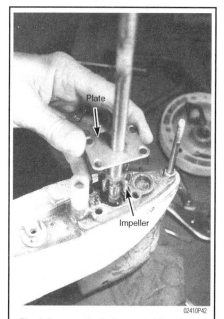

Fig. 8 Remove the fasteners holding the water pump housing and/or cover in place

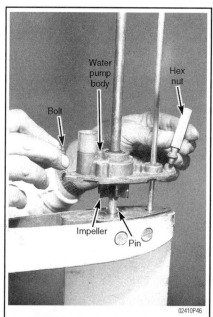

Fig. 9 Remove the housing and/or cover and impeller up and slide off the driveshaft

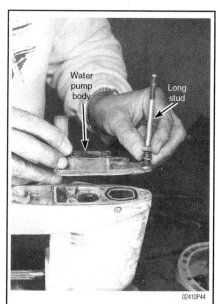

Fig. 10 Remove water pump body by removing the stud and lifting the body up and off the shaft

20. Lubricate the driveshaft splines on the sides only. Grease on the top of the shaft could cause it not to be able to fully seat into the crankshaft.
21. Lift the lower unit from the holder or vise.
22. Align the driveshaft with the tube or cavity and push the lower unit up into the mid housing.

➡It may be necessary to rotate the driveshaft or flywheel to align the driveshaft splines.

23. Be certain that the water tube is aligned with the water pump outlet and slide the unit into place.
24. Install the attaching hardware.
25. Connect the shift linkage with the fastener removed earlier.
26. Attach the spark plug leads to the spark plugs.
27. Supply the engine with an appropriate water supply.
28. Start the engine and test the water pump output.

✱✱ WARNING

If water does not appear to be flowing through the engine immediately shut the engine off. There could be a water supply problem or a misalignment in the water tube.

Over 15 HP with Internal Type Connection

▶ See Figure 11

1. Disconnect the spark plug leads.
2. Remove the fasteners holding the lower unit in place.
3. Carfully split the lower unit from the mid section to gain access to the internal type shift rod connection.
4. Disconnect the shift rod.

LUBRICATION AND COOLING 6-5

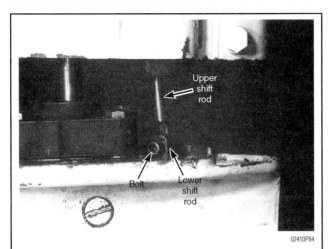

Fig. 11 With the lower unit unbolted, shift the unit into reverse gear and start to remove the unit to expose the shift connection

5. Carefully pull the lower unit down to remove.
6. Support the unit in a lower unit holder or a vise.
7. Remove the fasteners holding the water pump upper housing in place.
8. Remove the housing and the water pump impeller up and slide off the driveshaft.
9. Remove the impeller wear plate and gaskets.

To install:

➡ A brush on gasket sealer would be recommended in salt water applications or when the housings are in marginal condition.

10. Install the water pump plate and gasket(s).
11. Slide the water pump impeller down the driveshaft.
12. Place the impeller key into the slot on the driveshaft.
13. Align the slot in the water pump impeller with the key and slide down into place.
14. Lower the water pump housing down the driveshaft.
15. With a clockwise twisting motion of the driveshaft push the pump housing over the impeller.
16. Lubricate and install the fasteners that hold the pump housing in place.
17. Lubricate the driveshaft splines on the sides only. Grease on the top of the shaft could cause it not to be able to fully seat into the crankshaft.

18. Lift the lower unit from the holder or vise.
19. Align the driveshaft with the tube or cavity and push the lower unit up into the mid housing.

➡ It may be necessary to rotate the driveshaft or flywheel to align the driveshaft splines.

20. Be certain that the water tube is aligned with the water pump outlet and slide the unit into place leaving about a one inch gap.
21. Connect the shift linkage with the fastener removed earlier.
22. Move the lower unit up flush with the mid housing.
23. Install the attaching hardware.
24. Attach the spark plug leads to the spark plugs.
25. Supply the engine with an appropriate water supply.
26. Start the engine and test the water pump output.

✴✴ WARNING

If water does not appear to be flowing through the engine immediately shut the engine off. There could be a water supply problem or a misalignment in the water tube.

Over 15 HP with Threaded Shift Coupling

♦ See Figures 12, 13 and 14

1. Disconnect the spark plug leads.
2. Place an identifying mark on the shift rod to ensure proper location for assembly.
3. Disconnect the shift rod.
4. Remove the fasteners holding the lower unit in place.
5. Carefully pull the lower unit down to remove.
6. Support the unit in a lower unit holder or a vise.
7. Remove the fasteners holding the water pump housing in place.
8. Remove the housing and impeller up and slide off the driveshaft.
9. Remove the impeller wear plate and gaskets.

To install:

➡ A brush on gasket sealer would be recommended in salt water applications or when the housings are in marginal condition.

10. Install the water pump plate and gasket(s).
11. Slide the water pump impeller down the driveshaft.
12. Place the impeller key into the slot on the driveshaft.
13. Align the slot in the water pump impeller with the key and slide down into place.

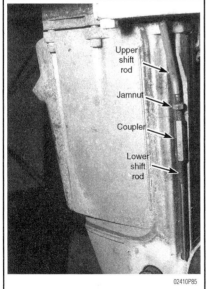

Fig. 12 Hold the coupler and remove the jam nut allowing the coupler to be threaded off the shift rod disconnecting the shift rod

Fig. 13 Remove the bolts holding the pump housing in place

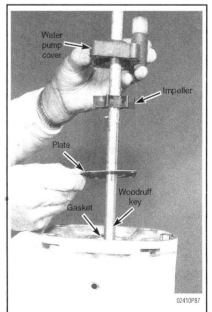

Fig. 14 The proper order of pump parts as they are removed or installed on the drive shaft

6-6 LUBRICATION AND COOLING

14. Lower the water pump housing down the driveshaft.
15. With a clockwise twisting motion of the driveshaft push the pump housing over the impeller.
16. Lubricate and install the fasteners that hold the pump housing in place.
17. Lubricate the driveshaft splines on the sides only. Grease on the top of the shaft could cause it not to be able to fully seat into the crankshaft.
18. Lift the lower unit from the holder or vise.
19. Align the driveshaft with the tube or cavity and push the lower unit up into the mid housing.

➥It may be necessary to rotate the driveshaft or flywheel to align the driveshaft splines.

20. Be certain that the water tube is aligned with the water pump outlet and slide the unit into place leaving enough space to thread together the shift connector.
21. Connect the shift linkage by threading the connector on to the shift rod
22. Secure the connector with the jam nut.
23. Move the lower unit up flush with the mid housing.
24. Install the attaching hardware.
25. Attach the spark plug leads to the spark plugs.
26. Supply the engine with an appropriate water supply.
27. Start the engine and test the water pump output.

※※ WARNING
If water does not appear to be flowing through the engine immediately shut the engine off. There could be a water supply problem or a misalignment in the water tube.

Over 15 HP with Shift Shaft Pin Connection "Aluminum Pump Housing"

▶ See Figures 15 thru 20

1. Disconnect the spark plug leads.
2. Place an identifying mark on the shift rod to ensure proper location for assembly.

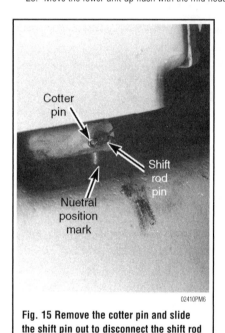
Fig. 15 Remove the cotter pin and slide the shift pin out to disconnect the shift rod

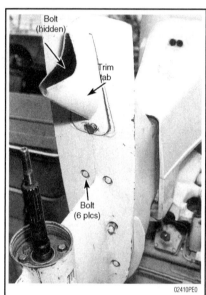
Fig. 16 Remove the remaining fasteners holding the lower unit in place

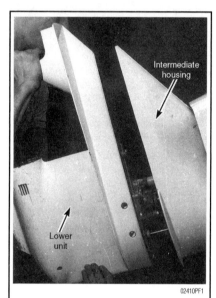
Fig. 17 Carefully pull the lower unit down to remove

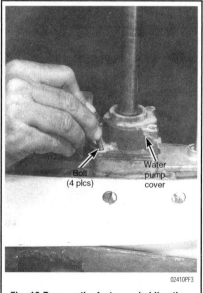
Fig. 18 Remove the fasteners holding the water pump housing in place

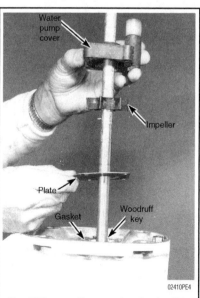
Fig. 19 Remove the pump housing by sliding it up the driveshaft

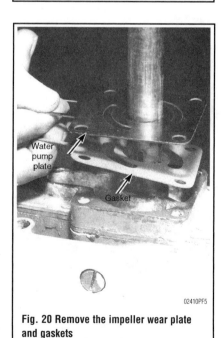
Fig. 20 Remove the impeller wear plate and gaskets

LUBRICATION AND COOLING 6-7

3. Remove the cotter pin and slide the shift pin out to disconnect the shift rod.
4. Remove the trim tab from the trailing end of the ventilation plate.
5. Remove the fastener from under the trim tab.
6. Remove the remaining fasteners holding the lower unit in place.
7. Carefully pull the lower unit down to remove.
8. Support the unit in a lower unit holder or a vise.
9. Remove the fasteners holding the water pump housing in place.
10. Remove the pump housing by sliding it up the driveshaft.
11. Remove the water pump impeller and slide off the driveshaft.
12. Remove the impeller wear plate and gaskets.

To install:

➡ A brush on gasket sealer would be recommended in salt water applications or when the housings are in marginal condition.

13. Install the water pump plate and gasket(s).
14. Place the impeller key into the slot on the driveshaft.
15. Slide the water pump impeller down the driveshaft.
16. Align the slot in the water pump impeller with the key and slide down into place.
17. Lower the water pump housing down the driveshaft.
18. With a clockwise twisting motion of the driveshaft push the pump housing over the impeller.
19. Lubricate and install the fasteners that hold the pump housing in place.
20. Lubricate the driveshaft splines on the sides only. Grease on the top of the shaft could cause it not to be able to fully seat into the crankshaft.
21. Lift the lower unit from the holder or vise.
22. Align the driveshaft with the tube or cavity and push the lower unit up into the mid housing.

➡ It may be necessary to rotate the driveshaft or flywheel to align the driveshaft splines.

23. Be certain that the water tube is aligned with the water pump outlet
24. Move the lower unit up flush with the mid housing.
25. Install the attaching hardware.
26. Install the trim tab with its hardware.
27. Connect the shift linkage by installing the shift pin into the connector and installing a new cotter pin.
28. Attach the spark plug leads to the spark plugs.
29. Supply the engine with an appropriate water supply.
30. Start the engine and test the water pump output.

✵✵ WARNING

If water does not appear to be flowing through the engine immediately shut the engine off. There could be a water supply problem or a misalignment in the water tube.

Over 15 HP with Shift Shaft Pin Connection "Plastic Pump Housing"

▶ See Figures 21, 22 and 23

1. Disconnect the spark plug leads.
2. Place an identifying mark on the shift rod to ensure proper location for assembly.
3. Remove the cotter pin and slide the shift pin out to disconnect the shift rod.
4. Remove the trim tab from the trailing end of the ventilation plate.
5. Remove the fastener from under the trim tab.
6. Remove the remaining fasteners holding the lower unit in place.
7. Carefully pull the lower unit down to remove.
8. Support the unit in a lower unit holder or a vise.
9. Remove the fasteners holding the water pump housing in place.
10. Remove the housing and the water pump impeller up and slide off the driveshaft.
11. Remove the impeller wear plate and gaskets.

To install:

➡ A brush on gasket sealer would be recommended in salt water applications or when the housings are in marginal condition.

12. Install the water pump plate and gasket(s).
13. Place the impeller key into the slot on the driveshaft.
14. Slide the water pump impeller down the driveshaft.
15. Align the slot in the water pump impeller with the key and slide down into place.
16. Lower the water pump housing down the driveshaft.
17. With a clockwise twisting motion of the driveshaft push the pump housing over the impeller.
18. Lubricate and install the fasteners that hold the pump housing in place.
19. Lubricate the driveshaft splines on the sides only. Grease on the top of the shaft could cause it not to be able to fully seat into the crankshaft.
20. Lift the lower unit from the holder or vise.
21. Align the driveshaft with the tube or cavity and push the lower unit up into the mid housing.

Fig. 21 Remove the fasteners holding down the plastic water pump housing

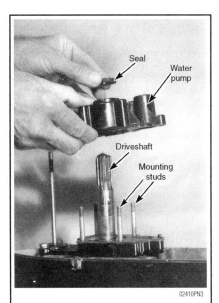

Fig. 22 Lift the pump housing up and off of the driveshaft with the impeller in the housing

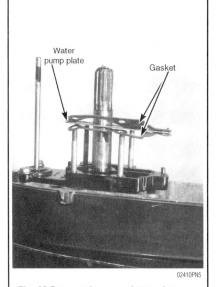

Fig. 23 Remove the wear plate and gaskets and inspect the lower housing for damage

6-8 LUBRICATION AND COOLING

➥It may be necessary to rotate the driveshaft or flywheel to align the driveshaft splines.

22. Be certain that the water tube is aligned with the water pump outlet
23. Move the lower unit up flush with the mid housing.
24. Install the attaching hardware.
25. Install the trim tab with its hardware.
26. Connect the shift linkage by installing the shift pin into the connector and installing a new cotter pin.
27. Attach the spark plug leads to the spark plugs.
28. Supply the engine with an appropriate water supply.
29. Start the engine and test the water pump output.

✱✱ WARNING

If water does not appear to be flowing through the engine immediately shut the engine off. There could be a water supply problem or a misalignment in the water tube.

Over 15 HP with Splined Shift Shaft Connections

♦ See Figures 24, 25, 26, 27 and 28

1. Disconnect the spark plug leads.
2. Shift the unit into forward gear.
3. Remove the fasteners holding the lower unit in place.
4. Carefully pull the lower unit down to remove.
5. Support the unit in a lower unit holder or a vise.
6. Remove the fasteners holding the water pump housing in place.

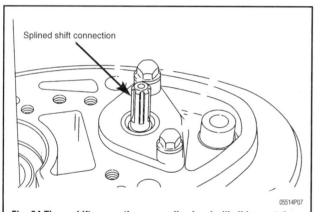

Fig. 24 These shift connections are splined and will slide apart during lower unit removal

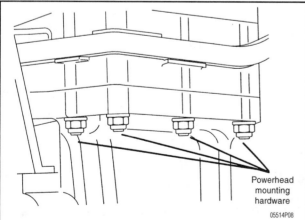

Fig. 25 Remove the fasteners holding the lower unit in place remembering to support the unit when removing the final fastener

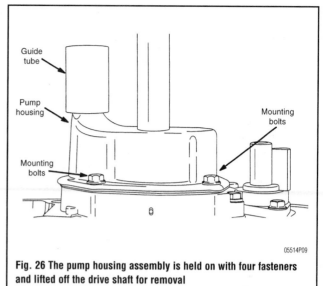

Fig. 26 The pump housing assembly is held on with four fasteners and lifted off the drive shaft for removal

7. Remove the housing and impeller up and slide off the driveshaft.
8. Remove the impeller wear plate and gaskets.

To install:

➥A brush on gasket sealer would be recommended in salt water applications or when the housings are in marginal condition.

9. Install the water pump plate and gasket(s).
10. Place the impeller key into the slot on the driveshaft.
11. Slide the water pump impeller down the driveshaft.
12. Align the slot in the water pump impeller with the key and slide down into place.
13. Lower the water pump housing down the driveshaft.
14. With a clockwise twisting motion of the driveshaft push the pump housing over the impeller.
15. Lubricate and install the fasteners that hold the pump housing in place.
16. Lubricate the driveshaft splines on the sides only. Grease on the top of the shaft could cause it not to be able to fully seat into the crankshaft.
17. Lift the lower unit from the holder or vise.
18. Align the driveshaft with the tube or cavity and push the lower unit up into the mid housing.

➥It may be necessary to rotate the driveshaft or flywheel to align the driveshaft splines.

19. Be certain that the water tube is aligned with the water pump outlet and shift splines are aligned with the shift connection then slide the unit into place.
20. Move the lower unit up flush with the mid housing.
21. Install the attaching hardware.
22. Attach the spark plug leads to the spark plugs.
23. Supply the engine with an appropriate water supply.
24. Start the engine and test the water pump output.

✱✱ WARNING

If water does not appear to be flowing through the engine immediately shut the engine off. There could be a water supply problem or a misalignment in the water tube.

CLEANING & INSPECTION

♦ See Figure 29

1. Clean the gasket surface on the pump housing and base housing.
2. Inspect the pump housing for pitting wear and warping.

LUBRICATION AND COOLING

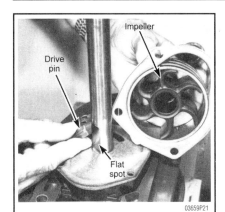

Fig. 27 Remove the water pump housing with the impeller off the driveshaft. Be sure to secure the drive pin for reinstallation

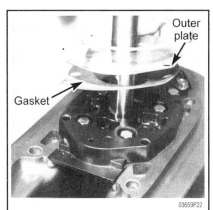

Fig. 28 Install the wear plate gasket and outer wear plate after the lower housing has been cleaned and inspected

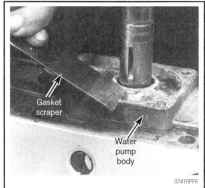

Fig. 29 Carefully clean the gasket surfaces with a scraper and inspect for a smooth surface. A leak in this area will cause an overheat condition that could be hard to find

3. Inspect the wear plate for distortion or wear.
4. Inspect any rubber grommets or tubes for pitting, deterioration or holes.
5. Check the water inlet ports for blockage i.e.: mud, corrosion, barnacles, mussels.
6. Inspect the driveshaft splines for wear and corrosion.

➡It is highly recommended that the water pump parts be replaced and not reused.

Thermostat

REMOVAL & INSTALLATION

♦ See Figure 30

Thermostats are located in the cylinder head above the number one cylinder.
1. Remove the fasteners holding the thermostat cover.
2. Carfully pry up the cover and remove the thermostat.

➡Remember the direction that the thermostat was in when it was removed. If it was to be installed backwards the engine will have overheat problems.

3. Clean the gasket surfaces on the cylinder head and thermostat cover.
4. Insert the thermostat into the powerhead with the spring facing out the bypass slot facing up.

5. Install the thermostat cover with a good gasket.
6. Clean, lubricate and install the hardware to hold the thermostat cover in place.

※※ WARNING

The engine should not be run without a thermostat in good working order. The engine running at a proper temperature will help prevent carbon build up and cold seizure scoring.

TESTING

♦ See Figures 31 and 32

The cause of a malfunctioning thermostat is often foreign matter stuck to the valve seat. Inspect the thermostat to make sure it is clean and free of foreign matter. If necessary, test the removed thermostat for operation using the procedure below:
1. With a piece of string pinched by the valve, suspend the thermostat in a pot in such a way that the thermostat floats above the bottom of the pot.
2. Raise the water temperature.
3. If the valve opens (the thermostat releases the string and drops to the bottom of the pot) at the temperature specified on the thermostat, the thermostat is functioning correctly.
4. If the valve does not open (the thermostat remains hanging) at the temperature specified on the thermostat, the thermostat is faulty and should be replaced.

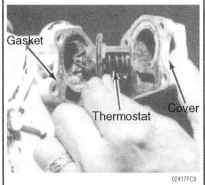

Fig. 30 A typical thermostat being removed from a four cylinder engine

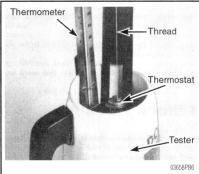

Fig. 31 Test thermostats by heating it to temperature and checking for ease of movement open then cooling and closing

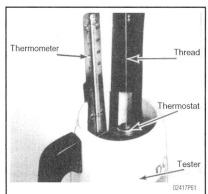

Fig. 32 Thermostat ready for testing as described in the text

6-10 LUBRICATION AND COOLING

WARNING SYSTEMS

Description and Operation

The only warning system function equipped on these outboard engines is the overheat warning. There is a temperature sensor mounted in the cylinder head that is set to short to ground once a certain temperature is met. It then completes a circuit and allows a warning horn to sound at the control station to alert the operator of an overheat condition.

Troubleshooting the Warning Systems

▶ See Figure 33

To test the warning horn turn the ignitions key to the **ON** position then short the temperature sensor to ground and listen for the horn to sound. If the horn sounds it is good.

Fig. 33 With the ignition key turned on the warning horn can be tested by grounding the overheat sensor with a screwdriver a listening for the horn to sound

Cylinder Head Temperature Sensor

The temperature sending unit (thermoswitch), was identified by color: Red, White, or Black. These colors indicated the temperature range of the unit.

The manufacturer recently decided to use only one sending unit (thermoswitch) with a slightly higher temperature range. This new unit is identified with the color Red. The manufacturer recommends all sending units be replaced with the new Red unit.

REMOVAL & INSTALLATION

There are three different kinds of holders for the temperature switches. Some use a bracket to hold in into place, some use a snap ring to hold it into a recessed area in the cylinder head and the others are threaded and screw into the cylinder head.

Remove the sensor as necessary making sure not to damage it if it is reusable.

Install the sensor in the same fashion that it had been removed.

➡ Be sure that the end of the sensor being inserted into the cylinder head is clean and can make a good connection to ground if activated.

TESTING

1. Clean the metal surface of the temperature sender (the area contacting the cylinder head). Clean the terminal rings at the ends of the temperature sender wires. Clean terminals will ensure a good electrical contact. The electrical switch in the temperature sender is designed to close if the temperature of the powerhead reaches a predetermined point. When the sender switch closes, a circuit is completed and an alarm horn is sounded.
2. Obtain a thermostat tester or similar device, multimeter. Turn the multimeter to the R x 1000 range.
3. Connect a test lead to each temperature sender wire. The meter must indicate no continuity. This means the sender switch is in the normally open position. If a reading on the meter indicates continuity the sender must be replaced.
4. Leave the meter connected to the sender and suspend the meter and a thermometer in the thermostat tester. Do not allow the sender or the thermometer to touch the bottom or sides of the tester.
5. Fill the thermostat tester with water to cover the sender unit. Connect the thermostat tester to an electrical source.
6. Observe the temperature at which the switch in the temperature sender "closes". The multimeter must have a scale reading of 0 ohms. The switch must "close" when the temperature reaches 180° to 200°F.
7. Disconnect the tester from the electrical source.
8. Allow the water in the tester to cool. Observe the temperature at which the sender "opens". The multimeter must have a scale reading of infinity (no continuity). The switch must "open" when the temperature cools to 160° to 180°F.
9. Replace the temperature sender if the switch fails to "open" or "close" within the specified temperature range.

ENGINE MECHANICAL 7-2
THE TWO-STROKE CYCLE 7-2
POWERHEAD 7-2
    REMOVAL & INSTALLATION 7-2
REED VALVE 7-6
    REMOVAL & INSTALLATION 7-6
    INSPECTION & CLEANING 7-8
POWERHEAD RECONDITIONING 7-8
DETERMINING POWERHEAD
  CONDITION 7-8
    COMPRESSION TEST 7-8
BUY OR REBUILD? 7-9
POWERHEAD OVERHAUL TIPS 7-9
    TOOLS 7-9
    CAUTIONS 7-10
    CLEANING 7-10
    REPAIRING DAMAGED
      THREADS 7-10
POWERHEAD PREPARATION 7-11
CYLINDER BLOCK AND HEAD 7-11
    GENERAL INFORMATION 7-11
    INSPECTION 7-12
CYLINDER BORES 7-13
    GENERAL INFORMATION 7-13
    INSPECTION 7-13
    REFINISHING 7-13
PISTONS 7-14
    GENERAL INFORMATION 7-14
    INSPECTION 7-14
PISTON PINS 7-15
    GENERAL INFORMATION 7-15
    INSPECTION 7-15
PISTON RINGS 7-16
    GENERAL INFORMATION 7-16
    INSPECTION 7-16
CONNECTING RODS 7-17
    GENERAL INFORMATION 7-17
    INSPECTION 7-17
CRANKSHAFT 7-18
    GENERAL INFORMATION 7-18
    INSPECTION 7-18
BEARINGS 7-19
    GENERAL INFORMATION 7-19
    INSPECTION 7-20
FLYWHEEL TIMING MARKS 7-20
    USING THE FACTORY TIMING
      GAUGE 7-20
    ALTERNATE METHOD 7-21
**POWERHEAD EXPLODED
  VIEWS 7-22
TORQUE SEQUENCE
  DIAGRAMS 7-29**

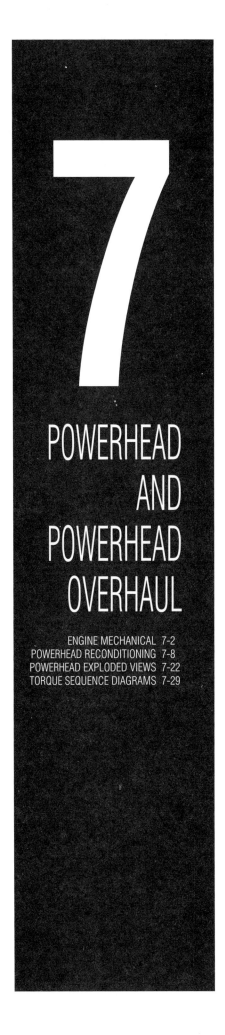

7
POWERHEAD
AND
POWERHEAD
OVERHAUL

ENGINE MECHANICAL 7-2
POWERHEAD RECONDITIONING 7-8
POWERHEAD EXPLODED VIEWS 7-22
TORQUE SEQUENCE DIAGRAMS 7-29

7-2 POWERHEAD AND POWERHEAD OVERHAUL

ENGINE MECHANICAL

The Two-Stroke Cycle

The two-stroke engine can produce substantial power for its size and weight. But why is a two-stroke so much smaller and lighter than a four-stroke? Well, there is no valve train. Camshafts, valves and push rods can really add weight to an engine. A two-stroke engine doesn't use valves to control the air and fuel mixture entering and exiting the engine. There are holes, called ports, cut into the cylinder, which allow for entry and exit of the fuel mixture. The two-stroke engine also fires on every second stroke of the piston, which is the primary reason why so much more power is produced than a four-stroke.

Since two-stroke engines discharge approximately one fourth of their fuel unburned, they have come under close scrutiny by environmentalists. Many states have tightened their grip on two-strokes and most manufacturers are hard at work developing new efficient models that can meet the tough emissions standards. Check out your state's regulations before you buy any two-stroke outboard.

The two-stroke engine is able to function because of two very simple physical laws. The first, gases will flow from an area of high pressure to an area of lower pressure. A tire blowout is an example of this principle. The high-pressure air escapes rapidly if the tube is punctured. Second, if a gas is compressed into a smaller area, the pressure increases, and if a gas expands into a larger area, the pressure is decreased. If these two laws are kept in mind, the operation of the two-stroke engine will be easier understood.

Two-stroke engines utilize an arrangement of port openings to admit fuel to the combustion chamber and to purge the exhaust gases after burning has been completed. The ports are located in a precise pattern in order for them to be opened and closed at an exact moment by the piston as it moves up and down in the cylinder. The exhaust port is located slightly higher than the fuel intake port. This arrangement opens the exhaust port first as the piston starts downward and therefore; the exhaust phase begins a fraction of a second before the intake phase.

Actually, the intake and exhaust ports are spaced so closely together that both open almost simultaneously. For this reason, the pistons of most two-stroke engines have a deflector-type top. This design of the piston top serves two purposes very effectively. First, it creates turbulence when the incoming charge of fuel enters the combustion chamber. This turbulence results in more complete burning of the fuel than if the piston top were flat. Second, it forces the exhaust gases from the cylinder more rapidly.

Beginning with the piston approaching top dead center on the compression stroke, the intake and exhaust ports are closed by the piston, the reed valve is open, the spark plug fires, the compressed air/fuel mixture is ignited, and the power stroke begins. The reed valve was open because as the piston moved upward, the crankcase volume increased, which reduced the crankcase pressure to less than the outside atmosphere.

As the piston moves downward on the power stroke, the combustion chamber is filled with burning gases. As the exhaust port is uncovered, the gases, which are under great pressure, escape rapidly through the exhaust ports. The piston continues its downward movement. Pressure within the crankcase increases, closing the reed valves against their seats. The crankcase then becomes a sealed chamber. The air/fuel mixture is compressed ready for delivery to the combustion chamber. As the piston continues to move downward, the intake port is uncovered. A fresh air/fuel mixture rushes through the intake port into the combustion chamber striking the top of the piston where it is deflected along the cylinder wall. The reed valve remains closed until the piston moves upward again.

When the piston begins to move upward on the compression stroke, the reed valve opens because the crankcase volume has been increased, reducing crankcase pressure too less than the outside atmosphere. The intake and exhaust ports are closed and the fresh fuel charge is compressed inside the combustion chamber.

Pressure in the crankcase decreases as the piston moves upward and a fresh charge of air flows through the carburetor picking up fuel. As the piston approaches top dead center, the spark plug ignites the air/fuel mixture, the power stroke begins and one full cycle has been completed.

The exact time of spark plug firing depends on engine speed. At low speed the spark is retarded, fires later than when the piston is at or beyond top dead center. Engine timing is built into the unit at the factory.

At high speed, the spark is advanced and fires earlier than when the piston is at top dead center. On all but the smallest horsepower outboards the timing can be changed adjusted to meet advance and retard specifications.

Because of the design of the two-stroke engine, the fuel passing through the engine must deliver lubrication of the piston and cylinder walls. Since gasoline doesn't make a good lubricant, oil must be added to the fuel and air mixture. The trick here is to add just enough oil to the fuel to provide lubrication. If too much oil is added to the fuel, the spark plug can become "fouled" because of the excessive oil within the combustion chamber. If there is not enough oil present with the air/fuel mixture, the piston can "seize" within the cylinder. What usually happens in this case is the piston and cylinder become scored and scratched, from lack of lubrication. In extreme cases, the piston will turn to liquid and eventually disintegrate within the cylinder.

Most two-stroke engines require that the fuel and oil be mixed before being poured into the fuel tank. This is known as "pre-mixing" the fuel. This can become a real hassle. You must be certain that the ratio is correct. Too little oil in the fuel could cause the piston to seize to the cylinder, causing major engine damage and completely ruining your weekend. Most modern two-stroke engines have an oil injection system that automatically mixes the proper amount of oil with the fuel as it enters the engine.

Powerhead

REMOVAL & INSTALLATION

When removing any powerhead, it is a good idea to make a sketch or take an instant picture of the location, routing and positioning of electrical harnesses, brackets and component locations for installation reference.

➡**Sometimes when attempting to remove the powerhead it won't come loose from the adapter. The gasket may hold the powerhead. Rock the powerhead back and forth or give it a gentile nudge with a pry bar. If the gasket breaks loose and the powerhead still will not come loose, then the driveshaft is seized to the crankshaft at the splines.**

The following procedures assume that tiller handle controlled outboards have been removed from the boat and placed on a suitable work stand. If a remote controlled engines powerhead is being removed be certain to remove the fuel line and the battery cables prior to disassembly.

On some powerheads it will be necessary to remove attached components if the powerhead is to be overhauled. Refer to the specific sections covering these components for removal and installation information.

In the event that the power head is to be removed for a service procedure other than a rebuild the total removal of the bolt on components is not necessary.

3 to 4 HP

♦ **See Figure 1**

1. Remove engine cover.
2. Remove the manual starter pull handle to allow the starter rope to be removed from the engine cover then tie a knot in the starter rope so that it does not retract all the way into the recoil assembly.
3. Remove the fasteners holding the starter in place.
4. Lift the manual starter off the engine.
5. Remove the engines fuel tank.
6. Remove fuel shut off handle and valve assembly.
7. Remove the E-clip holding the choke mechanism in place.
8. Remove the carburetor.
9. Disconnect the shut off switch wires.
10. Remove the front plate that houses the choke knob and stop button.
11. Remove the lower motor pan.
12. Remove the fasteners holding the powerhead in place.
13. Remove the powerhead.
14. Place on a clean suitable workbench.

To install:

15. Using a new gasket place the powerhead on the exhaust housing aligning the driveshaft splines with the crankshaft.
16. Install the fasteners holding the powerhead in place.
17. Install the lower motor pan.

POWERHEAD AND POWERHEAD OVERHAUL   7-3

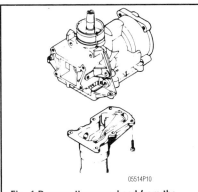

Fig. 1 Remove the powerhead from the exhaust housing

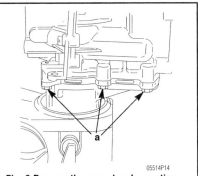

Fig. 2 Remove the powerhead mounting bolts and lift powerhead off of the exhaust housing

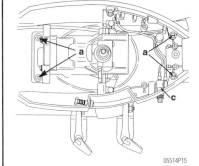

Fig. 3 To remove the lower cowl the mounting hardware and stop switch wiring need to be removed

18. Install the front plate that houses the choke knob and stop button.
19. Connect the shut off switch wires.
20. Install the carburetor.
21. Install the E-clip holding the choke mechanism in place.
22. Install the engines fuel tank.
23. Install the manual starter and fasten in place.
24. Pull the starter rope out and untie the knot. Put the pull rope through the engine cover and install the pull handle. Secure the rope with a knot.
25. Install the motor cover.

5 HP

♦ See Figures 2 and 3

1. Remove engine cover.
2. Remove the engines fuel tank, If equipped.
3. Disconnect the engines fuel supply connector, If equipped.
4. Remove the flywheel.
5. Remove the stator plate.
6. Remove the fuel pump.
7. Remove the carburetor.
8. Disconnect the stop switch wires.
9. Remove the lower engine pan.
10. Remove the fasteners holding the powerhead in place.
11. Remove the powerhead.
12. Place on a clean suitable workbench.

To install:

13. Using a new gasket place the powerhead on the exhaust housing aligning the driveshaft splines with the crankshaft.
14. Install the fasteners holding the powerhead in place.
15. Install the lower motor pan.
16. Connect the shut off switch wires
17. Install the carburetor.
18. Install the fuel pump.
19. Install the stator plate.
20. Install the flywheel.
21. Install the engines fuel tank, If equipped.
22. Install the fuel connector, If equipped.
23. Install engine cover.

7.5 to 15 HP

1. Remove the engine cover.
2. Disconnect the engines fuel supply connector.
3. Remove the spark plugs
4. Remove the manual starter pull handle to allow the starter rope to be removed from the engine cover then tie a knot in the starter rope so that it does not retract all the way into the recoil assembly.
5. Remove the fasteners holding the starter in place.
6. Lift the manual starter off the engine.
7. Remove the fuel pump.
8. Remove the carburetor.
9. Remove the flywheel.
10. Remove the stator plate.

11. Disconnect the neutral interlock arm that connects the shifter to the throttle stop.
12. Disconnect the wiring for the stop and emergency stop switches.
13. Remover the six fasteners holding the powerhead into place.
14. Remove the powerhead.
15. Place on a clean suitable workbench.

To install:

16. Using a new gasket place the powerhead on the exhaust housing aligning the driveshaft splines with the crankshaft.
17. Install the fasteners holding the powerhead in place.
18. Connect the shut off switch wires
19. Connect the neutral interlock arm assembly.
20. Install the stator plate.
21. Install the flywheel.
22. Install the carburetor.
23. Install the fuel pump.
24. Install the manual starter.
25. Pull the starter rope out and untie the knot. Put the pull rope through the engine cover and install the pull handle. Secure the rope with a knot.
26. Install the fuel connector.
27. Install engine cover.

25 HP

1. Remove the engine cover.
2. Disconnect the engines fuel supply connector.
3. Remove the spark plugs
4. Remove the fasteners holding the starter in place.
5. Lift the manual starter off the engine.
6. Remove the fuel pump.
7. Remove the carburetor.
8. Remove the flywheel.
9. Remove the stator plate.
10. Disconnect the neutral interlock arm that connects the shifter to the throttle stop.
11. Disconnect the wiring for the stop and emergency stop switches.
12. Remover the six fasteners holding the powerhead into place.
13. Remove the powerhead.
14. Place on a clean suitable workbench.

To install:

15. Using a new gasket place the powerhead on the exhaust housing aligning the driveshaft splines with the crankshaft.
16. Install the fasteners holding the powerhead in place.
17. Connect the wiring for the emergency stop and shut off switches.
18. Connect the neutral interlock arm assembly.
19. Install the stator plate.
20. Install the flywheel.
21. Install the carburetor.
22. Install the fuel pump.
23. Install the manual starter.
24. Install the spark plugs.
25. Install the fuel connector.
26. Install engine cover.

7-4 POWERHEAD AND POWERHEAD OVERHAUL

1984–1987 35 HP

1. Remove the engine cover.
2. Disconnect the battery leads from the battery, negative first then the positive.
3. Disconnect the engines fuel supply connector.
4. Remove the spark plugs.
5. Remove the flywheel.
6. Remove the charging system components/starter solenoid and terminal board components.
7. Remove the stator/ magneto plate and the remaining ignition components.
8. Remove the control linkages for the throttle.
9. Disconnect the choke rod and fuel supply hose.
10. Remove the carburetor.
11. Remove the rear half of the exhaust mid housing cover.
12. Remove the nuts from the shock mounts holding the front exhaust cover in place to be able to access the front two powerhead mounting screws.
13. Remove the lower engine mounts to allow enough room to slide down the front cover plate and be able to reach all the powerhead mounting screws.
14. Remove the powerhead mounting screws.

✱✱ WARNING

Some early engines have a powerhead mounting screw hidden under a removable exhaust muffler adapter. Be certain that if the engine has this fastener remove it before attempting to lift the powerhead from the exhaust housing.

15. Install a lifting ring on to the crankshaft.
16. Remove the powerhead.
17. Place on a clean suitable workbench.

To install:

18. Using a new gasket place the powerhead on the exhaust housing aligning the driveshaft splines with the crankshaft.
19. Install the fasteners holding the powerhead in place.
20. Install the lower engine mounts.
21. Install the nuts holding the shock mounts on the front exhaust cover.
22. Install the rear half of the exhaust mid housing cover.
23. Install the carburetor.
24. Connect the choke rod and fuel supply hose.
25. Install the control linkages for the throttle.
26. Install the stator/ magneto plate and the remaining ignition components.
27. Install the flywheel.
28. Install the spark plugs.
29. Connect the engines fuel supply connector.
30. Connect the battery leads from the battery. "Positive first than the negative"
31. Install the engine cover.

1988–91 35 To 50 HP and 1985 60 HP

♦ See Figure 4

The models included in this section are grouped because of very similar removal procedures. The 35HP, 40HP, 50HP and the 1985 60HP are based on a two cylinder design with comparable features.

1. Remove the engine cover.
2. Disconnect the engines fuel supply connector.
3. Disconnect the battery leads from the battery, negative first than the positive.
4. Remove the spark plugs.
5. Disconnect the main harness connections from the terminal board.
6. Remove the flywheel.
7. Remove the stator and trigger assemblies.
8. Disconnect the circuit breaker and terminal board.
9. Remove the starter solenoid.
10. Remove the CD module, rectifier and shift inter lock switch.
11. Disconnect the choke solenoid wire.
12. Remove the electrical/ coil bracket.
13. Remove the starter.
14. Remove the control linkages for the throttle including the tower shaft and throttle cam link rod.

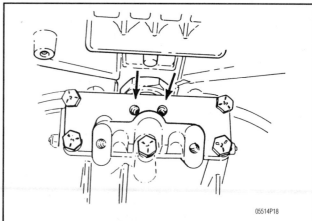

Fig. 4 The upper shock mount bolts are located in the front of the engine under the intake manifold

15. Remove the carburetor.
16. Remove the choke solenoid.
17. Remove the fuel pump assembly.
18. Disconnect the shift rod.
19. Remove the bolts holding the powerhead including the two 12 point fasteners in the front adjacent to the upper cowl shock mounts.
20. Remove the negative battery cable from the lower motor pan.
21. Remove the nuts securing the upper cowl shock mounts.
22. Install a lifting ring on to the crankshaft.
23. Remove the powerhead.
24. Place on a clean suitable workbench.

To install:

25. Using a new gasket place the powerhead on the exhaust housing aligning the driveshaft splines with the crankshaft.
26. Install the bolts that hold the powerhead including the two 12 point fasteners in the front adjacent to the upper cowl shock mounts.
27. Install the nuts that secure the upper cowl shock mounts.
28. Remove a lifting ring on to the crankshaft.
29. Connect the shift rod.
30. Install the fuel pump assembly.
31. Install the choke solenoid.
32. Install the carburetor.
33. Install the control linkages for the throttle including the tower shaft and throttle cam link rod.
34. Install the starter.
35. Install the electrical/ coil bracket.
36. Connect the choke solenoid wire.
37. Install the CD module, rectifier and shift inter lock switch.
38. Install the starter solenoid.
39. Connect the circuit breaker and mount the terminal board.
40. Install the stator and trigger assemblies.
41. Install the flywheel.
42. Connect the main harness connections from the terminal board.
43. Install the spark plugs.
44. Connect the battery leads from the battery, positive first than the negative.
45. Connect the engines fuel supply connector.
46. Install the engine cover.

1992–99 40 and 50 HP

1. Remove the engine cover.
2. Disconnect the battery leads from the battery, negative first than the positive.
3. Disconnect the engines fuel supply connector.
4. Remove the spark plugs.
5. Disconnect the remote control cables and main harness plug.
6. Remove battery cable connections at the engine block and starter solenoid.
7. Remove the ground terminal for the power trim & tilt system.
8. Remove the lower engine cowl covering the powerhead mounting hardware.

POWERHEAD AND POWERHEAD OVERHAUL 7-5

9. Disconnect the shift rod.
10. Remove the bolts holding the power head in place.
11. Lift the power head off the exhaust housing with an appropriate lifting device.
12. Place on a clean suitable workbench.

➡ Once the powerhead is placed on the workbench it can have the bolt on components removed as described in the following steps. The reason the newer power heads are removed complete is because the lifting ring is designed to thread into the flywheel therefore it can not be removed before the powerhead is removed.

13. Remove the flywheel.
14. Remove the stator and trigger assemblies.
15. Disconnect the circuit breaker and terminal board, as applicable.
16. Remove the starter solenoid.
17. Remove the CD module, Ignition coils and/or cdm modules.
18. Disconnect the choke solenoid or fuel primer wire.
19. Remove the electrical/ coil bracket including the starter solenoid rectifier etc.
20. Remove the starter.
21. Remove the control linkages for the throttle including the tower shaft and throttle cam link rod.
22. Remove the carburetor.
23. Remove the choke solenoid or primer valve.
24. Remove the fuel pump assembly.

To install:
25. Install the bolt on components in the following order then install the powerhead.
26. Install the fuel pump assembly.
27. Install the choke solenoid or primer valve.
28. Install the carburetor.
29. Install the control linkages for the throttle including the tower shaft and throttle cam link rod.
30. Install the starter.
31. Install the electrical/ coil bracket including the starter solenoid rectifier etc.
32. Connect the choke solenoid or fuel primer wire.
33. Install the CD module, Ignition coils and/or cdm modules.
34. Connect the circuit breaker and terminal board, as applicable.
35. Install the stator and trigger assemblies.
36. Install the flywheel.
37. Install a lifting ring on to the flywheel.
38. Using a new gasket place the powerhead on the exhaust housing aligning the driveshaft splines with the crankshaft.
39. Install the bolts holding the power head in place.
40. Connect the shift rod.
41. Install the lower engine cowl covering the powerhead mounting hardware.
42. Install the ground terminal for the power trim & tilt system.
43. Install battery cable connections at the engine block and starter solenoid.
44. Connect the remote control cables and main harness plug.
45. Install the spark plugs.
46. Connect the engines fuel supply connector.
47. Connect the battery leads from the battery, positive first than the negative.
48. Install the engine cover.

1984–96 70 to 150 HP

1. Remove the engine cover.
2. Disconnect the engines fuel supply connector.
3. Disconnect the battery leads from the battery, negative first than the positive.
4. Remove the spark plugs.
5. Disconnect the main harness connections from the terminal board.
6. Remove the flywheel.
7. Disconnect the remaining electrical and ignition components, as follows:
    a. Remove the stator and trigger assemblies or distributor and ignition coil.
    b. Remove the mounting hardware and lift the CD coil bracket from the engine block.
    c. Remove the shock mounts for the CD coil bracket from the engine block.
    d. Disconnect and remove the starter solenoid.
    e. Remove the engine ground wire from the cylinder block.
    f. Disconnect and remove the starter and rectifier assemblies.
    g. Disconnect the choke solenoid or primer valve.
    h. Disconnect the temperature warning sensor wire from the terminal strip.
    i. Remove the fasteners holding the terminal strip and lift it off the engine block.
    j. Remove the shift interlock switch.
8. Lay the electrical system components carefully on the workbench.
9. Remove the fuel system.
    a. Remove the fuel pump or pumps.
    b. Remove the air box cover or individual carb covers as applicable.
    c. Remove and secure the fuel supply hoses.
    d. Disconnect the carburetor throttle and choke plate link rods.
    e. Remove the bolts holding the carburetors in place.
10. Lay the fuel system components carefully on the workbench.
11. Remove the choke solenoid.
12. Remove the gearshift arm connection for the interlock cam located below the intake manifold.
13. Remove the exhaust housing cover to gain access to the powerhead mounting bolts.
14. Remove the six nuts that are exposed by removing the exhaust cover.
15. Remove the three bolts on each side of the front of the block holding the spacer plate.
16. Lift the power head off the exhaust housing with an appropriate lifting device.
17. Place on a clean suitable workbench.
18. Install an appropriate lifting device on the crankshaft.

To install:
19. Using a new gasket place the powerhead on the exhaust housing aligning the driveshaft splines with the crankshaft.
20. Install the three bolts on each side of the front of the block holding the spacer plate.
21. Install the six nuts that are exposed by removing the exhaust cover.
22. Install the exhaust housing cover to gain access to the powerhead mounting bolts.
23. Install the gearshift arm connection for the interlock cam located below the intake manifold.
24. Install the choke solenoid.
25. Install the fuel system.
26. Install the shift interlock switch.
27. Install the fasteners holding the terminal strip and lift it off the engine block.
28. Connect the temperature warning sensor wire from the terminal strip.
29. Connect the choke solenoid or primer valve.
30. Connect and remove the starter and rectifier assemblies
31. Install the engine ground wire from the cylinder block.
32. Connect and remove the starter solenoid.
33. Install the shock mounts for the CD coil bracket from the engine block.
34. Install the mounting hardware and lift the CD coil bracket from the engine block.
35. Install the stator and trigger assemblies or distributor and ignition coil.
36. Install the flywheel.
37. Connect the main harness connections from the terminal board.
38. Install the spark plugs.
39. Connect the battery leads from the battery, positive first than the negative.
40. Connect the engines fuel supply connector.
41. Install the engine cover.

1997–99 70 to 150 HP

1. Remove the engine cover.
2. Disconnect the battery leads from the battery, negative first than the positive.
3. Disconnect the engines fuel supply connector.
4. Remove the spark plugs.
5. Disconnect the remote control cables and main harness plug.
6. Remove battery cable connections at the engine block and starter solenoid.

7-6 POWERHEAD AND POWERHEAD OVERHAUL

7. Remove the ground terminal for the power trim & tilt system.
8. Disconnect the power trim wires from the solenoids, as applicable.
9. Remove the lower engine cowl covering the powerhead mounting hardware.
10. Disconnect the shift rod.
11. Remove the bolts holding the power head in place.
12. Lift the power head off the exhaust housing with an appropriate lifting device.
13. Place on a clean suitable workbench.

➡ Once the powerhead is placed on the workbench it can have the bolt on components removed as described in the following steps. The reason the newer power heads are removed complete is because the lifting ring is designed to thread into the flywheel therefore it can not be removed before the powerhead is removed.

14. Remove the flywheel.
15. Remove the stator and trigger assemblies.
16. Disconnect the circuit breaker and terminal board, as applicable.
17. Remove the starter solenoid.
18. Remove the CD module, Ignition coils and/or cdm modules.
19. Disconnect the choke solenoid or fuel primer wire.
20. Remove the electrical/ coil bracket including the starter solenoid rectifier etc.
21. Remove the starter.
22. Remove the control linkages for the throttle including the tower shaft and throttle cam link rod.
23. Remove the carburetor.
24. Remove the choke solenoid or primer valve.
25. Remove the fuel pump assembly.

To install:

26. Install the bolt on components in the following order then install the powerhead.
27. Install the fuel pump assembly.
28. Install the choke solenoid or primer valve.
29. Install the carburetor.
30. Install the control linkages for the throttle including the tower shaft and throttle cam link rod.
31. Install the starter.
32. Install the electrical/ coil bracket including the starter solenoid rectifier etc.
33. Connect the choke solenoid or fuel primer wire.
34. Install the CD module, Ignition coils and/or cdm modules.
35. Connect the circuit breaker and terminal board, as applicable.
36. Install the stator and trigger assemblies.
37. Install the flywheel.
38. Install a lifting ring on to the flywheel.
39. Using a new gasket place the powerhead on the exhaust housing aligning the driveshaft splines with the crankshaft.
40. Install the bolts holding the power head in place.
41. Connect the shift rod.
42. Install the lower engine cowl covering the powerhead mounting hardware.
43. Install the ground terminal for the power trim & tilt system.
44. Connect the power trim wires from the solenoids, as applicable.
45. Install battery cable connections at the engine block and starter solenoid.
46. Connect the remote control cables and main harness plug.
47. Install the spark plugs.
48. Connect the engines fuel supply connector.
49. Connect the battery leads from the battery. "Positive first than the negative"
50. Install the engine cover.

Reed Valve

The reed valves operate in response to changes in crankcase pressure. Located between the intake manifold and the crankcase, the reed valves admit the air-fuel mixture into the crankcase and during the scavenging stroke, act as a one-way valve to prevent the mixture from flowing back into the intake manifold. The travel of the reed itself is limited by the reed stop. By this action, the scavenging action is improved and the engine will produce greater power.

REMOVAL & INSTALLATION

3 and 4 HP

♦ See Figure 5

1. Remove the carburetor.
2. Remove the four fasteners holding the intake manifold in place.
3. Carefully remove the intake manifold.
4. Remove the reed stop plate and reed valves by loosening the two screws holding the assembly.

To install:

5. Install the reed stop plate and reed valves by placing the reed stop plate screws through the stop and reed valves.

➡ It is highly recommended that a screw locking compound such as Loctite® be used on the retaining screws

6. Install the intake manifold using a new gasket.
7. Install carburetor.

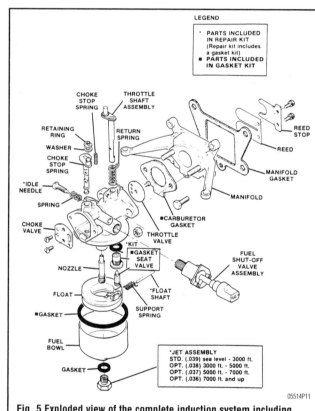

Fig. 5 Exploded view of the complete induction system including reed valve location

5 HP

♦ See Figures 6 and 7

1. Remove the carburetor.
2. Remove the six fasteners holding the intake manifold in place.
3. Carefully remove the intake manifold.
4. Remove the reed stop plate and reed valves by loosening the two screws holding the assembly.

To install:

5. Install the reed stop plate and reed valves by placing the reed stop plate screws through the stop and reed valves.

➡ It is highly recommended that a screw locking compound such as Loctite® be used on the retaining screws

6. Install the intake manifold using a new gasket.
7. Install carburetor.

POWERHEAD AND POWERHEAD OVERHAUL   7-7

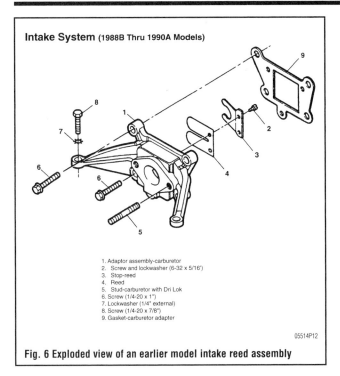

Fig. 6 Exploded view of an earlier model intake reed assembly

1. Adaptor assembly-carburetor
2. Screw and lockwasher (6-32 x 5/16")
3. Stop-reed
4. Reed
5. Stud-carburetor with Dri Lok
6. Screw (1/4-20 x 1")
7. Lockwasher (1/4" external)
8. Screw (1/4-20 x 7/8")
9. Gasket-carburetor adapter

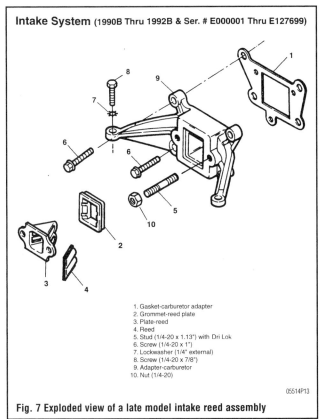

Fig. 7 Exploded view of a late model intake reed assembly

1. Gasket-carburetor adapter
2. Grommet-reed plate
3. Plate-reed
4. Reed
5. Stud (1/4-20 x 1.13") with Dri Lok
6. Screw (1/4-20 x 1")
7. Lockwasher (1/4" external)
8. Screw (1/4-20 x 7/8")
9. Adapter-carburetor
10. Nut (1/4-20)

7.5–15 HP

◆ See Figure 8

1. Remove the carburetor.
2. Remove the six fasteners that hold the intake to the cylinder block.
3. Carefully remove the intake manifold.
4. Remove the reed stop plate and reed valves by loosening the two screws holding each assembly.

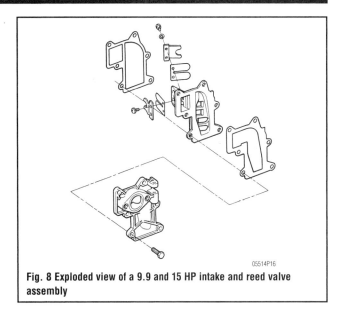

Fig. 8 Exploded view of a 9.9 and 15 HP intake and reed valve assembly

To install:

5. Install the reed stop plate and reed valves by placing the reed stop plate screws through the stop and reed valves.

➡ It is highly recommended that a screw locking compound such as Loctite® be used on the retaining screws

6. Install the intake manifold using a new gasket.
7. Install carburetor.

25 HP

◆ See Figure 9

1. Remove the carburetors.
2. Remove the six fasteners that hold the intake to the cylinder block.
3. Carefully remove the intake manifold.

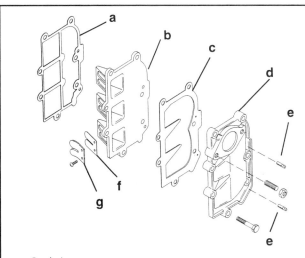

a - Gasket
b - Reed Plate
c - Gasket
d - Manifold
e - Check Valve
f - Reed
g - Reed Stop

Fig. 9 Exploded view of a 3 cylinder 25 HP intake and reed valve assembly

7-8 POWERHEAD AND POWERHEAD OVERHAUL

4. Remove the reed stop plate and reed valves by loosening the two screws holding each assembly.

To install:

5. Install the reed stop plate and reed valves by placing the reed stop plate screws through the stop and reed valves.

➡ It is highly recommended that a screw locking compound such as Loctite® be used on the retaining screws

6. Install the intake manifold using a new gasket.
7. Install carburetor.

35–150 HP

1. Remove the carburetor(s).
2. Remove the fasteners that hold each intake to the cylinder block.
3. Carefully remove each intake manifold.
4. Remove the reed box assemblies by loosening the mounting screws on the front of the intake plate.
5. Remove the reed stop plate and reed valves by loosening the screws holding each assembly.

To install:

6. Install the reed stop plate and reed valves by placing the reed stop plate screws through the stop and reed valves.
7. Using a new gasket install the reed box assemblies by tightening the mounting screws on the front of the intake plate securing the reed box assemblies.

➡ It is highly recommended that a screw locking compound such as Loctite® be used on the retaining screws for the reed stops and reed box assemblies.

8. Install the intake manifold using a new gasket.
9. Install carburetor.

INSPECTION & CLEANING

▶See Figures 10, 11 and 12

Always handle the reeds with the utmost care. Rough treatment will result in the reeds becoming distorted and will affect their performance.

Wash the reeds in solvent, and blow them dry with compressed air from the back side only. Do not blow air through the reed from the front side. Such action would cause the reed to open and fly up against the reed stop. Wipe the front of the reed dry with a lint free cloth.

Clean the reed box thoroughly by removing any old gasket material.

Secure the reed blocks to the reed plate with the screws tightened to a torque value of 25–35 Ft. lbs. (3–4Nm).

Check for chipped or broken reeds. Observe that the reeds are not preloaded or standing open. Satisfactory reeds will not adhere to the reed block surface, but still there is not more than the specified clearance between the reed and the block surface.

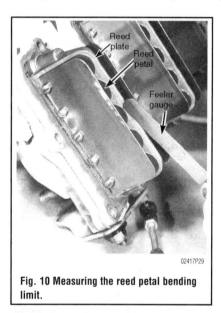

Fig. 10 Measuring the reed petal bending limit.

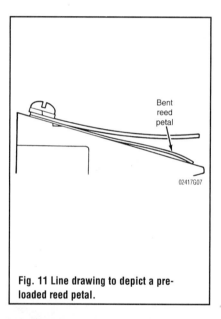

Fig. 11 Line drawing to depict a preloaded reed petal.

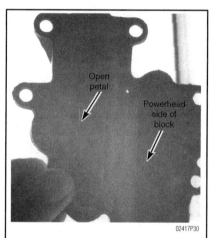

Fig. 12 If daylight can be detected when the reed block is held to the sunlight, the reed petal is distorted and the reeds must be replaced.

POWERHEAD RECONDITIONING

Determining Powerhead Condition

Anything that generates heat and/or friction will eventually burn or wear out (for example, a light bulb generates heat, therefore its life span is limited). With this in mind, a running powerhead generates tremendous amounts of both; friction is encountered by the moving and rotating parts inside the powerhead and heat is created by friction and combustion of the fuel. However, the powerhead has systems designed to help reduce the effects of heat and friction and provide added longevity. The oil injection system combines oil with the fuel to reduce the amount of friction encountered by the moving parts inside the powerhead, while the cooling system reduces heat created by friction and combustion. If either system is not maintained, a break-down will be inevitable. Therefore, you can see how regular maintenance can affect the service life of your powerhead.

There are a number of methods for evaluating the condition of your powerhead. A secondary compression test can reveal the condition of your pistons, piston rings, cylinder bores and head gasket(s). A primary compression test can determine the condition of all engine seals and gaskets. Because the 2-stroke powerhead is a pump, the crankcase must be sealed against pressure created on the down stroke of the piston and vacuum created when the piston moves toward top dead center. If there are air leaks into the crankcase, insufficient fuel will be brought into the crankcase and into the cylinder for normal combustion.

COMPRESSION TEST

▶ See Figure 13

The actual pressure measured during a secondary compression test is not as important as the variation from cylinder to cylinder. On multi-cylinder powerheads, a variation of 15 psi or more is considered questionable. On single cylinder powerheads, a drop of 15 psi from the normal compression pressure you established when it was new is cause for concern (you did do a compression test on it when it was new, didn't you?).

➡ If the powerhead been in storage for an extended period, the piston rings may have relaxed. This will often lead to initially low and misleading readings. Always run an engine to operating temperature to ensure that the reading you get is accurate.

POWERHEAD AND POWERHEAD OVERHAUL   7-9

Fig. 13 Check the secondary compression with an appropriate gauge, allow the engine to crank over enough to get three compression pulses

Fig. 14 The question of whether or not a powerhead is worth rebuilding is largely a subjective matter and one of personal worth. This powerhead is not worth much in its present condition

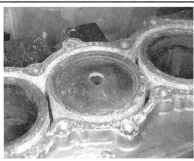

Fig. 15 A burned piston like this one will be replaced during an overhaul. The condition, which caused the hole in the top of the piston, must be identified and corrected or the same thing will happen again

1. Disable the ignition system by removing the lanyard clip. If you do not have a lanyard, take a wire jumper lead and connect one end to a good engine ground and the other end to the metal connector inside the spark plug boot, using one jumper for each plug wire. Never simply disconnect all the plug wires.

※※ CAUTION

Removing all the spark plugs and cranking over the powerhead can lead to an explosion if raw fuel/oil sprays out of the plug holes. A plug wire could spark and ignite this mix outside of the combustion chamber if it isn't grounded to the engine.

2. Remove all the spark plugs and be sure to keep them in order. Carefully inspect the plugs, looking for any inconsistency in coloration and for any sign of water or rust near the tip.
3. Thread the compression gauge into the No. 1 spark-plug hole, taking care too not cross thread the fitting.
4. Open the throttle to the wide open throttle position and hold it there.

➡ Some engines allow only minimal opening if the gearshift is in neutral, to guard against over-revving.

5. Crank over the engine an equal number of times for each cylinder you test, zeroing the gauge for each cylinder.
6. If you have electric start, count the number of seconds you count. On manual start, pull the starter rope four to five times for each cylinder you are testing.
7. Record your readings from each cylinder. When all cylinders are tested, compare the readings and determine if pressures are within the 15 psi criterion.
8. If compression readings are lower than normal for any cylinders, try a "wet" compression test, which will temporarily seal the piston rings and determine if they are the cause of the low reading.
9. Using a can of fogging oil, fog the cylinder with a circular motion to distribute oil spray all around the perimeter of the piston. Retest the cylinder.
   a. If the compression rises noticeably, the piston rings are sticking. You may be able to cure the problem by decarboning the powerhead.
   b. If the dry compression was low and no change is evident during the wet test, the cylinder is dead. The piston and/or rings are worn beyond specification and a powerhead overhaul or replacement is necessary.
10. If two adjacent cylinders on a multi-cylinder engine give a similarly low reading then the problem may be a faulty head gasket. This should be suspected if there was evidence of water or rust on the spark plugs from these cylinders.

Buy or Rebuild?

♦ See Figures 14 and 15

Now that you have determined that your powerhead is worn out, you must make some decisions. The question of whether or not a powerhead is worth rebuilding is largely a subjective matter and one of personal worth. Is the powerhead a popular one, or is it an obsolete model? Are parts available? Is the outboard it's being put into worth keeping? Would it be less expensive to buy a new powerhead, have your powerhead rebuilt by a pro, rebuild it yourself or buy a used powerhead? Or would it be simpler and less expensive to buy another outboard? If you have considered all these matters and more and have still decided to rebuild the powerhead, then it is time to decide how you will rebuild it.

➡ **The editors at Seloc® feel that most powerhead machining should be performed by a professional machine shop. Don't think of it as wasting money, rather, as insurance that the job has been done right the first time. There are many expensive and specialized tools required to perform such tasks as boring and honing a powerhead. Even inspecting the parts requires expensive micrometers and gauges to properly measure wear and clearances. Also, a machine shop can deliver to you clean and ready to assemble parts, saving you time and aggravation. Your maximum savings will come from performing the removal, disassembly, assembly and installation of the powerhead and purchasing or renting only the tools required to perform the above tasks. Depending on the particular circumstances, you may save 40 to 60 percent of the cost doing these yourself.**

A complete rebuild or overhaul of a powerhead involves replacing or reconditioning all of the moving parts (pistons, rods, crankshaft, etc.) with new or remanufactured ones and machining the non-moving wearing surfaces of the block and heads. Unfortunately, this may not be cost effective. For instance, your crankshaft may have been damaged or worn, but it can be machined for a minimal fee.

So, as you can see, you can replace everything inside the powerhead, but, it is wiser to replace only those parts that are really needed and, if possible, repair the more expensive ones.

Powerhead Overhaul Tips

♦ See Figure 16

Most powerhead overhaul procedures are fairly standard. In addition to specific parts replacement procedures and specifications for your individual powerhead, this section is also a guide to acceptable rebuilding procedures. Examples of standard rebuilding practice are given and should be used along with specific details concerning your particular powerhead.

Competent and accurate machine shop services will ensure maximum performance, reliability and powerhead life. In most instances it is more profitable for the do-it-yourself mechanic to remove, clean and inspect the component, buy the necessary parts and deliver these to a shop for actual machine work.

Much of the assembly work (crankshaft, bearings, pistons, connecting rods and other components) is well within the scope of the do-it-yourself mechanic's tools and abilities. You will have to decide for yourself the depth of involvement you desire in a powerhead repair or rebuild.

TOOLS

The tools required for a powerhead overhaul or parts replacement will depend on the depth of your involvement. With a few exceptions, they will be

7-10 POWERHEAD AND POWERHEAD OVERHAUL

Fig. 16 Much of the assembly work (crankshaft, bearings, pistons, connecting rods and other components) is well within the scope of the average do-it-yourself mechanic's tools and abilities

the tools found in an average do it yourselfer's tool kit. More in-depth work will require some or all of the following:
- A dial indicator (reading in thousandths) mounted on a universal base
- Micrometers and telescope gauges
- Jaw and screw-type puller
- Scraper
- Ring groove cleaner
- Piston ring expander and compressor
- Ridge reamer
- Cylinder hone or glaze breaker
- Plastigage®
- Powerhead stand

The use of most of these tools is illustrated in this section. Many can be rented for a one-time use from a local parts store or tool supply house.

Occasionally, the use of special tools is necessary. See the information on Special Tools and the Safety Notice in the front of this book before substituting another tool.

CAUTIONS

Aluminum is extremely popular for use in powerheads, due to its low weight. Observe the following precautions when handling aluminum parts:
- Never hot tank aluminum parts, the caustic hot tank solution will eat the aluminum
- Remove all aluminum parts (identification tag, etc.) from powerhead parts prior to hot tanking
- Always coat threads lightly with oil or anti-seize compounds before installation, to prevent seizure
- Never over tighten bolts or spark plugs especially in aluminum threads

When assembling the powerhead, any parts that will be exposed to frictional contact must be prelubed to provide lubrication at initial start-up. Any product specifically formulated for this purpose can be used.

When semi-permanent (locked, but removable) installation of bolts or nuts is desired, threads should be cleaned and coated with Loctite® or another similar, commercial non-hardening sealant.

CLEANING

Before the powerhead and its components are inspected, they must be thoroughly cleaned. You will need to remove any varnish, oil sludge and/or carbon deposits from all of the components to insure an accurate inspection. A crack in the block or cylinder head can easily become overlooked if hidden by a layer of sludge or carbon.

Most of the cleaning process can be carried out with common hand tools and readily available solvents or solutions. Carbon deposits can be chipped away using a hammer and a hard wooden chisel. Old gasket material and varnish or sludge can usually be removed using a scraper and/or cleaning solvent.

Extremely stubborn deposits may require the use of a power drill with a wire brush. Always follow any safety recommendations given by the manufacturer of the tool and/or solvent. You should always wear eye protection during any cleaning process involving scraping, chipping or spraying of solvents.

➡If using a wire brush, use extreme care around any critical machined surfaces (such as the gasket surfaces, bearing saddles, cylinder bores, etc.). Use of a wire brush is NOT RECOMMENDED on any aluminum components.

An alternative to the mess and hassle of cleaning the parts by yourself is to drop them off at a local machine shop. They will, more than likely, have the necessary equipment to properly clean all of the parts for a nominal fee.

✱✱ CAUTION

Always wear eye protection during any cleaning process involving scraping, chipping or spraying of solvents.

Remove any plugs or pressed-in bearings and carefully wash and degrease all of the powerhead components including the fasteners and bolts. Small parts should be placed in a metal basket and allowed to soak. Use pipe cleaner type brushes and clean all passageways in the components.

Use a ring expander to remove the rings from the pistons. Clean the piston ring grooves with a ring groove cleaner or a piece of broken ring. Scrape the carbon off of the top of the piston. You should never use a wire brush on the pistons. After preparing all of the piston assemblies in this manner, wash and degrease them again.

REPAIRING DAMAGED THREADS

▶ See Figures 17 thru 21

Several methods of repairing damaged threads are available. Heli-Coil®, Keenserts® and Microdot® are among the most widely used. All involve basically the same principle—drilling out stripped threads, tapping the hole and installing a prewound insert—making welding, plugging and oversize fasteners unnecessary.

Two types of thread repair inserts are usually supplied: a standard type for most inch coarse, inch fine, metric course and metric fine thread sizes and a

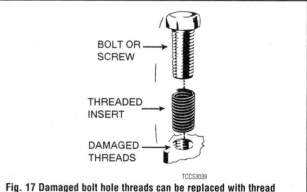

Fig. 17 Damaged bolt hole threads can be replaced with thread repair inserts

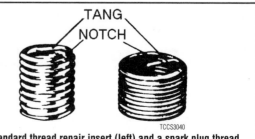

Fig. 18 Standard thread repair insert (left) and a spark plug thread insert

POWERHEAD AND POWERHEAD OVERHAUL

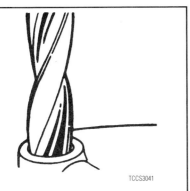

Fig. 19 Drill out the damaged threads with the specified size bit. Be sure to drill completely through the hole or to the bottom of a blind hole

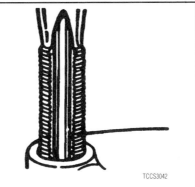

Fig. 20 Using the kit, tap the hole to receive the thread insert. Keep the tap well oiled and back it out frequently to avoid clogging the threads

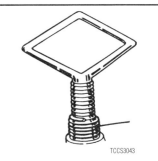

Fig. 21 Screw the insert onto the installer tool until the tang engages the slot. Thread the insert into the hole until it is ¼–½ turn below the top surface, then remove the tool and break of the tang using a punch

spark lug type to fit most spark plug port sizes. Consult the individual tool manufacturer's catalog to determine exact applications. Typical thread repair kits will contain a selection of prewound threaded inserts, a tap (corresponding to the outside diameter threads of the insert) and an installation tool. Spark plug inserts usually differ because they require a tap equipped with pilot threads and a combined reamer/tap section. Most manufacturers also supply blister-packed thread repair inserts separately in addition to a master kit containing a variety of taps and inserts plus installation tools.

Before attempting to repair a threaded hole, remove any snapped, broken or damaged bolts or studs. Penetrating oil can be used to free frozen threads. The offending item can usually be removed with locking pliers or using a screw/stud extractor. After the hole is clear, the thread can be repaired, as shown in the series of accompanying illustrations and in the kit manufacturer's instructions.

Powerhead Preparation

To properly rebuild a powerhead, you must first remove it from the outboard, then disassemble and inspect it. Ideally you should place your powerhead on a stand. This affords you the best access to the components. Follow the manufacturer's directions for using the stand with your particular powerhead.

Now that you have the powerhead on a stand, it's time to strip it of all but the necessary components. Before you start disassembling the powerhead, you may want to take a moment to draw some pictures, fabricate some labels or get some containers to mark and hold the various components and the bolts and/or studs which fasten them. Modern day powerheads use a lot of little brackets and clips which hold wiring harnesses and such and these holders are often mounted on studs and/or bolts that can be easily mixed up. The manufacturer spent a lot of time and money designing your outboard and they wouldn't have wasted any of it by haphazardly placing brackets, clips or fasteners. If it's present when you disassemble it, put it back when you assemble it, you will regret not remembering that little bracket which holds a wire harness out of the path of a rotating part.

You should begin by unbolting any accessories attached to the powerhead. Remove any covers remaining on the powerhead. The idea is to reduce the powerhead to the bare necessities (cylinder head(s), cylinder block, crankshaft, pistons and connecting rods), plus any other 'in block' components.

Cylinder Block and Head

GENERAL INFORMATION

▶ See Figures 22 thru 28

The cylinder block is made of aluminum and may have cast-in iron cylinder liners. It is the major part of the powerhead and care must be given to this part when service work is performed. Mishandling or improper service procedures performed on this assembly may make scrap out of an otherwise good casting. The cylinder assembly casting and other major castings on the outboard are expensive and need to be cared for accordingly.

There are three parts to the cylinder assembly, the cylinder block, the cylinder head and the crankcase half. The cylinder block and crankcase half are married together and line bored to receive the crankshaft bearings, reed blocks and on some powerheads sealing rings. After this operation they are treated as one casting.

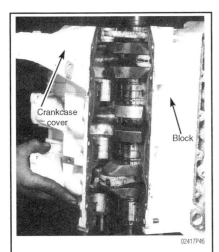

Fig. 22 The cylinder block and crankcase cover being separated to access the crankshaft.

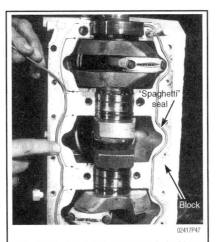

Fig. 23 The block halves are sealed with a spaghetti seal. Inspect that the seal groove is in good condition, never reuse spaghetti seal

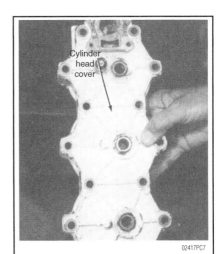

Fig. 24 A typical three cylinder head and cover assembly with the thermostat removed

7-12 POWERHEAD AND POWERHEAD OVERHAUL

Fig. 25 The cylinder block and crankcase half are machined to fit together perfectly. They provide a cradle for the spinning crankshaft

Fig. 26 To seal the top of the cylinder assembly around the crankshaft a gasket is installed around the end cap and neoprene seals are installed inside the cap and seal against the crankshaft

Fig. 27 The lower main bearing housing assembly incorporates the bearing and lower crankshaft seal

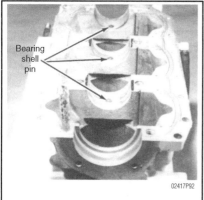

Fig. 28 Inspect the bearing alignment pins, they must be in place to align the crankshaft during assembly

Fig. 29 Inspect exhaust covers for warpage, cracks and rotted areas

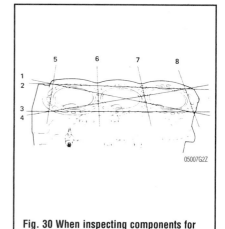

Fig. 30 When inspecting components for warping, check in multiple directions

➡Remember that anything done to the mating surfaces during service work will change the inner bore diameter for the main bearings, reed blocks and sealing rings and possibly prevent the block and crankcase mating surfaces from sealing.

The only service work allowed on the mating surface is a lapping operation to remove nicks from the service. Carefully guard this surface when other service work is being performed. The different sealing materials used to seal the mating surfaces are sealing strips, sealing compound and Loctite®.

Since the 2-stroke powerhead operates like a pump with one inlet and one outlet for each cylinder, special sealing features must be designed into the cylinder assembly to seal each individual cylinder in a multi-cylinder powerhead. Each inlet manifold must be completely sealed both for vacuum and pressure. One way of doing this internally is with a labyrinth seal, which is located between two adjacent cylinders next to the crankshaft. It may be of aluminum or brass, formed in the assembly and machined with small circular grooves running very close to a machined area on the crankshaft. The tolerance is so close that fuel residue puddling in the seal effectively completes the seal between the cylinder block and crankcase halves against the crankshaft. Crankcase pressures are therefore retained to each individual cylinder. No repair of the labyrinth seal is made. If damage has occurred to the seal, the main bearings have allowed the crankshaft to run out and rub.

Another method of internal sealing between the crankcases is with seal rings. These rings are installed in grooves in the crankshaft. When the crankshaft is installed, the sealing rings mate up to and seal against the web in the cylinder block crankcase halves and crankshaft. Sealing rings of different thickness are available for service work. The side tolerance is close, so puddled fuel residue will effectively complete the seal between crankcases and crankshaft.

To seal the ends of the cylinder assembly around the crankshaft, O-rings are installed around the end caps and neoprene seals are installed inside the cap and seal against the crankshaft.

INSPECTION

◆ See Figures 29, 30, 31 and 32

Everytime the cylinder head is removed, the cylinder head and cylinder block deck should be checked for warping. Do this with a straight edge or a surface block. If the cylinder head or cylinder block deck is warped, the surface should be machined flat by a competent machine shop. Using emery paper in a figure eight motion on a surface block until the surface is true may cure minor warpage.

Inspect the cylinder head and cylinder block for cracks and damage to the bolt holes caused by galvanic corrosion. On models that do not use a cylinder head, check the cylinder dome for holes or cracks caused by overheating and pre-ignition. The spark plug threads may also be damaged by over torquing the spark plug.

Quite often the small bolts around the cylinder block sealing area are seized by corrosion. If white powder is evident around the bolts, stop. Galvanic corrosion is probably seizing the shank of the bolt and possibly the threads as well. Putting a wrench on them may just twist the head off, creating one big mess. Know the strength of the bolt and stop before it breaks. If it does break, don't reach for an easy out, it won't work.

A good way to service these seized bolts is with localized heat (from a heat gun, not a torch) and a good penetrating oil. Heat the aluminum casting, not the

POWERHEAD AND POWERHEAD OVERHAUL 7-13

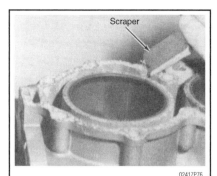

Fig. 31 Scrape the old gasket from the cylinder block and cylinder head, inspect for any gouges or areas that might have failed to seal with burn trace or water marks where the gasket should have been

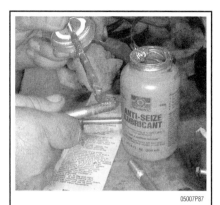

Fig. 32 To help prevent bolts from seizing due to corrosion, coat threads with a good antiseize compound.

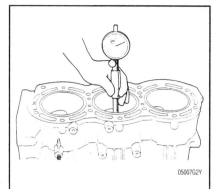

Fig. 33 The roundness of the cylinder diameter and the straightness of the cylinder wall should be inspected using a dial bore gauge

bolt. This releases the bolt from the corrosive grip by creating clearance between the bolt, the corrosion and the aluminum casting. Be careful because too much heat will melt the casting. Many bolts can be released in this way, preventing drilling out the total bolt and heli-coiling the hole or tapping the hole for an oversize bolt.

To help prevent bolts from seizing due to corrosion, coat threads with a good antiseize compound.

Cylinder Bores

GENERAL INFORMATION

The purpose of the cylinder bore is to help lock in combustion gases, provide a guided path for the piston to travel within, provide a lubricated surface for the piston rings to seal against and transfer heat to the cooling system. These functions are carried out through all engine speeds. To function properly the cylinder has to have a true machined surface and must have the proper finish installed on it to retain lubricant.

INSPECTION

▶ See Figures 33 and 34

The roundness of the cylinder diameter and the straightness of the cylinder wall should be inspected carefully. Micrometer readings should be taken at several points to determine the cylinder condition. Start at the bottom using an outside micrometer or dial bore gauge. By starting at the bottom, below the area of ring travel, cylinder bore diameter can be determined and a determination can be made if the powerhead is standard or has been bored oversize. Take the second measurement straight up from the first in the area of the ports and note that the cylinder is larger here. This is the area where the rings ride and it has worn slightly. Take the third measurement within a half inch of the top of the cylinder, straight up from where the second measurement was taken. These three measurements should be repeated with the measuring instrument turned 90° clockwise.

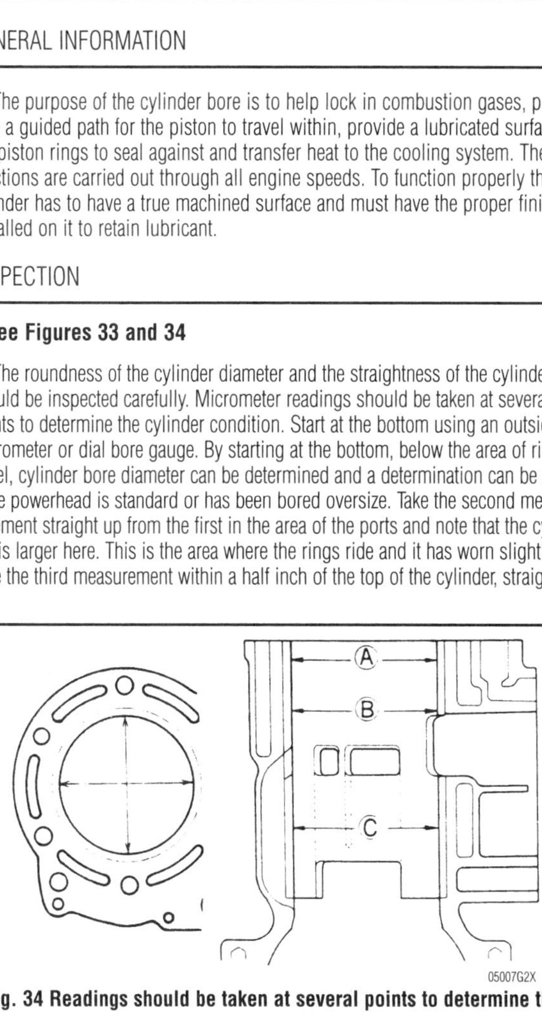

After the readings are taken, you will have enough information to access the cylinder condition. This will tell you if the rings can simply be replaced or if the cylinder will need to be over bored. While measuring the cylinder, you should also be noting if there is a cross-hatched pattern on the cylinder walls. Also note any scuffing or deep scratches.

REFINISHING

▶ See Figure 35

If the cylinder is out of round, worn beyond specification, scored or deeply scratched, reboring will be necessary. If the cylinder is within specification, it can be deglazed with a flex hone and new rings installed.

➡ Some cylinders are chrome plated and require special service procedures. Consult a qualified machine shop when dealing with chrome plated cylinders.

Almost all engine block refinishing must be performed by a machine shop. If the cylinders are not to be rebored, then the cylinder glaze can be removed with a ball hone. When removing cylinder glaze with a ball hone, use a light or penetrating type oil to lubricate the hone. Do not allow the hone to run dry as this may cause excessive scoring of the cylinder bores and wear on the hone. If new pistons are required, they will need to be installed to the connecting rods. This should be performed by a machine shop as the pistons must be installed in the correct relationship to the rod or engine damage can occur.

When deglazing, it is important to retain the factory surface of the cylinder wall. The cross-hatched patter on the cylinder wall is used to retain oil and seal the rings. As the piston rings move up and down the wall, a glaze develops. The hone is used to remove this glaze and reestablish the basket weave pattern. The pattern and the finish have a satin look and make an excellent surface for good retention of 2-stroke oil on the cylinder wall.

There is nothing magic about the crosshatch angle but there should be one similar to what the factory used. (Approximately 20–40°). Too steep an angle or too flat a pattern is not acceptable and as it is not good for ring seating. Since the hone reverses as it is being pushed down and pulled up the cylinder wall, many different angles are created. Multiple criss-crossing angles are the secret for longevity of the cylinder and the rings. The pattern allows 2-stroke oil to flow under the piston ring bearing surface and prevents a metal-to-metal contact between the cylinder wall and piston ring. The satin finish is necessary to prevent early break-in scuffing and to seat the ring correctly.

After the cylinder hone operation has been completed, one very important job remains. The grit that was developed in the machining process must be thoroughly cleaned up. Grit left in the powerhead will find its way into the bearings and piston rings and become embedded into the piston skirts, effectively grinding away at these precision parts. Relate this to emery paper applied to a piece of steel or steel against a grinding stone. The effect is removal of material from the steel. Grit left in the powerhead will damage internal components in a very short time.

Wiping down the cylinder bores with an oil or solvent soaked rag does not remove grit. Cleaning must be thorough so that all abrasive grit material has

Fig. 34 Readings should be taken at several points to determine the cylinder condition. Start at the bottom and work your way to the top

7-14 POWERHEAD AND POWERHEAD OVERHAUL

Fig. 35 If the cylinder is within specification, it can be deglazed with a flex hone and new rings installed

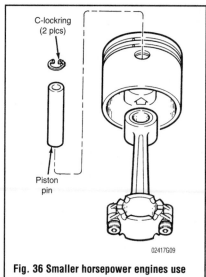

Fig. 36 Smaller horsepower engines use C-clips to hold the piston pins in place

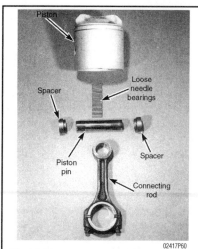

Fig. 37 The arrangement of the piston and its related parts from a larger horsepower engine

been removed from the cylinders. It is important to use a scrub brush and plenty of soapy water. Remember that aluminum is not safe with all cleaning compounds, so use a mild dish washing detergent that is designed to remove grease. After the cylinder is thought to be clean, use a white paper towel to test the cylinder. Rub the paper towel up and down on the cylinder and look for the presence of gray color on the towel. The gray color is grit. Re-scrub the cylinder until it is perfectly clean and passes the paper towel test. When the cylinder passes the test, immediately coat it with 2-stroke oil to prevent rust from forming.

➡Rust forms very quickly on clean, oil free metal. Immediately coat all clean metal with 2-stroke oil to prevent the formation of rust.

Pistons

GENERAL INFORMATION

♦ See Figures 36, 37, 38 and 39

Pistons are the moveable ends of a cylinder. The cylinder bore provides a guided path for the piston allowing a small clearance between the piston skirt and cylinder wall. This clearance allows for piston expansion and controls piston rock within the cylinder.

Modern piston design is such that the head of the piston directs incoming fuel toward the top of the cylinder and outgoing exhaust to the exhaust port in the cylinder wall. This design is called a deflector type piston head. The deflector dome deflects the incoming fuel upward to the spark plug end of the cylinder, partially cooling the cylinder and spark plug tip. It also purges the spent gases from the cylinder. In essence, the incoming fuel charge is chasing out the exhaust gases from the cylinder.

Not all piston designs are of the deflector head type. Other pistons have a small convex crown on the piston head. In this case, port design aids in directing the incoming fuel upward. The piston head bears the brunt of the combustion force and heat. Most of the heat is transferred from the piston head through the rings to the cylinder wall and then on to the cooling system.

The piston design can be round, cam ground or barrel shaped. The cam ground design allows for expansion of the piston in a controlled manner. As the piston heats up, expansion takes place and the piston moves out along the piston pin becoming more round as it warms up. Barrel shaped pistons rock very slightly in the bore, which helps to keep the rings free.

The piston has machined ring grooves in which the rings are installed. They are carried along with the piston as it travels up and down the cylinder wall. There is one small pin in each ring groove to prevent the ring from rotating. The piston skirt is the bearing area for thrust and rides on the cylinder wall oil film. The side thrust of the piston is dependent upon piston pin location. If the pin is in the center of the piston, then there will be more thrust. If the pin is offset a few thousandths of an inch from the center of the piston, there will be less thrust. A used piston will have one side of the piston skirt show more signs of wear than the opposite side. The side showing wear is the major thrust side.

Thrust is caused by the pendulum action of the rod following the crankshaft rotation, which pulls the rod out from under the piston. The combustion pressure therefore pushes and thrusts the piston skirt against the cylinder wall. Some heat is also transferred at this point. The other skirt receives only minor pressure. Some pistons have small grooves circling the skirts to retain oil in the critical area between the skirt and the cylinder wall.

INSPECTION

♦ See Figures 40 and 41

The piston needs to be inspected for damage. Check the head for erosion caused by excessive heat, lean mixtures and out of specification timing/syn-

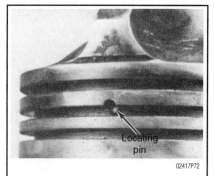

Fig. 38 There is one small pin in each ring groove to prevent the ring from rotating

Fig. 39 Piston diameter should be measured at a specific position on the piston that the manufacturer will specify

Fig. 40 This piston is severely scored from lack of lubrication and should not be reused

POWERHEAD AND POWERHEAD OVERHAUL   7-15

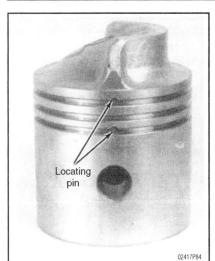

Fig. 41 The piston must be installed with the intake side "the right side from the locating pins" towards the intake side of the engine block

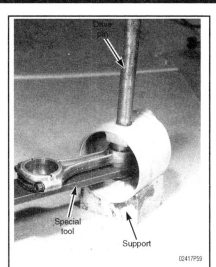

Fig. 42 The piston pins are removed and installed by using a press and a special tool as not to crush the piston as the pin slides in or out of the piston

Fig. 43 Measuring the piston pin bore inside diameter. This reading will be compared with the piston pin outside diameter to determine pin-to-bore clearance

chronization. Examine the ring land area to see if it is flat and not rounded over. Also look for burned through areas caused by preignition. Check the skirt for scoring caused by a break through of the oil film, excessive cylinder wall temperatures, incorrect timing/synchronization or inadequate lubrication.

To measure the piston diameter, place an outside micrometer on the piston skirt at the specified location. All pistons in a given powerhead should read the same. Check the specifications for placement of the micrometer when measuring pistons. Generally there is a specific place on the piston. This is especially true of barrel shaped pistons that are larger in the middle than they are at the top and bottom.

If the piston looks reasonably good after cleaning, take a close look at the ring lands. Wear may develop on the bottom of the ring lands. This wear is usually uneven, causing the ring to push on the higher areas and loads the ring unevenly when inertia is the greatest. Such uneven support of the ring will cause ring breakage and the piston will need to be replaced.

When installing a new ring in the groove, measure the ring side clearance against specification. Also check the see if the ring pins are there and that they have not loosened. Measure the skirt to see if the piston is collapsed.

Piston Pins

GENERAL INFORMATION

▶ See Figure 42

A hole placed in the side of the piston, commonly referred to as the piston boss, is used to mount the piston to the piston pin. The combustion pressure is transferred to the piston pin and connecting rod bearing, then on to the crankshaft where it is converted to rotary motion. The pin is fitted to the piston bosses. The piston pin is the inner bearing race for the bearing mounted in the small end of the connecting rod. This transfers the combustion pressures into the connecting rod and allows the rod to swing with a pendulum-like action.

Piston pins are secured into both piston bosses. All have retainers and in addition some use a press fit to secure the pin. There are some models that use a slip fit. These may require special installation techniques.

Another type of pin fitting is loose on one side and tight on the other. This type aids in removal of the pin without collapsing the piston. With this design, always press on the pin from the loose boss side. The piston is marked on the inside of the piston skirt with the word "loose" to identify the loose boss. Always press with the loose side up and press the pin all the way through and out. When installing, press with the loose side up.

In all pressing operations, set the piston in a cradle block to support the piston. Some pistons require heating to expand the piston bosses so the pin can be pressed out without collapsing the piston. Other pistons just have a slip fit.

INSPECTION

▶ See Figures 43, 44, 45 and 46

Check the piston pin retainer grooves for evidence of the retainers moving as they may have been distorted. Always replace the retainers once they have been removed. If there is evidence of wear in any of these areas, the piston should be replaced.

Inspect piston pin for wear in the bearing area. Rust marks caused by water will leave a needle bearing imprint. Chatter marks on the pin indicate that the

Fig. 44 Measure the piston pin outside diameter with a micrometer and check the measurement against the specification chart

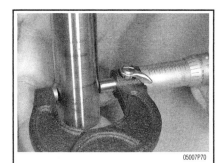

Fig. 45 Measure the piston pin on the bearing surface and compare the measurement to the outside diameter measurement to check for wear

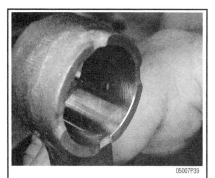

Fig. 46 The small end bore in the connecting rod must be perfectly round to prevent bearing troubles

7-16 POWERHEAD AND POWERHEAD OVERHAUL

piston pin should be replaced. If these marks are not too heavy, they may possibly be cleaned with emery paper for loose needle bearings or crocus cloth for caged bearings.

If the piston pin checks out visually, measure its outside diameter and compare that measurement with the inside diameter of the piston pin bore. Proper clearance is vital to providing enough lubrication.

Piston Rings

GENERAL INFORMATION

▶ See Figure 47

The piston ring seals the piston to the cylinder bore, just as other seals are used on the crankshaft and lower unit. To perform correctly, the rings must conform to the cylinder wall and maintain adequate pressure to insure their sealing action at required operating speeds and temperatures. There are different designs used throughout the outboard industry. A given manufacturer will select a ring design that meets the operating requirements of the powerhead. This may be a standard ring, a pressure back (Keystone) ring or a combination of rings.

The functions of the piston ring include sealing the combustion gases so they cannot pass between the piston and the cylinder wall into the crankcase upsetting the pulse and maintaining an oil film in conjunction with the cylinder wall finish throughout the ring travel area. The rings also transfer heat picked up by the piston during combustion. This heat is transferred into the cylinder wall and thus to the cooling system. There are either two or three rings per piston, which perform these functions.

➡ An oil control ring is not used on 2-stroke engines.

All piston rings used are of the compression type. This means that they are for sealing the clearance between piston and cylinder wall. They are not allowed to rotate on the piston as automotive piston rings do. They are prevented from rotating by a pin n the piston ring groove. If the ring was allowed to turn, a ring end could snap into the cylinder port and become broken. The ring ends are specially machined to compensate for the pin. As the rings warm up in a running powerhead they expand, thereby requiring a specific end gap between the ring ends for expansion. This ring gap decreases upon warm-up, effectively limiting blowing gases (from the combustion process) from going into the crankcase. The rings ride in a piston ring groove with minimal side clearance, which gives them support as they move up and down the cylinder wall. With this support, combustion gas pressure and oil effectively seal the piston ring against the ring land and the cylinder wall. As long as the oil mix is correct and temperatures remain where they should, the rings will provide service for many hours of operation.

INSPECTION

▶ See Figures 48 thru 53

One of the first indications of ring trouble is the loss of compression and performance. When compression has been lost or lowered because of the ring not sealing, the ring is either broken or stuck with carbon, gum or varnish. Improper oil mixing and stale gasoline provide the carbon, gum and varnish that cause the rings to stick. Low octane fuel, improperly adjusted timing/synchronization and lean fuel mixtures can damage the ring land, causing the ring to stick or break.

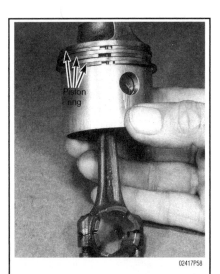

Fig. 47 The piston ring seals the piston to the cylinder bore

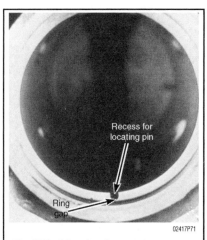

Fig. 48 To determine ring gap, use a feeler gauge to measure the expansion space between the ring ends with the ring installed in the cylinder

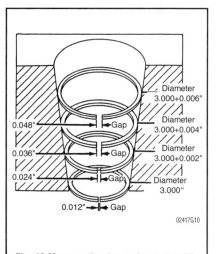

Fig. 49 Measure the ring end gap at multiple locations down the cylinder bore to check for cylinder taper

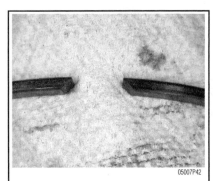

Fig. 50 Some rings are square shaped and are use together or as a pair with other shaped rings

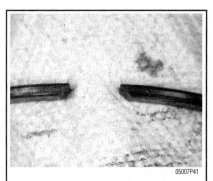

Fig. 51 Some rings are tapered and are generally used as a top ring or one of two rings on a particular piston

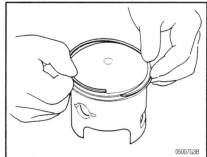

Fig. 52 When installing rings, place the ring in the groove and work it around the piston using a spiral motion until the ring is properly seated

POWERHEAD AND POWERHEAD OVERHAUL 7-17

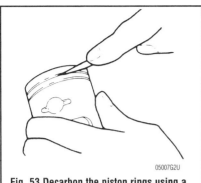

Fig. 53 Decarbon the piston rings using a ring groove cleaner or a broken piece of piston ring

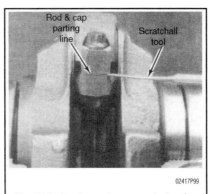

Fig. 54 Rod ends are freeze cracked and must be assembled as a matched set

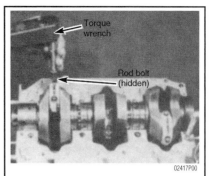

Fig. 55 Once the rod caps are perfectly matched they are torqued down evenly in one third increments

➡Running the outboard out of the water for even a few seconds can have damaging effects on the rings, pistons, cylinder walls and water pump.

To determine if the rings fit the cylinder and piston, two measurements are taken; ring gap and ring side clearance. To determine these measurements, the ring is pushed into the cylinder bore using the piston skirt, so it will be square. Position each ring, one at a time, at the bottom of the cylinder (the smallest diameter) and using a feeler gauge, measure the expansion space between the ring ends. This is known as the ring gap measurement. Compare this measurement against the specifications. If the measurement is too small, the ring must be filed to increase the gap. If it is too large, either the bore is too large or the ring is not correct for the powerhead.

After ring end gap has been determined, position each ring in the piston ring groove and using a feeler gauge, measure between the ring and the piston ring land. Compare this measurement against the specifications. If the measurement is too small, the ring groove may be compressed. Inspect the ring groove and ring land condition. If it is too large the ring may not correct for the powerhead.

Connecting Rods

GENERAL INFORMATION

The connecting rod transfers the combustion pressure from the piston pin to the crankshaft, changing the vertical motion into rotary motion. In doing so, the connecting rod swings back and forth on the piston pin like a pendulum while it is traveling up and down. It goes down by combustion pressure and goes up by flywheel momentum and/or other power strokes on a multi-cylinder powerhead. The connecting rod can be of aluminum on smaller horsepower fishing outboards or of steel on larger horsepower models.

Most connecting rod designs use a steel liner with needle bearings in the large end and a pressed-in needle bearing in the small end.

The steel rod is a bearing race at both the large and small ends of the rod. It is hardened to withstand the rolling pressures applied from the loose or caged needle bearings. Unlike many connecting rod designs, these rods do not use two piece caps. The connecting rod big end is one piece. This requires the crankshaft to be pressed together to form a rotating assembly with the connecting rods.

The connecting rods are mist lubricated. Some of the rods have a trough design in the shank area. Oil holes may be drilled into the bearing area at both ends of this trough. Oil mist that falls out of the fuel will settle into the rod trough and collect. As the rod moves in and out, the oil is sloshed back and forth in the trough and out the oil holes into the rod or piston pin bearings. This provides sufficient lubrication for these bearings. When the rod is equipped with oil holes, the oil holes have to be placed in the upward position toward the tapered end of the crankshaft when reassembled.

INSPECTION

◆ See Figures 54, 55, 56, 57 and 58

Damage to the connecting rod can be caused by lack of lubrication and will result in galling of the bearing and eventual seizing to the crankshaft. Over speeding of the powerhead may also cause the upper shank area of the rod to stretch and break near the piston pin.

Steel rods are inspected in the bearing areas, much like you would inspect a roller bearing. Look for scoring, pit marks, chatter marks, rust and color change. A blue color indicates overheating of the bearing surface. Minor rust marks or scoring may be cleaned up using crocus cloth for caged needle bearings or emery paper for loose needle bearings. A piece of round stock, cut with a slot in one end to accept a small piece of emery paper and mounted in a drill motor, can be used to clean up the rod ends.

The rod also needs to be checked to see if it is bent or has a twist in it. To do this, remove the piston and place the rod on a surface plate or a piece of flat glass (automotive widow). Using a flash light behind the rod and looking from in front of the rod, check for any light which can be seen under the rod ends. If light can be seen shining under the rod ends, the rod is bent and it must be replaced. You can also use a .002 feeler gauge. See if it will start under the

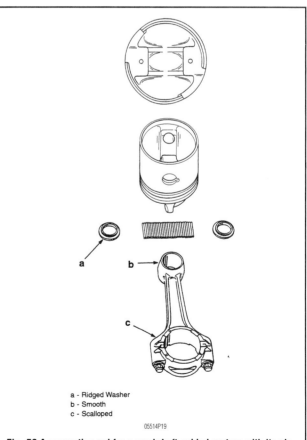

a - Ridged Washer
b - Smooth
c - Scalloped

Fig. 56 A connecting rod for a crankshaft guided system with its piston and bearings

7-18 POWERHEAD AND POWERHEAD OVERHAUL

machined area of the rod. If it will, the rod is bent. Examine the rod bolts and studs for damage and replace the nuts where used. Always reinstall the rod back on the same journal from which it was removed. The needle bearings, rod bearing surface and crankshaft journal are all mated to each other once the powerhead has been run.

➡ When installing the connecting rod, the long sloping side must be installed toward the exhaust side of the cylinder assembly and if there is a hole in the connecting rod, position the oil hole upward. Some piston designs are marked with the word "UP". This side should be placed toward the tapered end of the crankshaft.

Crankshaft

GENERAL INFORMATION

The crankshaft is used to convert vertical motion received from the mounted connecting rod into rotary motion, which turns the driveshaft. It mounts the flywheel, which imparts a momentum to smooth out pulses between power strokes. It also provides sealing surfaces for the upper and lower seals and provides a surface for the labyrinth seal to hold oil against and a groove in which sealing rings are installed to seal pressures into each crankcase. Mounted main bearings control the axial movement of the crankshaft as it accomplishes these functions. The crankshaft bearing journals are case hardened to be able to withstand the stresses applied by the floating needle bearings used for connecting rod and main bearings. In essence, the crankshaft journals are the inner bearing races for the needle bearings.

INSPECTION

▶ See Figures 59, 60, 61 and 62

Pressure from the power stoke applied to the crankshaft rod journal by the needle bearings has a tendency to wear the journal on one side. During crankshaft inspection, the journals should also be measured with a micrometer to determine if they are round and straight. They should also be inspected for scoring, pitting, rust marks, chatter marks and discoloration caused by heat.

Check the sealing surfaces for grooves worn in by the upper and lower crankshaft seals. Take a look at the splined area that receives the driveshaft. Inspect the side of the splines for wear. This wear can be caused by lack of lubrication or improper lubricant applied during a seasonal service. An exhaust housing/lower unit that has received a sudden impact can be warped and this can also cause spline damage in the crankshaft.

The crankshaft cannot be repaired because of the case hardening and the possibility of changing the metallurgical properties of the material during the welding and machine operation. Also, there are no oversized bearings available. Repairs are limited to cleaning up the journal surface with 320 emery paper when loose needle bearings are run on the journal. Where caged roller bearings are run, the journal may be polished with crocus cloth.

The tapered end of the crankshaft has a spline or a keyway and keys, which times the flywheel to the crankshaft. Inspect the spline or key and keyway for damage. The crankshaft taper should be clean and free of scoring, rust and lubrication. The taper must match the flywheel hub. If someone has hit the fly-

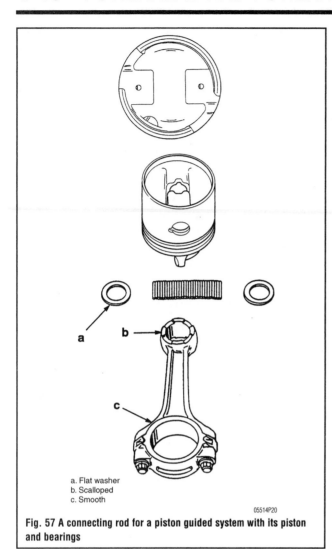

Fig. 57 A connecting rod for a piston guided system with its piston and bearings

a. Flat washer
b. Scalloped
c. Smooth

Fig. 58 Rod bearings are free floating bearings use only all new or all used bearings never mix

Fig. 59 Check the main bearing journals for specification and taper wear

Fig. 60 Check the connecting rod bearing journals for specification and taper wear

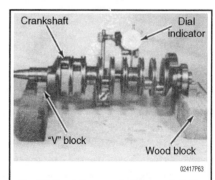

Fig. 61 Check the runout of the crankshaft, If bent or twisted it would need to be replaced

POWERHEAD AND POWERHEAD OVERHAUL 7-19

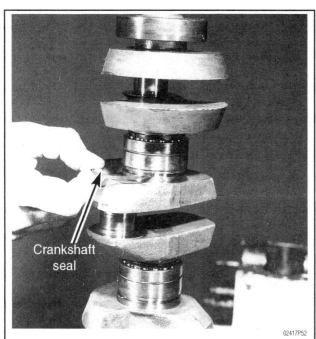

Fig. 62 Remove and inspect the crankshaft seals a bad seal will cause a loss of engine idle speed

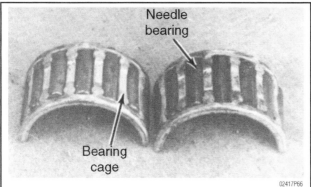

Fig. 63 Caged bearings from a connecting rod that are damaged beyond use

wheel with a heavy hammer or has used an improper puller to remove the flywheel, the flywheel and hub may be warped. Place the flywheel on the tapered end of the crankshaft and check the fit. If there is any rocking indication a distorted hub, replace the flywheel. Always use a puller that pulls from the bolt pattern or threaded inner hub of the flywheel. Never use a puller on the outside of the flywheel.

The taper is used to lock the flywheel hub to the crankshaft. When mounting the flywheel to the crankshaft the taper on the crankshaft an in the flywheel hub must be cleaned with a fast evaporating solvent. No lubrication on the crankshaft taper or flywheel hub should be done. The flywheel nut must be torqued to specification to obtain a press fit between the flywheel hub and the crankshaft taper. If the nut is not brought to specifications, the flywheel may spin on the crankshaft causing major damage.

The flywheel key is for alignment purposes and sets the flywheel's relative position to the crankshaft. Check the key for partial shearing on the side. If there is any indication of shearing, replace the key. Also check the keyway in the flywheel and crankshaft for damage. If there is damage that will allow incorrect positioning of the flywheel, the powerhead timing will be off.

Bearings

GENERAL INFORMATION

♦ See Figures 63, 64, 65, 66 and 67

Needle bearings are used to carry the load that is applied to the piston and rod. This load is developed in the combustion process and the bearings reduce the friction between the crankshaft and the connecting rod. They roll with little effort and at times have been referred to as anti-friction bearings, as they reduce friction by reducing the surface area that is in contact with the crankshaft and the connecting rod. These needle bearings are of two types, loose and caged. When loose bearings are used, there can be upwards to 32 loose bearings floating between the rod journal of the crankshaft and the connecting rod. These bearings are aided in rolling by the movement of the crank pin journal and the connecting rod pendulum action. The surface installed on the journal and rod encourages needle rotation because of its relative roughness. If the journal and rod surface was polished with crocus cloth, the loose needle bearings would have the tendency to scoot, wearing both surfaces. So, journals and rods which uses the loose needle bearings are cleaned up sing 320 grit emery paper.

Caged needle bearings used a reduced number of needles and the needles are kept separated and are encouraged to roll by the cage. The cage also controls end movement of the bearings. Because of the cage, the journal and rod surfaces can be smoother, so these surfaces are polished with crocus cloth.

Main bearings are used to mount and control the axial movement of the crankshaft. They are ball, needle or split race needle bearings. The split race needle bearings are held together with a ring and are sandwiched between the

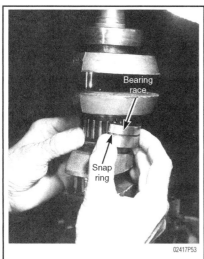

Fig. 64 Remove the snap ring and main bearing race

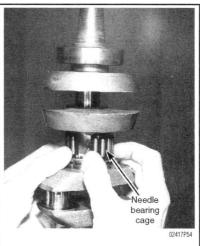

Fig. 65 Remove the needle bearing cage and set aside with the bearing race they are to be kept as a matched set

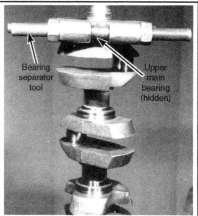

Fig. 66 Remove the upper bearing with a puller as shown, This is the bearing that keeps the crankshaft from riding up and down and acts as a thrust bearing

7-20 POWERHEAD AND POWERHEAD OVERHAUL

Fig. 67 The lower main bearing and seal assembly should slide off the end of the crankshaft

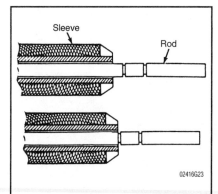

Fig. 68 Cross section drawing of the special timing gauge mentioned in the text. The exposed end of the rod is extended (top), to the second groove indicating TDC. At a specified number of degrees BTDC the exposed end only shows a single groove (bottom).

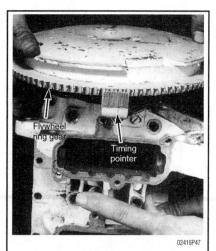

Fig. 69 If the decal is missing, or is illegible, the text outlines detailed instructions for establishing a new TDC mark.

crankcase and cylinder assembly. The split race bearings are commonly used as center main bearings, as this is the only type of bearing that can be easily installed in this location. The ball bearings may be mounted as top or bottom mains on the crankshaft.

The bearing is made up of three parts- the inner race, needle and the outer race. In most industrial applications, the outer or inner race of a needle bearing assembly is held in a fixed position by a housing or shaft. The connecting rod needle bearings in the outboard powerhead have the same basic parts, but differ in that both inner and outer races are in motion. The outer race-the connecting rod-is swinging like a clock pendulum. The inner race-the crankshaft-is rotating and the needle bearing are floating between the two races.

INSPECTION

When the powerhead is disassembled and inspection of the parts is made, then by necessity along with examining the needle bearings, the crankshaft main bearing journal, rod journal and connecting rod bearing surfaces are also examined. The surfaces of all three of these parts can give a tremendous amount of information and the examination will determine if the parts are reusable.

Surfaces should be examined for scoring, pitting, chatter marks, rust marks and discoloration from overheating of the bearing surfaces. Minor scoring or pitting and rust marks may be cleaned up and the surfaces brought back to a satisfactory condition. This is done using crocus cloth for caged needle bearings and 320 grit emery paper for loose needle bearings. This is not a metal removing process, rather just a clean up of the surfaces.

Needle bearings are used as main bearings and are inspected for the same conditions as listed above. There are no oversized bearings available for rod or main bearings. Because of the hardness of the crankshaft (a bearing race), it should not be turned or welded up in order to bring it back to standard size. The welding process may stress the metallurgical properties of the crankshaft, developing cracks.

The caged rod bearings and split race main bearings are inspected for the same condition as loose needle bearings, plus the cage is examined for wear, cracks and breaks.

Ball bearings are used for top and bottom main bearings in some powerheads. These may be pressed onto the crankshaft or pressed into the end cap. To examine these bearings, wash, dry, oil and check them on the crankshaft or in the bearing cap. Turn the bearing by hand and feel if there is any roughness or catching. Try to wobble the bearing by grasping the outer race, (inner race) checking for looseness of the bearing. Replace the bearing if any of these conditions are found. If the bearing is pressed off (out) the bearing will probably be damaged and should be replaced.

If new or used bearings are contaminated with grit or dirt particles at the time of installation, abrasion will naturally follow. Many bearing failures are due to the introduction of foreign material into the internal parts of the bearing during assembly. Misalignment of the rod cap, torque of the rod bolts and lack of proper lubrication also cause failures. Bearing failure is usually detected by a gradual rise in operating noise, excessive looseness (axial) in the bearing and shaft deflection. Keep the work area clean and use needle bearing grease or multipurpose grease to hold the bearings in place. This grease will dissipate quickly as the fuel mixture comes in contact with it. Do not use a wheel bearing or chassis grease, as this will cause damage to the bearings. Oil the ball bearings with 2-stroke oil upon installation. Remember to keep them clean.

Flywheel Timing Marks

Some engine models prior to 1990 have a timing decal on the flywheel. After a powerhead rebuild it is good practice to check the TDC marks that they still line up or if the decal that was on the flywheel has come loose or is missing. Follow these procedures to remark the flywheel for timing.

USING THE FACTORY TIMING GAUGE

♦ See Figures 68 and 69

The manufacturer suggests using a special tool, known as a timing gauge, inserted into the No. 1 spark plug opening to check the location of timing decals on larger horsepower units. This gauge is also used when establishing piston location on units without a timing decal.

For all ignition systems used on outboards covered in this manual, the gauge to use has P/N T2937-1.

The timing gauge is used to determine the exact position of the piston in the cylinder bore. This information is then used to precisely time the spark in the cylinder.

The timing gauge must be used if the timing decal is missing. Some midrange horsepower units are not equipped with a timing decal from the factory and the manufacturer recommends the use of this gauge each time the powerhead is timed.

If the timing gauge is not available, an alternate method of establishing piston position is presented later in this section. This alternate method is not as accurate as the gauge and involves measurements and calculations.

The timing gauge consists of a hollow sleeve and movable rod arrangement. One end of the sleeve has threads the same size as a spark plug. The rod passes through and can move inside the sleeve. One end will make contact with the piston crown while the other end extends out the end of the sleeve.

Two sets of sleeves and rods are available. One set is used to determine 28° BTDC. The other sleeve and rod is used for all other settings—30°, 32°, and 34° BTDC. Each sleeve and rod has its own part number.

A pair of grooves are embossed on the rod near both ends. The grooves at one end are calibrated for models 25 to 55hp and the other end is calibrated for models 70hp and above.

The groove closest to the sleeve will indicate when the piston crown has reached TDC (top dead center). The other groove in the pair, the groove closest to the end of the rod, will indicate when the piston crown is at 28°, 30°, 32°, or 34° BTDC, depending on the sleeve and rod being used.

POWERHEAD AND POWERHEAD OVERHAUL

The procedure is as follows.
1. Disconnect the battery cables and remove all the spark plugs.
2. Thread the sleeve into the No. 1 cylinder spark plug opening until only half the threads on the sleeve are visible.
3. Install the rod with the appropriate horsepower range visible and facing outward.
4. Hold the end of the rod against the piston dome.
5. Slowly rotate the flywheel clockwise.

✶✶ WARNING

Never rotate the flywheel counterclockwise such action may damage the water pump impeller.

6. As the piston begins to reach TDC the rod will emerge from the sleeve to a maximum length and then retract as the piston dome moves downward.
7. Lightly hold the rod and rotate the sleeve of the tool threading it in or out of the spark plug opening until the edge of the sleeve is aligned with the first groove closest to the sleeve.
8. The piston is now at the TDC position.
9. Two permanent marks will now be made. One mark on the vertical surface of the flywheel just above the ring gear. The other will be a matching mark in a convenient location on the block under the flywheel where both marks may be observed when aligned.
10. Identify this first flywheel mark as "TDC 1" (top dead center for No. 1 cylinder). The mark on the block will always be referred to as "the timing mark" throughout the remainder of these procedures.
11. While maintaining a slight pressure on the rod against the piston crown slowly rotate the flywheel clockwise until the edge of the sleeve is aligned with the second groove on the rod (the first groove will have disappeared inside the sleeve).
12. The piston is now in the advanced spark position, either at 28° BTDC, 30° BTDC, 32° BTDC, or 34° BTDC, depending on the model being serviced.
13. Make another mark on the flywheel aligned with the timing mark made on the block. Identify this second mark to indicate it is the appropriate BTDC mark.

ALTERNATE METHOD

▶ See Figure 70

Obtain a screwdriver with a shank at least 4" (10cm).
Insert the screwdriver blade into the No. 1 spark plug opening. Keep the shank at right angles to the cylinder head cover, and at the same time, slowly rotate the flywheel clockwise.
Determine the exact point when the screwdriver is at its maximum lift.

✶✶ WARNING

Do not rotate the flywheel counterclockwise. Such action may cause damage to the water pump impeller vanes.

1. Do not disturb the flywheel and remove the screwdriver.
2. Two permanent marks will now be made. One mark on the vertical surface of the flywheel just above the ring gear. The other mark will be a matching mark in a convenient location on the block under the flywheel where both marks may e observed when aligned.
3. Identify this first flywheel mark as "TDC 1" (top dead center for No. 1 cylinder). The mark on the block will always be referred to as 'the timing mark" throughout the remainder of these procedures.

The maximum timing mark must now be calculated and measured using the following steps and formula.
This fraction of the circumference of the flywheel can be determined with a single accurate measurement and a simple calculation. In fact, the calculation can be made even simpler if the fraction is changed into a decimal, thus:
- $28/360 = 0.077$
- $30/360 = 0.083$
- $32/360 = 0.088$
- $34/360 = 0.094$

This decimal remains constant when working on any diameter flywheel.
Begin by making as accurate a measurement as possible of the flywheel perimeter (distance around—circumference). A fabric tape measure is ideal for this task. Make the measurement from the TDC mark already scribed in either direction around the flywheel in the same plane as the mark
Measure to the closest 1/16". If the fabric tape measure is only marked in 1/8" intervals, estimate to the nearest 1/16". As a double check, mark the circumference on the fabric tape measure and then lay the fabric tape measure out next to a metal one and compare.
To change the inch fraction into a decimal, use the following table.
- $1/16$" = 0.062"
- $1/8$" = 0.125"
- $3/16$" = 0.187"
- $1/4$" = 0.250"
- $5/16$" = 0.312"
- $3/8$" = 0.375"
- $7/16$" = 0.437"
- $1/2$" = 0.573"
- $9/16$" = 0.562"
- $5/8$" = 0.625"
- $11/16$" = 0.687"
- $3/4$" = 0.750"
- $13/16$" = 0.812"
- $7/8$" = 0.875"
- $15/16$" = 0.937"

The following formula will give the figure in inches for the distance to be measured clockwise (to the left when facing the flywheel), on the circumference of the flywheel, between the "TDC 1" mark and the required number of degrees advance.

Units With 28° BTDC Advance

From the conversion table $28/360 = 0.077$. Multiply this figure by the flywheel circumference thus:
0.077 X Flywheel circumference = Distance to be measured on the flywheel from the TDC 1 mark. This distance will locate the 28° BTDC mark.

Units With 30° BTDC Advance

From the conversion table $30/360 = 0.083$. Multiply this figure by the flywheel circumference thus:
0.083 x Flywheel circumference = distance to be measured on the flywheel from the TDC/1 mark. This distance will locate the 30° BTDC mark.

Units With32°BTDC Advance

From the conversion table $32/360 = 0.088$. Multiply this figure by the flywheel circumference thus:
0.088 x Flywheel circumference = Distance to be measured on the flywheel from the TDC 1 mark. This distance will locate the 32° BTDC mark.

Units With 34°BTDC Advance

From the conversion table $34/360 = 0.094$. Multiply this figure by the flywheel circumference thus:
0.094 x Flywheel circumference = Distance to be measured on the flywheel from the TDC 1 mark. This distance will locate the 34° BTDC mark.

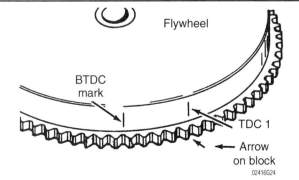

Fig. 70 Simplified drawing to indicate how the TDC 1 and BTDC are scribed on the flywheel just above the ring gear.

7-22 POWERHEAD AND POWERHEAD OVERHAUL

Example

Model with 32° BTDC advance: Flywheel circumference at the TDC 1 mark equals 28 3/16".
Referring to the table above: 28 3/16" is equivalent to 28.187".
Using the standard formula given:
- 0.088 X Flywheel circumference = Distance
- 0.088 X 28.187" = 2.48" = Distance

Therefore, the 32°BTDC mark is located 2.48" to the left (clockwise around the flywheel perimeter), from the TDC 1 mark on the flywheel.
Again referring to the table above, the inch decimal can be changed back into an inch fraction:
- 0.48" is between 7/16" and 1/2"

Therefore, the mark must be made between 2 7/16" and 2 1/2" to the left (clockwise around the flywheel perimeter) from the TDC 1 mark.

POWERHEAD EXPLODED VIEWS

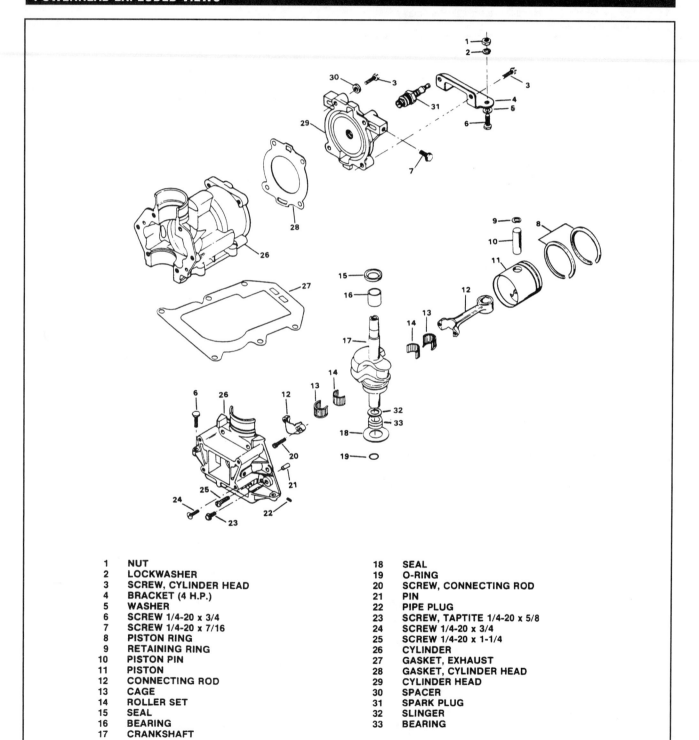

1	NUT	18	SEAL
2	LOCKWASHER	19	O-RING
3	SCREW, CYLINDER HEAD	20	SCREW, CONNECTING ROD
4	BRACKET (4 H.P.)	21	PIN
5	WASHER	22	PIPE PLUG
6	SCREW 1/4-20 x 3/4	23	SCREW, TAPTITE 1/4-20 x 5/8
7	SCREW 1/4-20 x 7/16	24	SCREW 1/4-20 x 3/4
8	PISTON RING	25	SCREW 1/4-20 x 1-1/4
9	RETAINING RING	26	CYLINDER
10	PISTON PIN	27	GASKET, EXHAUST
11	PISTON	28	GASKET, CYLINDER HEAD
12	CONNECTING ROD	29	CYLINDER HEAD
13	CAGE	30	SPACER
14	ROLLER SET	31	SPARK PLUG
15	SEAL	32	SLINGER
16	BEARING	33	BEARING
17	CRANKSHAFT		

Fig. 71 Exploded view of the cylinder block, crankshaft and piston assembly—3, 4 and 5 HP

POWERHEAD AND POWERHEAD OVERHAUL 7-23

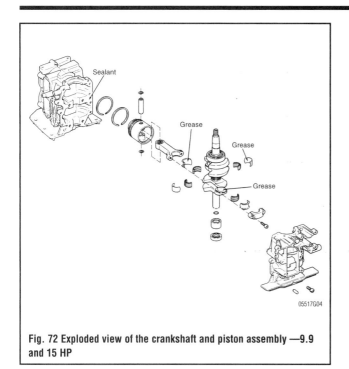

Fig. 72 Exploded view of the crankshaft and piston assembly—9.9 and 15 HP

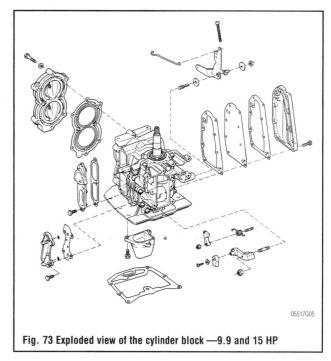

Fig. 73 Exploded view of the cylinder block—9.9 and 15 HP

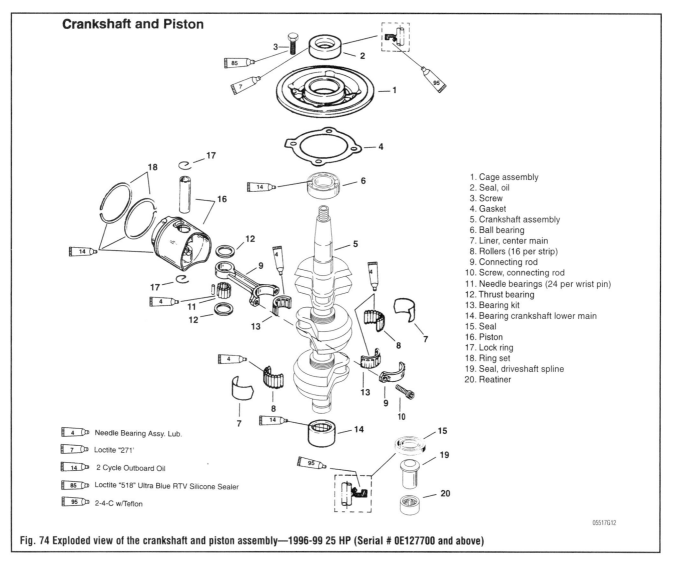

Fig. 74 Exploded view of the crankshaft and piston assembly—1996-99 25 HP (Serial # 0E127700 and above)

7-24 POWERHEAD AND POWERHEAD OVERHAUL

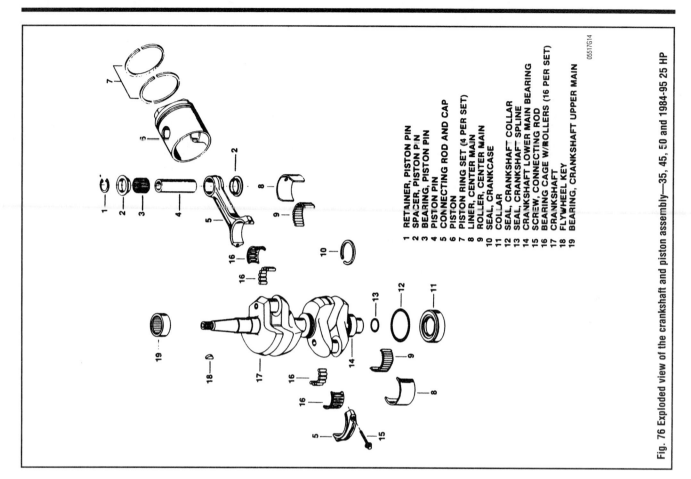

Fig. 76 Exploded view of the crankshaft and piston assembly—35, 45, 50 and 1984-95 25 HP

1. RETAINER, PISTON PIN
2. SPACER, PISTON PIN
3. BEARING, PISTON PIN
4. PISTON PIN
5. CONNECTING ROD AND CAP
6. PISTON
7. PISTON RING SET (4 PER SET)
8. LINER, CENTER MAIN
9. ROLLER, CENTER MAIN
10. SEAL, CRANKCASE
11. COLLAR
12. SEAL, CRANKSHAFT COLLAR
13. SEAL, CRANKSHAFT SPLINE
14. CRANKSHAFT LOWER MAIN BEARING
15. SCREW, CONNECTING ROD
16. BEARING CAGE W/ROLLERS (16 PER SET)
17. CRANKSHAFT
18. FLYWHEEL KEY
19. BEARING, CRANKSHAFT UPPER MAIN

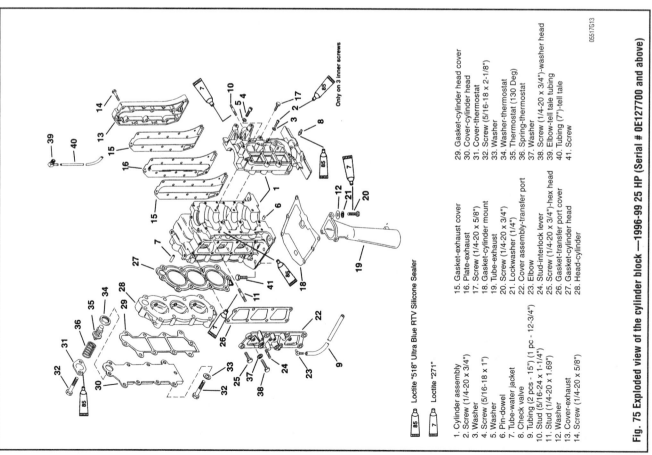

Fig. 75 Exploded view of the cylinder block—1996-99 25 HP (Serial # 0E127700 and above)

- Loctite "518" Ultra Blue RTV Silicone Sealer
- Loctite "271"

1. Cylinder assembly
2. Screw (1/4-20 x 3/4")
3. Washer
4. Screw (5/16-18 x 1")
5. Washer
6. Pin-dowel
7. Tube-water jacket
8. Check valve
9. Tubing (2 pcs - 15") (1 pc - 12-3/4")
10. Stud (5/16-24 x 1-1/4")
11. Stud (1/4-20 x 1.69")
12. Washer
13. Cover-exhaust
14. Screw (1/4-20 x 5/8")
15. Gasket-exhaust cover
16. Plate-exhaust
17. Screw (1/4-20 x 5/8")
18. Gasket-cylinder mount
19. Tube-exhaust
20. Screw (1/4-20 x 3/4")
21. Lockwasher (1/4")
22. Cover assembly-transfer port
23. Elbow
24. Stud-interlock lever
25. Screw (1/4-20 x 3/4")-hex head
26. Gasket-transfer port cover
27. Gasket-cylinder head
28. Head-cylinder
29. Gasket-cylinder head cover
30. Cover-cylinder head
31. Cover-thermostat
32. Screw (5/16-18 x 2-1/8")
33. Washer
34. Washer-thermostat
35. Thermostat (130 Deg)
36. Spring-thermostat
37. Washer
38. Screw (1/4-20 x 3/4")-washer head
39. Elbow-tell tale tubing
40. Tubing (7")-tell tale
41. Screw

POWERHEAD AND POWERHEAD OVERHAUL 7-25

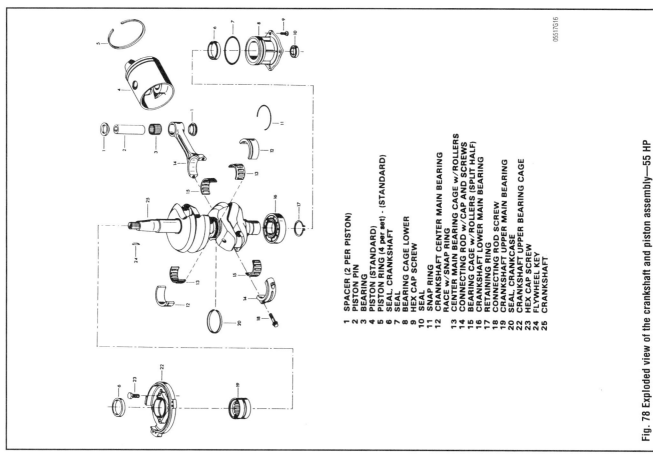

Fig. 78 Exploded view of the crankshaft and piston assembly—55 HP

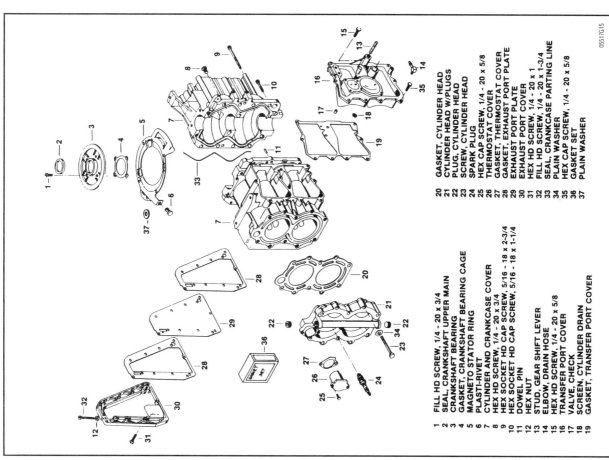

Fig. 77 Exploded view of the cylinder block—35, 45, 50 and 1984-95 25 HP

7-26 POWERHEAD AND POWERHEAD OVERHAUL

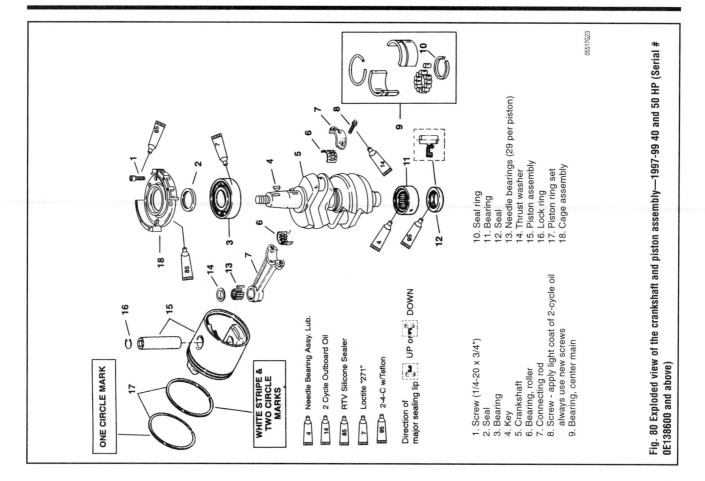

Fig. 80 Exploded view of the crankshaft and piston assembly—1997-99 40 and 50 HP (Serial # 0E138600 and above)

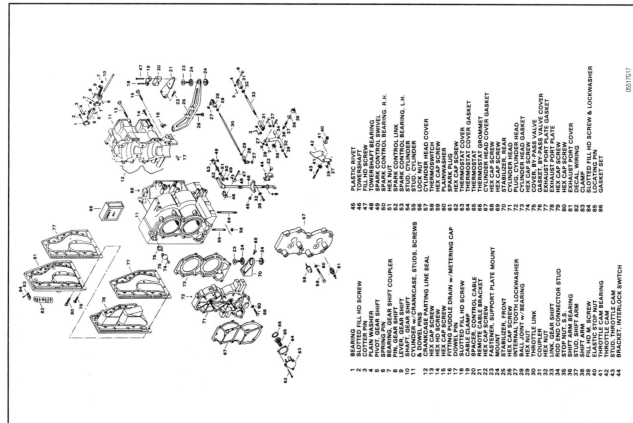

Fig. 79 Exploded view of the cylinder block—55 HP

POWERHEAD AND POWERHEAD OVERHAUL 7-27

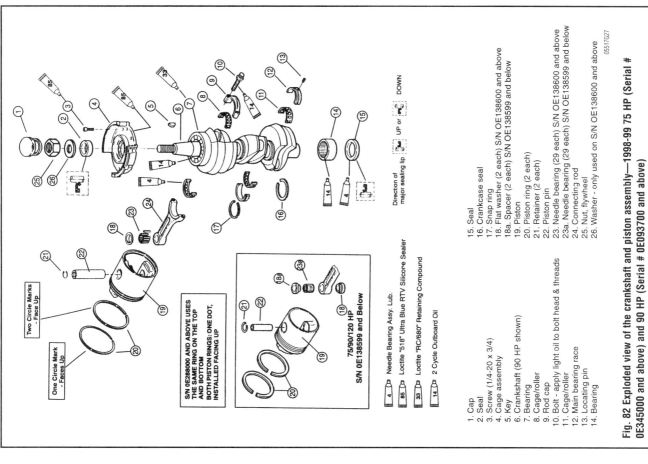

Fig. 82 Exploded view of the crankshaft and piston assembly—1998-99 75 HP (Serial # 0E345000 and above) and 90 HP (Serial # 0E093700 and above)

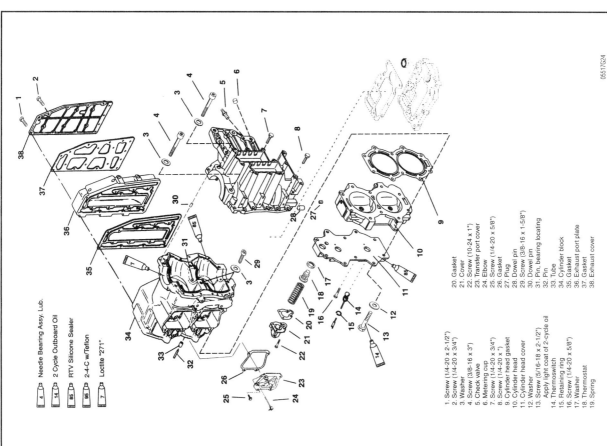

Fig. 81 Exploded view of the cylinder block—1997-99 40 and 50 HP (Serial # 0E138600 and above)

7-28 POWERHEAD AND POWERHEAD OVERHAUL

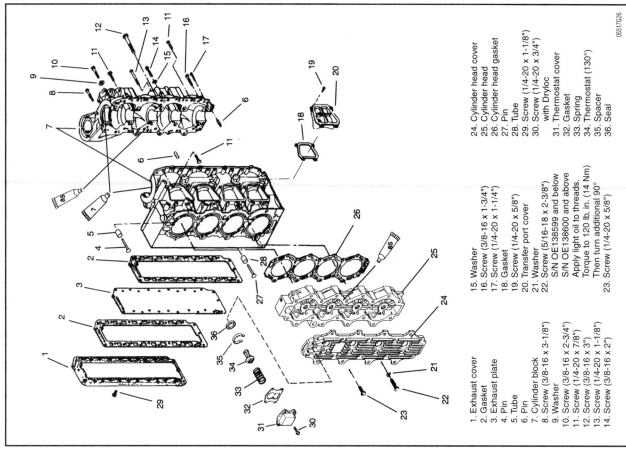

Fig. 84 Exploded view of the cylinder block —1998-99 120 HP (Serial # 0E093700 and above)

1. Exhaust cover
2. Gasket
3. Exhaust plate
4. Pin
5. Tube
6. Pin
7. Cylinder block
8. Screw (3/8-16 x 3-1/8")
9. Washer
10. Screw (3/8-16 x 2-3/4")
11. Screw (1/4-20 x 7/8")
12. Screw (3/8-16 x 3")
13. Screw (1/4-20 x 1-1/8")
14. Screw (3/8-16 x 2")
15. Washer
16. Screw (3/8-16 x 1-3/4")
17. Screw (1/4-20 x 1-1/4")
18. Gasket
19. Screw (1/4-20 x 5/8")
20. Transfer port cover
21. Washer
22. Screw (5/16-18 x 2-3/8")
    S/N OE138599 and below
    S/N OE138600 and above
    Apply light oil to threads.
    Torque to 120 lb. in. (14 Nm)
    Then turn additional 90°.
23. Screw (1/4-20 x 5/8")
24. Cylinder head cover
25. Cylinder head
26. Cylinder head gasket
27. Pin
28. Tube
29. Screw (1/4-20 x 1-1/8")
30. Screw (1/4-20 x 3/4") with Dryloc
31. Thermostat cover
32. Gasket
33. Spring
34. Thermostat (130°)
35. Spacer
36. Seal

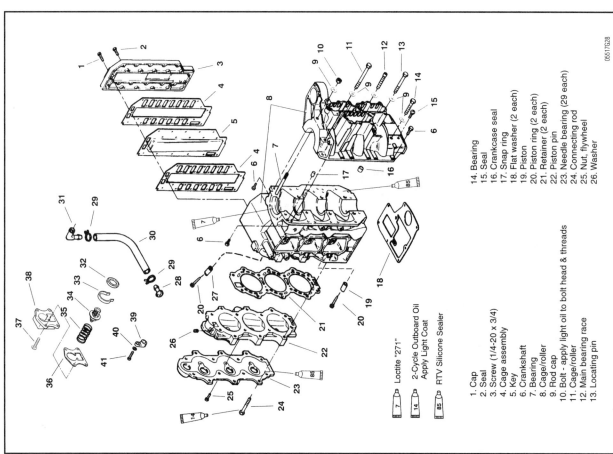

Fig. 83 Exploded view of the cylinder block —1998-99 75 HP (Serial # 0E345000 and above) and 90 HP (Serial # 0E093700 and above)

1. Cap
2. Seal
3. Screw (1/4-20 x 3/4)
4. Cage assembly
5. Key
6. Crankshaft
7. Bearing
8. Cage/roller
9. Rod cap
10. Bolt - apply light oil to bolt head & threads
11. Cage/roller
12. Main bearing race
13. Locating pin
14. Bearing
15. Seal
16. Crankcase seal
17. Snap ring
18. Flat washer (2 each)
19. Piston
20. Piston ring (2 each)
21. Retainer (2 each)
22. Piston pin
23. Needle bearing (29 each)
24. Connecting rod
25. Nut, flywheel
26. Washer

- 7 — Loctite "271"
- 14 — 2-Cycle Outboard Oil Apply Light Coat
- 85 — RTV Silicone Sealer

POWERHEAD AND POWERHEAD OVERHAUL 7-29

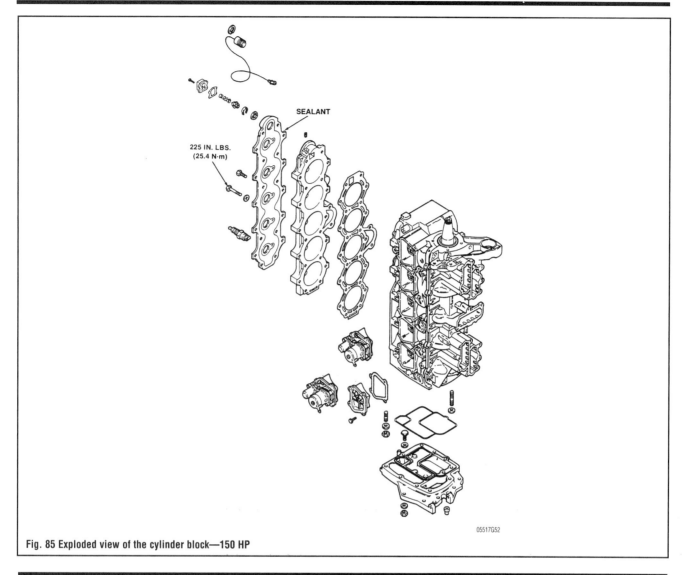

Fig. 85 Exploded view of the cylinder block—150 HP

TORQUE SEQUENCE DIAGRAMS

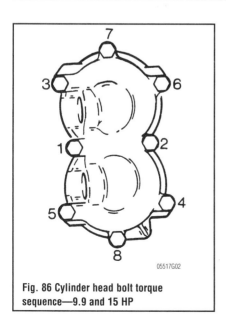

Fig. 86 Cylinder head bolt torque sequence—9.9 and 15 HP

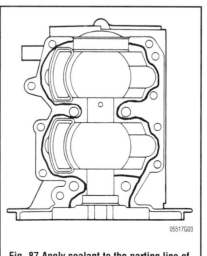

Fig. 87 Apply sealant to the parting line of the crankcase as illustrated—9.9 and 15 HP

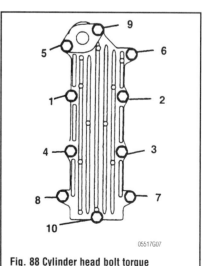

Fig. 88 Cylinder head bolt torque sequence—1996-99 25 HP (Serial # 0E127700 and above)

7-30 POWERHEAD AND POWERHEAD OVERHAUL

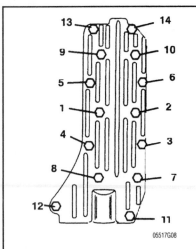

Fig. 89 Exhaust manifold bolt torque sequence—1996-99 25 HP (Serial # 0E127700 and above)

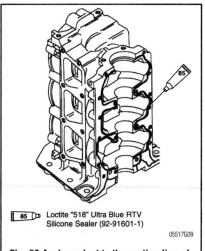

Fig. 90 Apply sealant to the parting line of the crankcase as illustrated—1996-99 25 HP (Serial # 0E127700 and above)

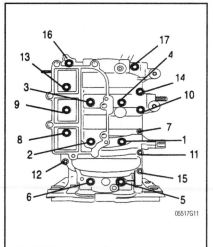

Fig. 91 Crankcase bolt torque sequence—1996-99 25 HP (Serial # 0E127700 and above)

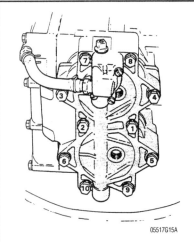

Fig. 92 Cylinder head bolt torque sequence—35, 45, 50, 55 and 1984-95 25 HP

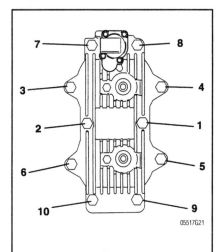

Fig. 93 Cylinder head bolt torque sequence—1997-99 40 and 50 HP (Serial # 0E138600 and above)

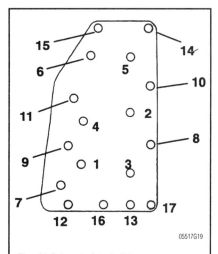

Fig. 94 Exhaust plate bolt torque sequence—1997-99 40 and 50 HP (Serial # 0E138600 and above)

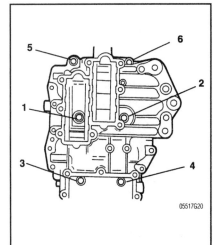

Fig. 95 Crankcase bolt torque sequence—1997-99 40 and 50 HP (Serial # 0E138600 and above)

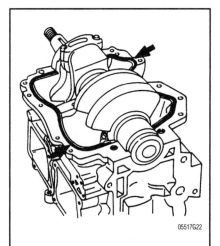

Fig. 96 Apply sealant to the parting line of the crankcase as illustrated—1997-99 40 and 50 HP (Serial # 0E138600 and above)

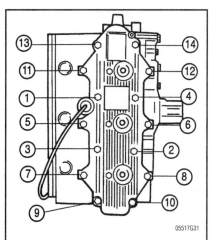

Fig. 97 Cylinder head bolt torque sequence—1998-99 75 HP (Serial # 0E345000 and above) and 90 HP (Serial # 0E093700 and above)

POWERHEAD AND POWERHEAD OVERHAUL 7-31

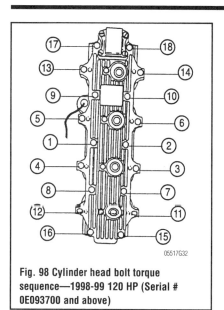

Fig. 98 Cylinder head bolt torque sequence—1998-99 120 HP (Serial # 0E093700 and above)

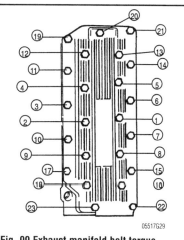

Fig. 99 Exhaust manifold bolt torque sequence—1998-99 75 HP (Serial # 0E345000 and above) and 90 HP (Serial # 0E093700 and above)

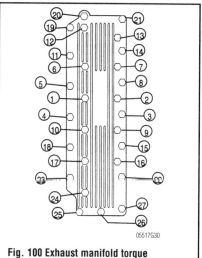

Fig. 100 Exhaust manifold torque sequence—1998-99 120 HP (Serial # 0E093700 and above)

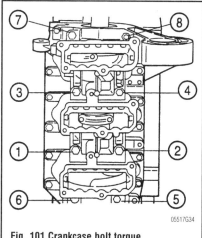

Fig. 101 Crankcase bolt torque sequence—1998-99 75 HP (Serial # 0E345000 and above) and 90 HP (Serial # 0E093700 and above)

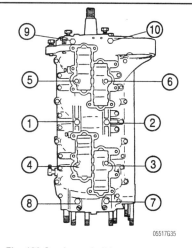

Fig. 102 Crankcase bolt torque sequence—1998-99 120 HP (Serial # 0E093700 and above)

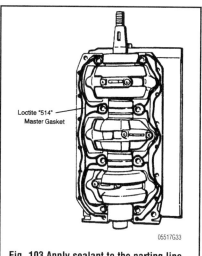

Fig. 103 Apply sealant to the parting line of the crankcase as illustrated—1998-99 120 HP (Serial # 0E093700 and above)

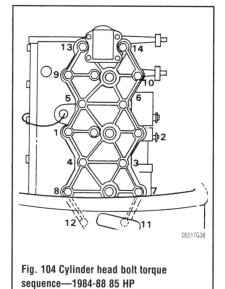

Fig. 104 Cylinder head bolt torque sequence—1984-88 85 HP

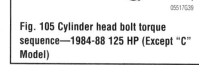

Fig. 105 Cylinder head bolt torque sequence—1984-88 125 HP (Except "C" Model)

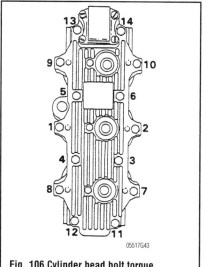

Fig. 106 Cylinder head bolt torque sequence—1989-99 85 and 90 HP L-Drive

7-32 POWERHEAD AND POWERHEAD OVERHAUL

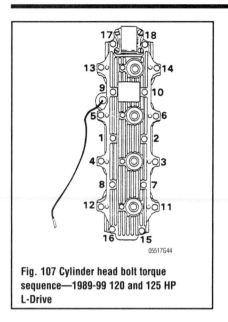

Fig. 107 Cylinder head bolt torque sequence—1989-99 120 and 125 HP L-Drive

Fig. 108 Apply sealant to the parting line of the crankcase as illustrated—1989-99 85 and 90 HP L-Drive

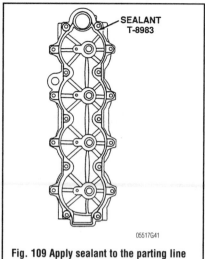

Fig. 109 Apply sealant to the parting line of the crankcase as illustrated—1989-99 120 and 125 HP L-Drive

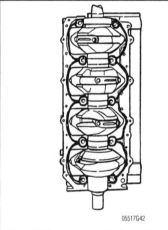

Fig. 110 Apply sealant to the parting line of the crankcase as illustrated—1989-99 120 and 125 HP L-Drive (85 and 90 HP L-Drive similar)

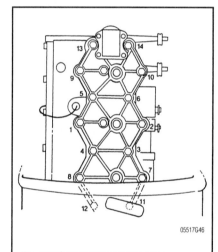

Fig. 111 Cylinder head bolt torque sequence—1988-91 85 and 90 HP ("A" Model)

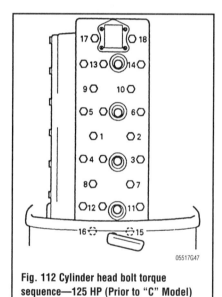

Fig. 112 Cylinder head bolt torque sequence—125 HP (Prior to "C" Model)

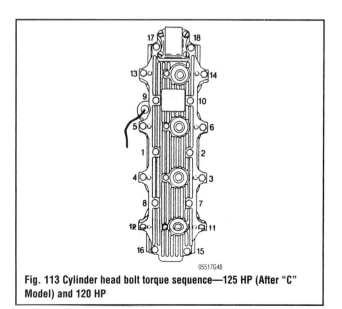

Fig. 113 Cylinder head bolt torque sequence—125 HP (After "C" Model) and 120 HP

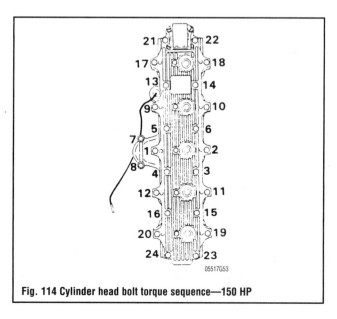

Fig. 114 Cylinder head bolt torque sequence—150 HP

POWERHEAD AND POWERHEAD OVERHAUL

Engine Rebuilding Specifications — 3, 4, 5 and 7.5 HP

Component	Standard (in.)	Standard Metric (mm)	Service Limit Standard (in.)	Service Limit Metric (mm)
Crankshaft				
Except 7.5 HP				
Upper main bearing journal	0.813	20.65	0.8125	20.64
Lower main bearing journal	0.7878	20.01	0.7874	20
Connecting rod journal	0.7496	19.04	0.7493	19.03
Upper seal surface	0.719	18026	0.7185	18.25
Lower seal surface	0.75	19.05	0.749	19.02
7.5 HP				
Upper main bearing journal	0.9853	25.03	0.9849	25.02
Center main bearing journal	0.8005	20.33	0.8	20.32
Lower main bearing journal	0.75	19.05	0.7495	19.04
Connecting rod journal	0.7496	19.04	0.7493	19.03
Upper seal surface	0.8135	20.66	0.8122	20.63
Lower seal surface	0.75	19.05	0.749	19.02
Connecting rod				
Wrist pin end	0.6245	15.86	0.625	15.88
Crank shaft end				
1988-69 3-5 HP	0.9399	23.87	0.9403	23.88
1990-99 5 HP	0.9409	23.9	0.9413	23.91
7.5 HP	0.9399	23.87	0.9403	23.88
Cylinder head distortion			0.012	0.76
Cylinder distortion			0.0005	0.01
Cylinder taper			0.002	0.051
Cylinder bore				
3 HP	1.85	47.6	1.85	47.6
Except 3 HP	2	50.9	2.001	50.83
Piston diameter				
Except 1990-99 5 HP	1.996	50.7	1.9965	50.71
1990-98 5 HP	2.122	53.9	2.123	53.92
Piston pin diameter	0.43765	11.12	0.4375	11.11
Piston pin hole diameter	0.4373	11.11	0.4376	11.12
Piston ring end gap				
Except 1990-99 5 HP	0.006	0.15	0.011	0.28
1990-99 5 HP	0.009	0.23	0.009	0.23
Maximum reed stop opening				
7.5 HP	0.2	5.1	0.22	5.6
All others	0.24	6.1	0.26	6.6
Reed seat overlap			0.03	0.76
Reed stop opening				
7.5 HP			0.012	0.76
All others			0.006	0.15

Engine Rebuilding Specifications — 9.9 and 15HP

Component	Standard (in.)	Standard Metric (mm)	Service Limit Standard (in.)	Service Limit Metric (mm)
Crankshaft				
Upper main bearing journal	0.9853	25.03	0.9849	25.02
Center main bearing journal	0.8005	20.33	0.8	20.32
Lower main bearing journal	0.75	19.05	0.7495	19.04
Connecting rod journal	0.7501	19.05	0.7496	19.04
Upper seal surface	0.8135	20.66	0.8122	20.63
Center seal surface	1.059	26.9	1.0585	26.89
Lower seal surface	0.75	19.05	0.7495	19.02
Connecting rod				
Wrist pin end	0.687	17.45	0.6875	17.46
Crank shaft end	0.9399	23.87	0.9403	23.88
Cylinder head distortion			0.012	0.76
Cylinder distortion			0.0005	0.01
Cylinder taper			0.002	0.051
Cylinder bore				
1984 9.9 HP	2.1875	55.56	2.1885	55.59
Except 1984 9.9 HP	2.251	57.18	2.252	57.2
Piston diameter				
1984 9.9 HP	2.185	55.5	2.1845	55.49
Except 1984 9.9 HP	2.2475	57.09	2.247	57.07
Piston pin diameter	0.50015	12.7	0.5	12.7
Piston pin hole diameter	0.5001	12.7	0.4998	12.69
Piston ring end gap				
1984 9.9 HP	0.006	0.15	0.016	0.41
Except 1984 9.9 HP	0.004	0.1	0.014	0.36
Maximum reed stop opening	0.24	6.1	0.26	6.6
Reed seat overlap			0.03	0.76
Reed stop opening			0.006	0.15

POWERHEAD AND POWERHEAD OVERHAUL

Engine Rebuilding Specifications— 25 (2 CYL) and 35 HP

Component	Standard (in.)	Standard Metric (mm)	Service Limit Standard (in.)	Service Limit Metric (mm)
Crankshaft				
Upper main bearing journal	1.378	35	1.3774	34.99
Center main bearing journal	1.3451	34.17	1.3446	34.15
Lower main bearing journal	0.9853	25.03	0.9849	25.02
Connecting rod journal	1.1395	28.94	1.1391	28.93
Upper seal surface	1.065	27.05	1.063	27
Lower seal surface	0.935	23.75	0.9345	23.74
Connecting Rod				
Wrist pin end				
1986-87	0.852	21.64	0.856	21.74
1988-91	0.8762	22.26	0.8767	22.27
Crank shaft end				
1986-87	1.4523	36.89	1.4528	36.9
1988-91	1.4528	36.9	1.4533	36.91
Cylinder head distortion			0.012	0.76
Cylinder distortion			0.0005	0.01
Cylinder taper			0.002	0.051
Cylinder bore				
25 HP	2.8125	71.44	2.8135	71.46
35 HP	3.001	76.23	3.003	76.28
Piston diameter				
25 HP				
Piston diameter measuring position ①	2.8084	71.33	2.8064	71.28
Cylinder measuring position ②	2.8059	71.27	2.8039	71.22
35 HP				
Piston diameter measuring position ①	2.995	76.07	2.9935	76.03
Cylinder measuring position ②	2.9915	75.98	2.9885	75.91
Piston pin diameter	0.50015	12.7	0.5	12.7
Piston pin hole diameter	0.6874	17.46	0.6877	17.47
Piston ring end gap				
25 HP	0.007	0.18	0.017	0.43
35 HP				
1st	0.006	0.15	0.016	0.41
2nd	0.004	0.1	0.014	0.36
Maximum reed stop opening	0.29	7.4	0.27	0.69
Reed seat overlap			0.04	1.02
Reed stop opening			0.01	0.25

① 90 Degrees to piston pin bore
② Parallel to the piston pin bore

Engine Rebuilding Specifications— 40 and 50HP

Component	Standard (in.)	Standard Metric (mm)	Service Limit Standard (in.)	Service Limit Metric (mm)
Crankshaft				
1984-95				
Upper main bearing journal	1.1821	30.03	1.1815	30.01
Center main bearing journal	1.1392	28.94	1.1388	28.93
Lower main bearing journal	1.1251	28.58	1.1245	28.56
Connecting rod journal	1.1395	28.94	1.1391	28.93
1996-99				
Upper main bearing journal	1.3793	35.03	1.3789	35.02
Center main bearing journal	1.2165	30.89	1.216	30.9
Lower main bearing journal	1.2505	31.76	1.25	31.75
Connecting rod journal	1.1818	30.02	1.1813	30.01
All				
Upper seal surface	1.1255	28.59	1.1245	28.56
Lower seal surface	1.1251	28.58	1.1245	28.56
Connecting rod				
Wrist pin end				
1984-95	0.8762	22.26	0.8767	22.27
1996-99	0.9568	24.3	0.9573	24.32
Crank shaft end				
1984-95	1.4528	36.9	1.4533	36.89
1996-99	1.4986	38.08	1.4991	38.06
Cylinder head distortion			0.012	0.76
Cylinder distortion			0.0005	0.01
Cylinder taper			0.002	0.051
Cylinder bore				
1984-89 50 HP	3.189	81	3.191	81.05
1989-92 50HP	3.313	84.15	3.3148	84.2
1992-99 40 and 50HP	3.3745	85.71	3.376	85.75
Piston diameter				
1984-87 50 HP				
Piston diameter measuring position ①	3.185	80.9	3.184	80.87
Cylinder measuring position ②	3.183	80.85	3.181	80.8
1988-89 50 HP				
Piston diameter measuring position ①	3.185	80.9	3.184	80.87
Cylinder measuring position ②	3.181	80.8	3.179	80.75
1989-92 50HP				
Piston diameter measuring position ①	3.309	84.05	3.308	84.02
Cylinder measuring position ②	3.305	83.95	3.304	83.92
1992-95 40 and 50HP				
Piston diameter measuring position ①	3.3628	85.42	3.3618	85.39
Cylinder measuring position ②	3.3608	85.36	3.3598	85.34
1996-99 40 and 50HP	3.3705	85.61	3.3695	85.59
Piston pin diameter	0.50015	12.7	0.5	12.7

POWERHEAD AND POWERHEAD OVERHAUL

Engine Rebuilding Specifications— 40 and 50HP

Component	Standard (in.)	Standard Metric (mm)	Service Limit (in.)	Service Limit Metric (mm)
Piston pin hole diameter				
1984-87 50 HP	0.6874	17.46	0.6877	17.47
1988-89 50 HP	0.6876	17.47	0.6879	17.5
1992-99 40 and 50HP	0.6878	17.48	0.6881	17.51
Piston ring end gap				
1984-89 50 HP	0.006	0.015	0.016	0.41
1989-92 50HP				
1st	0.01	0.25	0.02	0.51
2nd	0.006	0.015	0.016	0.41
1992-95 40 and 50HP	0.004	0.1	0.014	0.36
1996-99 40 and 50HP	0.01	0.25	0.2	0.51
Maximum reed stop opening	0.29	7.4	0.27	0.69
Reed seat overlap			0.04	1.02
Reed stop opening			0.01	0.25

① 90 Degrees to piston pin bore
② Parallel to the piston pin bore

Engine Rebuilding Specifications— 60HP

Component	Standard (in.)	Standard Metric (mm)	Service Limit (in.)	Service Limit Metric (mm)
Crankshaft				
Upper main bearing journal	1.375	34.93	1.3744	34.91
Center main bearing journal	1.3752	34.93	1.3748	34.92
Lower main bearing journal	1.3793	35.03	1.3789	35.02
Connecting rod journal	1.1827	30.04	1.1822	30.03
Upper seal surface	1.375	34.93	1.3744	34.91
Lower seal surface	1.376	34.95	1.374	34.9
Connecting rod				
Wrist pin end	0.9654	24.52	0.9659	24.53
Crank shaft end	1.5	38.11	1.5005	38.1
Cylinder head distortion			0.012	0.76
Cylinder distortion			0.0005	0.01
Cylinder taper			0.002	0.051
Cylinder bore	3.375	85.73	3.3762	85.76
Piston diameter				
Piston diameter measuring position ①	3.367	85.52	3.366	85.5
Cylinder measuring position ②	3.365	85.47	3.363	85.42
Piston pin diameter	0.7769	19.73	0.77675	19.73
Piston pin hole diameter	0.7769	19.73	0.7772	19.74
Piston ring end gap				
1st	0.004	0.1	0.014	0.36
2nd	0.006	0.015	0.016	0.41
Maximum reed stop opening	0.29	7.4	0.27	6.9
Reed seat overlap			0.04	1.02
Reed stop opening			0.01	0.25

① 90 Degrees to piston pin bore
② Parallel to the piston pin bore

7-36 POWERHEAD AND POWERHEAD OVERHAUL

Engine Rebuilding Specifications—1984-1995 70 to 150HP

Component	Standard (in.)	Standard Metric (mm)	Service Limit Standard (in.)	Service Limit Metric (mm)
Crankshaft				
Upper main bearing journal	1.3793	35.03	1.3789	35.02
Center main bearing journal	1.3752	34.93	1.3748	34.92
Lower main bearing journal	1.25	31.75	1.2495	31.74
Connecting rod journal	1.1395	28.94	1.1391	28.93
Upper seal surface	1.251	31.78	1.249	31.72
Lower seal surface	1.25	31.75	1.2495	31.74
Connecting rod				
Wrist pin end	0.8762	22.26	0.8767	22.27
Crank shaft end	1.4528	36.9	1.4533	36.91
Cylinder head distortion			0.012	0.76
Cylinder distortion			0.0005	0.01
Cylinder taper			0.002	0.051
Cylinder bore				
1984-90	3.313	84.15	3.3148	84.2
1990-95	3.375	85.73	3.3762	85.76
Piston diameter				
1984-89	3.3055	83.96	3.3045	83.93
1989-90				
Piston diameter measuring position ①	3.309	84.05	3.308	84.02
Cylinder measuring position ②	3.305	83.95	3.304	83.92
1990-95				
Piston diameter measuring position ①	3.37	85.6	3.369	85.57
Cylinder measuring position ②	3.368	85.55	3.367	85.52
Piston pin diameter	0.68765	17.47	0.6875	17.47
Piston pin hole diameter	0.6878	17.47	0.688	17.48
Piston ring end gap				
1984-89	0.006	0.15	0.016	0.41
1989-95				
1st	0.01	0.025	0.02	0.51
2nd	0.006	0.015	0.016	0.41
Maximum reed stop opening	0.29*	7.4*	0.27*	6.9*
Reed seat overlap			0.04*	1.02*
Reed stop opening			0.01*	0.25*

① 90 Degrees to piston pin bore
② Parallel to the piston pin bore
* Most of reed valves in these model engines are not servicable and must be replaced as assemblies

Engine Rebuilding Specifications—1996-1999 70 to 150HP

Component	Standard (in.)	Standard Metric (mm)	Service Limit Standard (in.)	Service Limit Metric (mm)
Crankshaft				
Upper main bearing journal	1.3793	35.03	1.3789	35.02
Center main bearing journal	1.3752	34.93	1.3748	34.92
Lower main bearing journal	1.25	31.75	1.2495	31.74
Connecting rod journal	1.1813	30.01	1.1818	30.02
Upper seal surface	1.376	34.95	1.374	34.9
Lower seal surface	1.25	31.75	1.2495	31.74
Connecting rod				
Wrist pin end	0.8762	22.26	0.8767	22.27
Crank shaft end	1.4986	38.06	1.4991	38.08
Cylinder head distortion			0.012	0.76
Cylinder distortion			0.0005	0.01
Cylinder taper			0.002	0.051
Cylinder bore	3.375	85.73	3.3762	85.76
Piston diameter	3.37005	85.6	3.3695	85.59
Piston pin diameter	0.77003	19.56	0.76993	19.56
Piston pin hole diameter	0.7769	19.73	0.7772	19.74
Piston ring end gap	0.01	0.025	0.02	0.51
Maximum reed stop opening	0.29*	7.4*	0.27*	6.9*
Reed seat overlap			0.04*	1.02*
Reed stop opening			0.01*	0.25*

* Most of reed valves in these model engines are not servicable and must be replaced as assemblies

POWERHEAD AND POWERHEAD OVERHAUL 7-37

Powerhead Torque Specifications

Component	Standard (ft. lbs.)	Metric (Nm)
Conventional Bolt/Nut		
6-32	9 in.lbs.	1.0
8-32	20 in.lbs.	2.3
10-24	30 in.lbs.	3.4
10-32	35 in.lbs.	4.0
12-24	45 in.lbs.	5.1
1/4 20	6	7.9
1/4 28	7	9.5
5/15 18	13	18.1
5/16 24	14	19.0
3/8 16	23	30.5
3/8 24	25	33.9
7/16 14	36	48.8
7/16 20	40	54.2
1/2 13	50	67.8
1/2 20	60	87.3
Connecting Rod		
3-15 HP	95 in.lbs.	10.7
Aluminum	95 in.lbs.	10.7
Steel	80 in.lbs.	9.0
25 HP (2-CYL)	180-190 in.lbs.	20.3-21.5
25 HP (3-CYL)	95 in.lbs.	10.7
1992-94 35, 40, 1988-94 50 HP	170 in.lbs.	19.2
1995-99 40-50 HP	120 in.lbs. ①	13.5 ①
60 HP	275 in.lbs.	31.0
70, 75, 90, 120 HP		
Serial# 093699 and later	120 in.lbs. ①	13.5 ①
Up to serial # 093699	170 in.lbs.	19.2
1984-87 85 and 125 HP	180-190 in.lbs.	20.3-21.5
1988-89 85 and 125 HP	170 in.lbs.	19.2
150 HP	170 in.lbs.	19.2
Cylinder Head		
3-15 HP	130 in.lbs.	14.7
25 HP (2-CYL) and 35 HP	190 in.lbs.	21.5
25 HP (3-CYL)	250 in.lbs.	28.2
40 and 60 HP, 1989-99 50 HP	225 in.lbs.	25.4
1984-89 50 HP	270 in.lbs.	30.5
70, 75, 90, 120 HP		
Serial # 093699 and later	120 in.lbs. ①	13.5 ①
Up to serial # 093699	225 in.lbs.	25.4
85 and 125 HP	225 in.lbs.	25.4
150 HP	225 in.lbs.	25.4

Powerhead Torque Specifications

Component	Standard (ft. lbs.)	Metric (Nm)
Drive Shaft Retainer		
70, 75, 90, 120 HP	75	101.7
85 and 125 HP	90	122.0
Engine Holder		
3, 7.5 HP	70 in.lbs.	7.9
9.9, 15 HP	160 in.lbs.	18.1
25 HP (3-CYL)	350 in.lbs.	39.5
25-60 HP	160 in.lbs.	18.1
1995-99 40 and 50 HP	55	74.6
70, 75, 90, 120 HP	270 in.lbs.	30.5
75 HP	37	50.0
85 and 125 HP	270 in.lbs.	30.5
150 HP	270 in.lbs.	30.5
Engine Holder Spacer Plate		
1990-94	270 in.lbs.	30.5
1995-99	30	340.7
Exhaust Cover		
7.5 HP	70 in.lbs.	7.9
9.9, 15 HP	90 in.lbs.	10.2
25 HP (3-CYL)	90 in.lbs.	10.2
25-60 HP	70 in.lbs.	7.9
1996-99 40 and 50 HP	100 in.lbs.	11.3
70, 75, 90, 120 HP	70 in.lbs.	7.9
75 HP	115 in.lbs.	13.0
85 and 125 HP	70 in.lbs.	7.9
1996-99 90 and 120 HP	80 in.lbs.	9.0
150 HP	70 in.lbs.	7.9
Flywheel		
3-5 HP	17	23.0
7.5 HP	40	54.0
9.9, 15 HP	50	67.8
25 HP (2-CYL)	60	81.3
25 HP (3-CYL)	45	61.0
35 HP	70	95.0
1995-99 40 and 50 HP	125	169.5
1984-94 40 and 50 HP	80	108.5
60 HP	90	122.0
70, 75, 90, 120 HP		
Serial # 093699 and later	125	169.5
Up to serial #OE093699	90	122.0
85 and 125 HP	125	169.5
150 HP	130	176.3

7-38 POWERHEAD AND POWERHEAD OVERHAUL

Powerhead Torque Specifications

Component	Standard (ft. lbs.)	Metric (Nm)
Gearcase Mounting		
3-15 HP	110 in.lbs.	12.4
9.9, 15 HP	130 in.lbs.	14.7
25-60 HP	270 in.lbs.	30.5
1995-99 40, 50 HP	40	54.2
70, 75, 90, 120 HP	40	54.2
85 and 125 HP	160 in.lbs.	18.1
150 HP	270 in.lbs.	30.5
Main Bearing		
3-15 HP	160 in.lbs.	18.1
25 and 35 HP	160 in.lbs.	18.1
40, 50 and 60 HP	270 in.lbs.	30.5
70, 75, 90, 120 HP	270 in.lbs.	30.5
85 and 125 HP	270 in.lbs.	30.5
150 HP	270 in.lbs.	30.5
Outer Cover Bolts		
3-15 HP	70 in.lbs.	7.9
25 HP (3-CYL)	90 in.lbs.	10.2
25-60 HP	70 in.lbs.	7.9
70, 75, 90, 120 HP	70 in.lbs.	7.9
85 and 125 HP	70 in.lbs.	7.9
150 HP	70 in.lbs.	7.9
Pinion Nut		
1995-99 40 and 50 HP	50	67.8
60 HP	85	115.2
1995-99 70, 75, 90, 120 HP	70	94.9
Except 1995-99 70, 75, 90, 120 HP	50	67.8
85 and 125 HP	85	115.2
150 HP	85	115.2
Pinion Screw		
25 HP (3-CYL)	150 in.lbs.	16.9
Propeller Nut		
25-60 HP	120 in.lbs.	13.5
70, 75, 90, 120 HP	55	74.6
85 and 125 HP	55	74.6
150 HP	55	74.6

Powerhead Torque Specifications

Component	Standard (ft. lbs.)	Metric (Nm)
Propeller Shaft Bearing Carrier		
3-15 HP	70 in.lbs.	7.9
25-60 HP	17	23.0
70, 75 HP (Except 1999 75 HP)	150 in.lbs.	16.9
1999 75 HP, 90, 120 HP	25	33.9
85 and 125 HP	160 in.lbs.	18.1
90, 120 HP Dual port Exhaust	130	176.3
150 HP	130	176.3
Upper Gearcase To Lower Gearcase		
3-60HP	70 in.lbs.	7.9
40, 50 HP 96-99	90 in.lbs.	10.2
70, 75, 90, 120 HP	70 in.lbs.	7.9
85 and 125 HP	70 in.lbs.	7.9
150 HP	70 in.lbs.	7.9
Upper Bearing Retainer		
3-15 HP	125 in.lbs.	14.1
25-60 HP	270 in.lbs.	30.5
Water Pump Adapter Plate		
25-60 HP	70 in.lbs.	7.9
70, 75, 90, 120 HP	60 in.lbs.	6.8
85 and 125 HP	70 in.lbs.	7.9
150 HP	70 in.lbs.	7.9
Water Pump Cover		
25-60 HP	70 in.lbs.	7.9
Water Pump Cover Dual Port Exhaust Long stud		
70, 75, 90, 120 HP	50 in.lbs.	5.6
150 HP	50 in.lbs.	5.6
Water Pump Cover Dual Port Exhaust Short stud		
70, 75, 90, 120 HP	15 in.lbs.	1.7
150 HP	15 in.lbs.	1.7
Water Pump Cover Single Port Exhaust Long stud		
70, 75, 90, 120 HP	70 in.lbs.	7.9
85 and 125 HP	70 in.lbs.	7.9
150 HP	70 in.lbs.	7.9
Water Pump Cover Single Port Exhaust Short Stud		
70-150 HP	40 in.lbs.	4.5

① Torque the bolt to specification then turn it an additional 1/4 turn.

LOWER UNIT 8-2
GENERAL INFORMATION 8-2
SHIFTING PRINCIPLES 8-2
   STANDARD ROTATING UNIT 8-2
TROUBLESHOOTING THE LOWER
 UNIT 8-2
PROPELLER 8-3
   REMOVAL & INSTALLATION 8-3
LOWER UNIT 8-4
   REMOVAL & INSTALLATION 8-4
LOWER UNIT OVERHAUL 8-7
UNITS UP TO 7.5 HP 8-7
   DISASSEMBLY 8-7
   CLEANING & INSPECTION 8-10
   ASSEMBLY 8-12
9.9 HP AND LARGER ENGINES WITH A
 SPLIT GEARCASE 8-12
   DISASSEMBLY 8-13
   CLEANING & INSPECTION 8-16
   ASSEMBLY 8-17
   SHIMMING 8-19
1984–91 70 TO 150 HP 8-19
   DISASSEMBLY 8-19
   CLEANING & INSPECTION 8-22
   ASSEMBLY & SHIMMING 8-24
1991–95 70 TO 150 HP 8-27
   DISASSEMBLY 8-27
   CLEANING & INSPECTION 8-33
   SHIMMING 8-35
   ASSEMBLY 8-37
1995–99 70 TO 120 HP 8-42
   DISASSEMBLY 8-42
   CLEANING & INSPECTION 8-45
   ASSEMBLY 8-47
   SHIMMING 8-49
SPECIFICATIONS CHARTS
   SHIM CHART—DRIVESHAFT 8-35

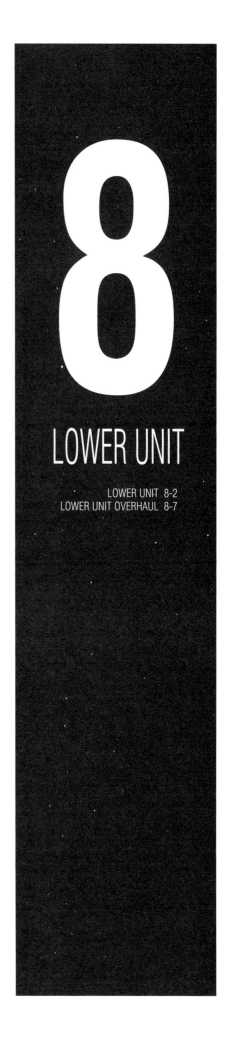

8
LOWER UNIT

LOWER UNIT 8-2
LOWER UNIT OVERHAUL 8-7

8-2 LOWER UNIT

LOWER UNIT

General Information

The lower unit consists of the driveshaft, water pump, and pinion gear, bearings, forward and reverse gears, propeller shaft. The shifting mechanism and the lower unit housing. The housing is bolted to the exhaust housing, which places the driveshaft and water tube(s) through the center of the exhaust housing. The water tube carries water from the water pump to the powerhead.

The driveshaft splines insert into the crankshaft, which then transmits power from the powerhead to the gearcase. A pinion gear on the driveshaft takes the power from the driveshaft and transfers the power to the propeller shaft. The powerhead driveshaft rotates in a clockwise direction continuously while the engine is running, propeller shaft direction is controlled by the gearcase shifting assembly.

The lower unit can be removed from the engine without removing the engine from it's mounting on the boat. The lower units in this chapter differ somewhat in their design and construction and require different servicing procedures.

The lower unit is normally trouble free until water enters the gearcase, the operator shifts incorrectly, or the oil is not changed regularly and corrosion enters the unit. Constant maintenance is required to prevent these problems. Because the unit is normally underwater, extra care must be taken to prevent problems.

Shifting the unit in and out of gear needs to be quick and positive to prevent rounding over the shift clutch dogs. Slow engagement will damage the parts. This problem is evident when the unit jumps out of gear.

Following the recommended oil change schedule allows the oil to be drained and checked for contamination, especially water intrusion. A milky looking oil is a sign of water has entered the gearcase and it must then be pressure tested to check the seals and the leaks repaired to prevent damage to the bearings and gears.

Shifting Principles

STANDARD ROTATING UNIT

Non-Reverse Type

Some of the smaller horsepower engines are equipped with a neutral, but no reverse gear. They utilize a spring-loaded clutch to shift between neutral and forward gear. To reverse the boat, the engine is simply rotated so that the propeller faces forward.

Reverse Type

♦ See Figures 1 and 2

On outboards equipped with a reverse gear, a sliding-type clutch engages the chosen gear in the gearcase housing. This clutch when engaged creates a direct connection that then moves the power flow from the pinion to the propeller shaft.

Power flow in the lower unit goes through the driveshaft into the pinion gear, which constantly turns the forward and reverse gears in opposite directions. The clutch dog is part of the shaft mechanism and is splined to the propeller shaft. The clutch dog is held in the central position (neutral) between the forward and reverse gears. When the shift shaft (rod) is moved, the shift cam (shifter) moves the follower (shifter shaft), which in turn, moves the clutch dog into mesh with the selected gear. Power is then transmitted from the gear through the clutch dog into the propeller shaft, and finally to the propeller.

Troubleshooting the Lower Unit

Once a season the lower unit needs to be dropped and lubrication on the driveshaft splines renewed. An extreme pressure monolithium lube is applied directly to the splines. Seals or an O-ring are used around the driveshaft to retain the lubrication and keep the water and exhaust from the splined joint area. This is a mandatory service to keep the splines from rusting together. Exhaust pressure and water are both present at the joint seal. If a seal failure occurs, water washes the lubricant from the splines and rusting will occur. The rust can be so severe that the two shafts will be rusted together. This will first be known there is an attempt to drop the lower unit. it will not separate from the powerhead.

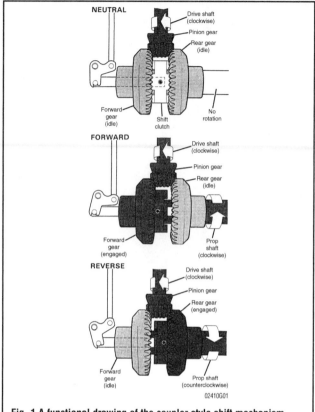

Fig. 1 A functional drawing of the coupler style shift mechanism

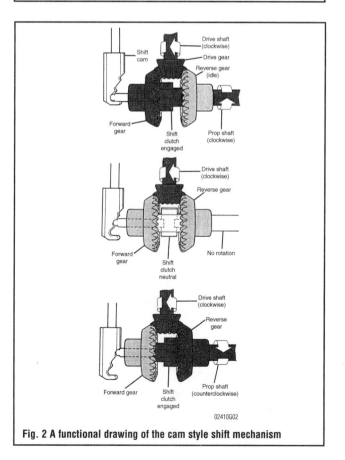

Fig. 2 A functional drawing of the cam style shift mechanism

LOWER UNIT 8-3

If rust has seized the joint, the drive shaft will have to be cut with a saw or cutting torch. Some driveshafts use stainless steel and others a good grade of steel. The best preventive service is not to let it happen. Service the spline joint each season along with a water pump impeller replacement. When water has been found in the lower unit oil, the unit should be pressure tested with air to 16–18 psi and the submerged under water or sprayed with a mixture of soapy water. Bubbles will come out of the areas that are leaking. Also, the unit should be subjected to 3–5 inches of vacuum. This will test the seals in the other direction. The pressure vacuum should hold for a few minutes.

After the shafts and housings are cleaned and inspected, special attention should be paid to the area's that are leaking. It is common for grit in the water to wear the shafts at the seal contact points or for corrosion to eat away at the housing. If the gearshift jumps out of gear, check the detent balls, spring, clutch dog, gears and shifter. If the gearshift won't shift, check the pivot pin, lever or cable adjustment shift rod connection, gearcase components and driveshaft. Replace any damaged or worn parts.

If the gearcase is seized, check the gearcase for lubricant. If lubricant is present, drain and disassemble the gearcase. Inspect all components for damage or corrosion. Replace broken or corroded components. Check for a distorted gearcase housing.

Propeller

REMOVAL & INSTALLATION

Shear Pin Style

♦ See Figures 3, 4 and 5

1. Disconnect the high tension lead/s from the spark plug/s to prevent the powerhead from starting accidentally.
2. Pull the cotter pin free of the propeller nut with a pair of pliers.
3. Place a block of wood between one blade of the propeller and the ventilation plate to prevent the shaft from turning.
4. Remove the propeller nut.
5. Remove the propeller lock pin.
6. Slide the propeller free of the propeller shaft.

To install:

7. Make certain the shaft is clean and free from corrosion.
8. Coat the propeller shaft with a liberal amount of water resistant grease to prevent the propeller from getting seized to the shaft.
9. Slide the propeller onto the shaft.

➡If the propeller shaft is equipped with splines and a pin be sure to install the propeller so that the pin holes in the propeller shaft and propeller align when installed.

10. Insert the pin through the propeller and propeller shaft.
11. Wedge a block of wood between one of the propeller blades and the ventilation plate to prevent the propeller from rotating.
12. Install the propeller nut.
13. Tighten the nut securely and then check to determine if the holes in the nut are aligned with the hole in the propeller shaft.
14. Adjust the nut accordingly to align the cotter pin hole.
15. Install the cotter pin.

Spline Driven

♦ See Figures 6, 7, 8 and 9

1. Disconnect the high tension leads from the spark plugs to prevent the powerhead from starting accidentally.
2. Pull the pin free of the propeller shaft with a pair of pliers.
3. Remove the cone from the end of the propeller shaft. If equipped

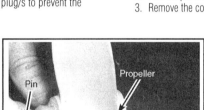

Fig. 3 Remove the cotter pin and prop cone to expose the shear pin and or nut seal

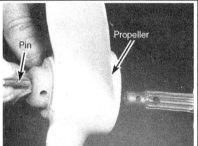

Fig. 4 Remove the shear pin and slide the propeller from the shaft

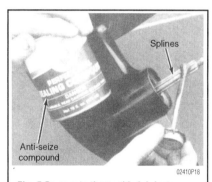

Fig. 5 Be sure to thoroughly lubricate a cleaned prop shaft before installing the propeller

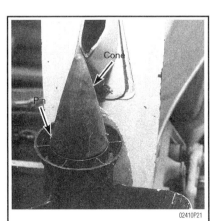

Fig. 6 Some units have a cone attached to the end of the propeller shaft covering the propeller nut

Fig. 7 Wedge the propeller with a piece of wood to hold it in place while removing the propeller nut

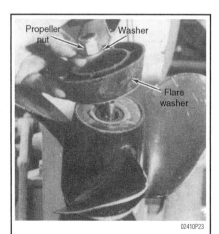

Fig. 8 Remove the propeller nut, washer, flare spacer then the propeller

8-4 LOWER UNIT

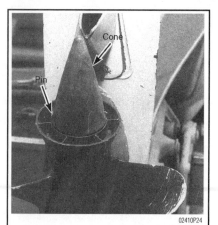

Fig. 9 Remove the propeller thrust washer and set it aside it must be reinstalled with the propeller housing damage will occur

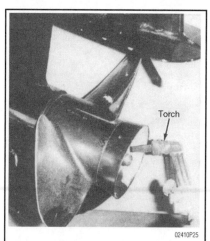

Fig. 10 It may be necessary to heat and melt the hub off the propeller

Fig. 11 The propeller hub and sleeve left behind after the propeller has been successfully melted off of the hub

4. Place a block of wood between one blade of the propeller and the ventilation plate to prevent the shaft from rotating.
5. Loosen the propeller nut.
6. Remove the block of wood and back the propeller nut free of the propeller shaft.
7. Slide the flat washer and the flare washer off the shaft.
8. Remove the propeller and the spacer.

To install:
9. Make certain the shaft is clean and free from corrosion.
10. Coat the propeller shaft with a liberal amount of water resistant grease to prevent the propeller from getting seized to the shaft.
11. Slide the spacer and then the propeller onto the shaft with the internal splines of the propeller indexing with the splines on the shaft.
12. Slide the flare washer and the flat washer onto the propeller shaft.
13. Thread the nut onto the propeller shaft.
14. Wedge a block of wood between one of the propeller blades and the ventilation plate to prevent the propeller from rotating.
15. Tighten the nut securely.
16. Slide the cone into place and insert the retaining pin.

Siezed Propeller

♦ See Figures 10, 11, 12 and 13

1. If the propeller is seized to the shaft heat must be applied to the shaft to melt out the rubber inside the hub. Using heat will destroy the hub, but there is no other way.

✱✱ CAUTION

As heat is applied, the rubber will expand and the propeller will actually be blown from the shaft. Therefore, stand clear to avoid personal injury.

2. Use a knife and cut the hub off the inner sleeve.
3. The sleeve can be removed by cutting it with a hacksaw or it can be removed with a puller. Again, if the sleeve is frozen, it may be necessary to apply heat. Remove the thrust hub from the propeller shaft.

Lower Unit

REMOVAL & INSTALLATION

Units Up To 7.5 HP

♦ See Figures 14, 15, 16, 17 and 18

1. Disconnect the spark plug leads.
2. Place an identifying mark on the shift rod to ensure proper location for assembly.
3. Disconnect the shift rod "Reference the particular type by the illustrations provided".
4. Remove the fasteners holding the lower unit in place.

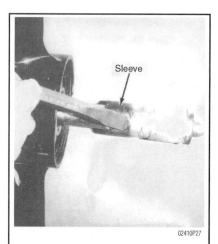

Fig. 12 Cut away the rubber portion of the hub and remove the sleeve by chiseling it or cutting it off the propeller shaft

Fig. 13 When all else fails a tool such as this one can be used to forcibly remove the propeller most likely leaving the propeller in an unusable condition

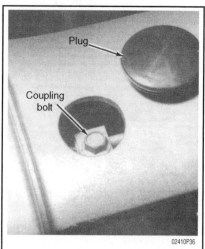

Fig. 14 Remove the cover plug to expose shift rod connection. Loosen to disconnect

LOWER UNIT 8-5

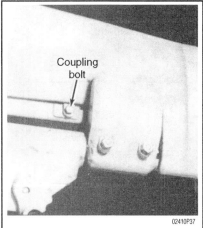
Fig. 15 Remove coupling bolt to disconnect shift rod

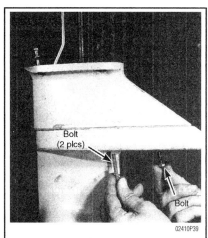
Fig. 16 Remove the fasteners holding the lower unit in place

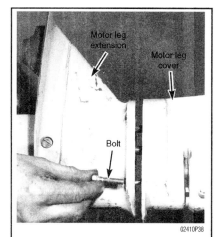
Fig. 17 Carefully pull the lower unit down to remove

5. Carefully pull the lower unit down to remove.
6. Support the unit in a lower unit holder or a vise.

To install:

➡ Prior to any assembly it is recommended that all fasteners be clean and greased to ease installation and help in the removal of the fasteners during future repairs or maintenance. A brush on gasket sealer would be recommended in salt water applications or when the housings are in marginal condition.

7. Lift the lower unit from the holder or vise.
8. Align the driveshaft with the tube or cavity and push the lower unit up into the mid housing.

➡ It may be necessary to rotate the driveshaft or flywheel to align the driveshaft splines.

9. Be certain that the water tube is aligned with the water pump outlet and slide the unit into place.

10. Install the attaching hardware.
11. Connect the shift linkage with the fastener removed earlier.
12. Attach the spark plug leads to the spark plugs.
13. Supply the engine with an appropriate water supply.
14. Start the engine and test the water pump output.

※※ WARNING

If water does not appear to be flowing through the engine immediately shut the engine off. There could be a water supply problem or a misalignment in the water tube.

Units With An Internal Type Connection

♦ See Figure 19

1. Disconnect the spark plug leads.
2. Remove the fasteners holding the lower unit in place.
3. Carfully split the lower unit from the mid section to gain access to the internal type shift rod connection.
4. Disconnect the shift rod "Reference the particular type by the illustrations provided".
5. Carefully pull the lower unit down to remove.
6. Support the unit in a lower unit holder or a vise.

To install:

➡ A brush on gasket sealer would be recommended in salt water applications or when the housings are in marginal condition.

7. Lubricate the driveshaft splines on the sides only. Grease on the top of the shaft could cause it not to be able to fully seat into the crankshaft.
8. Lift the lower unit from the holder or vise.
9. Align the driveshaft with the tube or cavity and push the lower unit up into the mid housing.

➡ It may be necessary to rotate the driveshaft or flywheel to align the driveshaft splines.

10. Be certain that the water tube is aligned with the water pump outlet and slide the unit into place leaving about a one inch gap.
11. Connect the shift linkage with the fastener removed earlier.
12. Move the lower unit up flush with the mid housing.
13. Install the attaching hardware.
14. Attach the spark plug leads to the spark plugs.
15. Supply the engine with an appropriate water supply.
16. Start the engine and test the water pump output.

※※ WARNING

If water does not appear to be flowing through the engine immediately shut the engine off. There could be a water supply problem or a misalignment in the water tube.

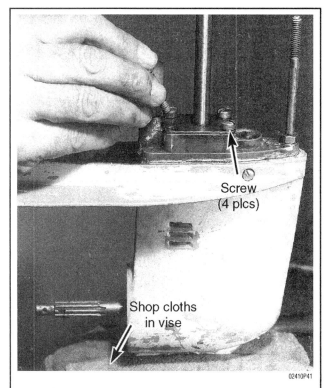
Fig. 18 Support the unit in a lower unit holder or a vise

8-6 LOWER UNIT

Units With Threaded Shift Coupling

♦ See Figure 20

1. Disconnect the spark plug leads.
2. Place an identifying mark on the shift rod to ensure proper location for assembly.
3. Disconnect the shift rod.
4. Remove the fasteners holding the lower unit in place.
5. Carefully pull the lower unit down to remove.
6. Support the unit in a lower unit holder or a vise.

To install:

➡A brush on gasket sealer would be recommended in salt water applications or when the housings are in marginal condition.

7. Lubricate the driveshaft splines on the sides only. Grease on the top of the shaft could cause it not to be able to fully seat into the crankshaft.
8. Lift the lower unit from the holder or vise.
9. Align the driveshaft with the tube or cavity and push the lower unit up into the mid housing.

➡It may be necessary to rotate the driveshaft or flywheel to align the driveshaft splines.

10. Be certain that the water tube is aligned with the water pump outlet and slide the unit into place leaving enough space to thread together the shift connector.
11. Connect the shift linkage by threading the connector on to the shift rod
12. Secure the connector with the jam nut.
13. Move the lower unit up flush with the mid housing.
14. Install the attaching hardware.
15. Attach the spark plug leads to the spark plugs.
16. Supply the engine with an appropriate water supply.
17. Start the engine and test the water pump output.

✳✳ WARNING

If water does not appear to be flowing through the engine immediately shut the engine off. There could be a water supply problem or a misalignment in the water tube.

Units With Shift Shaft Pin Connection

♦ See Figures 21, 22 and 23

1. Disconnect the spark plug leads.
2. Place an identifying mark on the shift rod to ensure proper location for assembly.

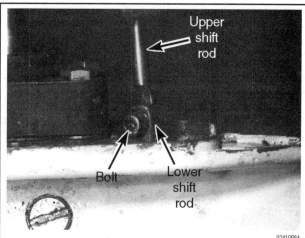

Fig. 19 With the lower unit unbolted, shift the unit into reverse gear start to remove the unit to expose the shift connection and remove the bolt

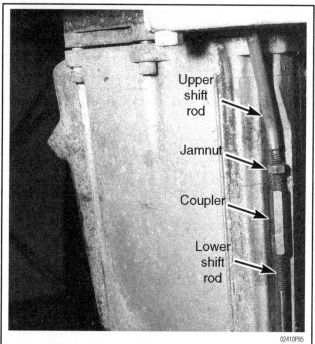

Fig. 20 Hold the coupler & remove the jam nut allowing the coupler to be threaded off the shift rod disconnecting the shift rod

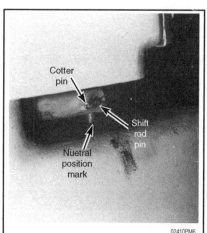

Fig. 21 Remove the cotter pin and slide the shift pin out to disconnect the shift rod

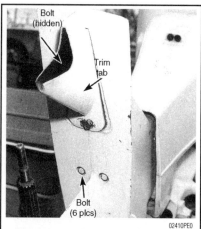

Fig. 22 Remove the remaining fasteners holding the lower unit in place

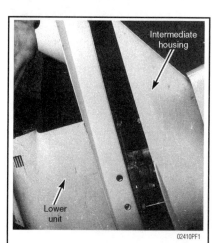

Fig. 23 Carefully pull the lower unit down to remove

LOWER UNIT 8-7

3. Remove the cotter pin and slide the shift pin out to disconnect the shift rod.
4. Remove the trim tab from the trailing end of the ventilation plate.
5. Remove the fastener from under the trim tab.
6. Remove the remaining fasteners holding the lower unit in place.
7. Carefully pull the lower unit down to remove.
8. Support the unit in a lower unit holder or a vise.

To install:

➡ A brush on gasket sealer would be recommended in salt water applications or when the housings are in marginal condition.

9. Lubricate the driveshaft splines on the sides only. Grease on the top of the shaft could cause it not to be able to fully seat into the crankshaft.
10. Lift the lower unit from the holder or vise.
11. Align the driveshaft with the tube or cavity and push the lower unit up into the mid housing.

➡ It may be necessary to rotate the driveshaft or flywheel to align the driveshaft splines.

12. Be certain that the water tube is aligned with the water pump outlet
13. Move the lower unit up flush with the mid housing.
14. Install the attaching hardware.
15. Install the trim tab with its hardware.
16. Connect the shift linkage by installing the shift pin into the connector and installing a new cotter pin.
17. Attach the spark plug leads to the spark plugs.
18. Supply the engine with an appropriate water supply.
19. Start the engine and test the water pump output.

✲✲ WARNING

If water does not appear to be flowing through the engine immediately shut the engine off. There could be a water supply problem or a misalignment in the water tube.

Units With Splined Shift Shaft Connections

1. Disconnect the spark plug leads.
2. Shift the unit into forward gear.
3. Remove the fasteners holding the lower unit in place.
4. Carefully pull the lower unit down to remove.
5. Support the unit in a lower unit holder or a vise.

To install:

➡ A brush on gasket sealer would be recommended in salt water applications or when the housings are in marginal condition.

6. Lubricate the driveshaft splines on the sides only. Grease on the top of the shaft could cause it not to be able to fully seat into the crankshaft.
7. Lift the lower unit from the holder or vise.
8. Align the driveshaft with the tube or cavity and push the lower unit up into the mid housing.

➡ It may be necessary to rotate the driveshaft or flywheel to align the driveshaft splines.

9. Be certain that the water tube is aligned with the water pump outlet and shift splines are aligned with the shift connection then slide the unit into place.
10. Move the lower unit up flush with the mid housing.
11. Install the attaching hardware.
12. Attach the spark plug leads to the spark plugs.
13. Supply the engine with an appropriate water supply.
14. Start the engine and test the water pump output.

✲✲ WARNING

If water does not appear to be flowing through the engine immediately shut the engine off. There could be a water supply problem or a misalignment in the water tube.

LOWER UNIT OVERHAUL

Units Up To 7.5 HP

DISASSEMBLY

♦ See Figures 24 thru 41

This section contains instructions for complete service of the lower unit with a one piece gear case and a cam shift arrangement. This lower unit section contains repair procedures for shifting and non-shifting units. The procedures are similar therefore only use the procedural steps applicable to the unit that is being repaired.

1. Remove all water pump parts.
2. Loosen the nut on the long stud securing the water pump body. On some models the nut is in the form of a hex bolt. Pry the water pump body from the lower unit.
3. Slowly raise the driveshaft straight up and out of the lower unit.

➡ Once the driveshaft is raised the pinion gear on the lower end of the driveshaft will come free of the shaft. If the gear were separated from the driveshaft the lower unit would most likely need to be completely disassembled to correct the condition.

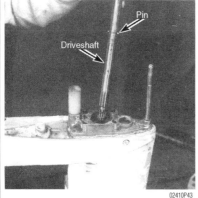

Fig. 24 The pinion gear on the lower end will separate from the lower end of the shaft. The water pump pin on the driveshaft may remain in place

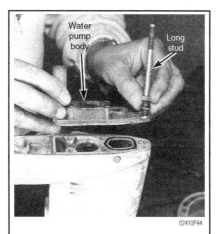

Fig. 25 Loosen the nut on the long stud securing the water pump body then pry the water pump body from the lower unit

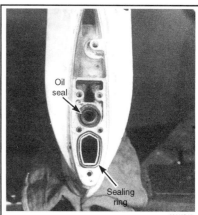

Fig. 26 Check the condition of the driveshaft oil seal and seal ring around the shift shaft pocket on a non-shifting unit

8-8 LOWER UNIT

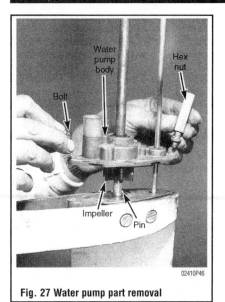
Fig. 27 Water pump part removal

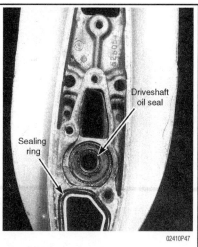
Fig. 28 Check the condition of the driveshaft oil seal and seal ring around the shift shaft pocket on a shifting unit

Fig. 29 If the retainer or the bearing carrier is seized in the lower unit a universal puller must be used along with possibly heating the lower unit housing with a torch

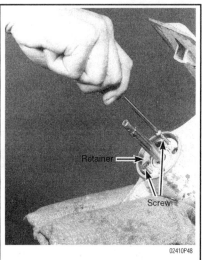
Fig. 30 Tighten the screws alternately and evenly to remove the retainer

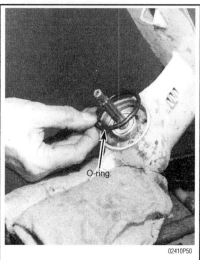
Fig. 31 Remove the O-ring seal between the retainer and bearing carrier

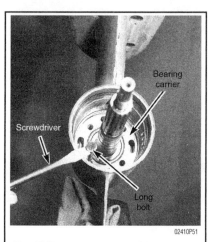
Fig. 32 Using a couple screwdrivers pry under both bolt heads at the same time with the edge of the lower unit opening as a fulcrum

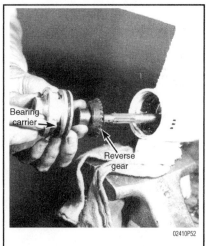
Fig. 33 Slowly slide the bearing carrier and the reverse gear off of the propeller shaft

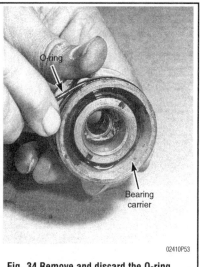
Fig. 34 Remove and discard the O-ring around the bearing carrier

Fig. 35 Inspect the condition of the propeller shaft oil seal

LOWER UNIT 8-9

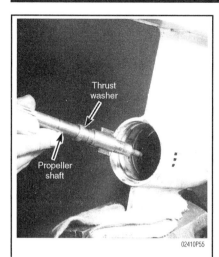

Fig. 36 Remove the propeller shaft from the lower unit. Slide the thrust washer free of the shaft

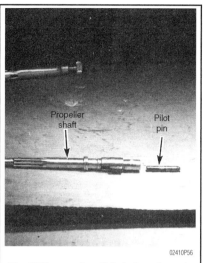

Fig. 37 Remove the pilot pin from the end of the propeller shaft

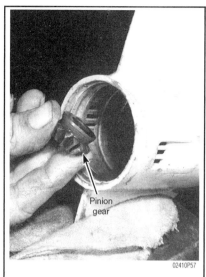

Fig. 38 Remove the pinion gear

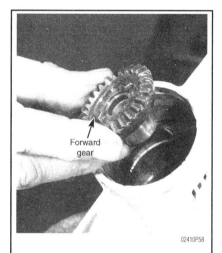

Fig. 39 Reach to the forward end of the lower unit cavity and remove the forward gear

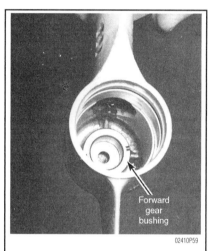

Fig. 40 Inspect the lower unit cavity and determine the condition of the forward gear bushing

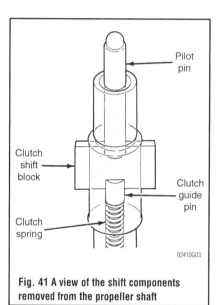

Fig. 41 A view of the shift components removed from the propeller shaft

✱✱ WARNING

Remove the shaft slowly to prevent the splines on the end of the shaft from damaging any oil seals.

4. Check the conditions of the driveshaft oil seal and seal ring around the shift shaft pocket. If the seal appears damaged pry the seal out with a screwdriver.
5. Rotate the shift rod counterclockwise to separate the rod from the shift cam.

➡ The shift cam will be held in place inside the lower unit by spring pressure on the pilot pin.

6. Lift the shift rod free of the backing plate and the water pump body.
7. Remove the backing plate and the driveshaft.
8. Use the appropriate wrench and remove the four screws from the propeller shaft retainer.
9. Observe the two other screw holes in the propeller shaft retainer.
10. Obtain two screws at least 2 in. (5cm) long with the same threads as in the retainer.
11. Thread the two screws into these two extra holes (not the mounting screw holes) until the screws make contact with the bearing carrier.
12. Tighten the screws alternately and evenly to remove the retainer.
13. Once the retainer is removed take out the two screws used for removal.

➡ If the retainer is seized in the lower unit and cannot be removed using the jacking bolts described in the previous step a universal puller must be used. Obtain a universal puller with long bolts of the proper size threads and remove the retainer.

14. Remove and discard the large O-ring.
15. Thread a couple of bolts into opposite holes in the carrier.
16. Using a couple screwdrivers pry under both bolt heads at the same time with the edge of the lower unit opening as a fulcrum.

➡ If the bearing carrier is seized in the lower unit and cannot be removed using the jacking bolts described in the previous step a universal puller must be used. Obtain a universal puller with long bolts of the proper size threads and remove the retainer.

17. Remove the bearing carrier from the lower unit cavity.
18. Slowly slide the bearing carrier aft and free of the propeller shaft.
19. Slide the reverse gear aft and free of the propeller shaft.
20. Remove and discard the O-ring around the bearing carrier.
21. Inspect the condition of the propeller shaft oil seal.
22. If the seal is no longer fit for service pry the seal free with a screwdriver.
23. Remove the propeller shaft from the lower unit.

8-10 LOWER UNIT

24. Slide the thrust washer free of the shaft.
25. Remove the pilot pin from the end of the propeller shaft.

➡This pin is slide fit and should free easily. If the pilot pin failed to remain with the propeller shaft as the shaft was removed reach into the lower unit cavity and retrieve the pin.

26. Remove the pinion gear.
27. Reach to the forward end of the lower unit cavity and remove the forward gear.
28. Inspect the lower unit cavity and determine the condition of the forward gear bushing.

➡The bushing cannot be removed so in the event that the bushing is damaged the lower unit casing must be replaced.

29. Remove the propeller shaft shift components by depressing the clutch spring with a small blade screwdriver inserted into the spring coils through the slot on one side of the propeller shaft.
30. With the spring depressed remove the clutch guide pin and shift clutch block from the propeller shaft. Remove the spring.

CLEANING & INSPECTION

♦ See Figures 42 thru 47

Inspect the oil seal housing for cracks and distortion.
Clean the beveled gears with solvent and then dry them with compressed air. Inspect the gear teeth for wear. Under normal conditions the gear will show signs of wear but it will be smooth and even.
Clean the bearing carrier with solvent and dry it with compressed air.
Check the clutch dogs to be sure they are not rounded off or chipped. Such damage is usually the result of poor operator habits and is caused by shifting too slowly or shifting while the engine is operating at high rpm. This type damage might also be caused by improper shift rod adjustments.
Clean the driveshaft with solvent and dry it with compressed air.
Inspect the driveshaft splines for excessive wear. Check the oil seal surfaces above and below the water pump drive pin or Woodruff key area for grooves. Replace the shaft if grooves are discovered.
Measure the length of the clutch spring. The spring should measure 2 7/16" (62mm) from end to end. This precise length is necessary to ensure the correct tension on the shift cam. Replace the spring if the measured length is not correct. Do not attempt to alter the length by forcibly stretching or compressing the spring. The attempt will fail and the spring will return to its natural length.
Inspect the two cutout portions of the clutch shift block for cracks. Also check for rounding at the clutch dog contact area on the forward and reverse gears as applicable. If rounding is discovered replace the block, this condition can only worsen and cause the lower unit to slip while in gear.
Metal particles broken from the clutch shift block, may find their way between the meshing teeth of the pinion gear and a mating gear. Metal particles between the gear teeth could chip gear teeth or cause even more serious damage in the lower unit.

Inspect the propeller shaft oil seal surface to be sure it is not pitted, grooved or scratched.
Inspect the propeller shaft splines for wear and corrosion damage. Check the propeller shaft for cracks distortion and be certain it is not bent.
Inspect the shift cam ramp for wear, corrosion or other signs of damage. Check the lower stop on the cam for cracks.
Check the gear housing for impact damage and clean off any metal particles from the drain plug magnet.
Check the trim tab. The trim tab must make a good ground inside the lower unit. Therefore, the trim tab and the cavity must not be painted. In addition to trimming the boat, the trim tab acts as a zinc electrode to prevent electrolysis

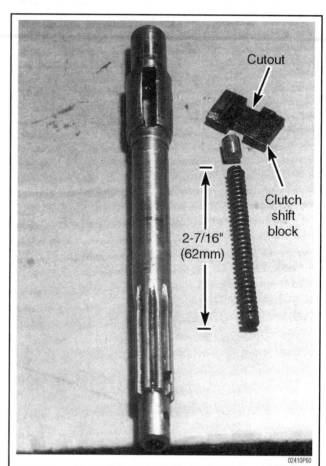

Fig. 42 Layout to indicate arrangement of parts for the cam shift mechanism

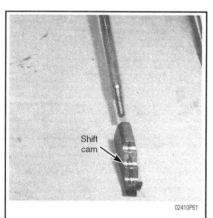

Fig. 43 Close-up views of the shift cam to show a normal wear pattern

Fig. 44 A rusted and corroded gear that has been water damaged and is not useable

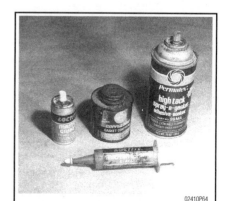

Fig. 45 Obtaining a good quality sealing product is well worth the extra cost to prolong satisfactory service of the lower unit

LOWER UNIT   8-11

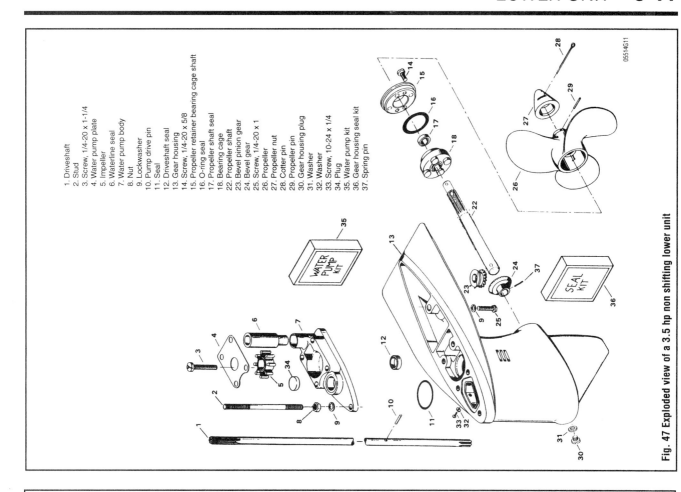

Fig. 47 Exploded view of a 3.5 hp non shifting lower unit

1. Driveshaft
2. Stud
3. Screw, 1/4-20 x 1-1/4
4. Water pump plate
5. Impeller
6. Waterline seal
7. Water pump body
8. Nut
9. Lockwasher
10. Pump drive pin
11. Seal
12. Driveshaft seal
13. Gear housing
14. Screw, 1/4-20 x 5/8
15. Propeller retainer bearing cage shaft
16. O-ring seal
17. Propeller shaft seal
18. Bearing cage
22. Bevel pinion gear
23. Bevel gear
24. Gear housing
25. Screw, 1/4-20 x 1
26. Propeller
27. Propeller nut
28. Cotter pin
29. Propeller pin
30. Gear housing plug
31. Washer
32. Washer
33. Screw, 10-24 x 1/4
34. Plug
35. Water pump kit
36. Gear housing seal kit
37. Spring pin

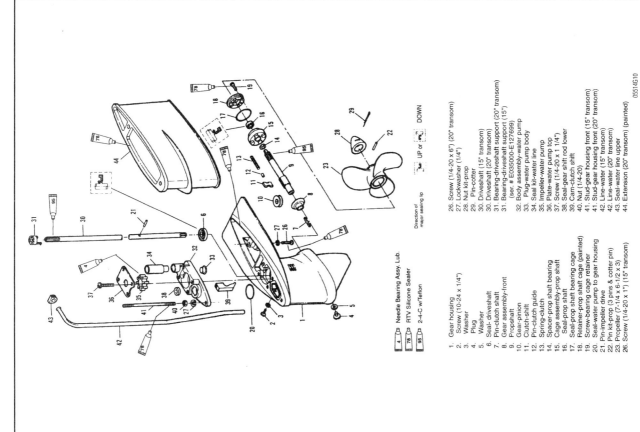

Fig. 46 Current model 5 hp exploded view

1. Gear housing
2. Screw (10-24 x 1/4")
3. Washer
4. Plug
5. Washer
6. Seal- driveshaft
7. Pin-clutch shaft
8. Gear assembly-front
9. Propshaft
10. Gear-pinion
11. Clutch-shift
12. Pin-clutch guide
13. Spring-clutch
14. Spacer-prop shaft bearing
15. Cage assembly-prop shaft
16. Seal-prop shaft
17. Seal-prop shaft bearing cage
18. Retainer-prop shaft cage (painted)
19. Screw-bearing cage retainer
20. Seal-water pump to gear housing
21. Pin-impeller drive
22. Pin kit-prop (3 pins & cotter pin)
23. Propeller (7-1/4 x 6-1/2 x 3)
26. Screw (1/4-20 x 1") (15" transom)
26. Screw (1/4-20 x 6") (20" transom)
27. Lockwasher (1/4")
28. Nut kit-prop
29. Pin-cotter
30. Driveshaft (15" transom)
30. Driveshaft (20" transom)
31. Bearing-driveshaft support (20" transom)
31. Bearing-driveshaft support (15")
    (ser. # E035000-E127699)
32. Body assembly-water pump
33. Plug-water pump body
34. Seal kit-water line
35. Impeller-water pump
36. Plate-water pump top
37. Screw (1/4-20 x 1 1/4")
38. Seal-gear shift rod lower
39. Cam-clutch shift
40. Nut (1/4-20)
41. Stud-gear housing front (15" transom)
41. Stud-gear housing front (20" transom)
42. Line-water (15" transom)
42. Line-water (20" transom)
43. Seal-water line upper
44. Extension (20" transom) (painted)

Direction of major sealing lip   UP or   DOWN

4 — Needle Bearing Assy. Lub.
78 — RTV Silicone Sealer
95 — 2-4-C w/Teflon

8-12 LOWER UNIT

from acting on more expensive parts. It is normal for the tab to show signs of erosion. The tabs are inexpensive and should be replaced frequently.

Clean the exterior surface of the unit thoroughly. Inspect the finish for damage or corrosion. Clean any damaged or corroded areas and apply primer and matching paint.

➡During the Assembly and Installation procedures, a sealant is applied to the lower unit, motor leg extension and the intermediate housing mating surfaces. The sealant recommended by the manufacturer is P/N T8983. An acceptable substitute is Permatex #27® which is a high temperature sealant.

The manufacturer also recommends the use of Mercury sealant P/N T8983 on bolt threads in the lower unit. For this application, a good substitute is Loctite ® or Lock N' Seal®. These products are non-hardening adhesives especially produced for application on the threads of load bearing fasteners.

This material helps prevent the bolt from loosening due to vibration, thread wear and corrosion. Loctite ® or Lock N' Seal® must only be applied to clean, dry parts.

Any sealant is always applied in a thin continuous bead to only one of the mating surfaces. After installation of the two mating surfaces, the sealant is compressed to form a watertight gasket between the mating surfaces.

If too much sealant is applied, the excess will ooze out both internally and externally and possibly restrict passages. If the sealant finds its way into a closed end threaded hole it will form a pool in the bottom of the cavity. When a bolt is later installed and tightened to a specific torque value, the bolt will hydraulically lock.

A hydraulic lock occurs when a bolt that has been installed into a threaded cavity, the bolt end contacts the liquid sealant in the bottom of the hole. Because a liquid cannot be compressed, the bolt cannot move further. Even though the reading on the torque wrench indicates the bolt has been tightened to the required torque value, the bolt remains seated against the fluid

ASSEMBLY

♦ See Figures 48, 49, 50 and 51

➡It is highly recommended that the installation of new O-rings and oil seals regardless of their appearance.

1. Pack the lip of the driveshaft oil seal with water resistant multi-purpose lubricant.
2. Using the proper size socket install the seal with the lip facing down.
3. Install a new sealing ring into the recess of the lower unit.
4. Insert the shift cam into the shift rod cavity with the stepped side facing aft.
5. Insert the clutch guide pin into the coils at one end of the spring.
6. Slide the spring and guide pin into the end of the propeller shaft with the empty end of the spring going in first.
7. Depress the spring with a small blade screwdriver, and at the same time move the shift clutch block through the slot in the propeller on top of the pin.
8. Center the block in the propeller shaft.
9. Release the pressure on the spring and remove the screwdriver.
10. Place the forward gear into the bushing at the forward end of the lower unit cavity.
11. Carefully lower the driveshaft down through the driveshaft oil seal
12. Hold the pinion gear in position inside the lower unit and push the driveshaft down with a slight twisting motion to align the splines in the pinion gear.
13. Push the pilot pin into the hole at the end of the propeller shaft, with the flat on the end of the pin going in first. The pin will rest against the clutch shift block.
14. Slide the thrust washer over the splined end of the propeller shaft.
15. Insert propeller shaft into the forward gear located at the forward end of the lower unit housing
16. Gently push the propeller shaft until the rounded end of the pilot rod rests against a stepped notch on the shift cam.
17. Pack the lip of the propeller shaft oil seal with water resistant multi-purpose lubricant.
18. Using the proper size socket, install the seal with the lip facing down.
19. Install a new O-ring around the bearing carrier.
20. Slide the reverse gear onto the propeller shaft followed by the bearing carrier.
21. Install the large O-ring over the carrier.
22. Position the propeller shaft retainer in place on the end of the bearing carrier.
23. Rotate the retainer until the holes in the retainer are aligned with the threaded holes in the bearing carrier.
24. Secure the retainer with the four Allen head cap screws.
25. Tighten the screws securely.
26. Install the water pump parts.
27. Adjust the shift rod height with the unit in neutral gear rotate the lower shift rod until the distance between the top surface of the motor leg extension and the bottom edge of the hole in the rod measures 6 5/32" (15.6cm).

9.9 HP and Larger Engines with a Split Gearcase

The following procedures give detailed illustrated instructions to completely disassemble service and assemble the split style lower unit. There are some slight variations in the units but all follow the same repair procedures in general. There will be procedural differences in two versions of this style lower unit.

The identifying factor in the repair procedure differences is that the older units have a stud that secures the lower assembly to the upper with the securing nut on the top of the unit in the exhaust chamber.

The newer units have a short stud securing the rear half of the unit with a nut protruding into the area just behind the bearing carrier.

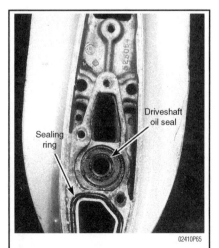

Fig. 48 Pack the lip of the driveshaft oil seal with water resistant multi-purpose lubricant for ease of installation

Fig. 49 Insert the shift cam into the shift rod cavity with the stepped side facing aft

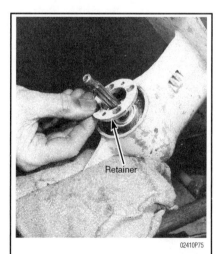

Fig. 50 Position the propeller shaft retainer in place on the end of the bearing carrier

LOWER UNIT   8-13

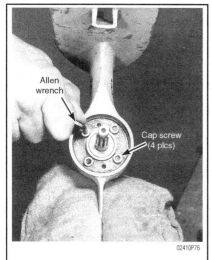

Fig. 51 Secure the retainer with the four Allen head cap screws

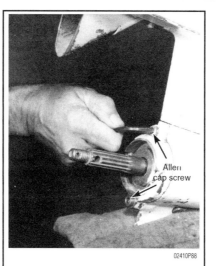

Fig. 52 Remove the two fasteners securing the bearing carrier in the lower unit

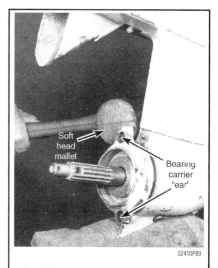

Fig. 53 Use a soft head mallet tap the ears of the bearing carrier to remove

➡The removal procedures for the bearings and races are listed as part of the disassembly and assembly procedures. In many cases the removal of the bearing will damage it beyond use. In the event the bearing is in good reusable condition skip that step and continue the procedure.

DISASSEMBLY

♦ See Figures 52 thru 74

Units having a weed guard installed in a groove at the aft end of the bearing carrier need to pry the ring from the carrier with a screwdriver before removal of the bearing carrier.

If the unit being serviced does not have the ears with a pair of snap ring pliers remove the internal snap ring used to secure the bearing carrier in the lower unit and remove using a puller

1. Remove the two fasteners securing the bearing carrier in the lower unit.
2. Use a soft head mallet and tap the upper ear of the bearing carrier counterclockwise until the upper and lower ear clear the lower unit.
3. Tap the upper and lower ears alternately and evenly on the forward side to move the carrier out of the housing.
4. Remove the bearing carrier from the propeller shaft.

➡If the bearing carrier is seized in the lower unit and cannot be removed using the method described in the previous step a universal puller must be used. Obtain a universal puller with long bolts of the proper size threads and remove the retainer.

5. If necessary the bearing set in the carrier assembly can be removed by using a slide hammer with jaw attachment.
6. Save any shim material installed behind the ball bearing or bearing race which ever is applicable.
7. Remove the propeller shaft seal by prying the old seal out with a screwdriver.
8. Remove and discard the bearing carrier O-ring.

✱✱ WARNING

Early production lower units that do not have a visible nut holding the lower and upper housings together hold the pinion gear to the drive shaft with a nut. Therefore the upper housing and driveshaft would be removed as an assembly.

9. Clamp the driveshaft in a vise equipped with soft jaws or a couple shop clothes
10. Using a block of wood to protect the upper surface of the lower unit hit the lower unit with a heavy hammer to free the pinion gear from the shaft.
11. Remove the shaft from the vise.
12. Place the lower unit in a vise and clamp in place holding the skeg.

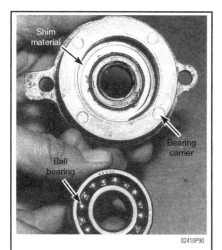

Fig. 54 A bearing carrier with the ball bearing removed. Note the location of the shims

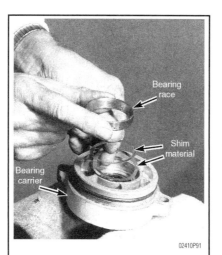

Fig. 55 A bearing carrier with the tapered bearing race removed. Note the location of the shims

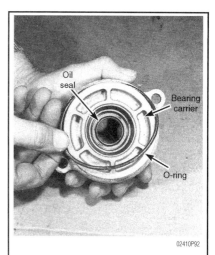

Fig. 56 The propeller shaft seal and bearing carrier O-ring should be replaced to insure a good seal after assembly

8-14 LOWER UNIT

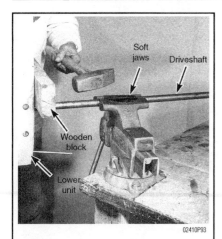

Fig. 57 Clamp the driveshaft in a vise equipped with soft jaws or a couple shop cloths to free the pinion gear from the driveshaft

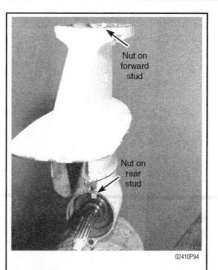

Fig. 58 Remove the nuts located as shown to split the upper and lower housings

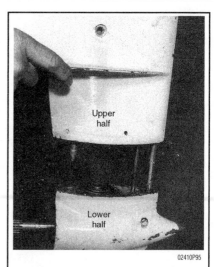

Fig. 59 Separate the two halves of the lower unit

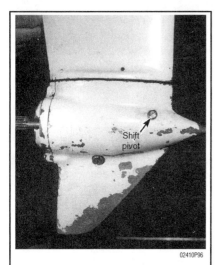

Fig. 60 Remove the pivot screw on the starboard side to release the shift coupler

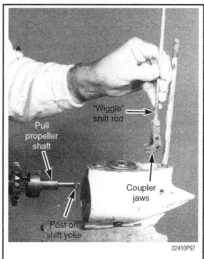

Fig. 61 Disassembly of a coupler style shift system

Fig. 62 Removal of the pinion and forward gear on a unit without a nut holding the pinion gear in place

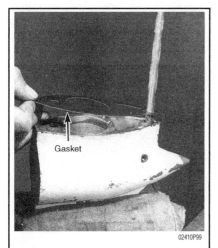

Fig. 63 Remove and discard the gasket from the mating surface of the lower unit halves

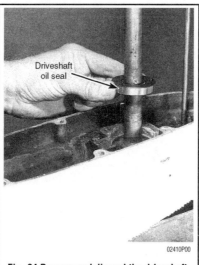

Fig. 64 Remove and discard the driveshaft oil seal

Fig. 65 Three long studs secure the upper and lower halves of the lower unit together on the style without the nut behind the bearing carrier

LOWER UNIT 8-15

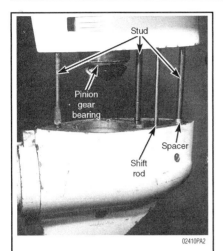

Fig. 66 Separate the lower unit upper and lower halves on the style without the nut behind the bearing carrier

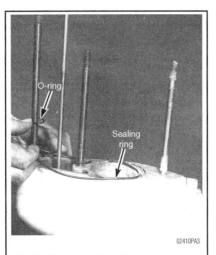

Fig. 67 The units with a three stud securing system use an O-ring and seal ring to seal the unit halves

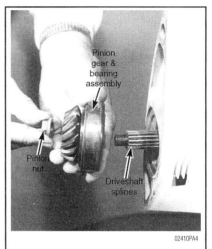

Fig. 68 With the driveshaft clamped in a vise use the correct size wrench and remove the pinion nut, as applicable

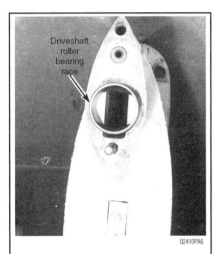

Fig. 69 If necessary remove the upper driveshaft tapered roller bearing race using the slide hammer

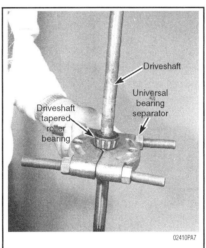

Fig. 70 A universal bearing separator is used to remove a driveshaft bearing that is no longer usable

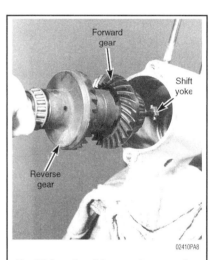

Fig. 71 Once the pinion gear is removed the propeller shaft assembly can be lifted out of the housing

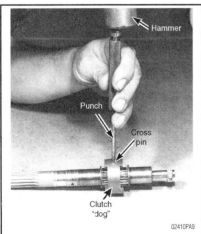

Fig. 72 Using a long pointed punch press out the cross pin to remove the shift arm or plunger and spring from the end of the propeller shaft

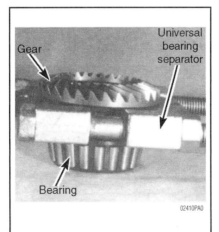

Fig. 73 To remove the bearing from the pinion of forward gear a universal bearing separator is installed as shown and the bearing pressed off of the gear

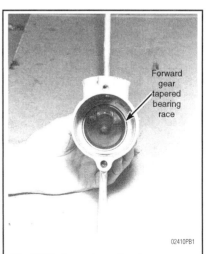

Fig. 74 The forward gear bearing race pressed into the lower unit housing can be removed with a slid hammer if necessary

8-16 LOWER UNIT

13. Pull the shaft up and out of the lower unit.
14. Remove the nut from behind the bearing carrier area holding the art section of the lower unit.
15. Remove the nut from the forward end of the lower unit in the area where the shift rod is located.
16. Separate the two halves of the lower unit.
17. If the unit being serviced is equipped with a pivot screw, remove the pivot screw, accessible from outside the lower unit on the starboard side.
18. Pull the propeller shaft straight back and out of the lower unit cavity.

※※ WARNING

If the attempt to pull the propeller shaft free fails the obstacle is in the shift mechanism. Wiggling the shift rod is usually enough to free the pivot connection. DO NOT USE FORCE UNLESS ABSOLUTELY NECESSARY.

19. Remove the shift rod.
20. Lift the pinion gear and driveshaft bearing assembly out of the upper cavity of the lower unit.
21. Remove the forward gear and tapered roller bearing assembly from the forward end of the lower unit.
22. Remove and discard the gasket from the mating surface of the lower unit halves.
23. Remove the upper driveshaft tapered roller bearing race using the slide hammer
24. Press the bearing free from the drive shaft using the proper size mandrel.

※※ WARNING

Take care not to bend or distort the driveshaft.

25. Remove the forward gear and thrust washer from the forward end of the propeller shaft.
26. Remove the reverse gear and thrust washer from the aft end of the shaft.
27. Using a long pointed punch press out the cross pin.
28. Remove the shift arm or plunger and spring from the end of the propeller shaft as applicable.

※※ WARNING

Observe how the clutch dog was installed. It must be installed in the same direction if it is to be reused.

29. Slide the clutch dog free of the shaft.

※※ WARNING

If the forward gear bearing or race is to be replaced, the bearing and the race must be replaced as a matched set.

➡ Three different special tools are required to properly install the bearing races for the pinion and/or the forward gear bearings. If the special tools are not available it is very difficult to install a new race without these tools.

30. Position a bearing separator between the forward gear and its bearing or between the pinion gear and its bearing.
31. Using a hydraulic press separate the gear from the bearing by pressing on the bearing shank.
32. Using a slide hammer and jaw attachment remove the bearing races as applicable.

CLEANING & INSPECTION

▶ See Figures 75, 76 and 77

Inspect the oil seal housing for cracks and distortion.
Clean the beveled gears with solvent and then dry them with compressed air. Inspect the gear teeth for wear. Under normal conditions the gear will show signs of wear but it will be smooth and even.
Clean the bearing carrier with solvent and dry it with compressed air.
Check the clutch dogs to be sure they are not rounded off or chipped. Such damage is usually the result of poor operator habits and is caused by shifting too slowly or shifting while the engine is operating at high rpm. This type damage might also be caused by improper shift rod adjustments.
Clean the driveshaft with solvent and dry it with compressed air.
Inspect the driveshaft splines for excessive wear. Check the oil seal surfaces above and below the water pump drive pin or Woodruff key area for grooves. Replace the shaft if grooves are discovered.
Inspect for rounding at the clutch dog contact area on the forward and reverse gears. If rounding is discovered replace the clutch dog. This condition can only worsen and cause the lower unit to slip while in gear.
Metal particles broken from the clutch shift block, may find their way between the meshing teeth of the pinion gear and a mating gear. Metal particles between the gear teeth could chip gear teeth or cause even more serious damage in the lower unit.
Inspect the propeller shaft oil seal surface to be sure it is not pitted, grooved or scratched.
Inspect the propeller shaft splines for wear and corrosion damage. Check the propeller shaft for cracks distortion and be certain it is not bent.
Inspect the shift cam ramp for wear, corrosion or other signs of damage. Check the lower stop on the cam for cracks.
Check the gear housing for impact damage and clean off any metal particles from the drain plug magnet.
Check the trim tab. The trim tab must make a good ground inside the lower unit. Therefore, the trim tab and the cavity must not be painted. In addition to

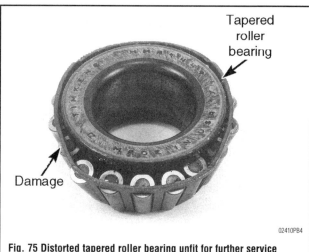

Fig. 75 Distorted tapered roller bearing unfit for further service

Fig. 76 Caged ball bearing set destroyed due to lack of lubrication, vibration, corrosion, metal particles or all of the above

LOWER UNIT 8-17

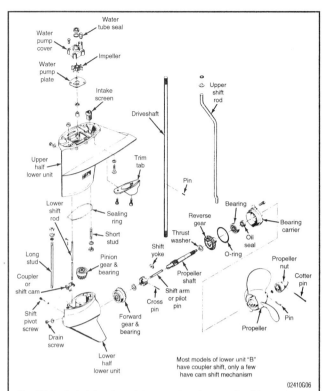

Fig. 77 Exploded drawing of the two piece lower unit with coupler shift or cam shift, and with forward, neutral, and reverse. Major parts are identified

trimming the boat, the trim tab acts as a zinc electrode to prevent electrolysis from acting on more expensive parts. It is normal for the tab to show signs of erosion. The tabs are inexpensive and should be replaced frequently.

Clean the exterior surface of the unit thoroughly. Inspect the finish for damage or corrosion. Clean any damaged or corroded areas and apply primer and matching paint.

➡During the assembly and installation procedures, a sealant is applied to the lower unit, motor leg extension and the intermediate housing mating surfaces. The sealant recommended by the manufacturer is P/N T8983. An acceptable substitute is Permatex #27® which is a high temperature sealant.

The manufacturer also recommends the use of sealant P/N T8983 on bolt threads in the lower unit. For this application, a good substitute is Loctite® or Lock N' Seal®. These products are non-hardening adhesives especially produced for application on the threads of load bearing fasteners.

This material helps prevent the bolt from loosening due to vibration, thread wear and corrosion. Loctite® or Lock N' Seal® must only be applied to clean, dry parts.

Any sealant is always applied in a thin continuous bead to only one of the mating surfaces. After installation of the two mating surfaces, the sealant is compressed to form a watertight gasket between the mating surfaces.

If too much sealant is applied, the excess will ooze out both internally and externally and possibly restrict passages. If the sealant finds its way into a closed end threaded hole it will form a pool in the bottom of the cavity. When a bolt is later installed and tightened to a specific torque value, the bolt will hydraulically lock.

A hydraulic lock occurs when a bolt that has been installed into a threaded cavity, the bolt end contacts the liquid sealant in the bottom of the hole. Because a liquid cannot be compressed, the bolt cannot move further. Even though the reading on the torque wrench indicates the bolt has been tightened to the required torque value, the bolt remains seated against the fluid

ASSEMBLY

♦ See Figures 78 thru 85

➡It is highly recommended that new O-rings and oil seals be installed, regardless of their appearance.

1. Position the tapered roller bearing over the gear.
2. Using a suitable mandrel, press on the inner race and install the bearing is flush against the shoulder of the gear.
3. Repeat this procedure for the forward and pinion gear bearings

➡There are special tools for these procedures but in the event these special tools are not available the old bearing race a small round plate or washer with equal diameter of the race or the old bearing and a drive rod will serve as substitutes.

4. Block the lower unit on a wooden work surface with the opening facing up.
5. Set the bearing race in position at the forward end of the lower unit with the large end of the race facing down.
6. Insert the tapered end of the race installer into the new race.
7. Place the driver handle over the race installer and slide the guide plate down the handle with the stepped side of the plate facing down.
8. Drive the bearing race into the lower unit by tapping on the driver handle with a hammer.
9. As soon as the race is fully seated the sound of the hammer striking the driver will change noticeably.
10. Place the new race into position over the lower unit.
11. Install the pinion gear bearing race into the upper housing.
12. Use the proper special tool or place the old bearing race on top of the new race with the small end of the race facing up.
13. As soon as the race is fully seated the sound of the hammer striking the driver will change noticeably.
14. Install the new driveshaft support bearing by first threading the pinion nut upside down onto the end of the driveshaft to protect the end of the shaft and the threads while the bearing is pressed into place.

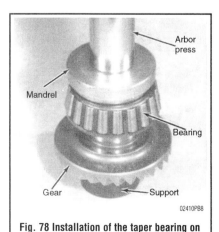

Fig. 78 Installation of the taper bearing on to the pinion or forward gear

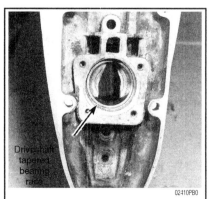

Fig. 79 Location of the driveshaft bearing race looking at the top of the upper section of the unit

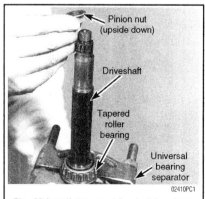

Fig. 80 Install the new driveshaft bearing as shown protecting the pinion nut thread area

8-18 Lower Unit

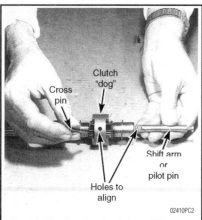

Fig. 81 Aligning the clutch dog, shift arm/pilot pin and cross pin are most easily done flat on a work bench as shown

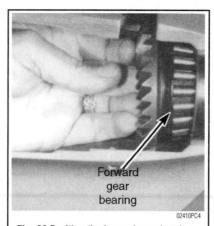

Fig. 82 Position the forward gear bearing assembly into place at the forward end of the lower unit cavity

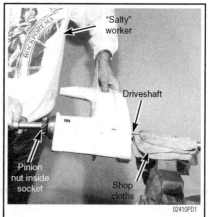

Fig. 83 Tighten the pinion nut as shown to achieve a minimal rolling torque to pre-load the pinion and driveshaft bearings

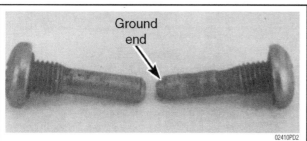

Fig. 84 Two shift pivot screws. The right screw has been ground to assist installation

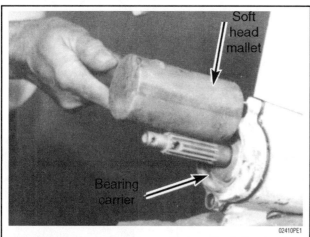

Fig. 85 Slide the bearing carrier over the propeller shaft and seat with a mallet

15. Press the new tapered roller bearing onto the driveshaft with the pinion end facing towards the tapered end of the bearing.

✳✳ WARNING

Because of the driveshaft length take care not to bend or distort the driveshaft while the bearing is being pressed on

16. Assmeble the propeller shaft shifting components by first sliding the clutch dog in place on the splines so that it can be aligned with the slot for the shift rod/pilot pin.
17. Slide the shift tension spring down into the propeller shaft, if applicable.
18. Install the shift rod or pilot pin into the end of the propeller shaft.
19. Press in or align as necessary to put the cross pin through the clutch dog and the shift rod /pilot pin.
20. Place the lower unit in a vise and clamp in place holding the skeg.
21. Position the forward gear and bearing assembly into place at the forward end of the lower unit cavity.
22. Units that do not have a nut holding the pinion gear in place install the pinion gear.
23. Place a small amount of fairly thick lubricant onto the end of the shift arm and place the shift yoke over the end of the shift arm on the cradle style shift units.
24. Lower the shift rod with shift rod coupler or cam attached into the lower unit with the jaws of the coupler or the stepped side of the cam must face aft.
25. At this time on a coupler shifted unit the pivot screw must be inserted into the threaded hole on the starboard side of the gear case and through the pivot hole in the coupler.
26. Install the propeller shaft.

➡ **Be certain to align the cradle shifter with the yoke at the end of the propeller shaft. The cam style systems are self aligning.**

27. Units having a nut holding the pinion gear on the driveshaft will now insert the driveshaft into the upper half of the lower unit.
28. Slide the pinion gear and bearing assembly onto the driveshaft with the internal splines of the gear indexing with the external splines on the driveshaft.
29. Install the pinion nut with Loctite®.
30. Clamp the driveshaft in a vise equipped with soft jaws or padded with a couple shop towels.
31. Tighten the pinion nut slowly and at the same time rotate the lower unit housing about the driveshaft.
32. Tighten the pinion gear nut until a slight amount of drag is felt as the lower unit is rotated.

➡ **There is no torque specification just a slight bearing preload is specified.**

33. Remove the driveshaft from the vise.
34. Install a new gasket or O-ring over the lower half of the unit.
35. Slide the upper half of the gear case over the lower half guiding the top of the shift rod slowly through the shift rod seal at the top of the upper half
36. Continue to install the upper half until both mating surfaces are seated against the gasket.
37. Install and hand tighten the lock washers and nuts on the studs holding the unit halves together.
38. Tighten the nuts alternately and evenly.
39. Lower the driveshaft down into the top of the lower unit if not already installed.
40. Rotate the propeller shaft to permit the internal splines of the pinion gear shank to index with the driveshaft splines.

LOWER UNIT   8-19

✱✱ WARNING

It may be necessary to tap the drive shaft into place on the units that do not have a nut holding the pinion gear. Driving the shaft down will most likely damage the drive shaft bearing and prematurely cause the lower unit to fail.

41. Install the reverse gear thrust washer as applicable.
42. Install the reverse gear and bearing.
43. Install a new seal in the bearing carrier by driving it into place with a special tool or a appropriate sized socket.
44. Use gasket sealer around the outside of the seal and install it with the seal lip facing in toward the gears.
45. Pack the lip of a new seal with multi-purpose water resistant lubricant.
46. Using a soft mallet install the bearing into the carrier with the same shims that had been removed.
47. Install the bearing carrier back over the propeller shaft and into the housing aligning the bolt holes as necessary.
48. Check the backlash shimming as described in the next section.
49. Install a new O-ring around the bearing carrier for final assembly.

SHIMMING

▶ See Figure 86

1. Install the bearing carrier without the O-ring.
2. Install and tighten one of the securing bolts to a torque value of 160 inch lbs. (18Nm).
3. Install the long bolt into the other threaded hole.
4. Push the propeller shaft into the lower unit as far as possible.
5. Rotate the shaft back and forth though a quarter turn to remove clearance.
6. Clamp a pair of vice grip type pliers around the propeller shaft with the vice grip handle in front of the bolt head.
7. Using a feeler gauge measure the gap between the vice grip handle and the bolt head.
8. Pull the propeller shaft out of the lower unit as far as possible.
9. Measure the gap between the handle and the bolt head.

The difference between the two measurements is the propeller shaft end play. If this difference exceeds 0.010" (0.25mm), the thickness of the thrust washer around the propeller shaft between the reverse gear and the clutch dog must be changed. A new thrust washer of the correct thickness must be purchased and installed.

10. Remove the bolt and vise grip pliers
11. Remove the bearing carrier.
12. Install a new O-ring around the bearing carrier.
13. Install the bearing carrier into the lower unit.
14. Install and tighten the two bolts alternately and evenly.

1984–91 70 to 150 HP

The following procedures give detailed illustrated instructions to completely disassemble service and assemble this style lower unit. There are some slight variations in the units but all follow the same repair procedures in general.

The visual identification of this unit is the water pump inlets. They are located on the round bullet portion of the lower unit on either side and behind the propeller.

➡The removal procedures for the bearings and races are listed as part of the disassembly and assembly procedures. In many cases the removal of the bearing will damage it beyond use. In the event the bearing is in good reusable condition skip that step and continue the procedure.

DISASSEMBLY

▶ See Figures 87 thru 110

1. Secure the lower unit assembly in a vise.
2. Drain the lower unit oil.
3. Remove the three bolts and washers securing the water pump body to the lower unit.
4. Pry the body up over the shift rod and driveshaft.
5. Remove and discard the O-ring and sealing ring from the water pump body.
6. Remove the shift shaft oil seal by prying the seal out with a screwdriver.
7. Remove and discard the driveshaft oil seal.
8. Remove and discard the crush ring.

✱✱ WARNING

The crush ring cannot be used a second time. Once installed the ring is crushed and deformed to provide proper clearance and positioning for driveshaft components when the pinion gear nut is tightened to the required torque value.

9. Remove the two screws securing the anode to the bearing carrier.
10. Remove the anode.
11. Obtain P/N T8948-1 or an equivalent puller and long bolts with the proper size threads.
12. Install the puller, and then take up on the bolts until the bearing carrier comes free.
13. Remove and discard the two O-rings from both ends of the bearing carrier.
14. Pry the oil seal from the carrier using a screwdriver.
15. Using a slide hammer with a pulling jaw attachment pull the needle bearing from the carrier.

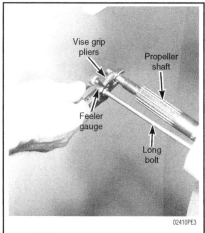

Fig. 86 Shimming set up as described. Be sure not to secure the vise grip on the propeller nut thread area and damage the threads

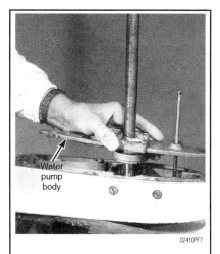

Fig. 87 Remove the bolts securing the water pump body to the lower unit and pry the body to remove

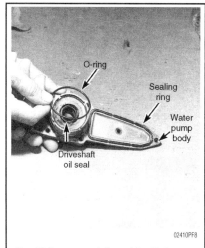

Fig. 88 Remove and discard the O-ring and sealing ring from the water pump body

8-20 LOWER UNIT

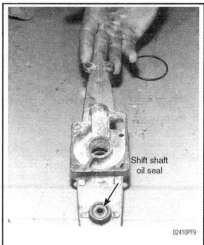

Fig. 89 Carefully inspect the shift shaft bushing and seal for possible leakage due to wear

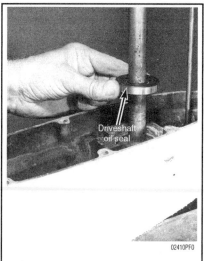

Fig. 90 Remove and discard the driveshaft oil seal

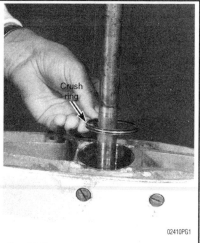

Fig. 91 The crush ring is not a reusable part it maintains a space that controls pinion height and bearing preload

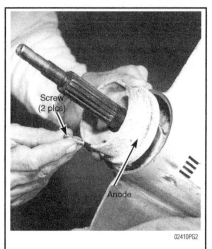

Fig. 92 The anode can be removed by taking out the two screws securing it to the bearing carrier

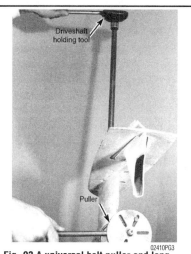

Fig. 93 A universal bolt puller and long bolts is used to remove both the inner and outer bearing carriers

Fig. 94 The outer bearing carrier removed showing the two sealing O-rings

Fig. 95 Inspecting the condition of the propeller shaft oil seal and needle bearing

Fig. 96 The two split rings must be removed prior to attempting to remove the inner bearing carrier

Fig. 97 The inner bearing carrier being removed from the lower unit

LOWER UNIT 8-21

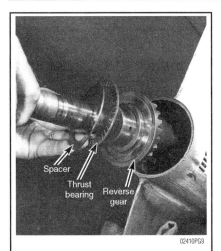

Fig. 98 Propeller shaft and internal gearcase components as they appear in there proper order

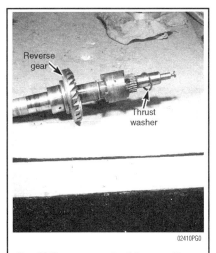

Fig. 99 The components of the propeller shaft including the reverse gear, note the location of the thrust washer

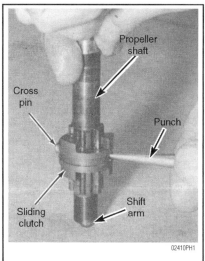

Fig. 100 A long pointed punch is used to press out the spring loaded cross pin

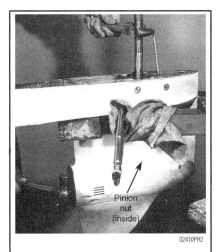

Fig. 101 A socket and ratchet being used to reach in and hold the pinion nut for removal

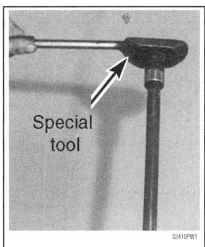

Fig. 102 A special tool is required on the upper end of the driveshaft while a socket and handle hold the pinion nut

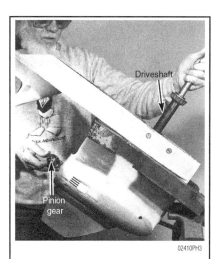

Fig. 103 After the pinion nut is removed the driveshaft and pinion gear can be lifted from the unit

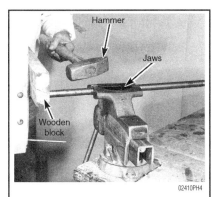

Fig. 104 It may be necessary to clamp the driveshaft in a vise equipped with soft jaws or a couple shop cloths. Using a block of wood to protect the upper surface of the lower unit and strike the lower unit with a heavy hammer to free the pinion gear from the shaft

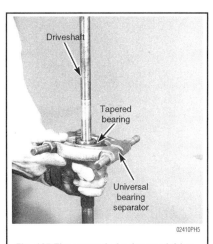

Fig. 105 The removal of a damaged driveshaft bearing can be accomplished using a hydraulic press and a bearing separator installed as shown

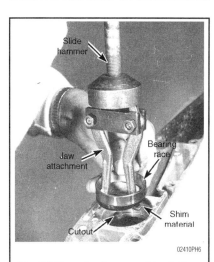

Fig. 106 Removing the upper driveshaft bearing race noting the location of the shims

8-22 LOWER UNIT

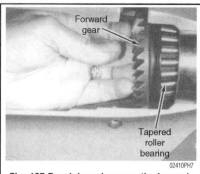

Fig. 107 Reach in and remove the forward gear and the forward gear tapered roller bearing

Fig. 108 A slide hammer is the proper tool to remove the forward gear bearing race

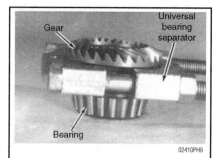

Fig. 109 Position a bearing separator between the forward gear and the tapered roller bearing. Using a hydraulic press separate the gear from the bearing

16. Insert a blade type screwdriver into the lower unit and hold the blade between the end of the retaining ring half that has a bevel cut and the housing.
17. Tap the end of the screwdriver with a soft head mallet to release the retaining ring from the groove.
18. Remove both halves of the ring from the lower unit.
19. Using a universal puller remove the inner bearing carrier.
20. Remove the inner bearing carrier from the lower unit.
21. Remove the spacer, thrust washer and reverse gear.
22. Remove the propeller shaft, thrust washer and the shift yoke from the lower unit.
23. Remove all the components from the propeller shaft.
24. Using a long pointed punch press out the cross pin.
25. Remove the shift arm and spring from the end of the propeller shaft.
26. Slide the clutch dog from the shaft observing how the clutch dog was installed.

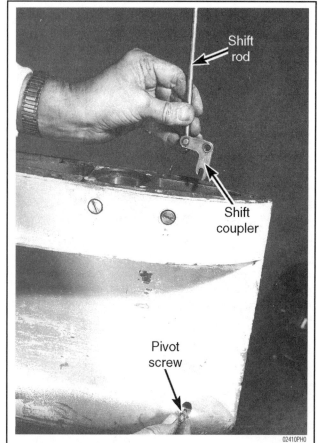

Fig. 110 Remove the pivot screw from the starboard side of the lower unit and lift the shift rod assembly out of the housing

27. The drive shaft must be removed next by first holding the driveshaft with a special spline socket.
28. Hold the pinion nut using the proper size socket to fit the pinion gear nut
29. With both tools in place rotate the driveshaft counterclockwise and loosen the pinion nut.
30. Remove the pinion nut.
31. Pull up on the driveshaft.
32. Pull the driveshaft up out of the lower unit housing.

➡It may be necessary to clamp the driveshaft in a vise equipped with soft jaws or a couple shop cloths. Using a block of wood to protect the upper surface of the lower unit and strike the lower unit with a heavy hammer to free the pinion gear from the shaft.

33. Remove the tapered roller bearing by pressing the bearing off the driveshaft using the proper size mandrel.

✱✱ WARNING

Because of its length, take care not to bend or distort the driveshaft.

34. Using a slide hammer and jaw attachment remove the upper driveshaft tapered roller bearing race and the pinion gear shim material using the slide hammer.

➡Save any shim material from behind the bearing it will be critical to obtain the correct backlash during installation.

35. Remove the pivot screw from the starboard side of the lower unit. Pull the shift rod and the shift rod coupler up and out of the lower unit.
36. Reach in and remove the forward gear and the forward gear tapered roller bearing.

➡Three different special tools are required to properly install the bearing race. If the special tools are not available it will be difficult to install a new race without them.

37. A slide hammer and jaw attachment will be used to remove the forward gear bearing race.
38. If it is necessary to replace the forward gear bearing, position a bearing separator between the forward gear and the tapered roller bearing.
39. Using a hydraulic press separate the gear from the bearing.

CLEANING & INSPECTION

♦ See Figures 111, 112 and 113

Good shop practice requires installation of new O-rings and oil seals regardless of their appearance.

Clean all water pump parts with solvent, and then dry them with compressed air. Inspect the water pump body and oil seal for cracks and distortion possibly caused from overheating. Inspect the water pump plate for grooves and/or rough surfaces. If possible, always install a new water pump impeller while the lower unit is disassembled. If the old impeller must be returned to service never install it in reverse to the original direction of rotation. Installation in reverse will cause premature impeller failure.

LOWER UNIT 8-23

If installation of a new impeller is not possible, check the seal surfaces. All must be in good condition to ensure proper pump operation. Check the upper, lower, and ends of the impeller vanes for grooves, cracking and wear. Check to be sure that the indexing notch of the impeller hub is intact and will not allow the impeller to slip.

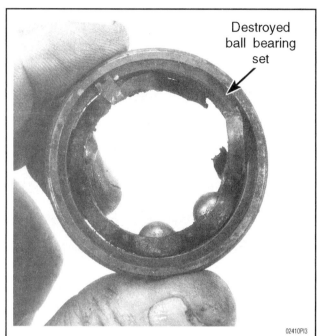

Fig. 111 Caged ball bearing set destroyed due to lack of lubrication, vibration, corrosion, metal particles, or all of the above

Clean around the Woodruff key. Clean all bearings with solvent, dry them with compressed air, and inspect them carefully. Be sure there is no water in the airline. Direct the air stream through the bearing. Never spin a bearing with compressed air. Such action is highly dangerous and may cause the bearing to score from lack of lubrication. After the bearings are clean and dry, lubricate them with clean lower unit lubricant. Do not lubricate tapered bearing cups until after they have been inspected.

Determine the condition of tapered bearing rollers and inner bearing race by inspecting the bearing cup for pitting, scoring, grooves, uneven wear, imbedded particles and discoloration caused from overheating. Always replace tapered roller bearings as a set.

Clean the forward gear with solvent, and then dry it with compressed air. Insect the gear teeth for wear. Under normal conditions the gear will show signs of wear but it will be smooth and even.

Clean the inner and outer bearing carriers with solvent, and then dry them with compressed air. Never spin bearings with compressed air. Such action is highly dangerous and may cause the bearing to score from lack of lubrication. Check the gear teeth of the reverse gear for wear. The wear should be smooth and even.

Check the clutch dogs to be sure they are not rounded-off or chipped. Such damage is usually the result of poor operator habits and is caused by shifting too slowly or shifting while the engine is operating at high rpm. Such damage might also be caused by improper shift rod adjustments.

Rotate the reverse gear and check for catches and roughness.

Inspect the roller bearing surface of the propeller shaft. Check the shaft surface for pitting, scoring, grooving, embedded particles, uneven wear and discoloration caused from overheating.

Clean the driveshaft with solvent, and then dry it with compressed air. Never spin bearings with compressed air. Such action is dangerous and could damage the bearing. Inspect the bearing for roughness, scratches, or side wear. If the bearing shows signs of such damage it should be replaced. If the bearing is satisfactory for further service coat it with oil.

Inspect the driveshaft splines for excessive wear. Check the oil seal surfaces above and below the water pump drive pin or Woodruff key area for grooves. Replace the shaft if grooves are discovered.

Inspect the drive shaft bearing surface above the pinion gear splines for pitting, grooves, scoring, uneven wear, embedded metal particles and discoloration caused by overheating.

Inspect the propeller shaft oil seal surface to be sure it is not pitted, grooved, or scratched. Inspect the roller bearing contact surface on the propeller shaft for pitting, grooves, scoring, uneven wear, embedded metal particles, and discoloration caused from overheating.

Inspect the propeller shaft splines for wear and corrosion damage. Check the propeller shaft for straightness.

Clean all parts with solvent and dry them with compressed air.
Inspect all bearing bores for loose fitting bearings.
Check the lower unit housing for impact damage.
Check the pinion nut corners for wear or damage.
Check the condition of the shift yoke for wear on the cutout portion and posts. This small inexpensive but vital part of the shifting mechanism should be replaced at the first sign of wear.

➡ During the Assembly and Installation procedures a sealant is applied to the lower unit motor leg extension and the intermediate housing mating surfaces. The sealant recommended by the manufacturer carries P/N T8983. An acceptable substitute is Permatex #27® which is a high temperature sealant.

The manufacturer also recommends the use of sealant P/N T8983 on bolt threads in the lower unit. For this application, a good substitute is Loctite® or Lock N' Seal®. These products are non-hardening adhesives especially produced for application on the threads of load bearing fasteners.

This material helps prevent the bolt from loosening due to vibration, thread wear and corrosion. Loctite® or Lock N' Seal® must only be applied to clean, dry parts.

Any sealant is always applied in a thin continuous bead to only one of the mating surfaces. After installation of the two mating surfaces, the sealant is compressed to form a watertight gasket between the mating surfaces.

If too much sealant is applied, the excess will ooze out both internally and externally and possibly restrict passages. If the sealant finds its way into a closed end threaded hole it will form a pool in the bottom of the cavity. When a bolt is later installed and tightened to a specific torque value, the bolt will hydraulically lock.

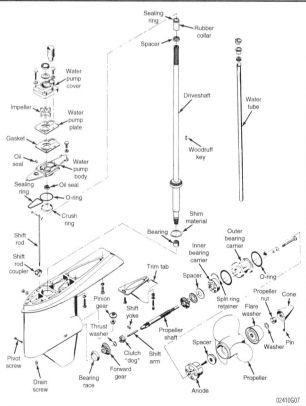

Fig. 112 Exploded drawing of one piece lower unit with coupler shift and forward, neutral and reverse. Major parts are identified

8-24 LOWER UNIT

Fig. 113 A new shift yoke (right) compared with one badly worn and unfit for service (left)

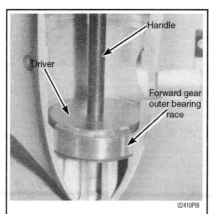

Fig. 114 A cutaway showing the forward gear bearing race being installed with the proper special tools

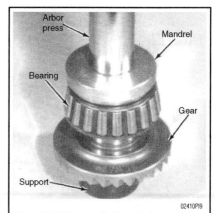

Fig. 115 Using a suitable mandrel press the bearing flush against the shoulder of the forward gear

A hydraulic lock occurs when a bolt that has been installed into a threaded cavity, the bolt end contacts the liquid sealant in the bottom of the hole. Because a liquid cannot be compressed, the bolt cannot move further. Even though the reading on the torque wrench indicates the bolt has been tightened to the required torque value, the bolt remains seated against the fluid

ASSEMBLY & SHIMMING

♦ See Figures 114 thru 126

Before assembling the lower unit be certain that all the parts have been properly cleaned, inspected and replacement items have been purchased and are on hand.

Good shop practice dictates new O-rings and seals be installed each time the lower unit is disassembled for service.

1. Install the lower unit housing in a vise of lower unit holder.
2. Install the forward gear tapered roller bearing race with special tools, Race Installer P/N T11204, Driver Handle P/N T8907, and Guide Plate P/N T11209.

➡If these special tools are not available the old bearing race a small round plate or washer with equal diameter as the race and a drive rod will be an acceptable replacement.

3. Block the lower unit on a wooden work surface with the opening facing up.
4. Set the bearing race in position at the forward end of the lower unit with the large end of the race facing down.
5. Using special tool: Insert the tapered end of the race installer into the new race.
6. Place the driver handle over the race installer and slide the guide plate down the handle with the stepped side of the plate facing down.
7. Drive the bearing race into the lower unit by tapping on the driver handle with a hammer.
8. Position the forward gear tapered roller bearing over the gear.
9. Useing a suitable mandrel, press the bearing flush against the shoulder of the forward gear.

WARNING

Always press on the inner race, never on the cage or on the rollers. This could damage the bearing and render it unusable.

10. Insert the assembled gear and bearing into the lower unit with the bearing sliding into the race.
11. Install the drive shaft taper roller bearing by first threading the pinion nut upside down onto the end of the driveshaft to protect the threads while the bearing is pressed into place.
12. Press a new tapered roller bearing onto the driveshaft with the pinion end facing towards the tapered end of the bearing.

WARNING

Because of the driveshaft length take care not to bend or distort the driveshaft while the bearing is being pressed on

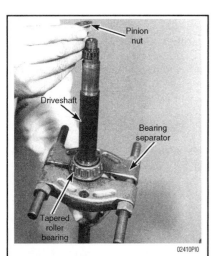

Fig. 116 Install the new driveshaft bearing as shown protecting the pinion nut thread area

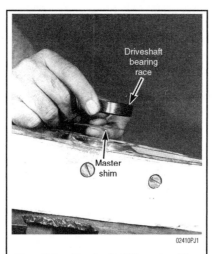

Fig. 117 Inserting the 0.050" master shim in place under the driveshaft tapered bearing race

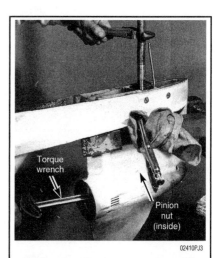

Fig. 118 It is essential to properly torque the pinion nut for shimming procedures and final assembly

LOWER UNIT 8-25

Fig. 119 Install the water pump body onto the surface of the lower unit for proper driveshaft alignment for pinion gear shimming

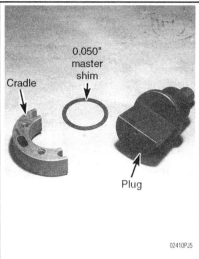

Fig. 120 Components parts of the pinion shimming tool kit

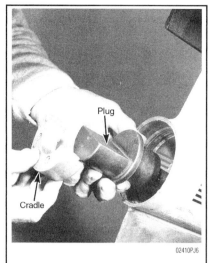

Fig. 121 Insert the shimming plugs into the lower unit to gauge the pinion height

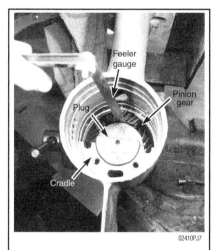

Fig. 122 The proper location of the shimming tools and feeler gauge checking the pinion height

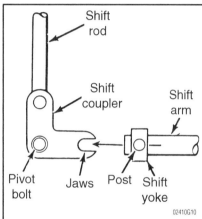

Fig. 123 Simple line drawing to depict operation of the shift coupler, as explained in the text. The coupler pivots on the pivot bolt. The coupler jaws must engage the posts on the shift yoke

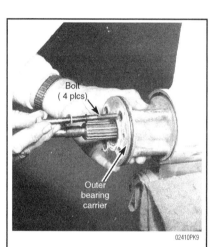

Fig. 124 Install the outer bearing carrier without the two O-rings with three of the mounting bolts for the backlash measurement procedure

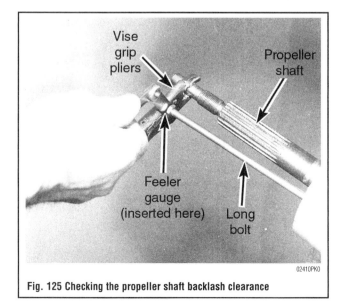

Fig. 125 Checking the propeller shaft backlash clearance

Fig. 126 Install a new crush ring over the driveshaft on top of the driveshaft roller bearing

8-26 LOWER UNIT

➡ In order to determine the amount of shim material required a special 3-piece tool set and a feeler gauge is necessary. The special tool is listed as P/N T8997A. If the special tool set is not available there is no way to determine the amount of shim material required meeting the manufacturer's recommendations. In the event the tool set is not available the best advice is to assemble the lower unit using the same amount of shim material removed during disassembling. If the unit is not being shimmed skip the following steps concerning the pinion shimming procedure.

13. Insert the 0.050" master shim, part of special tool P/N T8997, into the driveshaft cavity.
14. Install the driveshaft bearing race over the master shim.

➡ Do not use any of the shim material removed during disassembling.

15. Lower the driveshaft down through the upper end of the lower unit.
16. Raise the pinion gear to allow the driveshaft to pass through the pinion gear rotating it slightly to align the splines.
17. Install the pinion gear nut hand tight.
18. Torque the pinion gear nut by holding the splines of the driveshaft with driveshaft tool P/N T7048 or equivalent and a torque wrench.
19. Hold the pinion nut with an appropriate sized socket and ratchet
20. Tighten pinion nut to 85 ft. lbs. (115 Nm).
21. Install the water pump body onto the surface of the lower unit.
22. Secure the pump body with the attaching bolts.
23. Install the four bolts that would normally be installed to secure the water pump cover.
24. Insert the shimming plugs into the lower unit with the small end of the plug indexing into the forward gear and with the flat on the plug facing the pinion gear.
25. Slide the cradle under the "plug" to support the plug in the lower unit.
26. Pull up on the driveshaft to provide the maximum clearance between the lower end of the pinion gear and the diameter of the plug.
27. Insert the largest size feeler gauge possible between the pinion gear teeth and the upper surface of the plug.
28. Push down on the driveshaft and attempt to remove the feeler gauge.
29. The desired measurement is a slight drag when the feeler gauge is removed from between the pinion gear and plug while the driveshaft is being pushed down.
30. Record the final feeler gauge thickness reading.
31. The factory recommended clearance is 0.065" (1.65mm) for the 85 hp unit and 0.005" (0.13mm) for all other units. Subtract this figure from the feeler gauge size and subtract the answer from the 0.050" (1.23mm) master shim used. The final answer is the amount of shim material to be used under the driveshaft roller bearing race to provide proper pinion gear depth.

Example:
- Feeler gauge blade size 0.026"
- Factory recommended 0.005"
- Subtracted answer 0.021"
- Master shim 0.050"
- Answer from above 0.021"
- Shim material required 0.029"

Shim material is color coded in the following sizes:
- 0.003"—Brass
- 0.005"—Silver
- 0.010"—Black

➡ Individual pieces of shim material must be measured individually.

32. This concludes the pinion shimming procedure. All of the component parts that were assembled to perform the shimming procedure must be removed before the final assembly steps.
33. Lower the shift rod with shift rod coupler attached down into the lower unit with the jaws of the coupler facing aft.
34. Insert the pivot screw into the threaded hole on the starboard side of the gear case and through the pivot hole in the coupler.
35. Place the correct amount of shim material or the shims that were removed into the top of the lower unit underneath where the driveshaft roller bearing race is to be installed.
36. Install the driveshaft roller bearing race.
37. Lower the driveshaft down through the upper end of the lower unit.
38. Raise the pinion gear to allow the driveshaft to pass through the pinion gear with a slight twisting motion to align the splines of the drive shaft.
39. Be certain the teeth of the pinion gear mesh with the teeth of the forward gear and do not bind.
40. Install the pinion gear nut hand tight.
41. Torque the pinion gear nut by holding the splines of the driveshaft with driveshaft tool P/N T7048 or equivalent and a torque wrench.
42. Hold the pinion nut with an appropriate sized socket and ratchet
43. Tighten pinion nut to 85 ft. lbs. (115 Nm).
44. Assemble the propeller shaft components by first sliding the spring down into the propeller shaft

➡ Most model lower units do not have a spring. Skip this step when there is no spring installed

45. Insert a narrow screwdriver into the slot in the shaft and compress the spring until approximately ½" (12mm) is obtained between the top of the slot and the screwdriver.
46. Hold the spring compressed and slide the clutch dog over the splines of the propeller shaft.
47. Align the hole in the clutch dog with the slot in the shaft.
48. Slide the shift arm into the open end of the propeller.
49. Align the hole in the clutch dog, the slot in the propeller shaft and the hole in the shift arm
50. Install the cross pin into the clutch dog and through the propeller shaft slot and shift arm.
51. Center the pin and then remove the screwdriver allowing the spring to pop back into place.
52. Slide the thrust washer onto the forward end of the propeller shaft.
53. Slide the reverse gear, thrust washer and large spacer onto the propeller shaft.
54. Place a small amount of fairly thick lubricant onto the end of the shift arm.
55. Place the shift yoke over the end of the shift arm.

➡ The thick lubricant will help hold the shift yoke in place while the propeller shaft is installed. The posts on the sides of the yoke must index with the jaws of the coupler.

56. Install the propeller shaft being careful not to knock the shift yoke from the end of the shift arm.
57. Inspect to be certain that the shift yoke in fully engaged with the jaws of the shift coupler.
58. Install new propeller shaft needle bearings into the inner and outer bearing carriers as necessary using a press and a proper mandrel.
59. Install a new propeller shaft seal in the outer bearing carrier with the lip facing in on a one seal system and back to back on a two seal system.
60. Pack the lip of the oil seal with multi-purpose water resistant lubricant.
61. Slide the inner bearing carrier over the propeller shaft.
62. Seat the carrier against the spacer until the groove in the housing for the retaining ring is visible.
63. Insert the two halves of the retaining ring into the housing groove.
64. Install the outer bearing carrier without the O-ring.
65. Install and tighten three of the securing bolts to a torque value of 160 in lb. (18Nm).
66. Install the long bolt into the other threaded hole.
67. Push the propeller shaft into the lower unit as far as possible.
68. Rotate the shaft back and forth though a quarter turn to remove clearance.
69. Clamp a pair of vice grip type pliers around the propeller shaft with the vice grip handle in front of the bolt head.
70. Using a feeler gauge measure the gap between the vice grip handle and the bolt head.
71. Pull the propeller shaft out of the lower unit as far as possible.
72. Measure the gap between the handle and the bolt head.
73. The difference between the two measurements is the propeller shaft end "play". If this difference exceeds 0.010" (0.25mm), the thickness of the thrust washer around the forward end propeller shaft must be changed. A new thrust washer of the correct thickness must be purchased and installed.
74. Remove the bolt and vise grip pliers
75. Remove the bearing carrier.
76. Install a new O-ring around the bearing carrier.
77. Install the bearing carrier into the lower unit.
78. Install and tighten the four bolts alternately and evenly.
79. Install the anode. Tighten the attaching bolts securely.
80. Place a new crush ring over the driveshaft down onto the driveshaft roller bearing.
81. Pack the lip of the driveshaft oil seal with multi-purpose water resistant lubricant.

LOWER UNIT   8-27

82. Slowly slide a new driveshaft oil seal, with the lip facing down, over the upper splines of the driveshaft.
83. Pack the lip of the shift shaft oil seal with water resistant multi-purpose lubricant.
84. Position the seal into the water pump body with the seal lip facing down.
85. Use a suitable size socket and drive the seal until it seats in the body.
86. Pack the lip of the water pump body oil seal with water resistant multi-purpose lubricant.
87. Position the seal into the water pump body with the seal lip facing up (will be facing down following installation of the pump body).
88. Use a suitable size socket and drive the seal until it seats in the body.
89. Install a new O-ring and sealing ring into the grooves of the water pump body.
90. Carfully install the water pump body over the splines of the driveshaft and down the shaft.
91. As the pump body is lowered on the driveshaft feed the shift rod up through the shift shaft oil seal.
92. Tap the perimeter of the pump body lightly with a soft head mallet to seat it fully in the lower unit.
93. Install and tighten the three bolts securing the water pump body.
94. Pressure tests the unit with a pressure pump and gauge assembly to test the integrity of the unit.
95. Fill the lower unit with lower unit lubricant.

1991–95 70 to 150 HP

The following procedures give detailed illustrated instructions to completely disassemble service and assemble this style lower unit. There are some slight variations in the units but all follow the same repair procedures in general.

The visual identification of this unit is the water pump inlets. They are located on the round bullet portion of the lower unit on either side and behind the propeller.

➡The removal procedures for the bearings and races are listed as part of the disassembly and assembly procedures. In many cases the removal of the bearing will damage it beyond use. In the event the bearing is in good reusable condition skip that step and continue the procedure.

DISASSEMBLY

Single Exhaust Units

♦ See Figures 127 thru 149

1. The lower unit assembly once removed needs to be secured in a lower unit fixture or a suitable substitute.
2. Remove all water pump parts.

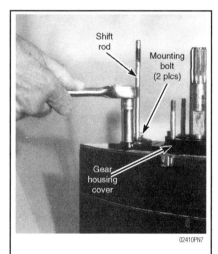

Fig. 127 Remove the two bolts located on either side of the shift rod that secure the gear housing cover to the lower unit

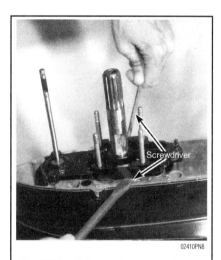

Fig. 128 Carefully pry the gear housing cover loose using equal pressure on both sides

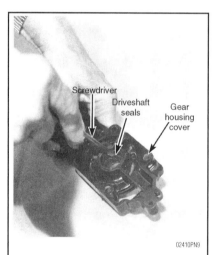

Fig. 129 Using a screwdriver push or pry the driveshaft seals out of the gear housing cover

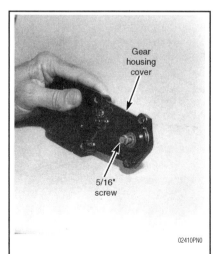

Fig. 130 Thread a 5/16" screw into the seal from the top side of the cover and press the seal out

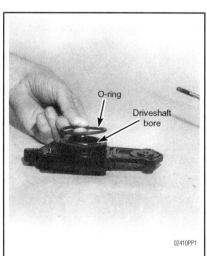

Fig. 131 Remove and discard the O-ring seal from the recess around the driveshaft bore

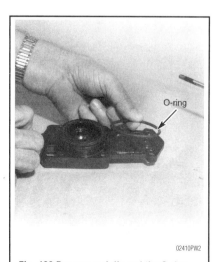

Fig. 132 Remove and discard the O-ring seal from the recessed groove in the aft end of the gear housing cover

8-28 LOWER UNIT

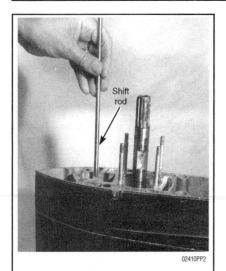

Fig. 133 Unthread and remove the shift rod from the lower unit

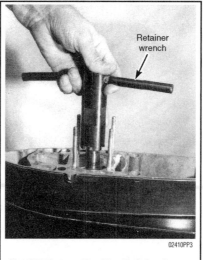

Fig. 134 Remove the driveshaft bearing retainer using the proper special tool

Fig. 135 Removing the sacrificial anode from the bearing carrier

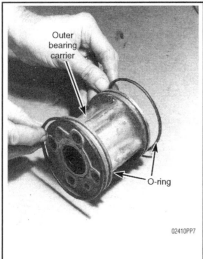

Fig. 136 Remove and discard the two O-rings from both ends of the bearing carrier

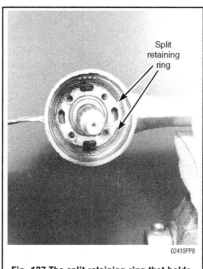

Fig. 137 The split retaining ring that holds the inner bearing carrier in place

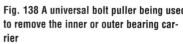

Fig. 138 A universal bolt puller being used to remove the inner or outer bearing carrier

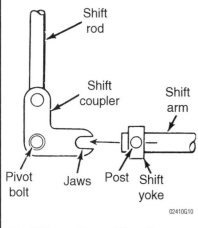

Fig. 139 Inspect the condition of the needle bearing. If the needle bearing worn or damaged the bearing carrier must be replaced

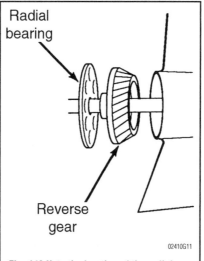

Fig. 140 Note the location of the radial bearing and reverse gear shims

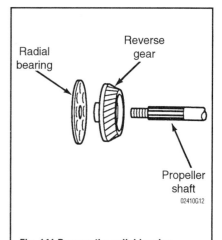

Fig. 141 Remove the radial bearing assembly and reverse gear from the aft end of the propeller shaft to gain access to the pinion gear and nut

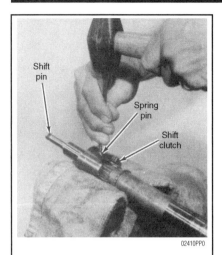

Fig. 142 A long punch of suitable size is used to drive the spring pin out of the shift clutch

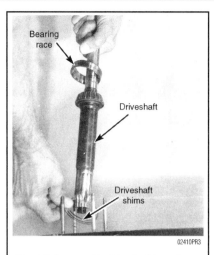

Fig. 143 Remove the driveshaft by pull upwards noting the location of the driveshaft shims

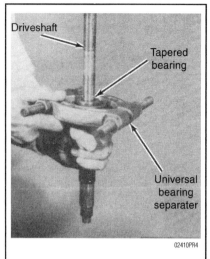

Fig. 144 A bearing separator in place around the driveshaft bearing for removal

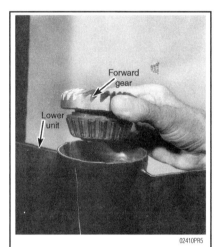

Fig. 145 Remove the forward gear from the propeller bore in the lower unit housing

Fig. 146 Make sure the spacer washer if equipped is retained with the forward gear

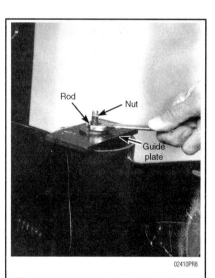

Fig. 147 Removing the forward gear bearing race with the factory special tool

3. Remove the tie wrap from the bellows cover, if equipped.
4. Remove the two bolts located on either side of the shift rod, which secure the gear housing cover to the lower unit.
5. Use two screwdrivers positioned near the center and on opposite sides of the cover and pry the cover frees from the lower unit.

✱✱ WARNING

The gear housing cover is bonded to the lower unit with RTV sealant. Some force may be required to remove the cover. Always pry from the middle of the housing to help avoid damage to the cover during removal.

6. Remove the gear housing cover.
7. Remove the seals by placing the gear housing cover on a pair of suitable support blocks supporting as much of the cover as possible.
8. Place the blade of the screwdriver against the seals and use the hammer to tap the screwdriver handle.

✱✱ WARNING

Using excessive force to drive the seals out may damage the cover.

9. Remove the shift rod seal.
   a. To remove a bellows type seal from the gear housing cover.
   b. Turn the cover upside down and drive the seal out using a drift pin punch through the holes on either side of the shift rod bore.
   c. To remove a lip style shift rod seal thread a 5/16" screw into the seal from the top side of the cover.
   d. Apply pressure to the point of the screw from the underside of the cover to press the seal out.
10. Remove and discard the first O-ring seal from the recess around the driveshaft bore.
11. Remove and discard the second O-ring seal from the recessed groove in the aft end of the gear housing cover.
12. Unscrew and remove the shift rod from the lower unit.

➥**Special tool P/N T-11293 or a suitable replacement is required to properly remove the driveshaft upper bearing race retainer.**

13. Slide the wrench down over the driveshaft and into the recess in the lower unit housing.
14. Make sure the four notches in the retainer wrench mate with the four tabs on the inside diameter of the race retainer.
15. Remove the retainer by turning it counter clockwise to loosen.
16. Remove the two screws securing the sacrificial anode to the bearing carrier.

8-30 LOWER UNIT

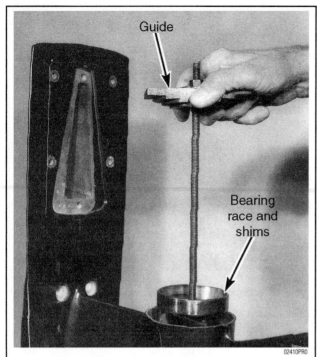

Fig. 148 Removing the guide, forward gear bearing race and shims from the lower unit housing

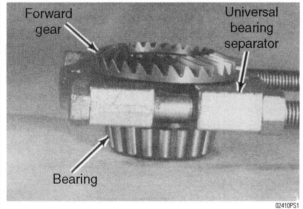

Fig. 149 A bearing separator positioned between the forward gear and the tapered roller bearing for disassembly

17. Remove the anode.
18. Remove the four fasteners with O-rings securing the bearing carrier in the lower unit.
19. Remove and discard the O-rings.

✱✱ WARNING

It may be necessary to use torch to heat the lower unit housing around the bearing carriers so that the housing expands to loosen up any salt and /or corrosion to allow the parts to be removed without damage.

20. Using puller P/N T-8948-1 and long bolts with ¼inch threads or an equivalent puller assembly remove the outer bearing carrier.
    a. Thread the puller center screw into the puller.
    b. Install the ¼inch bolts in the bearing carrier into the threaded holes for the sacrificial anode.
    c. Holding the puller square and stationary rotate the center screw clockwise until the bearing carrier is free of the lower unit housing.

21. Remove the puller and bearing carrier from the lower unit housing.
22. Remove the puller from the bearing carrier.
23. Remove and discard the two O-rings from both ends of the bearing carrier.
24. Insert a blade-type screwdriver into the lower unit housing and hold the blade between the end of the retaining ring half which has a bevel cut and the housing.
25. Tap the end of the screwdriver with a soft head mallet to release the retaining ring from the groove.
26. Remove both halves of the retaining ring from the lower unit housing.
27. Install the bearing carrier without the O-rings into the lower unit housing.
28. Secure the bearing carrier to the reverse gear bearing carrier with two mounting bolts only.
29. Repeat the removal procedure and pull the reverse gear bearing carrier out of the lower unit housing
30. Remove the propeller shaft oil seal by using a punch or a blade type screwdriver and drive the oil seal out of the bearing carrier.
31. Remove the spacer from the reverse gear bearing carrier.

➥**The bearing carrier needle bearing is not a separately replaceable part and would be replaced as apart of the bearing carrier assembly.**

32. Remove the propeller shaft assembly and reverse gear from the lower unit.
33. Remove the radial bearing assembly and reverse gear from the aft end of the propeller shaft.
34. Remove the shift yoke from the forward end of the propeller shaft.
35. Secure the propeller shaft in a vise protecting it with shop cloths.
36. Using a long punch of suitable size drive the spring pin out of the shift clutch.
37. Remove the shift pin and the shift clutch from the propeller shaft.
38. Remove the driveshaft by first holding the splined end with the special tool or an appropriate socket.
39. Using the proper size six point socket and breaker reach into the housing and hold the pinion nut.

✱✱ WARNING

The breaker bar handle must not directly contact the lower unit housing when loosening the pinion nut. Special tool P/N T-11209 would be inserted across the opening of the housing to protect the unit during disassembly. This tool is also essential to properly remove the forward gear bearing race during the shimming operation or bearing replacement.

40. Install Guide P/N T-11209 or other suitable holding fixture to protect the casing over the wrench handle.
41. Hold the pinion nut stationary when rotating the driveshaft until the pinion nut unthreads itself from the driveshaft.
42. Remove the pinion nut and washer from the lower unit housing.
43. Grasp the driveshaft firmly with both hands and pull upwards.
44. Remove the driveshaft from the lower unit housing.
45. Remove the pinion gear from the lower unit housing.
46. Remove the driveshaft shims.
47. If it is necessary to replace the driveshaft bearing, position a bearing separator behind the tapered roller bearing.
48. Using a hydraulic press separate the shaft from the bearing.
49. The lower unit may have one or more driveshaft shims, or no driveshaft shims at all. The driveshaft shims are used to obtain necessary clearance requirements during assembly. Do not discard the shims. Undamaged shims may be reused when assembling the lower unit. Additionally, if any shimming adjustments are required, it may be necessary to know the thickness of the previous shim pack.
50. Remove the forward gear and spacer washer "if equipped" from the propeller bore in the lower unit housing.
51. Remove the pivot pin and seal securing the shift coupler in the lower unit housing.
52. Remove the shift arm.

➥**Two special tools are required to remove the forward gear bearing race from the lower unit housing: Guide P/N T-11209—previously used to remove the driveshaft—and Cup Remover P/N T-16813. If these special tools are not available it will be very difficult to remove and then install the race properly without damaging the parts or the lower unit housing. But a slide hammer with jaws is an acceptable replacement for removal.**

LOWER UNIT 8-31

53. Place the Cup Remover P/N T-16813 flat against the backside of the forward gear bearing race.
54. Thread the rod from Guide P/N T-11209 into the cup remover.
55. Install the guide plate and nut over the rod.
56. Tighten the nut onto the rod until the forward gear bearing race is pulled free of the lower unit housing.
57. Remove the guide, the forward gear bearing race, and shims from the lower unit housing.
58. Position a bearing separator between the forward gear and the tapered roller bearing. Using a hydraulic press, separate the gear from the bearing.
59. Remove the driveshaft lower bearing by driving it down and out of the housing using special tools P/N T11209, T8907 & T11205 as an assembly or P/N T16433 on the two piece driveshaft units.

Dual Exhaust Units

▶ See Figures 150 thru 164

1. Remove the sacrificial anode from the lower gear housing, as necessary.
2. Use a blade-type screwdriver and a hammer to bend back the tab on the retainer ring lock ring
3. Remove the bearing carrier retainer ring with special tool P/N T-11275.

※※ WARNING

Removing this ring using a hammer and chisel would most likely damage the part.

➡ Once the retaining ring has been removed it can not be properly reinstalled and torqued into place without the proper tool.

4. Mate the tabs on the end of the wrench with the notches on the retainer ring.
5. Using a heavy-duty wrench of suitable size turn the Retainer Ring Wrench counter clockwise to loosen the retainer ring.
6. Remove the retainer ring and the lock ring from the lower unit.
7. Grasp the underside of the bearing carrier outer flange with a bolt head or hook shaped tool and slide the bearing carrier out of the lower unit housing.

※※ WARNING

The bearing carrier can seize in the lower unit housing due to salt and corrosion. It may be necessary to use a torch to heat the housing and use a puller as outlined in the next step

8. An alternate method for removing the bearing carrier from the lower unit. Obtain a long jaw puller (C-91-46086A1). Place the ends of the puller, inside

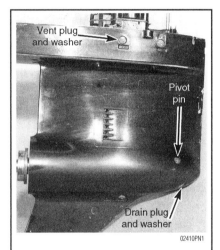

Fig. 150 Disassembly of the lower unit requires the gear housing lubricant to be drained

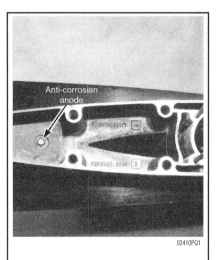

Fig. 151 Remove the sacrificial anode from the lower gear housing, as necessary

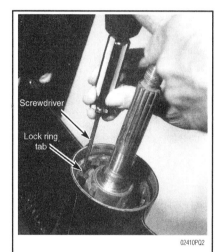

Fig. 152 Use a blade-type screwdriver and a hammer to bend back the tab on the retainer ring lock ring

Fig. 153 Removing the bearing carrier retaining ring with the proper factory special tool

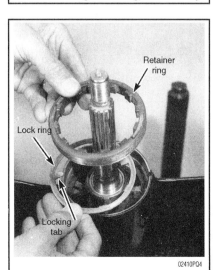

Fig. 154 Remove the retainer ring and the lock ring from the lower unit

Fig. 155 Grasp the underside of the bearing carrier outer flange and slide the bearing carrier out of the lower unit housing

8-32 LOWER UNIT

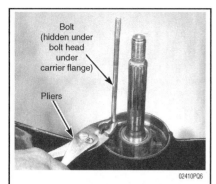

Fig. 156 Using the head of a bolt or a hook shaped tool and a pair of pliers use the edge of the lower unit as the fulcrum point for leverage to remove the bearing carrier

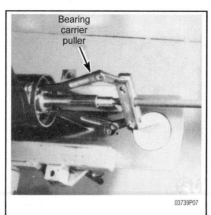

Fig. 157 Removing the bearing carrier with a jaw puller

Fig. 158 Remove the O-ring from the forward end of the bearing carrier

the lip on the bearing carrier. Screw the puller bolt up against the end of the propeller shaft. If needed, slide the propeller thrust hub (large splined washer), onto the propeller shaft to wedge the puller jaws tight against the lip of the bearing carrier.

9. Remove the O-ring from the forward end of the bearing carrier.

➡If the needle bearings are no longer fit for service the bearing carrier should be replaced.

10. Use a punch or a flat blade type screwdriver to drive the two propeller shaft oil seals out of the bearing carrier.

11. Remove the propeller shaft assembly and reverse gear from the lower unit.

12. Remove the bearing plate, radial bearing assembly, reverse gear, and spacer washer from the aft end of the propeller shaft.

13. Remove the shift yoke from the forward end of the propeller shaft.

14. Properly support the propeller shaft in a vise protecting it by using several shop cloths to prevent damage while driving out the spring pin.

15. Using a long punch of suitable size drive the spring pin out of the shift clutch.

16. Remove the shift pin and the shift clutch from the propeller shaft.

17. Remove the driveshaft by first holding the splined end with the special tool or an appropriate socket.

18. Using the proper size six point socket and breaker reach into the housing and hold the pinion nut.

Fig. 159 Removing the propeller shaft seals

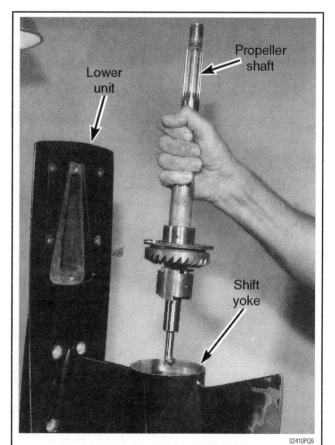

Fig. 160 Lift the propeller shaft assembly with reverse gear from the lower unit being careful not to loose the shift yoke on the forward end of the propeller shaft

LOWER UNIT 8-33

✽✽ WARNING

The breaker bar handle must not directly contact the lower unit housing when loosening the pinion nut. Special tool P/N T-11209 would be inserted across the opening of the housing to protect the unit during disassembly. This tool is also essential to properly remove the forward gear bearing race during the shimming operation or bearing replacement.

19. Install Guide P/N T-11209 or other suitable holding fixture to protect the casing over the wrench handle.
20. Hold the pinion nut stationary when rotating the driveshaft until the pinion nut unthreads itself from the driveshaft.
21. Remove the pinion nut and washer from the lower unit housing.
22. Grasp the driveshaft firmly with both hands and pull upwards.
23. Remove the driveshaft from the lower unit housing.
24. Remove the pinion gear from the lower unit housing.
25. Remove the driveshaft shims.
26. If it is necessary to replace the driveshaft bearing, position a bearing separator behind the tapered roller bearing.
27. Using a hydraulic press separate the shaft from the bearing.

The lower unit may have one or more driveshaft shims, or no driveshaft shims at all. The driveshaft shims are used to obtain necessary clearance requirements during assembly. Do not discard the shims. Undamaged shims may be reused when assembling the lower unit. Additionally, if any shimming adjustments are required, it may be necessary to know the thickness of the previous shim pack.

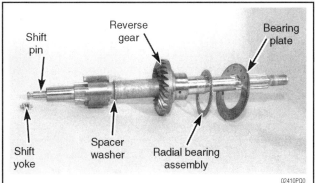

Fig. 161 Remove the bearing plate, radial bearing assembly, reverse gear, and spacer washer from the aft end of the propeller shaft

28. Remove the forward gear and spacer washer "if equipped" from the propeller bore in the lower unit housing.
29. Remove the pivot pin and seal securing the shift coupler in the lower unit housing.
30. Remove the shift arm.

➡Two special tools are required to remove the forward gear bearing race from the lower unit housing: Guide P/N T-11209—previously used to remove the driveshaft—and Cup Remover P/N T-16813. If these special tools are not available it will be very difficult to remove and then install the race properly without damaging the parts or the lower unit housing. But a slide hammer with jaws is an acceptable replacement for removal.

31. Place the Cup Remover P/N T-16813 flat against the backside of the forward gear bearing race.
32. Thread the rod from Guide P/N T-11209 into the cup remover.
33. Install the guide plate and nut over the rod.
34. Tighten the nut onto the rod until the forward gear bearing race is pulled free of the lower unit housing.
35. Remove the guide, the forward gear bearing race, and shims from the lower unit housing.
36. Position a bearing separator between the forward gear and the tapered roller bearing. Using a hydraulic press, separate the gear from the bearing.
37. Remove the driveshaft lower bearing by driving it down and out of the housing using special tools P/N T11209, T8907 & T11205 as an assembly or P/N T16433 on the two piece driveshaft units.

CLEANING & INSPECTION

♦ See Figures 165, 166 and 167

Good shop practice requires installation of new O-rings and oil seals regardless of their appearance.

Clean all water pump parts with solvent, and then dry them with compressed air. Inspect the water pump body and oil seal for cracks and distortion, possibly caused from overheating. Inspect the water pump plate for grooves and/or rough surfaces. If possible, always install a new water pump impeller while the lower unit is disassembled. A new impeller will ensure extended satisfactory service and give peace of mind to the owner. If the old impeller must be returned to service, never install it in reverse to the original direction of rotation. Installation in reverse will cause premature impeller failure.

If installation of a new impeller is not possible, check the seal surfaces. All must be in good condition to ensure proper pump operation. Check the upper, lower, and ends of the impeller vanes for grooves, cracking, and wear. Check to be sure the indexing notch of the impeller hub is intact and will not allow the impeller to slip.

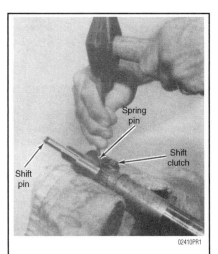

Fig. 162 Obtain a long punch of suitable size to drive the spring pin out of the shift clutch

Fig. 163 Hold the pinion nut with a breaker bar and the driveshaft with a spline socket and remove the pinion nut

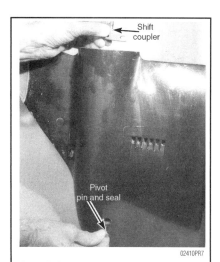

Fig. 164 Remove the pivot pin and seal securing the shift coupler in the lower unit housing to remove the shift arm

8-34 LOWER UNIT

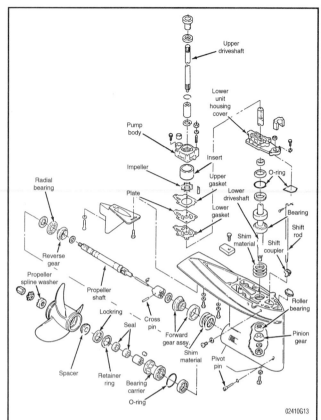

Fig. 165 Exploded drawing of a dual exhaust lower unit, with two-piece driveshaft. Major parts are identified

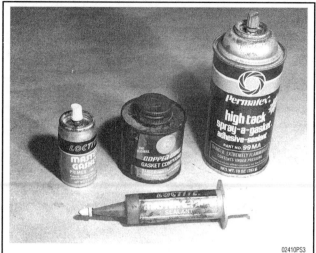

Fig. 166 Obtaining good quality sealing products is will worth the extra cost to prolong satisfactory service of the lower unit

Fig. 167 A new shift yoke (right), compared with one badly worn and unfit for further service (left)

Clean around the Woodruff key. Clean all bearings with solvent, dry them with compressed air, and inspect them carefully. Be sure there is no water in the air line. Direct the air stream through the bearing. Never spin a bearing with compressed air. Such action is highly dangerous and may cause the bearing to score from lack of lubrication. After the bearings are clean and dry, lubricate them with Formula 50 oil, or equivalent. Do not lubricate tapered bearing cups until after they have been inspected.

Determine the condition of tapered bearing rollers and inner bearing race, by inspecting the bearing cup for pitting, scoring, grooves, uneven wear, imbedded particles, and discoloration caused from overheating. Always replace tapered roller bearings as a set.

Clean the forward gear with solvent, and then dry it with compressed air. Inspect the gear teeth for wear. Under normal conditions the gear will show signs of wear, but it will be smooth and even.

Clean the inner and outer bearing carriers with solvent, and then dry them with compressed air. Never spin bearings with compressed air. Such action is highly dangerous and may cause the bearing to score from lack of lubrication. Check the gear teeth of the reverse gear for wear. The wear should be smooth and even.

Check the clutch dogs to be sure they are not rounded-off or chipped. Such damage is usually the result of poor operator habits and is caused by shifting tool slowly or shifting while the powerhead is operating at high rpm. Such damage might also be caused by improper shift rod adjustments.

Inspect the roller bearing surface of the propeller shaft. Check the shaft surface for pitting, scoring, grooving, embedded particles, uneven wear and discoloration caused from overheating.

Clean the driveshaft with solvent, and then dry it with compressed air. Never spin bearings with compressed air. Such action is dangerous and could damage the bearing. Inspect the bearing for roughness, scratches, or side wear. If the bearing shows signs of such damage, it should be replaced. If the bearing is satisfactory for further service, coat it with oil.

Inspect the driveshaft splines for excessive wear. Check the oil seal surfaces above and below the water pump drive pin or Woodruff key area for grooves. Replace the shaft if grooves are discovered.

Inspect the driveshaft bearing surface above the pinion gear splines for pitting, grooves, scoring uneven wear, embedded metal particles and discoloration caused by overheating.

Inspect the propeller shaft oil seal surface to be sure it is not pitted, grooved, or scratched. Inspect the roller bearing contact surface on the propeller shaft for pitting, grooves, scoring, uneven wear, imbedded metal particles, and discoloration caused by overheating.

Inspect the propeller shaft splines for wear and corrosion damage. Check the propeller shaft for straightness.

Clean all parts with solvent, and then dry them with compressed air.
Inspect all bearing bores for loose fitting bearings.
Check the lower unit housing for impact damage.
Check the pinion nut corners for wear or damage.
Check the condition of the shift yoke for wear on the cutout portion and posts. This small, inexpensive but vital part of the shifting mechanism should be replaced at the first sign of wear.

➡**During the Assembling and Installation procedures, a sealant is applied to the lower unit, motor leg extension, and the intermediate housing mating surfaces. The sealant recommended by the manufacturer carries P/N T8983. An acceptable substitute is Permatex #27 ®which is a high temperature sealant.**

The manufacturer also recommends the use of Mercury sealant P/N T8983 on bolt threads in the lower unit. For this application, a good substitute is Loctite ®or Lock N' Seal®. These products are non-hardening adhesives especially produced for application on the threads of load bearing fasteners.

This material helps prevent the bolt from loosening due to vibration, thread wear and corrosion. Loctite® or Lock N' Seal® must only be applied to clean, dry parts.

Any sealant is always applied in a thin continuous bead to only one of the mating surfaces. After installation of the two mating surfaces, the sealant is compressed to form a watertight gasket between the mating surfaces.

LOWER UNIT   8-35

If too much sealant is applied, the excess will ooze out both internally and externally and possibly restrict passages. If the sealant finds its way into a closed end threaded hole it will form a pool in the bottom of the cavity. When a bolt is later installed and tightened to a specific torque value, the bolt will hydraulically lock.

A hydraulic lock occurs when a bolt that has been installed into a threaded cavity, the bolt end contacts the liquid sealant in the bottom of the hole. Because a liquid cannot be compressed, the bolt cannot move further. Even though the reading on the torque wrench indicates the bolt has been tightened to the required torque value, the bolt remains seated against the fluid

SHIMMING

Shimming procedure for the lower unit actually consist of two separate and distinct operations:

- Shimming the forward gear.
- Shimming the driveshaft assembly.

1. The forward gear must be shimmed only if one of the following conditions exist:
   a. A new forward gear is being installed in the original lower unit housing.
   b. The original forward gear is being installed in a new lower unit housing.
2. Locate the forward gear housing code stamped in the lower unit housing.

➥Housing and gear codes will be either a number 0–9 or a letter A–I.

3. Locate the forward gear code on the tag attached to a new forward gear.
4. On a used forward gear to get the code measure the shims taken from behind the forward gear bearing race.

SHIM CHART -- DRIVESHAFT

STAMPED-IN GEAR HSG CODE	DRIVESHAFT CODE															
	5	6	7	8	9	10	11	12	13	14	15	16	17	18	19	20
0	.005	.006	.007	.008	.009	.010	.011	.012	.013	.014	.015	.016	.017	.018	.019	.020
1	.006	.007	.008	.009	.010	.011	.012	.013	.014	.015	.016	.017	.018	.019	.020	.021
2	.007	.008	.009	.010	.011	.012	.013	.014	.015	.016	.017	.018	.019	.020	.021	.022
3	.008	.009	.010	.011	.012	.013	.014	.015	.016	.017	.018	.019	.020	.021	.022	.023
4	.009	.010	.011	.012	.013	.014	.015	.016	.017	.018	.019	.020	.021	.022	.023	.024
5	.010	.011	.012	.013	.014	.015	.016	.017	.018	.019	.020	.021	.022	.023	.024	.025
6	.011	.012	.013	.014	.015	.016	.017	.018	.019	.020	.021	.022	.023	.024	.025	.026
7	.012	.013	.014	.015	.016	.017	.018	.019	.020	.021	.022	.023	.024	.025	.026	.027
8	.013	.014	.015	.016	.017	.018	.019	.020	.021	.022	.023	.024	.025	.026	.027	.028
9	.014	.015	.016	.017	.018	.019	.020	.021	.022	.023	.024	.025	.026	.027	.028	.029

STAMPED-IN GEAR HSG CODE	DRIVESHAFT CODE														
	21	22	23	24	25	26	27	28	29	30	31	32	33	34	35
0	.021	.022	.023	.024	.025	.026	.027	.028	.029	.030	.031	.032	.033	.034	.035
1	.022	.023	.024	.025	.026	.027	.028	.029	.030	.031	.032	.033	.034	.035	.036
2	.023	.024	.025	.026	.027	.028	.029	.030	.031	.032	.033	.034	.035	.036	.037
3	.024	.025	.026	.027	.028	.029	.030	.031	.032	.033	.034	.035	.036	.037	.038
4	.025	.026	.027	.028	.029	.030	.031	.032	.033	.034	.035	.036	.037	.038	.039
5	.026	.027	.028	.029	.030	.031	.032	.033	.034	.035	.036	.037	.038	.039	.040
6	.027	.028	.029	.030	.031	.032	.033	.034	.035	.036	.037	.038	.039	.040	.041
7	.028	.029	.030	.031	.032	.033	.034	.035	.036	.037	.038	.039	.040	.041	.042
8	.029	.030	.031	.032	.033	.034	.035	.036	.037	.038	.039	.040	.041	.042	.043
9	.030	.031	.032	.033	.034	.035	.036	.037	.038	.039	.040	.041	.042	.043	.044

SHIM CHART -- DRIVESHAFT

STAMPED-IN GEAR HSG CODE	DRIVESHAFT CODE															
	5	6	7	8	9	10	11	12	13	14	15	16	17	18	19	20
A	.004	.005	.006	.007	.008	.009	.010	.011	.012	.013	.014	.015	.016	.017	.018	.019
B	.003	.004	.005	.006	.007	.008	.009	.010	.011	.012	.013	.014	.015	.016	.017	.018
C	.002	.003	.004	.005	.006	.007	.008	.009	.010	.011	.012	.013	.014	.015	.016	.017
D	.001	.002	.003	.004	.005	.006	.007	.008	.009	.010	.011	.012	.013	.014	.015	.016
E	.000	.001	.002	.003	.004	.005	.006	.007	.008	.009	.010	.011	.012	.013	.014	.015
F		.000	.001	.002	.003	.004	.005	.006	.007	.008	.009	.010	.011	.012	.013	.014
G			.000	.001	.002	.003	.004	.005	.006	.007	.008	.009	.010	.011	.012	.013
H				.000	.001	.002	.003	.004	.005	.006	.007	.008	.009	.010	.011	.012
I					.000	.001	.002	.003	.004	.005	.006	.007	.008	.009	.010	.011
J						.000	.001	.002	.003	.004	.005	.006	.007	.008	.009	.010
K							.000	.001	.002	.003	.004	.005	.006	.007	.008	.009

STAMPED-IN GEAR HSG CODE	DRIVESHAFT CODE														
	21	22	23	24	25	26	27	28	29	30	31	32	33	34	35
A	.020	.021	.022	.023	.024	.025	.026	.027	.028	.029	.030	.031	.032	.033	.034
B	.019	.020	.021	.022	.023	.024	.025	.026	.027	.028	.029	.030	.031	.032	.033
C	.018	.019	.020	.021	.022	.023	.024	.025	.026	.027	.028	.029	.030	.031	.032
D	.017	.018	.019	.020	.021	.022	.023	.024	.025	.026	.027	.028	.029	.030	.031
E	.016	.017	.018	.019	.020	.021	.022	.023	.024	.025	.026	.027	.028	.029	.030
F	.015	.016	.017	.018	.019	.020	.021	.022	.023	.024	.025	.026	.027	.028	.029
G	.014	.015	.016	.017	.018	.019	.020	.021	.022	.023	.024	.025	.026	.027	.028
H	.013	.014	.015	.016	.017	.018	.019	.020	.021	.022	.023	.024	.025	.026	.027
I	.012	.013	.014	.015	.016	.017	.018	.019	.020	.021	.022	.023	.024	.025	.026
J	.011	.012	.013	.014	.015	.016	.017	.018	.019	.020	.021	.022	.023	.024	.025
K	.010	.011	.012	.013	.014	.015	.016	.017	.018	.019	.020	.021	.022	.023	.024

5. Refer to the associated Forward Gear/Driveshaft Shimming Chart. Locate the forward gear housing code in the vertical column on the left side of the chart. Locate the forward gear code in the horizontal column on the top of the chart.

6. Determine the appropriate shim thickness by locating the number on the shim chart where the forward gear housing code row and the forward gear code column intersect.

7. With a used forward gear take the shim total and match it with the original housing code and the chart follow that number to the left to get the gear code.

8. Install exactly this thickness of shims when assembling the lower unit. It may be necessary to use several shims in order to obtain the required thickness, or no shims at all may be required. Retain the proper thickness shim pack with the forward gear for use during assembling.

WARNING

When installing a new forward gear in the original lower unit housing, make sure all of the original forward gear shims have been

removed from the housing before installing the new forward gear. Also, make sure the new outer bearing race that came with the new forward gear is installed in the lower unit housing.

9. The driveshaft must be shimmed only is either of the two following conditions exist:
   a. A new driveshaft is being installed in the original lower unit housing.
   b. The original driveshaft is being installed in a new lower unit housing.
10. Locate the driveshaft housing code stamped in the lower unit housing.

➡Housing and gear codes will be either a number 0–9 or a letter A–I.

11. Locate the driveshaft code on the tag attached to a new driveshaft.
12. On a used driveshaft to get the code measure the shims taken from behind the upper driveshaft bearing race.
13. Refer to the associated Forward Gear/Driveshaft Shimming Chart. Locate the driveshaft housing code in the vertical column on the left side of the chart. Locate the driveshaft code in the horizontal column on the top of the chart.
14. Determine the appropriate shim thickness by locating the number on the shim chart where the driveshaft housing code row and the driveshaft code column intersect.
15. With a used driveshaft take the shim total and match it with the original housing code and the chart follow that number up to get the driveshaft code.

16. Install exactly this thickness of shims when assembling the lower unit. It may be necessary to use several shims in order to obtain the required thickness, or no shims at all may be required. Retain the proper thickness shim pack with the forward gear for use during assembling.

✲✲ WARNING

When installing a new driveshaft in the original lower unit housing, make sure all of the original shims have been removed from the housing before installing the new driveshaft. Also, make sure to use the new bearing race that came with the new driveshaft is installed in the lower unit housing.

ASSEMBLY

Single Exhaust Units

▶ See Figures 168 thru 178

The following procedure pickup the work after all parts have been cleaned and inspected. The proper shim material thickness has been determined following the shimming instructions in the previous section.
Replacement parts having been purchased and on hand.

➡Installing the lower driveshaft support bearing requires a special installer tool consisting of P/N T16496, T16496A and T16496B.

1. Apply some grease to the eighteen pinion gear roller bearings and place the bearings one at a time into the bearing outer race. The grease will help to hold the individual bearings in place.
2. Reach in and place the bearing race with the roller bearings and with the lettered side of the race facing up into position to be drawn up into the opening.
3. Hold the race in place and lower the rod part of the special installer down the driveshaft opening and through the bearings without disturbing the bearings.
4. Attach the lower end of the tool P/N 16496B with the nut to the end of the rod against the bearing race.
5. Install the top plate and nut onto the upper end of the rod.
6. Tighten the nut by hand to exert pressure on the bearing.
7. Hold the bottom nut from turning with a wrench and tighten the upper nut.
8. Tighten the nut on the rod will draw the bearing assembly up into the cavity until seated on the shoulder in the housing.

➡Install the forward gear tapered roller bearing race using special tool Race Installer P/N 16814, Driver Handle P/N T8907, and Guide Plate P/N T11209 or an acceptable equivalent such as an old bearing and race and a long punch.

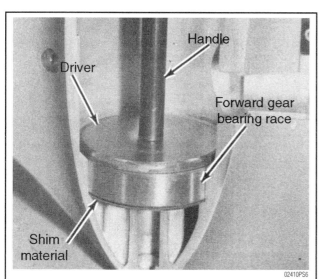

Fig. 168 Installation of the forward gear bearing race using the factory special tools

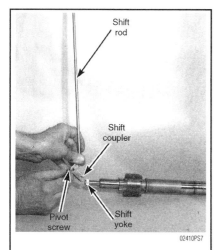

Fig. 169 An external view of the shifter assembly being mated together and properly aligned

Fig. 170 Apply some grease to the pinion gear roller bearings and insert the bearings into the race

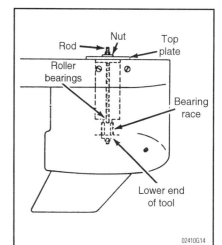

Fig. 171 The factory special tool showing the pinion support bearing being pulled up into the housing

8-38 LOWER UNIT

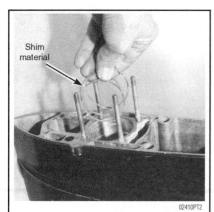

Fig. 172 Insert the same amount of shim material into the driveshaft bore and onto the ledge as was removed or as determined during Shimming procedures

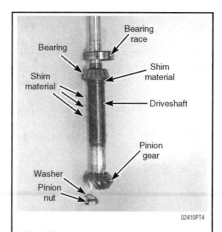

Fig. 173 An external view of the location of the components the are assembled with the driveshaft

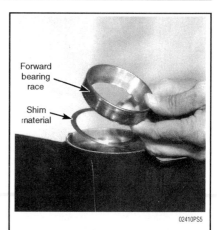

Fig. 174 Shim material and forward bearing race ready to be installed into the lower unit housing

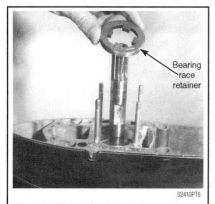

Fig. 175 Thread the driveshaft upper bearing race retainer into the lower unit housing

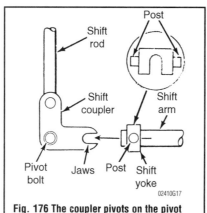

Fig. 176 The coupler pivots on the pivot bolt while the coupler jaws must engage the posts on the shift yoke

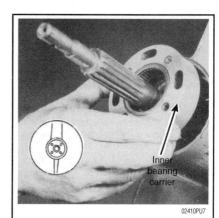

Fig. 177 Install the carrier with the bolt holes are roughly at 2 ,4 ,8 and 10 o'clock

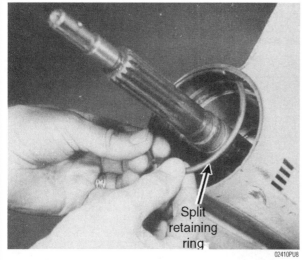

Fig. 178 Install the two halves of the retaining ring into the groove in the lower unit housing

9. Block the lower unit on a wooden work surface with the opening facing up.
10. Set the proper amount of shim material as determined from the shimming procedures in the previous section into the forward end of the lower unit.

11. Set the bearing race in position at the forward end of the lower unit with the large end of the race facing down—forward.
12. Insert the tapered end of the race installer into the new race.
13. Place the driver handle over the race installer and slide the guide plate down the handle with the stepped side of the plate facing down.
14. Drive the bearing race into the lower unit by tapping on the driver handle with a hammer.
15. As soon as the race is fully seated—the sound of the hammer striking the driver will change noticeably.
16. Insert the shift coupler into the forward end of the lower unit with the jaws facing aft.
17. Insert the pivot screw—with O-ring—into the threaded hole on the starboard side of the lower unit and through the pivot hole in the shift coupler.
18. Lower the shift rod down into the lower unit opening and thread it into the shift arm.
19. Thread the shift rod until it is fully seated.
20. Unthread the shift rod out four turns for the preliminary shift rod height adjustment.
21. Tighten the pivot screw.
22. Apply a coating of grease to the spacer washer and then insert it into the forward gear bore, if applicable.

➡The spacer washer should be the same one removed during disassembling. The propeller shaft end play can only be checked after the unit is fully assembled.

23. Insert the forward gear and bearing into the lower unit cavity with the bearing sliding into the forward bearing race.
24. Install the drive shaft taper roller bearing by first threading the pinion

LOWER UNIT   8-39

nut upside down onto the end of the driveshaft to protect the threads while the bearing is pressed into place.

25. Press a new tapered roller bearing onto the driveshaft with the small end of the bearing facing the splined end of the driveshaft.

※ WARNING

Because of the driveshaft length take care not to bend or distort the driveshaft while the bearing is being pressed on

26. Insert the same amount of shim material into the driveshaft bore and onto the ledge as was removed or as determined during Shimming procedures section under "Shimming".
27. Lower the driveshaft down into the lower unit housing gently so the pinion gear roller bearings stay in place.
28. Install the pinion gear on the drive shaft by lifting the driveshaft up just enough to put the gear in place against the forward gear.
29. Lower the driveshaft with a slight twisting motion to align the splines as the shaft and pinion gear fit into place.
30. Apply Loctite® or Lock N' Seal® to the threads of the pinion gear nut.
31. Thread the pinion nut on to the driveshaft with the pinion washer and the groove on the nut facing the gear.

※ WARNING

Always use a new pinion gear nut. Once the nut is tightened and released it loses its full locking ability and could come loose during normal operation

32. Tighten the pinion gear nut using the same tool arrangement as was used to remove it
33. Thread the driveshaft upper bearing race retainer into the lower unit housing.
34. Using the special tool tighten the retainer to a torque value of 90 ft. lbs. (122Nm).
35. The propeller shaft is next assembled by sliding the spring down into the propeller shaft.

➡ **Most model lower units do not have a spring. Skip this step when there is no spring installed**

36. Insert a narrow screwdriver into the slot in the shaft and compress the spring until approximately ½inch (12mm), is obtained between the top of the slot and the screwdriver.
37. Hold the spring compressed slide the clutch dog over the splines of the propeller shaft in the same direction as originally installed.
38. Align the hole in the dog with the slot in the shaft.
39. Slide the shift arm into the open end of the propeller.
40. Align the hole in the clutch dog the slot in the propeller shaft, and the hole in the shift arm
41. Insert the cross pin into the clutch dog and through the propeller shaft slot and shift arm.
42. Center the pin and then remove the screwdriver allowing the spring to pop back into place.
43. Coat the large spacer with a film of grease.
44. Slide the reverse gear its radial bearing assembly and large spacer onto the propeller shaft.
45. Place a small amount of fairly thick lubricant onto the end of the shift arm to help hold the shift yoke in place for assembly.
46. Place the shift yoke over the end of the shift arm with the open end of the yoke facing down.
47. Carefully install the propeller shaft and align the shift yoke with the shift cradle.

➡ **Be certain that the shifting mechanism is properly aligned and the shift rod when moved will move the clutch dog before continuing.**

48. Slide the inner bearing carrier onto the propeller shaft and into the lower unit.
49. Rotate the carrier until the bolt holes are roughly positioned at 2, 4, 8 and 10 o'clock.
50. Install the two halves of the retaining ring into the groove in the lower unit housing.

51. Install the propeller needle bearing and oil seal in the outer bearing carrier by positioning a new needle bearing over the outer bearing carrier.
52. Using the special tool or a suitable size socket drive the bearing into the carrier until the bearing is seated.
53. In the same manner install the propeller shaft seals.
54. Install the smaller diameter O-ring at the forward end of the bearing carrier.
55. Install the larger diameter O-ring at the aft end of the carrier.
56. Pack the lip of the oil seal with multi-purpose water resistant lubricant.
57. Install the outer bearing carrier into the lower unit.
58. Secure the bearing carrier with the four bolts with an O-ring on each bolt.
59. Tighten the bolts, in a criss-cross pattern.
60. Install a dial indicator on the lower unit housing to check the propeller shaft end play.
61. The total movement of the shaft should be between .005 in. and .020in.

➡ **If the propeller shaft end play does not fall within this range the lower unit need to be disassembled and the shim between the forward gear and the propeller shaft at the clutch dog splines**

62. Slide a new sacrificial anode onto the propeller shaft.
63. Install the driveshaft seals into the gear housing cover in a back to back orientation.
64. Install the O-rings around the driveshaft collar and shift shaft area.
65. If applicable, thread the intermediate shift rod into the lower shift rod until it bottoms out and back it out four full turns.
66. Install the lower unit housing cover over the driveshaft and shift rod assembly.
67. Secure the housing with its fasteners.
68. Install the water pump parts
69. Pressure tests the unit with a pressure pump and gauge assembly to test the integrity of the unit.
70. Fill the lower unit with lower unit lubricant.

Dual Exhaust Units

♦ **See Figures 179 thru 190**

1. The following procedure pickup the work after all parts have been cleaned and inspected. The proper shim material thickness has been determined following the shimming instructions in the previous section. Replacement parts having been purchased and on hand.

➡ **Installing the lower driveshaft support bearing requires a special installer tool consisting of P/N T16496, T16496A and T16496B.**

2. Apply some grease to the eighteen pinion gear roller bearings and place the bearings one at a time—into the bearing outer race. The grease will help to hold the individual bearings in place.

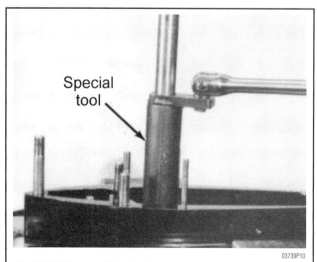

Fig. 179 Tighten the driveshaft bearing retainer with the proper special tool

8-40 LOWER UNIT

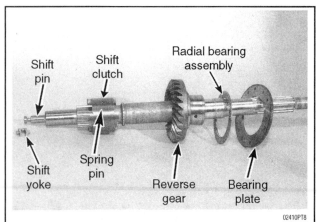

Fig. 180 Assemble the parts onto the propeller shaft in the sequence pictured

3. Reach in and place the bearing race with the roller bearings and with the lettered side of the race facing up into position to be drawn up into the opening.

4. Hold the race in place and lower the rod part of the special installer down the driveshaft opening and through the bearings without disturbing the bearings.

5. Attach the lower end of the tool P/N 16496B with the nut to the end of the rod against the bearing race.

6. Install the top plate and nut onto the upper end of the rod.

7. Tighten the nut by hand to exert pressure on the bearing.

8. Hold the bottom nut from turning with a wrench and tighten the upper nut.

9. Tighten the nut on the rod will draw the bearing assembly up into the cavity until seated on the shoulder in the housing.

→ Install the forward gear tapered roller bearing race using special tool Race Installer P/N 16814, Driver Handle P/N T8907, and Guide Plate P/N T11209 or an acceptable equivalent such as an old bearing and race and a long punch.

10. Block the lower unit on a wooden work surface with the opening facing up.

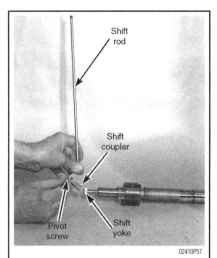

Fig. 181 As the propeller shaft is moved forward into the lower unit housing pull up on the shift rod to help align the shifter parts as depicted in this external view of the shifter components

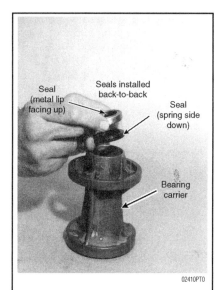

Fig. 182 Install the two seals into the bearing carrier in a back to back orientation with the outer seal metal lip facing up

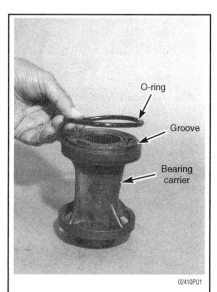

Fig. 183 Installing a new O-ring into the groove at the forward end of the bearing carrier unit

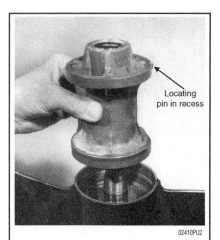

Fig. 184 Insert the bearing carrier into the lower unit with the cutout for the locating pin aligned with a matching cutout on the lower unit housing

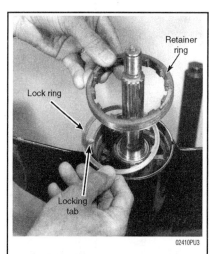

Fig. 185 Place the retainer ring into the housing over the lock ring with the word "off" facing outward

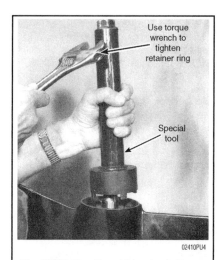

Fig. 186 Tighten the retainer ring in the lower unit to a torque value of 130 ft. lbs. (176m)

LOWER UNIT 8-41

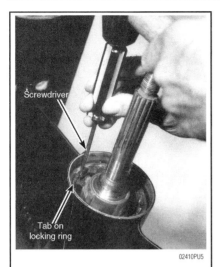

Fig. 187 Secure the lock tabs to hold the retainer in place

Fig. 188 Using an appropriate driver install the two seals back-to-back into the gear housing cover

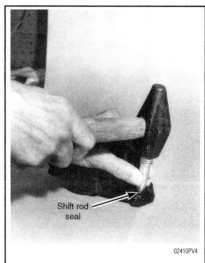

Fig. 189 Using an appropriate driver install the shift rod seal

Fig. 190 Install new O-rings around the driveshaft opening and into the groove around the shift shaft area

11. Set the proper amount of shim material as determined from the Shimming procedures in the previous section into the forward end of the lower unit.

12. Set the bearing race in position at the forward end of the lower unit with the large end of the race facing down forward.

13. Insert the tapered end of the race installer into the new race.

14. Place the driver handle over the race installer and slide the guide plate down the handle with the stepped side of the plate facing down.

15. Drive the bearing race into the lower unit by tapping on the driver handle with a hammer.

16. As soon as the race is fully seated the sound of the hammer striking the driver will change noticeably.

17. Insert the shift coupler into the forward end of the lower unit with the jaws facing aft.

18. Insert the pivot screw with O-ring into the threaded hole on the starboard side of the lower unit and through the pivot hole in the shift coupler.

19. Lower the shift rod down into the lower unit opening and thread it into the shift arm.

20. Thread the shift rod until it is fully seated.

21. Unthread the shift rod out four turns for the preliminary shift rod height adjustment.

22. Tighten the pivot screw.

23. Apply a coating of grease to the spacer washer and then insert it into the forward gear bore, if applicable.

➡The spacer washer should be the same one removed during disassembling. The propeller shaft end play can only be checked after the unit is fully assembled.

24. Insert the forward gear and bearing into the lower unit cavity with the bearing sliding into the forward bearing race.

25. Install the drive shaft taper roller bearing by first threading the pinion nut upside down onto the end of the driveshaft to protect the threads while the bearing is pressed into place.

26. Press a new tapered roller bearing onto the driveshaft with the small end of the bearing facing the splined end of the driveshaft.

✴✴ WARNING

Because of the driveshaft length take care not to bend or distort the driveshaft while the bearing is being pressed on

27. Insert the same amount of shim material into the driveshaft bore and onto the ledge as was removed or as determined during Shimming procedures section under "Shimming".

28. Lower the driveshaft down into the lower unit housing gently so the pinion gear roller bearings stay in place.

29. Install the pinion gear on the drive shaft by lifting the driveshaft up just enough to put the gear in place against the forward gear.

30. Lower the driveshaft with a slight twisting motion to align the splines as the shaft and pinion gear fit into place.

31. Apply Loctite® or Lock N' Seal® to the threads of the pinion gear nut.

32. Thread the pinion nut on to the driveshaft with the pinion washer and the groove on the nut facing the gear.

✴✴ WARNING

Always use a new pinion gear nut. Once the nut is tightened and released it loses its full locking ability and could come loose during normal operation

33. Tighten the pinion gear nut using the same tool arrangement as was used to remove it.

34. Thread the driveshaft upper bearing race retainer into the lower unit housing.

35. Using the special tool tighten the retainer to a torque value of 90 ft. lbs. (122Nm).

36. The propeller shaft is next assembled by sliding the spring down into the propeller shaft.

➡Most model lower units do not have a spring. Skip this step when there is no spring installed

37. Insert a narrow screwdriver into the slot in the shaft and compress the spring until approximately ½ inch (13mm), is obtained between the top of the slot and the screwdriver.

38. Hold the spring compressed slide the clutch dog over the splines of the propeller shaft in the same direction as originally installed.

39. Align the hole in the dog with the slot in the shaft.
40. Slide the shift arm into the open end of the propeller.
41. Align the hole in the clutch dog the slot in the propeller shaft, and the hole in the shift arm
42. Insert the cross pin into the clutch dog and through the propeller shaft slot and shift arm.
43. Center the pin and then remove the screwdriver allowing the spring to pop back into place.
44. Coat the large spacer with a film of grease.
45. Slide the reverse gear its radial bearing assembly, shims as applicable and large spacer onto the propeller shaft.
46. Place a small amount of fairly thick lubricant onto the end of the shift arm to help hold the shift yoke in place for assembly.
47. Place the shift yoke over the end of the shift arm with the open end of the yoke facing down.
48. Carefully install the propeller shaft and align the shift yoke with the shift cradle.
49. As the propeller shaft is moved forward into the lower unit housing, pull up on the shift rod. The open end of the shift coupler is facing aft toward the incoming propeller shaft. The open end of the shift coupler is to slide over the two posts of the shift yoke. The forward end of the shaft will index into the forward gear.

➡ Be certain that the shifting mechanism is properly aligned and the shift rod when moved will move the clutch dog before continuing.

50. Install the two seals into the bearing carrier in a back to back orientation with the outer
    seal metal lip facing up. The inside seal prevents lubricant from escaping and the outside seal prevents water from entering.
51. Install a new O-ring into the groove at the forward end of the bearing carrier.
52. Apply a light coating of grease to the propeller shaft seal in the bearing carrier.
53. Coat the bore of the lower unit and the O-ring in the bearing carrier with a light film of gear lubricant.
54. Insert the bearing carrier into the lower unit with the cutout for the locating pin aligned with a matching cutout on the lower unit housing.
55. Once the bearing carrier is in place push the locating pin into the cutouts to prevent the bearing carrier from rotating
56. Insert a new lock ring into the lower unit housing with the pointed locating tab of the ring aligned with the recess in the bearing carrier.
57. Place the retainer ring into the housing over the lock ring with the word OFF facing outward.
58. Using special tool P/N T11275 tighten the retainer ring in the lower unit to a torque value of 130 ft. lbs. (176m).
59. Install a dial indicator on the lower unit housing to check the propeller shaft end play.
60. The total movement of the shaft should be between .005 in. and .020in.

➡ If the propeller shaft end play does not fall within this range the lower unit need to be disassembled and the shim between the forward gear and the propeller shaft at the clutch dog splines

61. Select one square tab of locking ring centered between two internal cogs of the retainer ring.
62. Bend this tab outward aft between the cogs.
63. Bend the remaining three square tabs inward toward the bearing carrier.

➡ Bending these three tabs will permit improved exhaust flow exiting from the lower unit.

64. Using seal driver tool P/N T8985 or equivalent install the two seals back-to-back into the gear housing cover.
65. Using seal driver tool P/N T8947 or the proper size socket and drive the shift rod seal into place.
66. Install a new O-ring type seal around the driveshaft opening.
67. Install an O-ring type seal into the groove around the shift shaft area.
68. If applicable, thread the intermediate shift rod into the lower shift rod until it bottoms out and back it out four full turns.
69. Install the lower unit housing cover over the driveshaft and shift rod assembly.
70. Secure the housing with its fasteners.
71. Install the water pump parts
72. Pressure tests the unit with a pressure pump and gauge assembly to test the integrity of the unit.
73. Fill the lower unit with lower unit lubricant.

1995–99 70 to 120 HP

DISASSEMBLY

♦ See Figures 191 thru 205

1. The reverse gear-to-pinion gear backlash and the forward gear-to-pinion gear backlash should be checked prior to disassembly. The reverse gear backlash can be checked only at a point of propeller shaft rotation where it is not possible to shift from neutral gear into reverse, as outlined in the following procedure.
2. Three hands are necessary for the backlash check, therefore, obtain the help of an assistant.
3. Slowly rotate the propeller shaft and at the same time attempt to shift into the reverse gear position. Once this position is reached:
   a. Pull up on the driveshaft.
   b. Pull outward on the propeller shaft.
   c. Hold pressure on the lower shift shaft toward reverse.
   d. Lightly rotate the propeller shaft clockwise and counterclockwise.

The amount of free play felt is the reverse gear-to-pinion gear backlash. For all powerheads covered in this manual, the correct amount of backlash allowable is 0.040–0.060 in. (1.0–1.5mm). Record the amount of backlash felt because it may affect shimming of the reverse gear during assembly.

4. Repeat the previous step to check the backlash of the forward gear, except push in on the propeller shaft. Allowable backlash for the forward gear varies, depending on the model being serviced. Therefore, consult the specification chart.
5. Remove the two bolts and washers securing the bearing carrier in the

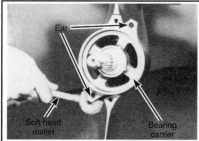

Fig. 191 Using a soft head mallet tap the ears of the bearing carrier to offset the carrier from the housing and tap opposite ears alternately and evenly on the backside to remove

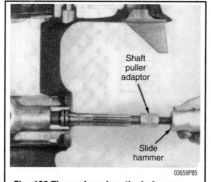

Fig. 192 The preferred method of removing the bearing carrier is to use a shaft puller adapter and a slide hammer

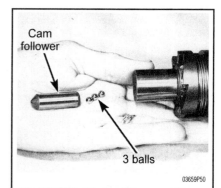

Fig. 193 Slide the cam follower and the three very small balls from the forward end of the propeller shaft

LOWER UNIT   8-43

lower unit. Use a soft head mallet and tap the ears of the bearing carrier to offset the carrier from the housing, as shown in the accompanying illustration. Now, tap opposite "ears" alternately and evenly on the backside to remove the carrier from the lower unit housing.

If the carrier is frozen and refuses to budge, one of three methods may prove successful to remove the carrier. The first and safest method is to use a simple adapter with a slide hammer. The second method involves the use of a slide hammer and long puller jaws to remove the carrier from the lower unit. The third and least desirable method is to use a mallet to actually drive the lower unit off the carrier.

In all three methods, heat carefully applied to the outside of the lower unit, will assist in removing the bearing carrier.

6. The first method of removing the bearing carrier is to use a shaft puller adapter and a slide hammer, as shown in the accompanying illustration.

7. Thread the adapter onto the propeller shaft as far as possible. Next, thread the slide hammer into the adapter and the shaft, clutch dog, reverse gear, and bearing carrier are ready to be removed as a unit. Operate the slide hammer to pull the parts mentioned.

8. As explained, this step is an alternate method for removing the bearing carrier from the lower unit. Obtain Slide Hammer (C-91-34569A1) with long Puller Jaws (C-91-46086A1). Use the propeller thrust hub to maintain an outward pressure on the puller jaws. Remove the propeller shaft from the lower unit. Set the unit aside for disassembly later.

9. The third and least desirable method is as follows: Clamp the propeller shaft in a vise equipped with soft jaws in a horizontal position. Use a mallet and strike the lower unit with quick sharp blows midway between the ventilation plate and the propeller shaft. This action will drive the lower unit off the bearing carrier. Take care not to drop the lower unit when the unit finally comes free of the carrier.

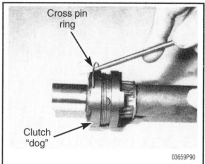

Fig. 194 Insert a thin blade screwdriver under the first coil of the cross pin ring and rotate the propeller shaft to unwind the spring from the sliding clutch

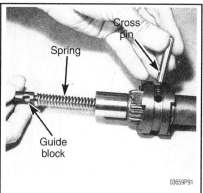

Fig. 195 Push the cross pin clear of the clutch dog and the propeller shaft

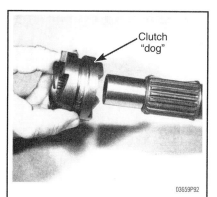

Fig. 196 Slide the clutch dog from the propeller shaft

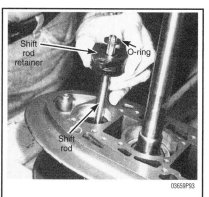

Fig. 197 Pull the shift rod boot up and over the shift rod

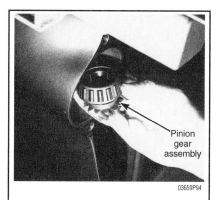

Fig. 198 Remove the pinion gear from the housing

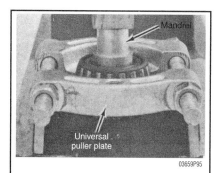

Fig. 199 Remove the bearing from the gear using a Universal Puller Plate (C-91-37241) between the pinion gear and the tapered roller bearing

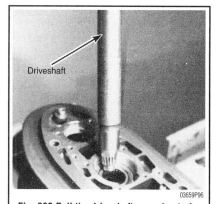

Fig. 200 Pull the driveshaft up and out of the lower unit housing

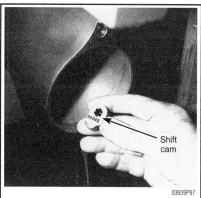

Fig. 201 Reach into the housing and pull out the shift cam

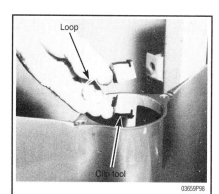

Fig. 202 Insert the E-clip tool against the race with the spring loop in the propeller shaft cavity

8-44 LOWER UNIT

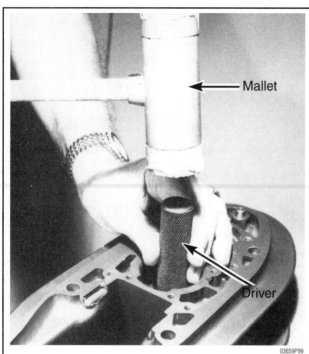

Fig. 203 Place the driver down through the top of the lower unit to index into the E-clip tool

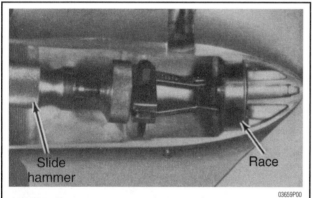

Fig. 204 Remove the forward gear tapered roller bearing race using a slide hammer

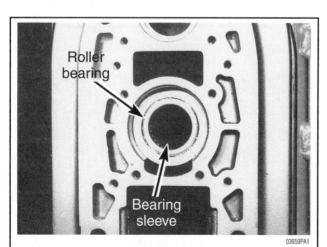

Fig. 205 Inspect the condition of the upper driveshaft roller bearing

➡ If the lower unit refuses to move, it may be necessary to carefully apply heat to the lower unit in the area of the carrier and at the same time attempt to move it off the carrier.

10. Remove the propeller shaft and bearing carrier from the vise. Slide the propeller shaft out of the bearing carrier from the rear to the front. Set the bearing carrier assembly aside for disassembly.

11. Inspect the two seals and the condition of the needle bearing at the rear end of the carrier. If the seals have failed and have allowed water to enter the lower unit, both needle bearings are no longer fit for further service. If the bearing does not roll freely or shows any sign of corrosion, clamp the carrier in a vise equipped with soft jaws. Obtain and use Slide Hammer (C-91-34569A1) to pull the bearing free of the carrier.

12. Use a screwdriver and remove both seals. These seals should have been installed back to back. The outer seal prevents water from entering and the other seal prevents oil from escaping.

13. Obtain Bearing Removal and Installation Kit (C-91-31229A-5). Obtain Mandrel (C-91-36569) and Driver Rod (C-91-37323), or a suitable substitute mandrel. Insert the removal tools into the forward end of the carrier and press the needle bearing out the rear end of the carrier.

14. Remove and discard the large O-ring around the carrier.

15. Tilt the forward end of the propeller shaft up slightly to prevent loss of the cam follower and very small parts behind the follower into the depths of the housing. Pull the shaft free of the housing. The reverse gear, a thrust bearing, and a thrust washer will come out with the shaft.

➡ Note the absence of any shim material behind the reverse gear. The manufacturer gives a backlash specification. However, an adjustment is not possible by adding or removing shim material behind the reverse gear.

16. Slide the cam follower, and then the three very small balls from the forward end of the propeller shaft.

17. Insert a thin blade screwdriver or an awl under the first coil of the cross pin ring, and then rotate the propeller shaft to unwind the spring from the sliding clutch. Take care not to over-stretch the spring.

18. Push the cross pin clear of the clutch dog and the propeller shaft. Be prepared to catch the guide block and spring when they fly free from the end of the propeller shaft.

19. Slide the clutch dog from the propeller shaft.

20. Pull the shift rod boot up and over the shift rod. Pry the circlip off the rod or remove the coupler and nylon spacer, depending on which items are used.

21. Remove the two attaching bolts and lift the shift rod and shift rod retainer straight up and out of the lower unit. Remove and discard the O-ring around the retainer.

22. Select a box end wrench the same size as the pinion nut. This tool will be used to prevent the nut from turning as the driveshaft is rotated. Obtain Driveshaft. Holding Tool (C-91-56775).

23. Install the holding tool onto the end of the driveshaft. Now, with the box end wrench on the pinion gear nut, use an appropriate wrench and rotate the tool and driveshaft counterclockwise until the pinion gear nut is free.

24. Remove the nut and the pinion gear assembly.

➡ Set aside the old pinion gear nut. The manufacturer recommends the nut not be used a second time. Once the nut has been removed it loses its designed locking characteristics. The old pinion nut will be temporarily installed on the driveshaft for shimming and backlash calculations, and then removed and discarded.

25. Obtain and position Universal Puller Plate (C-91-37241) between the pinion gear and the tapered roller bearing. Place the puller plate and gear, with the gear on the bottom, in an arbor press. Use a suitable mandrel and press the gear free of the bearing. The mandrel must contact the gear collar, but clear the bearing cage.

➡ Once the bearing has been removed, it cannot be used a second time. The roller cage will be distorted during the removal process.

26. Pull the driveshaft up and out of the lower unit housing.

27. Remove and discard the sealing ring around the driveshaft.

28. Inspect the condition of the wear sleeve at the lower end of the driveshaft. If the sleeve is worn or distorted it will allow water to enter the lower unit.

LOWER UNIT 8-45

29. To remove the wear sleeve, support the driveshaft in a universal bearing separator tool, resting over an open vice.

30. Carefully, using a soft head mallet, tap the upper splined driveshaft end to force the wear sleeve up and free of the driveshaft.

31. After the pinion gear and driveshaft have been removed, the forward gear assembly and associated bearing can be lifted out of the housing.

➥**The forward gear tapered roller bearing is pressed onto the short shaft of the gear assembly.**

32. The needle bearing supports the forward end of the propeller shaft. If the needle bearing is no longer fit for further service, it can be removed by tapping, with a blunt punch and hammer, around the bearing cage from the aft end the gear end of the bearing. This action will destroy the bearing cage. Therefore, be sure a replacement bearing is on hand, before attempting to remove the defective bearing.

33. Reach into the housing and pull out the shift cam. If the attempt to remove the shift shaft failed during earlier disassembly procedures, the shaft will still be engaged with the cam. Attempt to "wiggle" the cam free of the shaft. An application of penetrating oil dribbled down the shaft and sprayed into the housing may help the cam come free.

34. If the forward gear tapered roller bearing or the pinion gear tapered roller bearing was removed, the race of the affected bearing must also be removed. The bearing and the race must be installed as a set. A wear pattern will have been worn in the race by the old bearing. Therefore, if the race is not replaced, the new bearing will be quickly worn by the old worn race.

35. The upper driveshaft bearing and oil sleeve need not be disturbed for this procedure. Obtain Bearing Race Tool (C-91-14308A1). Insert the E-clip tool against the race with the spring loop in the propeller shaft cavity.

36. Place the driver down through the top of the lower unit to index into the E-clip tool.

37. Tap the driver with a hammer or mallet until the race is driven free of the housing.

38. Save the shim material from behind the race. The same amount of shim material will most likely be required during assembly.

39. Remove the forward gear tapered roller bearing race using a slide hammer.

40. SAVE any shim material from behind the race after the race is removed. The same amount of shim material will probably be used during assembly.

41. Inspect the condition of the upper driveshaft roller bearing—not the tapered bearing. The bearing is a one piece unit. If defective, it can be removed using a slide hammer with puller jaw attachment.

42. The roller bearing is pressed into a bearing sleeve, then the sleeve is pressed into the driveshaft bore.

43. To remove the bearing and sleeve, hook the puller jaws first around the roller bearing and pull it from the sleeve.

44. Then hook the jaws under the sleeve and pull the sleeve from the driveshaft.

45. If necessary, remove the oil sleeve behind the bearing using the same tool.

CLEANING & INSPECTION

◆ **See Figures 206 thru 215**

1. Clean all water pump parts with solvent, and then dry them with compressed air.

2. Inspect the water pump cover and base for cracks and distortion, possibly caused from overheating.

3. Inspect the face plate and water pump insert for grooves and/or rough surfaces. If possible, always install a new water pump impeller while the lower unit is disassembled. A new impeller will ensure extended satisfactory service and give peace of mind to the owner. If the old impeller must be returned to service, never install it in reverse to the original direction of rotation. Installation in reverse will cause premature impeller failure.

4. Inpsect the impeller side seal surfaces and the ends of the impeller blades for cracks, tears, and wear. Check for a glazed or melted appearance, caused from operating without sufficient water. If any question exists, and as previously stated, install a new impeller if at all possible.

5. Clean all bearings with solvent, dry them with compressed air, and inspect them carefully. Be sure there is no water in the air line. Direct the air stream through the bearing. Never spin a bearing with compressed air. Such action is highly dangerous and may cause the bearing to score from lack of lubrication. After the bearings are clean and dry, lubricate them with Quicksilver Formula 50-D lubricant or equivalent. Do not lubricate tapered bearing cups until after they have been inspected.

6. Inspect all ball bearings for roughness, catches, and bearing race side wear. Hold the outer race, and work the inner bearing race in-and-out, to check for side wear.

7. Determine the condition of tapered bearing rollers and inner bearing race, by inspecting the bearing cup for pitting, scoring, grooves, uneven wear, imbedded particles, and discoloration caused from overheating. Always replace tapered roller bearings as a set.

8. Inspect the bearing surface of the shaft roller bearing support. Check the shaft surface for pitting, scoring, grooving, imbedded particles, uneven wear and discoloration caused from overheating. The shaft and bearing must be replaced as a set if either is unfit for continued service.

Inspect the sliding clutch of the propeller shaft. Check the reverse gear side clutch dogs". If the dogs are rounded one of three causes may be to blame
  a. Improper shift cable adjustment.
  b. Running engine at too high a rpm while shifting.
  c. Shifting from neutral to reverse gear too quickly.

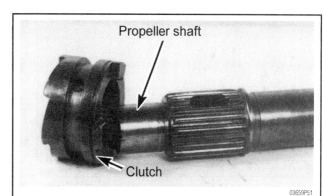

Fig. 206 The clutch dog and matching splines on the propeller shaft should be closely inspected

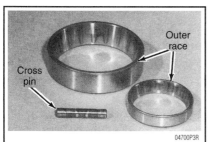

Fig. 207 Grooves on the cross pin or marked wear patterns on the outer roller bearing races are evidence of premature failure of these or associated parts

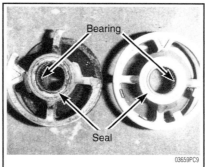

Fig. 208 Comparison of a worn bearing carrier (left) with a new one (right)

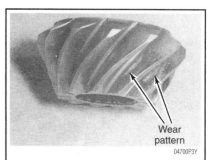

Fig. 209 Unacceptable pinion gear wear pattern, probably caused by inadequate lubrication in the lower unit

8-46 LOWER UNIT

9. Inspect the cam follower and replace it and the shift cam if there is any evidence of pitting, scoring, or rough surfaces.

10. Check the straightness of the propeller shaft with a set of V-blocks. Rotate the propeller shaft on the blocks and at the same observe the propeller shaft splined area for a bend. Inspect the seal surfaces, the area of the shaft where the bearing carrier oil seals make contact. If any type of groove is discovered in this area the shaft should be replaced. Inspect the propeller shaft spline area for corrosion damage or wear.

11. Inspect the propeller shaft roller bearing surfaces for pitting, rust marks, uneven wear, imbedded metal particles or signs of overheating caused by lack of adequate lubrication.

➡ **Good shop practice requires installation of new O-rings and oil seals regardless of their appearance.**

12. Clean the bearing carrier, pinion gear, drive gear clutch spring, and the propeller shaft with solvent. Dry the cleaned parts with compressed air.

13. Check the pinion gear and the drive gear for abnormal wear. Apply a coating of light-weight oil to the roller bearing. Rotate the bearing and check for cracks or catches.

14. Inpsect the propeller shaft oil seal surface to be sure it is not pitted, grooved, or scratched. Inspect the roller bearing contact surface on the propeller shaft for pitting, grooves, scoring, uneven wear, imbedded metal particles, and discoloration caused from overheating.

Fig. 210 This lower unit housing was destroyed because the bearing carrier was frozen so badly drastic action was required. Even so, all working components inside the unit were saved for further service

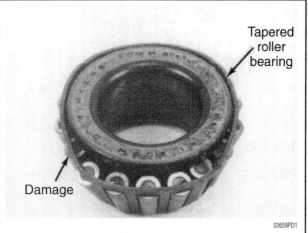

Fig. 211 Distorted tapered roller bearing unfit for further service

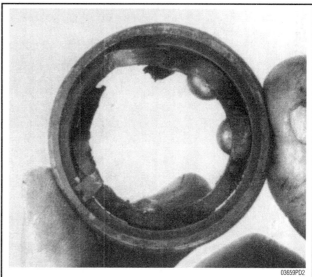

Fig. 212 Caged ball bearing set destroyed due to lack of lubrication, vibration, corrosion, metal particles, or all of the above

Fig. 213 A rusted and corroded gear. Water was allowed to enter the lower unit through a bad seal and cause this damage to the gear and other expensive parts

LOWER UNIT 8-47

ASSEMBLY

♦ See Figures 216 thru 221

✱✱ WARNING

Before beginning the installation work, count the number of teeth on the pinion gear and on the reverse gear. Not necessary to count the forward gear. Knowing the number of teeth on the pinion and reverse gear will permit obtaining the correct tool setup for the shimming procedure.

1. If the wear sleeve was removed, obtain Wear Sleeve Installation Tool (C-91-14310A1). Insert the new wear sleeve into the sleeve holder. Slide the bottom end of the driveshaft into the sleeve and holder. Slide the long collar portion of the installation tool over the top end of the driveshaft.

2. Move the assembled tool and driveshaft to an arbor press. Position the base of the sleeve holder onto a suitable support which will allow the bottom

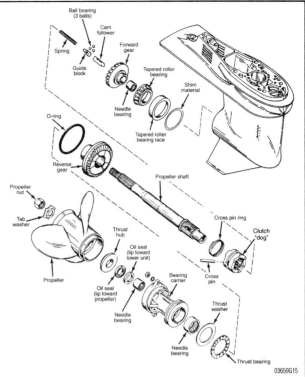

Fig. 215 Exploded drawing of the 70–120 hp models propeller shaft and shifting mechanism assemblies

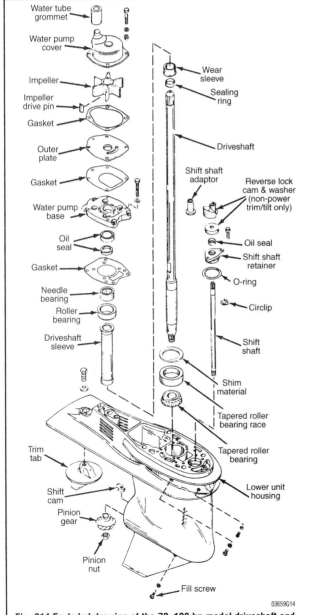

Fig. 214 Exploded drawing of the 70–120 hp model driveshaft and water pump assemblies

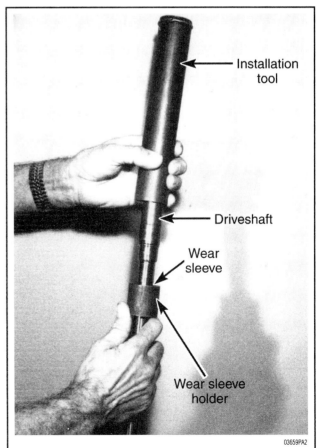

Fig. 216 Using the Wear Sleeve Installation Tool (C-91-14310A1) insert the new wear sleeve into the sleeve holder

8-48 Lower Unit

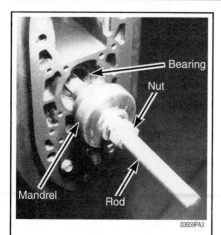

Fig. 217 Installing the upper driveshaft bearing

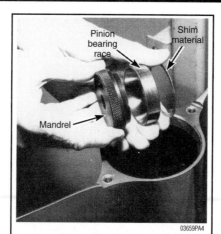

Fig. 218 Install the pinion bearing race and the appropriate amount of shims

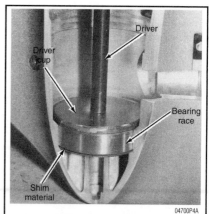

Fig. 219 Install the forward bearing race with the same amount of shim material saved during disassembly

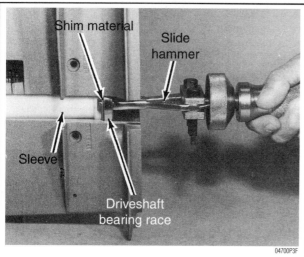

Fig. 220 Place the forward gear on a press with the gear teeth down and press the forward gear tapered bearing over the gear

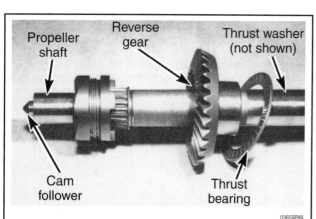

Fig. 221 Slide the reverse gear and thrust bearing onto the propeller shaft

4. Set the driveshaft aside for later installation.
5. Apply a light coating of Quicksilver Needle Bearing Lubricant to the inner surface of the bearing sleeve.
6. Identify the end of the new roller bearing with the embossed numbers. Identify the end of the bearing sleeve with a slight taper. Place the tapered end down on the arbor press support plate.
7. Insert the new roller bearing down into the larger end of the bearing sleeve with the numbered end facing upward. Obtain a suitable mandrel and press the bearing in until it is flush with the sleeve.
8. If the oil sleeve was removed from the driveshaft bore, install the sleeve into the bore with the tab on the upper portion of the sleeve facing aft. This sleeve must be in place before the upper driveshaft bearing is installed.
9. To install the upper driveshaft bearing, obtain the following special tools: Two mandrels from Bearing Installation Tool Kit (C-91-14309A1), a long threaded rod (C-91-31229), and a special nut (C-11-24156).
10. Thread the nut onto the rod until the nut is about ⅔ of the way up. Hold the shorter of the two mandrels inside the lower unit under the driveshaft bore.
11. Insert the threaded rod into the bore and thread the mandrel onto the rod until the mandrel is secure. Place the assembled roller bearing and sleeve over the threaded rod, into the cavity, and with the flush/numbered side facing upward.
12. Slide the other mandrel over the threaded rod with the shoulder side down to fit inside the sleeve.
13. Thread the nut on the rod down against the upper mandrel until all slack (clearance) is removed from the setup. With the proper size wrench on the top of the rod, prevent the rod from rotating, and at the same time rotate the nut clockwise down against the upper mandrel. As the nut is rotated the two mandrels will come closer together. This action will seat the bearing and sleeve assembly. When the shoulder of the upper mandrel makes contact with the upper face of the lower unit, the bearing/sleeve assembly is correctly positioned in the driveshaft bore.
14. Remove the special tools.
15. Using the same special tool setup employed in the previous step, the pinion gear bearing race can be installed. Set the pinion gear bearing race onto the shorter mandrel with the taper toward the mandrel. Place the same amount of shim material, removed during disassembly, on top of the race.

➡If the shim material was misplaced or not recorded, begin with material totaling 0.025 in. (0.635mm). Remember, shim material must be measured individually and the thickness of each added to arrive at the total thickness. Never use a micrometer to measure the total thickness of several pieces of shim material. Such a procedure results in a very inaccurate measurement and will lead to incorrect shimming of the lower unit.

16. Insert the threaded rod through the shim material, the bearing race, and into the mandrel. Assemble the tool as described in the previous step. The upper driveshaft bearing has already been installed. Install the pinion gear bearing race in the same manner.

end of the driveshaft to pass through. Press on the top surface of the long collar until the bottom surface of the collar seats against the top surface of the sleeve holder.

3. Install a new sealing ring around the driveshaft, and then apply a light coating of Loctite® Grade A around the ring.

LOWER UNIT 8-49

※※ WARNING

The bearing carrier is used as a pilot while installing the forward gear bearing race in the next step. Therefore, the bearing carrier must be temporarily assembled to include at least the propeller shaft roller bearing.

17. Place the same amount of shim material saved during disassembly into the lower unit. If the shim material was lost, or if a new lower unit is being used, begin with approximately 0.010 in. (0.25mm) material. Coat the forward bearing race bore with Quicksilver Formula 50-D lubricant, or equivalent.

18. Position the tapered bearing race squarely over the bearing bore in the front portion of the lower unit. Obtain Bearing Driver cup tool (C-91-31106). Place the tool over the tapered bearing race.

19. Insert the propeller shaft into the hole in the center of the bearing race. Lower the bearing carrier assembly down over the propeller shaft, and then lower it into the lower unit. The bearing carrier will serve as a pilot to ensure proper bearing race alignment.

20. Use a mallet and drive the propeller shaft against the bearing driver cup until the tapered bearing race is seated against the shim material. Withdraw the propeller shaft and bearing carrier, then lift out the driver cup.

21. Place the cam into the forward portion of the lower unit between the cast webbing. The ramps on the shift cam must be visible from the rear of the lower unit and the longer side, the reverse ramp, must be toward the left side, the fill screw hole side. The numbered face of the shift cam must face upward.

22. Install a new O-ring around the shift rod retainer. Place the lower shift shaft, the short spline end, into the shift shaft cavity. Rotate the shift shaft to allow the shaft splines to index with the cam splines.

23. Tap the retainer evenly into the shift shaft cavity. Apply Loctite® Grade A to the threads of both securing bolts. Install and tighten the bolts to a torque value of 60 inch lbs. (6.8Nm).

24. Install the coupler and nylon spacer, or boot, depending on the item used.

25. Place the forward gear on a press with the gear teeth down. Position the forward gear tapered bearing over the gear.

26. Press the bearing onto the gear with a suitable mandrel until the bearing is firmly seated.

➡ Press on the inner bearing race only. Pressing on the bearing cage will distort the bearing.

27. Check for clearance (gap) between the inner bearing race and the shoulder of the gear. There should be zero clearance.

28. Position the roller bearing over the center bore of the forward gear with the numbered side of the bearing facing up. Use a suitable mandrel and press the roller bearing is seated against the shoulder.

29. Insert the forward gear assembly into the forward gear bearing race.

30. Place the pinion gear on a press with the gear teeth down. Position the pinion gear tapered bearing over the gear.

31. Press the bearing onto the gear with a suitable mandrel until the bearing is firmly seated.

➡ Press on the inner bearing race only. Pressing on the bearing cage will distort the bearing.

32. Check for clearance (gap) between the inner bearing race and the shoulder of the gear. There should be zero clearance.

33. With one hand, lower the driveshaft into the top of the lower unit.

34. With the other hand, hold the pinion gear up in the lower unit below the driveshaft cavity and with the teeth of the gear indexed (meshed) with the teeth of the forward gear.

35. Rotate and insert the driveshaft until the driveshaft splines align and engage with the splines of the pinion gear. Continue to insert the driveshaft into the pinion gear until the tapered bearing is against the bearing race.

➡ Use the old pinion nut for this step, because if a new nut is used, it will lose its locking characteristics. Start the nut onto the end of the driveshaft.

36. Install the old pinion gear nut onto the driveshaft. Hold the pinion nut with a socket wrench. Pad the area where the socket wrench flex handle will contact the lower unit while the pinion nut is being tightened.

37. Obtain Drive Shaft Nut Wrench (C-91-56775). Place the special wrench over the upper end of the driveshaft. Use a torque wrench and socket to tighten the nut to a torque value of 70 ft. lbs. (95Nm).

※※ WARNING

After the pinion gear depth and the forward gear backlash have been set, the old nut must be replaced with a new pinion gear nut. Once the nut is tightened and removed, the nut loses some of its locking abilities.

38. Align the hole through the clutch dog with the slot in the propeller shaft, and then slide the dog onto the clutch with the square teeth facing the propeller end of the shaft. The teeth with the slanted ramps face forward.

39. Insert the spring into the end of the propeller shaft. Insert the guide block, stepped end, into the front end of the propeller shaft with the cross-pin hole aligned with the cross-pin hole in the sliding clutch.

40. Temporarily insert the flat end of the cam follower into the front end of the propeller shaft.

41. Position the cam follower against a solid object and push against the cam follower to compress the spring.

42. Hold the propeller shaft in this position and at the same time, use a punch to align the guide block cross-pin opening with the sliding clutch cross-pin hole.

43. Remove the punch and insert the cross-pin.

44. Release pressure on the spring.

45. Remove the cam follower.

46. Install the cross-pin ring over the sliding clutch. Take care not to over stretch the spring.

47. Insert just a little water resistant multipurpose lubricant into the end of the propeller shaft. Insert the three very small steel balls. Set the propeller shaft aside ready for later installation.

➡ When handling the assembled propeller shaft, Take care to keep the forward end tilted slightly upward to keep the small balls and other parts in place inside the shaft.

48. Position the smaller propeller shaft roller bearing into the aft end of the bearing carrier with the numbered side toward the aft end. Press the needle bearing into the bearing carrier with mandrel (C-91-37263) or other suitable mandrel.

49. Coat the outer diameter of the propeller shaft oil seals with Loctite® Type A.

50. Obtain Oil Seal Driver (C-91-31108). Place one seal on the longer shoulder side of the driver tool with the lip of the seal away from the shoulder.

51. Press the seal into the bearing carrier until the seal driver bottoms against the bearing carrier. Place the second seal on the short shoulder side of the seal driver with the lip of the seal toward the shoulder.

52. Press the seal into the bearing carrier until the seal driver bottoms against the bearing carrier. Clean excess Loctite® from the seals.

53. Position the larger propeller shaft roller bearing into the forward end of the bearing carrier with the numbered side toward the forward end.

54. Press the roller bearing into the bearing carrier using mandrel (C-91-13945) or equivalent substitute.

55. Install a new O-ring around the carrier. Apply a coating of Special Lubricant 101 (C-92-13872A1) onto the installed O-ring.

56. Slide the reverse gear and thrust bearing onto the propeller shaft. Check to be sure the cam follower and the three very small steel balls are still in place inside the shaft.

57. Insert the assembled propeller shaft, reverse gear and thrust bearing into the lower unit. Rotate the bearing carrier, if necessary, to position the word "TOP" facing upward after installation.

58. Following final installation, coat the threads of the attaching bolts with Loctite® Grade A.

59. Install and tighten the bolts to a torque value of 150 inch lbs. (17Nm). On some models, nuts on studs are used in place of bolts.

60. Install and tighten the nuts to a torque value of 275 inch lbs. (31.5Nm).

SHIMMING

♦ See Figures 222, 223, 224, 225 and 226

1. Obtain Bearing Preload Tool (C-91-14311A1). This tool consists of a spring, which seats against the lower unit and is used to simulate the upward driving force on the pinion gear bearing. Install the following, one by one, over the driveshaft.

8-50 LOWER UNIT

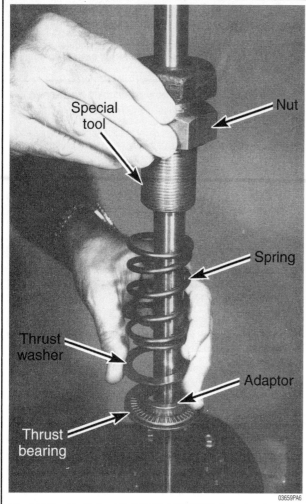

Fig. 222 Bearing Preload Tool (C-91-14311A1) being set up to preload the pinion bearings for shimming

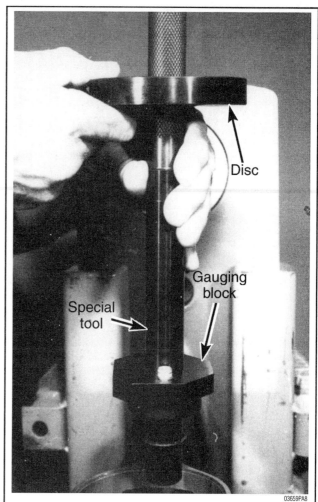

Fig. 224 Installing the proper special tool to check the pinion gear depth

Fig. 223 Align the two holes in the sleeve and tighten the Allen screws to hold the driveshaft

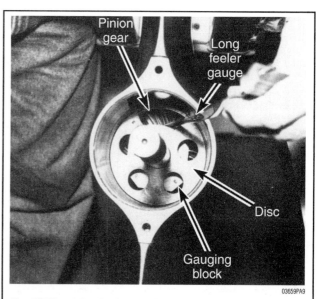

Fig. 225 The pinion depth is the distance between the bottom of the pinion gear teeth and the flat surface of the special tool

LOWER UNIT 8-51

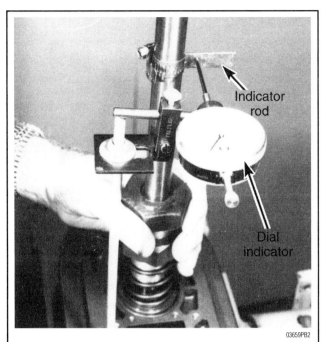

Fig. 226 Using a dial indicator and the back lash indicator rod check the forward gear back lash

a. Adaptor, with the flat face down against the lower unit
b. Thrust bearing
c. Thrust washer
d. Spring
e. Special large bolt, with the nut on the bolt threaded all the way to the top of the threads.
f. Sleeve, with two holes

2. Align the two holes in the sleeve with the two Allen set screws in the bolt head. These two set screws grip the driveshaft and prevent the tool from slipping on the driveshaft. Tighten the two set screws securely with the Allen key provided with the tool kit.

3. Measure the distance between the bottom surface of the bolt head and the top surface of the nut. Record this measurement.

4. Use a wrench to hold the bolt head steady and another wrench to tighten the nut down against the spring. Continue tightening the nut until the distance between the bottom surface of the bolt head and the top surface of the nut equals 1.00 in. (25.4mm) plus the distance recorded when the tool was installed.

5. The driveshaft has now been forced upwards placing an upward preload on the pinion gear bearing.

6. Obtain Pinion Gear Locating Tool (C-91-12349A-2) for gear ratios 13/30 and 14/29 or (C-91-817057A-1) for a gear ratio of 14/23.

7. Insert the tool into the installed forward gear assembly. If positioned correctly, one of the flats on the gauging block should be positioned directly under the pinion gear teeth.

8. Notice the gauging block on tool (C-91-12349A-2) has a series of flat sides, identified by the numbers 1 thru 8. The eight different numbered faces are needed for use with different combinations of gear ratios. Only two gear ratios are used on units covered in this manual. At the beginning of this section, the assembler was asked to count and record the number of teeth on the pinion and reverse gears. The reason for the count is as follows:

If the pinion gear has 13 teeth and the reverse gear has 30 teeth, the gear ratio is 13/30. With this gear ratio, face No. 8 is positioned closest to the pinion gear.

If the pinion gear has 14 teeth and the reverse gear has 29 teeth, the gear ratio is 14/29. With this gear ratio, face No. 2 is positioned closest to the pinion gear.

➡ No other gear ratio combinations are possible for units covered in this manual. If another "odd-ball" gear ratio is encountered, somewhere, sometime, someone has purposely changed the gear ratio to suit their own requirements, possibly for operation of the outboard at different altitudes. Therefore, factory specifications would no longer apply. Should this occur, consult your local dealer for advice. He will probably advise you to put the lower unit back together as you found it, wait for it to fail and replace the failed parts to meet with a factory specified gear ratio.

Let's assume everything is as it should be and the appropriate tool face to use is determined. Position this face (either No. 8 or No. 2) under the pinion gear teeth. Install the locating disc over the tool shank with the access hole facing upward. Push the disc all the way into the lower unit until it rests against a machined shoulder of the bore. This disc supports the tool shank in the lower unit bore.

9. The pinion depth is the distance between the bottom of the pinion gear teeth and the flat surface of the gauging block.

10. Measure this distance by inserting a long feeler gauge into the access hole and between the pinion gear teeth and gauging block flat face.

For all units covered in this section, the correct clearance is 0.025 in. (0.64mm). If the clearance is correct, remove only the Pinion Gear Locating Tools from the inside of the gearcase. Leave the Pinion Gear Bearing Preload Tool in place on the driveshaft for further shimming procedures.

If the clearance is not correct, add or subtract shim material from behind the pinion gear bearing race to lower or raise the gear.

If the pinion gear depth was found to be greater than 0.025 in. (0.64mm), then shim material must be added to lower the gear.

If the pinion gear depth was found to be less than 0.025 in. (0.64mm), then shim material must be removed to raise the gear.

The difference between 0.025 in. (0.64mm) and the actual clearance measured is the amount of shim material needed to correct the pinion gear depth.

➡ To check the forward gear backlash, the propeller shaft is temporarily installed for this procedure.

Final installation of the shaft must be performed later because an old pinion gear nut was installed on the driveshaft to determine the pinion gear depth.

➡ Obtain a piece of PVC sprinkler pipe, 6 in. (15.2cm) long by 1¼–1½in. (3.2-3.8cm) diameter and a washer large enough to slide over the propeller shaft threads. Before installing the assembled bearing carrier into the lower unit, install the PVC pipe over the propeller end of the shaft, followed by the large washer and propeller nut to hold it all together. Hand tighten the nut. The pipe will exert a pressure on the bearing carrier, thrust washer and reverse gear and will allow each component to be held in alignment while the shaft is installed. Once the shaft is well seated inside the lower unit, the propeller nut, washer and PVC pipe is removed.

11. Insert the assembled propeller shaft, reverse gear and thrust bearing into the lower unit. Rotate the bearing carrier, if necessary, to position the word "TOP" facing upward after installation.

12. Following final installation, coat the threads of the attaching bolts with Loctite® Grade "A".

13. Install and tighten the bolts to a torque value of 150 inch lbs. (17Nm). On some models, nuts on studs are used in place of bolts. Install and tighten the nuts to a torque value of 275 inch lbs. (31.5Nm).

14. To shim the forward gear to correct backlash, install a bearing carrier puller with the arms of the puller on the carrier and the center bolt on the end of the propeller shaft. Tighten the puller center bolt to a torque value of 45 inch lbs. (5Nm). This action places a preload on the forward gear, pushing the forward gear into the forward gear, pushing the forward gear into the forward gear bearing.

15. Rotate the driveshaft about ten full revolutions to properly seat the forward gear tapered roller bearing. Then recheck the torque value on the center puller bolt.

16. Attach the Backlash Dial Indicator Rod (C-91-83155) for 75 hp and 90 hp units or (C-91-78473) for 100 hp and larger units to the driveshaft. Position the dial indicator shaft to the line marked "4" on the dial indicator rod for a pinion gear to forward gear ratio of 2.30:1 13 teeth on the pinion gear. Use the line marked ".448" on the indicator rod for a pinion gear to forward gear ratio of 2.07:1 14 teeth on the pinion gear. Use the line marked "3" on the indicator rod for a pinion gear to forward gear ration of 1.64:1 also 14 teeth on the pinion gear.

8-52 LOWER UNIT

17. Rotate the driveshaft back-and-forth and observe movement of the dial indicator. Total movement is the forward gear backlash. Check the listings in the specifications chart for proper amount of backlash permissible for the unit being serviced. If the backlash is too great, add shim material behind the forward gear bearing race. If the backlash is too small, remove shim material from behind the forward gear bearing race. For each 0.001 in. (0.025mm) of backlash, add or remove 0.00125 in. (0.032mm) shim material.

18. Remove all special tools. Remove the bearing carrier and propeller shaft.
19. Replace the old pinion nut with a new nut.
20. Tighten the new nut to a torque value of 70 ft. lbs. (95Nm) for 75 to 125 hp.
21. Install the propeller shaft and bearing carrier.

MANUAL TILT 9-2
DESCRIPTION AND OPERATION 9-2
   SERVICING 9-2
CENTER MOUNTED PISTON
TRIM/TILT SYSTEM 9-3
DESCRIPTION AND OPERATION 9-3
TROUBLESHOOTING THE CENTER
  MOUNTED PISTON TRIM/TILT
  SYSTEM 9-4
TRIM/TILT PUMP 9-4
   TESTING 9-4
   REMOVAL & INSTALLATION 9-5
   DISASSEMBLY 9-5
   CLEANING & INSPECTION 9-6
   ASSEMBLY 9-6
   SYSTEM BLEEDING 9-6
TRIM/TILT MOTOR 9-6
   TESTING 9-6
   REMOVAL & INSTALLATION 9-7
   DISASSEMBLY 9-7
   CLEANING & INSPECTION 9-7
   ASSEMBLY 9-7
TRIM/TILT CYLINDER 9-8
   TESTING 9-8
   REMOVAL & INSTALLATION 9-8
   DISASSEMBLY 9-9
   CLEANING & INSPECTION 9-10
   ASSEMBLY 9-10
TRIM/TILT SWITCH 9-10
   TESTING 9-10
TRIM/TILT RELAY 9-11
   TESTING 9-11
   REMOVAL & INSTALLATION 9-11
SINGLE RAM INTEGRAL POWER
TRIM/TILT SYSTEM 9-11
DESCRIPTION AND OPERATION 9-11
TRIM/TILT PUMP 9-11
   REMOVAL & INSTALLATION 9-11
   HYDRAULIC SYSTEM
    BLEEDING 9-12
TILT/TRIM MOTOR 9-13
   TESTING 9-13
   REMOVAL & INSTALLATION 9-13
   CLEANING & INSPECTION 9-14
TILT/TRIM CYLINDER 9-15
   REMOVAL & INSTALLATION 9-15
   CLEANING & INSPECTION 9-17
TILT/TRIM SWITCH 9-18
   TESTING 9-18
TILT/TRIM RELAY 9-18
   TESTING 9-18

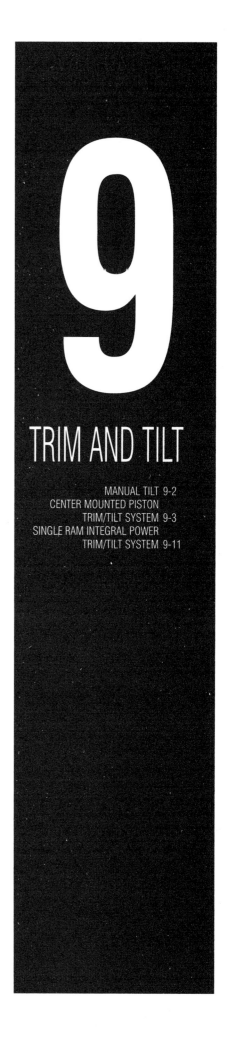

TRIM AND TILT

MANUAL TILT 9-2
CENTER MOUNTED PISTON
TRIM/TILT SYSTEM 9-3
SINGLE RAM INTEGRAL POWER
TRIM/TILT SYSTEM 9-11

9-2 TRIM AND TILT

MANUAL TILT

Description and Operation

All outboard installations are equipped with some means of raising or lowering (pivoting) the outboard for efficient operation under various load, boat design, water conditions, and for trailering to and from the water. By pivoting the outboard, the correct trim angle can be achieved to ensure maximum performance and fuel economy as well as a more comfortable ride for the crew and passengers.

The manual tilt mechanism is used to tilt the outboard in relation to the stern bracket. To adjust the outboard angle, the tilt lever is moved to the tilt position. This disengages the release rod and causes the reverse lock to disengage from the adjusting pin. The unit can be set at varying angles by raising or lowering the outboard. To release the tilt mechanism, return the tilt lever to the normal position, then raise and lower the extension case slightly.

➡ **The tilt lever should be kept in the normal position whenever the outboard is operating.**

SERVICING

◆ See Figures 1 and 2

Service procedures for the manual tilt system are confined to general lubrication and inspection. If individual components should wear or break, replacement of the defective components is necessary.

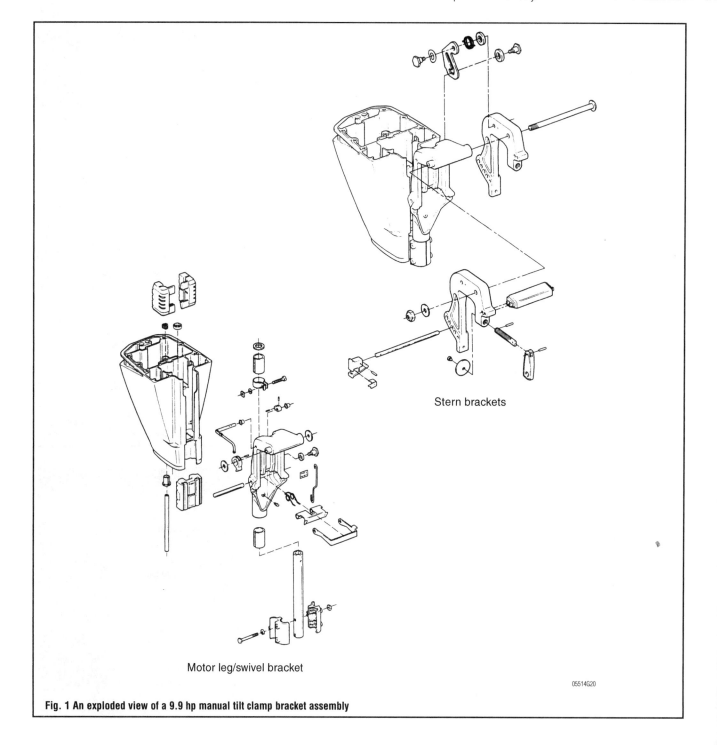

Fig. 1 An exploded view of a 9.9 hp manual tilt clamp bracket assembly

TRIM AND TILT

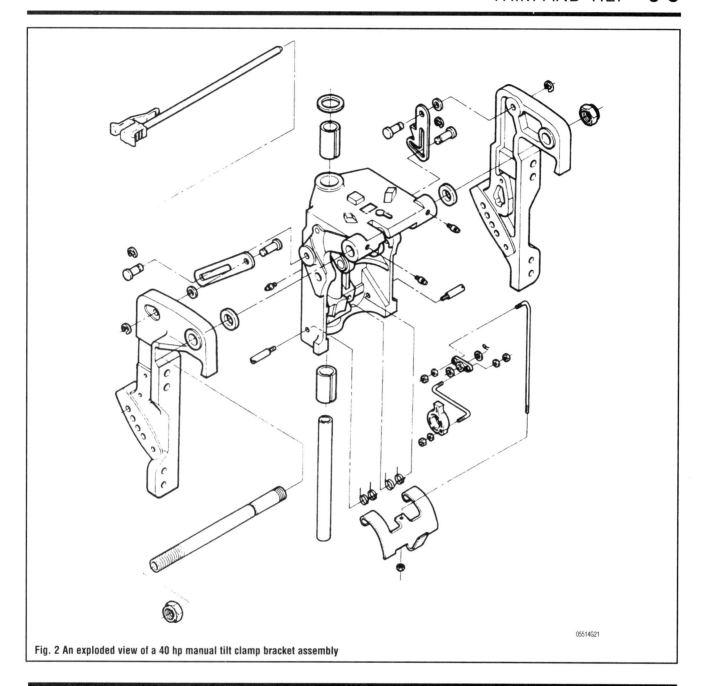

Fig. 2 An exploded view of a 40 hp manual tilt clamp bracket assembly

CENTER MOUNTED PISTON TRIM/TILT SYSTEM

Description and Operation

The power trim/tilt systems consist of a housing with an electric motor, a gear driven hydraulic pump, hydraulic reservoir and on or more trim/tilt cylinders. The cylinders perform a double function as trim/tilt cylinders and also as a shock absorbers, should the lower unit strike an underwater object while the boat is underway.

The necessary valves, check valves, relief valves, and hydraulic passageways are incorporated internally and externally for efficient operation. A manual release valve is provided to permit the outboard unit to be raised or lowered should the battery fail to provide the necessary current to the electric motor or if a malfunction should occur in the hydraulic system.

The gear driven pump operates in much the same manner as an oil circulation pump installed on motor vehicles. The gears rotate in either direction, depending on the desired cylinder movement. One side of the pump is considered the suction side, and the other the pressure side, when the gears rotate in a given direction. These sides are reversed, the suction side becomes the pressure side and the pressure side becomes the suction side when gear movement is changed to the opposite direction.

Depending on the model, up to two relays may be used for the electric motor. The relays are usually located at the bottom cowling pan, where they are fairly well protected from moisture.

➡ As a convenience, on some models an auxiliary trim/tilt switch is installed on the exterior cowling.

When the up portion of the trim/tilt switch is depressed, the up circuit, through the relay, is closed and the electric motor rotates in a clockwise direction. Pressurized oil from the pump passes through a series of valves to the lower chamber of the trim cylinders, the pistons are extended and the outboard unit is raised. The fluid in the upper chamber of the pistons is routed back to the reservoir as the piston is extended. When the desired position for trim is obtained, the switch on the control handle is released and the outboard is held stationary.

If the trim cylinder pistons should become fully extended, such as in a tilt up situation, fluid pressure in the lower chamber of the trim cylinders increases. This increase in pressure opens an up relief valve and the fluid is routed to the reservoir. The sound of the electric motor and the pump will have a noticeable change.

When the down portion of the trim/tilt switch is depressed, the down circuit, through the relay, is closed and the electric motor rotates in a counterclockwise direction. The pressure side of the pump now becomes the suction side and the original suction side becomes the pressure side. Pressurized oil from the pump passes through a series of valves to the upper chamber of the trim cylinders, the pistons are retracted and the outboard unit is lowered. The fluid in the lower chamber of the pistons is routed back to the reservoir as the retracted is extended. When the desired position for trim is obtained, the switch on the control handle is released and the outboard is held stationary.

If the trim cylinder pistons should become fully retracted, such as in a tilt down situation, fluid pressure in the upper chamber of the trim cylinders increases. This increase in pressure opens an up relief valve and the fluid is routed to the reservoir. The sound of the electric motor and the pump will have a noticeable change.

In the event the outboard lower unit should strike an underwater object while the boat is underway, the tilt piston would be suddenly and forcibly extended, moved upward. For this reason, the lower end of the tilt piston is capped with a free piston. This free piston normally moves up and down with the tilt piston.

The free piston also moves upward but at a much slower rate than the tilt piston. The action of the tilt piston separating from the free piston causes two actions. First, the hydraulic fluid in the upper chamber above the piston is compressed and pressure builds in this area. Second, a vacuum is formed in the area between the tilt piston and the free piston.

This vacuum in the area between the two pistons sucks fluid from the upper chamber. The fluid fills the area slowly and the shock of the lower unit striking the object is absorbed. After the object has been passed the weight of the outboard unit tends to retract the piston. The fluid between the tilt piston and the free piston is compressed and forced through check valves to the reservoir until the free piston reaches its original neutral position.

A manual relief valve, located on the stern bracket, allows easy manual tilt of the outboard should electric power be lost. The valve opens when the screw is turned counterclockwise, allowing fluid to flow through the manual passage. When the relief valve screw is turn fully clockwise, the manual passage is closed and the outboard lock in position.

A thermal valve is used to protect the trim/tilt motor and allow it to maintain a designated trim angle. Oil in the upper chamber is pressurized when force is applied to the outboard from the rear while cruising. Oil is directed through the right side check valve and activates the thermal valve to release oil pressure and lessen the strain on the motor and pump.

Troubleshooting The Center Mounted Piston Trim/Tilt System

Any time a problem develops in the power trim/tilt system the first step is to determine whether it is electrical or hydraulic in nature. After the determination is made, then the appropriate steps can be taken to remedy the problem.

The first step in troubleshooting is to make sure all the connectors are properly plugged in and that all the terminals and wires are free of corrosion. The simple act of disconnecting and connecting a terminal may sometimes loosen corrosion that preventing a proper electrical connection. Inspect each terminal carefully and coat each with dielectric grease to prevent corrosion.

The next step is to make sure the battery is fully charged and in good condition. While checking the battery, perform the same maintenance on the battery cables as you did on the electrical terminals. Disconnect the cables (negative side first), clean and coat them and then reinstall them. If the battery is past its useful life, replace it. If it only requires a charge, charge it.

Check the power trim/tilt fuse, as appropriate. Many systems will have a fuse to prevent large current draws from damaging the system. If this fuse is blown, the system will cease to function. This is a good indicator that you may have problems elsewhere in the electric system. Fuses don't blow without cause.

After inspecting the electrical side of the system, check the hydraulic fluid level and top it off as necessary. Remember to position the motor properly (full tilt up or down) to get an accurate measurement of fluid level. A slight decrease in the level of hydraulic fluid may cause the system to act sporadically.

Finally, make sure the manual release valve is in the proper position. A slightly open manual release valve may prevent the system from working properly and mimic other more serious problems.

Just remember to check the simple things first. If these simple tests do not diagnose the cause of the problem, then it is time to investigate more deeply. Perform the hydraulic pressure tests in this section to determine if the pump is making adequate pressure. Inspect the entire power trim/tilt electrical harness with a multimeter, checking for excessive resistance and proper voltage.

Trim/Tilt Pump

The trim and tilt unit is a combined assembly of an electric motor and hydraulic pump with reservoir. In the event that the entire unit is to be worked on the trim/tilt pump and trim/tilt motor sections will need to be combined as necessary to repair the entire unit.

TESTING

▶ See Figures 3 and 4

1. There are no pressure tests or pressure specifications for this style system. When a problem needs to be diagnosed a simple test to narrow down the failed component is performed. The repair is made through a process of elimination disassembling the most common failed part first.
2. Before testing the hydraulic system it must be properly bleed and the fluid level be at the proper height.
3. To test the down circuit of the hydraulic system
   a. Disconnect the lower hydraulic trim cylinder line from the hydraulic pump base.
   b. Cap the line and plug the opening in the pump to prevent loss of fluid and contaminants from entering the system.
   c. Move the control toggle switch to the up position.
   d. The tilt cylinder lifts the outboard on its own with the trim cylinder bypassed for this test.
   e. Remove the cap and plug from the line and pump.
   f. Connect the trim cylinder line.
4. If the outboard descends slowly the problem is in the tilt cylinder or the up circuit of the valve body. If there is no external leak the problem will be either the tilt cylinder or the hydraulic valve body.
5. Follow the procedures to remove and disassemble the tilt cylinder. If there are no visible defects with the cylinder assemble it and replace the hydraulic pump.
6. To test the up circuit of the hydraulic system
   a. With the outboard in the full down position disconnect the lower hydraulic tilt cylinder line from the hydraulic pump base.
   b. Cap the line and plug the opening in the pump to prevent loss of fluid and contaminants from entering the system.
   c. Move the control toggle switch to the up position.

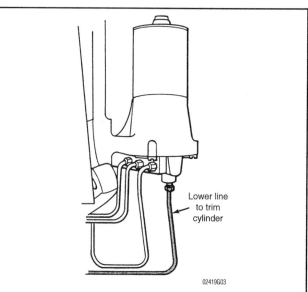

Fig. 3 The location of the trim cylinder pressure line. Newer units will have the same location but will be turned clockwise by 90°

TRIM AND TILT 9-5

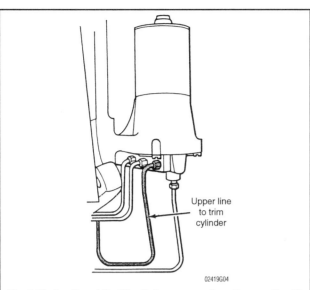

Fig. 4 The location of the tilt cylinder pressure line. Newer units will have the same location but will be turned clockwise by 90°

  d. The trim cylinder lifts the outboard on its own with the tilt cylinder bypassed for this test.
  e. Remove the cap and plug from the line and pump.
  f. Connect the tilt cylinder line.
7. If the outboard descends slowly the problem is in the trim cylinder or the hydraulic valve body. If there is no external leak the problem will be either the trim cylinder or the hydraulic valve body.
8. Follow the procedures to remove and disassemble the trim cylinder. If there are no visible defects with the cylinder assemble it and replace the hydraulic pump.

REMOVAL & INSTALLATION

1. Disconnect the wires connecting the trim motor to the harness or its solenoids.

✱✱ CAUTION

The hydraulic fluid in this system can be under extreme pressure. Safety glasses and proper protection should be worn before loosening the pressure lines in this system.

2. With a shop cloth around the fitting area loosen the four hydraulic lines with an appropriate sized wrench.
3. Remove the fasteners holding the pump assembly to the clamp bracket.
4. With the hydraulic lines removed lift the assembly off the clamp bracket.
To install:
5. Support the hydraulic pump assembly by the clamp bracket.

➙Leaving the pump assembly loose allows more freedom to install the hydraulic lines leaving less of a possibility of them getting cross threaded.

6. Thread the hydraulic lines in to there proper locations.
7. Bolt the pump assembly on to the clamp bracket.
8. Tighten the hydraulic lines.
9. Connect the pump motor wires on the proper wire connections or solenoid terminals.

DISASSEMBLY

▸ See Figures 5, 6 and 7

1. Before disassembling the unit mark the following locations indexing the unit for alignment during assembly.
   a. Make a mark across the cap and motor frame.
   b. Make a mark across the motor frame and reservoir.
   c. Make a mark across the reservoir and valve body.
2. With the assembly in the upright position remove the two through bolts.
3. Separate the motor frame from the fluid reservoir.
4. Remove the armature and ball bearing.
5. Drain the trim and tilt fluid from the reservoir.
6. Set aside the cap and motor frame, testing procedures for these parts are found in the tilt motor section.
7. Invert the reservoir and remove the four screws securing the valve body to the reservoir.
8. Separate the reservoir from the valve body.
9. Remove and the O-ring that seals the valve body to the reservoir.
10. Hold the valve body in the upright position to avoid disturbing the valve springs, check balls, and pump gears while the pump body and the valve body are separated.
11. Remove the four screws securing the upper pump body to the lower pump body.
12. Separate the two pump body halves.
13. Remove the drive gear (center) and the idle gear (off set).

➙Note the arrangement and location of the three springs and the spring retainers in the upper pump body.

14. Remove the three spring retainers and the springs.
15. Remove the two poppet valves and the three check balls from the lower pump body.

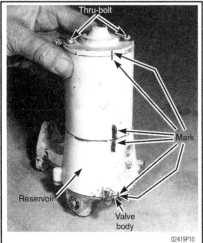

Fig. 5 Make three sets of matching marks on the assembly to index the unit before disassembly

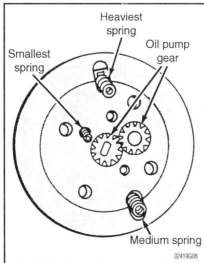

Fig. 6 Note the location and size of the springs in relation to the oil pump gear set

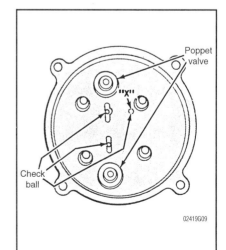

Fig. 7 The location of the two poppet valves and the three check balls in the lower pump body

9-6 TRIM AND TILT

➡ The pump and valve body assembly cannot be serviced except for cleaning. Individual parts are not available. The pump and valve bodies share a single part number and are replaced as an assembly.

CLEANING & INSPECTION

Make every effort to keep the work area clean to prevent contaminating parts to be installed.

Clean all parts thoroughly with solvent and carefully blow them dry with low pressure compressed air.

Inspect the three check balls for possible out of roundness. An out of round condition could prevent a check ball from seating properly. If a check ball fails to seat properly hydraulic fluid may be able to pass when the circuit should be closed.

Inspect the idle and drive gear teeth for chips or cracks. Check the springs and poppet valves for excessive wear.

Remember that individual parts are not available for replacement. If any part is found to be defective and not reusable the complete pump and electric motor must be replaced as an assembly

Coat all internal parts with power trim and tilt fluid prior to assembling.

ASSEMBLY

1. Install the two poppet valves and the three check balls into their original locations in the lower valve body.
2. Install the three springs and their retainers into the upper pump body.
3. Install the drive gear with the slotted center into the center recess.
4. Install the idle gear with the round center into the offset recess in the upper valve body.
5. Position the upper valve body over the lower valve body.

➡ The smallest spring should align over the check ball marked with an "X".

6. Install and tighten the four attaching screws holding the two pump halves together.
7. Install a new O-ring over the assembled valve body.
8. Position the reservoir over the valve body.
9. Align the marks made prior to disassembly.
10. Install and tighten the four attaching screws.

SYSTEM BLEEDING

1. Move the outboard to the full up position, and then tap the toggle switch down just once to relieve excess pressure in the hydraulic lines. The outboard unit must be in the full up position when checking or adding fluid because the pistons are extended and the fluid level in the reservoir will be at it highest.
2. Clean the area around the fill plug to prevent contaminants entering the system when the plug is removed.
3. Remove the fill plug.
4. Check the fluid level in the reservoir. The fluid should reach the lower edge of the fill opening.
5. Replenish the system using non detergent 30-weight automotive engine oil.

✳✳ WARNING

Do not use automatic transmission fluid because such fluid contains a high degree of detergent agents that could cause the internal seal to prematurely wear and fail.

➡ Brand name oils should never be mixed, except in an emergency situation. The complete system should then be drained at the first opportunity and filled with only one name brand fluid.

6. With the fill plug removed, tilt the outboard up and down three to five times to purge (bleed) trapped air from the hydraulic system. Bubbles will appear at the fill opening.
7. Replenish the reservoir as necessary.
8. Install and tighten the fill plug.

Trim/Tilt Motor

TESTING

3 Wire Trim Motor

▸ See Figure 8

1. This simple procedure will quickly determine if the pump motor requires service.
2. Disconnect the pump motor wire harness from the system wire harness.
3. Put the outboard in the down position.
4. Using a fully charged 12 volt battery connect the Black electric motor lead to the negative battery terminal.
5. Connect the Blue motor lead to the positive battery terminal.
6. The motor should run and the trim cylinder should begin to extend.
7. Keeping the Black motor lead connected to the negative battery terminal.
8. Disconnect the Blue wire and connect Green motor lead to the positive battery terminal.
9. The trim cylinder should retract.
10. If the outboard hesitates in either direction the motor lead may have a loose connection or be grounded inside the electric motor.
11. If the trim motor fails to run during the two tests the problem is inside the electric motor.
12. Remove the two through bolts securing the upper cap to the electric motor.

Using an ohmmeter set on the Rx1000 scale make contact with the Red meter lead to the switch lead attached to the bimetallic switch and the Black meter lead to the brush lead between the bimetallic switch and the brush set.

The meter needle should register continuity. If the meter needle registers a high resistance reading or flickers back and forth clean the contacts and repeat the test.

13. Keep the leads in place and separate the contacts with a toothpick (or other non-conducting material). The meter should register no continuity. If the meter continues to register continuity the switch is shorted.

✳✳ WARNING

The bimetallic switch is a safety device and should be replaced when shorted and or damaged in any way. Making the switch work by bending or deliberately shorting it can cause component damage and or possible personal injury.

2 Wire Trim Motor

1. This simple procedure will quickly determine if the pump motor requires service.
2. Put the outboard in the down position.
3. Disconnect the pump motor wire harness from the system wire harness.
4. Connect the pump motor Blue wire to the battery positive terminal and the Green wire to the battery negative terminal.

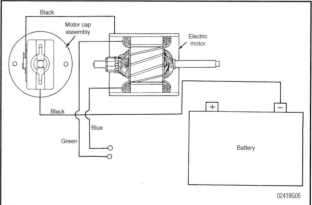

Fig. 8 A line drawing showing the wire connections for a 3 wire motor

TRIM AND TILT 9-7

5. The motor should operate in one direction.
6. Reverse the two wires connected to the battery and pump.
7. The pump should operate in the opposite direction.
8. If the motor fails to run in one, or both directions, disassemble the motor and check for defective components.
9. Check the armature for a short circuit by placing it on a growler and holding a hack saw blade over the armature core while the armature is rotated. If the saw blade vibrates, the armature is shorted. Clean between the armature bars, and then check again on the growler. If the saw blade still vibrates, the armature must be replaced. Occasionally carbon dust from the brushes will short the armature. Therefore, blow the slots in the armature clean with compressed air.
10. Make contact with one probe of an ohmmeter on the armature core or shaft. Make contact with the other probe on the commutator. If the meter fails to indicate continuity, the armature is grounded and must be replaced.
11. Check the commutator for run out. Inspect the armature shaft and both bearings for scoring.
12. Turn the commutator in a lathe if it is out-of-round by more than 0.005" (0.13mm).
13. Never undercut the mica because the brushes are harder than the insulation. Undercut the insulation between the commutator bars 1/32" (0.80mm) to the full width of the insulation and flat at the bottom. A triangular groove is not satisfactory. After the undercutting work is completed, clean out the slots carefully to remove dirt and copper dust. Sand the commutator lightly with No.600 sandpaper to remove any burrs left from the undercutting.
14. Make contact with the ohmmeter leads to any two commutator bars. The ohm meter should indicate continuity. Check the armature a second time on the growler for possible short circuits.

Brush Testing

3 WIRE TRIM MOTOR

1. Make contact with a meter lead to one of the negative grounded brushes. Make contact with the other meter lead to the positive terminal on the exterior of the motor frame. The meter should register no continuity. If the meter registers continuity, the terminal stud, or one or both of the positive brushes is grounded to the motor frame. Inspect the connections and repair or replace as necessary.
2. Check the field brush connections and lead insulation. A brush kit and a contact kit are available at your local marine dealer, but all other assemblies must be replaced rather than repaired.
3. The armature, fields, and brush holders must be checked before assembling the cranking motor.

2 WIRE TRIM MOTOR

1. Test the brushes by connecting one meter lead to one positive brush and the other meter lead to the second positive brush. The ohmmeter must indicate continuity between the brushes. If the meter indicates any resistance, check the lead to the brush and the lead to the positive terminal solder connection. If the connection cannot be repaired, the brushes must be replaced.
2. Connect a meter lead to one of the positive brushes, and make contact with the other meter lead to the end cap. The ohmmeter must register no continuity.
3. Repeat this test for the other positive brush. If the meter registers continuity in either test, the positive brush is shorted to the end cap. Inspect the connections and repair or replace as necessary.
4. Slide the positive brushes from the end cap. Make contact with one meter lead to the motor frame on an unpainted surface. Make contact with the other meter lead to the positive brush. The meter should register no continuity. If the meter registers continuity, the positive brush is shorted to the motor frame. Inspect the connections and repair or replace as necessary.
5. Make contact with one meter lead to one of the negative brushes. Make contact with the other meter lead to the end cap. The meter should indicate continuity. If the meter registers no continuity, the negative brush has an open in the circuit. Inspect the connections and repair or replace as necessary.
6. Make contact with a meter lead to one of the positive insulated brushes. Make contact with the other meter lead to the positive terminal on the exterior of the motor housing. The meter should register continuity. Repeat the test for the other positive brush. If the meter registers no continuity in either test, the positive brush has an open in the circuit. Inspect the connections and repair or replace as necessary.
7. Make contact with a meter lead to an unpainted portion of the motor frame. Make contact with the other meter lead to the positive terminal on the exterior of the motor housing. The meter should register no continuity. If the meter registers continuity, the terminal stud, or one or both of the positive brushes is grounded to the motor frame. Inspect the connections and repair or replace as necessary.

REMOVAL & INSTALLATION

1. Disconnect the wires connecting the trim motor to the harness or its solenoids.

※※ CAUTION

The hydraulic fluid in this system can be under extreme pressure. Safety glasses and proper protection should be worn before loosening the pressure lines in this system.

2. With a shop cloth around the fitting area loosen the four hydraulic lines with an appropriate sized wrench.
3. Remove the fasteners holding the pump assembly to the clamp bracket.
4. With the hydraulic lines removed lift the assembly off the clamp bracket.

To install:
5. Support the hydraulic pump assembly by the clamp bracket.

➡**Leaving the pump assembly loose allows more freedom to install the hydraulic lines leaving less of a possibility of them getting cross threaded.**

6. Thread the hydraulic lines in to there proper locations.
7. Bolt the pump assembly on to the clamp bracket.
8. Tighten the hydraulic lines.
9. Connect the pump motor wires on the proper wire connections or solenoid terminals.

DISASSEMBLY

1. Disconnect the trim/tilt motor wires.
2. Remove the two through bolts securing the upper cap to the electric motor.
3. Lift the entire motor assembly off of the pump housing.
4. Lift the upper cap off of the motor housing.
5. Slide the armature out of the motor housing

CLEANING & INSPECTION

The electric motor is very similar to other electric motors used on the outboard unit. Brush replacement, armature testing, and commutator repair may be made in the same manner. Therefore, the motor can be checked, tested, and serviced following the procedures outlined in
Inspect the brushes in the end cap for wear. If the brushes are worn down to less than 3/8in. inch replace the brushes.
Inspect the ends of the pump driveshaft for worn or damaged splines. If the shaft is worn or damaged replace the shaft.

ASSEMBLY

1. Slide the armature into the motor housing.
2. Support the armature and housing and in a padded vise.
3. Carefully place the upper cap on the motor housing while holding the brushes back to clear the commutator.
4. Place the entire motor assembly on the pump housing.
5. Align the notch in the armature shaft with the hydraulic pump gear.
6. Align the two through bolts and secure the upper cap and electric motor assembly..
7. Connect the trim/tilt motor wires.

9-8 TRIM AND TILT

Trim/Tilt Cylinder

TESTING

1. There are no pressure tests or pressure specifications for this style system. When a problem needs to be diagnosed a simple test to narrow down the failed component is performed. The repair is made through a process of elimination disassembling the most common failed part first.
2. Before testing the hydraulic system it must be properly bleed and the fluid level be at the proper height.
3. To test the down circuit of the hydraulic system
    a. Disconnect the lower hydraulic trim cylinder line from the hydraulic pump base.
    b. Cap the line and plug the opening in the pump to prevent loss of fluid and contaminants from entering the system.
    c. Move the control toggle switch to the up position.
    d. The tilt cylinder lifts the outboard on its own with the trim cylinder bypassed for this test.
    e. Remove the cap and plug from the line and pump.
    f. Connect the trim cylinder line.
4. If the outboard descends slowly the problem is in the tilt cylinder or the up circuit of the valve body. If there is no external leak the problem will be either the tilt cylinder or the hydraulic valve body.
5. Follow the procedures to remove and disassemble the tilt cylinder. If there are no visible defects with the cylinder assemble it and replace the hydraulic pump.
6. To test the up circuit of the hydraulic system
    a. With the outboard in the full down position disconnect the lower hydraulic tilt cylinder line from the hydraulic pump base.
    b. Cap the line and plug the opening in the pump to prevent loss of fluid and contaminants from entering the system.
    c. Move the control toggle switch to the up position.
    d. The trim cylinder lifts the outboard on its own with the tilt cylinder bypassed for this test.
    e. Remove the cap and plug from the line and pump.
    f. Connect the tilt cylinder line.
7. If the outboard descends slowly the problem is in the trim cylinder or the hydraulic valve body. If there is no external leak the problem will be either the trim cylinder or the hydraulic valve body.
8. Follow the procedures to remove and disassemble the trim cylinder. If there are no visible defects with the cylinder assemble it and replace the hydraulic pump.

REMOVAL & INSTALLATION

▶ See Figures 9, 10 and 11

1. With the outboard unit in the full down position, disconnect both cables at the battery terminals.
2. Disconnect the electric motor harness leads.
3. Obtain a suitable container and position it under the hydraulic pump base.
4. Loosen, but do not remove, each of the four hydraulic lines at the hydraulic pump.
5. Once the fluid stops draining from the fittings disconnect, cap or plug the four lines to prevent contamination from entering the system.
6. Manually raise and lock the outboard unit in the full up position.

※※ WARNING

The lower pivot shaft must be removed before the upper pivot shaft. If the upper shaft is removed first, the tilt cylinder or the shock absorber will fall forward and very possibly bend or distort the hydraulic lines.

7. Remove the lock nut on one end of the lower pivot shaft by holding each end of the shaft with the appropriate sized socket or wrench.
8. Remove the upper pivot pin.

※※ CAUTION

The pivot pin passing through the assembly and the cross brace is the only means of supporting the trim assembly. Therefore, use adequate measures to support the trim cylinder assembly, to prevent it from falling when the pivot pin is driven out.

9. Remove the washer and then drive the pivot shaft through the clamp bracket and cross brace using a long thin blunt drift punch.

To install:

10. Before installing the assembly be certain that it is cleaned and lubricated to ease installation.

➡ An alignment tool such as a brass pin or a broomstick with a slightly smaller diameter than the pivot rod makes the task of installing the upper and lower pivot shafts easier.

11. Hold the tilt cylinder in place and pass the alignment tool through the starboard clamp bracket into the tilt cylinder bushing.

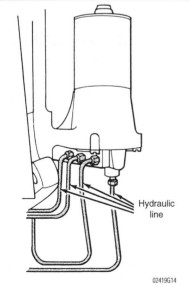

Fig. 9 Loosen the hydraulic lines to relieve the pressure in the system before disassembling the cylinders

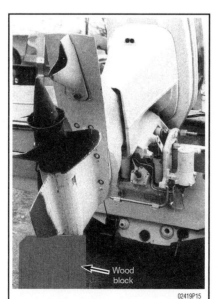

Fig. 10 Supporting the lower unit with a block of wood is a good safety tip when working on the trim and tilt unit

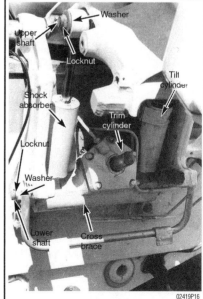

Fig. 11 Part location on the center mounted trim piston system

TRIM AND TILT

12. Hold the shock absorber in place and slide the alignment tool into the shock absorber bushing, and into the port clamp bracket.
13. Push the dowel through until the end of the dowel is flush with the outer edge of the starboard clamp bracket.

with a soft head mallet Use the upper pivot shaft to gently push the alignment tool out of the bushing area

14. Stop tapping on the shaft when the amount of shaft extending beyond the starboard and port clamp brackets is equal.
15. Install a cupped washer onto each end of the pivot shaft, with the cupped side facing inward.

Install and tighten a lock nut onto each end of the pivot shaft, with approximately the same number of threads showing on the shaft at each end.

16. Hold the cross brace in place and use the alignment tool to align parts for installation of the lower pivot shaft.
17. Tighten the lock nuts in the same manner used to tighten the lock nuts on the upper pivot shaft.
18. Remove the caps from the hydraulic lines and connect the four lines to the hydraulic pump.
19. Connect the electric motor harness leads.

DISASSEMBLY

▶ See Figures 12 and 13

Tilt Cylinder

1. Disconnect the upper and lower hydraulic lines from the tilt cylinder.
2. Clamp the base of the tilt cylinder in a vice equipped with soft jaws.

➡A spanner wrench is required to remove the tilt cylinder end cap. Removing the end cap can be a very difficult task, especially if the unit has been used in contaminated or salt water. If the end cap can not be removed or is severely damaged during removal the cylinder should be replaced.

3. Using a spanner wrench inserted into the two opposing holes of the end cap, loosen the end cap.
4. Pull the tilt piston out of the cylinder.
5. Remove the O-ring around the threaded end cap and the O-ring in the piston groove.

➡Save the O-rings for comparison when purchasing new rings.

6. Further disassembly of the tilt cylinder is not possible.

Trim Cylinder

1. Clamp the cross brace in a vice equipped with soft jaws.
2. Disconnect the hydraulic lines from the trim cylinder.
3. Use a 12-point socket and remove the five cap screws from the trim cylinder cover.

✲✲ WARNING

Heat, from a light source, but not a flame source, may have to be applied to free the cap screws. An open flame could damage the seals or warp the housings.

➡If the cap screws cannot be removed, the cylinder cannot be serviced. The only solution is to replace the trim cylinder.

4. Lift the cover off the cylinder
5. Pull the piston free of the cylinder.
6. Remove the O-ring from around the piston.
7. Remove the two cover O-rings.
8. Remove the scraper oil seal.
9. Slide the liner out of the trim cylinder.
10. Using a tool with a small hook remove the O-ring from the bottom of the cylinder.

➡Save all of the O rings and seal to match them with there replacements. The trim and tilt systems have been superseded a number of times and the seals and O rings could be wrong if not properly matched.

Having the old ring for comparison when purchasing a new O-ring will help ensure a proper fit.

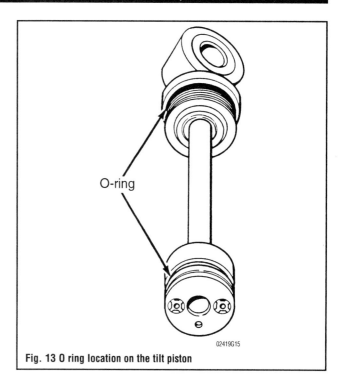
Fig. 13 O ring location on the tilt piston

Fig. 12 Once the tilt cylinder cover is loosened the inner shaft lifts out for seal inspection or replacement

9-10 TRIM AND TILT

CLEANING & INSPECTION

Keep the work area as clean as possible to prevent contamination by foreign material entering the system on parts to be installed.

Clean all parts with solvent and then blow them dry with low pressure compressed air.

Carefully inspect the tilt piston for any sign of damage.

Purchase, if available, new O-rings and discard the old items, but only after the replacement O-ring is verified as correct for the intended installation.

Inspect the bushing at the top of the shaft. Remove and replace the bushing, if the bushing shows signs of excessive wear.

Use an awl and push down the check valve balls in the piston. Two are located at the bottom end and one at the top end. Check the return action of each check ball. If a ball fails to seat, push down on the ball and at the same time flush with solvent. Dry in and around the ball with compressed air, while pushing down on the ball. Repeat this procedure for each ball.

ASSEMBLY

Tilt Cylinder

♦ See Figure 14

➡ Make every effort to keep the parts as clean as possible during the installation work to prevent contaminating the system. Good shop practice dictates new O-rings be installed anytime the unit is disassembled and the rings are exposed.

1. Coat each O-ring with nondetergent 30-weight automotive oil prior to installation.
2. Install two new O-rings, one around the piston groove and the other around the threaded end cap.
3. Push the tilt piston into the cylinder.
4. Thread the end cap onto the cylinder.
5. Using a spanner wrench inserted into the two opposing holes of the end cap tighten the cap securely.

➡ Take extra care not to elongate the spanner wrench holes.

6. Connect the upper and lower hydraulic lines.
7. Tighten the fittings securely using the flare or brake line wrenches.

Trim Cylinder

➡ Make every effort to keep the parts as clean as possible during the installation work to prevent contaminating the system. Good shop practice dictates new O-rings be installed anytime the unit is disassembled and the rings are exposed.

1. Coat each O-ring with nondetergent 30-weight automotive oil prior to installation.
2. Install the O-ring into the bottom of the trim cylinder.
3. Slide the trim cylinder liner into the cylinder with the light colored band toward the upper end of the cylinder.
4. Install a new seal with the seal lip facing downward from the top side of the cover.

Pack the seal lip with water resistant multipurpose lubricant.

5. Install the two new O-rings into the trim cylinder cover.
6. Install a new O-ring around the trim piston.
7. Slide the piston into the cylinder liner.

Install the trim cylinder cover over the piston with the opening for the hydraulic line at the 11 o'clock position.

8. Secure the cover in place with the five cap screws.

Tighten the screws using a 12-point socket to a torque value of 95 in lbs (10.7Nm).

Connect the upper and lower hydraulic lines.

Tighten the fittings securely using the flare or brake line wrenches.

Trim/Tilt Switch

TESTING

♦ See Figures 15 and 16

1. Using an ohmmeter set on the Rx1000 scale connect the Red meter lead to the Red lead from the toggle switch.
2. Connect the Black meter lead to the Blue lead from the toggle switch.
3. Depress the toggle switch to the up position.
4. Depress the toggle switch to the down position.
5. Next connect the Black meter lead to the Green lead from the toggle switch.
6. Depress the toggle switch to the down position.
7. Depress the toggle switch to the up position.

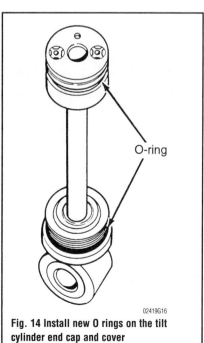

Fig. 14 Install new O rings on the tilt cylinder end cap and cover

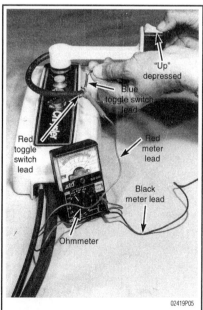

Fig. 15 Test for continuity shown with switch pressed in the up direction

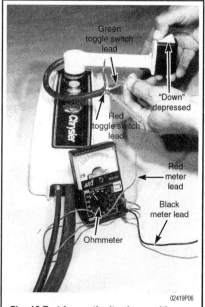

Fig. 16 Test for continuity shown with switch pressed in the down direction

TRIM AND TILT

8. If the meter registers continuity in one direction only the switch passes the continuity test. In the event that switch has continuity in both they up and down directions or no continuity at all the switch is bad and need to be replaced.

Trim/Tilt Relay

TESTING

Early Models

RESISTANCE TEST

1. Connect one test lead of a multimeter to each of the large solenoid terminals.
2. Connect the positive (+) lead from a fully charged 12-volt battery to the small solenoid terminal marked S.
3. Momentarily make contact with the ground lead from the battery to the small solenoid terminal marked "I".
4. If a loud click sound is heard and the multimeter indicates continuity, the solenoid is in serviceable condition.
5. If, however a click sound is not heard and/or the multimeter does not indicate continuity, the solenoid is defective and must be replaced and only with a marine solenoid.

VOLTAGE TEST

1. Connect the voltmeter between the common powerhead ground and the No. 1 point.
2. Activate the trim switch in the up or down depending upon which solenoid is being tested position.
3. Observe the voltmeter. If there is no reading, the trim solenoid is defective and must be replaced. If a reading is indicated and a click sound is heard, the solenoid is functioning properly.

Late Models

1. The single cylinder trim/tilt system uses a set of relays for the pump motor operation direction and control. Previous trim/tilt systems have used a set of solenoids to perform the same function. The relays are compact, sealed units and are easily changed with their plug-in feature.
2. If the trim system operates in one direction, but fails in the opposite direction, a quick and easy test can be performed to determine if one of the relays has failed and which one has failed.
3. Mark both relays, the top relay is the Up relay and the lower one is the Down relay.
4. Unplug both relays from the plug-in connectors. Plug each relay back into the opposite connector from which it was removed.
5. Depress the trim/tilt switch in the inoperative direction. If the trim/tilt system now operates, the relay is good.
6. Press the trim/tilt switch in the opposite direction. If the trim/tilt system fails to operate in the previously known good direction, the failed relay must be replaced.

REMOVAL & INSTALLATION

Early Models

Early models are usually equipped with a standard type solenoid in the trim/tilt system. This solenoid can be serviced by unbolting it from the lower cowl. Make sure to label the wires, marking them for proper installation.

Late Models

On late model units, service is as simple as removing the fasteners holding the solenoid in place and unplug the electrical the connector.

SINGLE RAM INTEGRAL POWER TRIM/TILT SYSTEM

Description and Operation

The single cylinder power tilt/trim system consists of an electric motor, pump, pressurized fluid reservoir, one large trim/tilt cylinder and electrical components.

The single cylinder extends very slowly through the first 20° of movement, this is considered the trim range, and then accelerates to move the outboard to the desired tilt position for trailering or shallow water operation.

The hydraulically powered single cylinder tilt/trim system permits changing the tilt angle of the outboard unit from the helm position. Controls and indicators for the system are located on or about the remote control handle. A set of switches is also installed on the side of the powerhead on the lower cowling.

Trim/Tilt Pump

REMOVAL & INSTALLATION

♦ See Figures 17, 18, 19 and 20

➡As a safety measure to prevent accidental movement of the outboard while work is being performed, it is strongly recommended a few minutes be used to make a safety support tool as shown in the accompanying illustration.

1. The tool may be made from any metal bar stock or small channel iron of suitable size, with a ⅜in. (9.53 mm) hole drilled through at each end and 14 in. (35.6 cm) apart, as shown. Cut-off the head of a ⅜in. bolt about 2-½in. (6.4 cm) long. Drill a hole through each bolt for a cotter pin. Secure the bolts through the holes made in the bar stock with two nuts, one on each side of the bar. The tool is now ready for installation, one end through the clamp bracket and the other end through the tilt stop bracket. Secure each end of the tool in place with a washer and cotter pin. The trim/tilt system may now be serviced or other work performed with confidence and in safety.
2. Raise the outboard unit to the full up position. If the hydraulic system is inoperative, first rotate the manual release valve counterclockwise about three complete turns, and then manually lift the unit to the full up position.
3. Set the tilt lock lever in place. Install the safety support tool as described at the beginning of this section. Secure one end of the tool to the clamp bracket with a washer and cotter pin. Secure the other end to the tilt stop bracket in a similar manner, as shown.
4. Disconnect the trim/tilt motor electrical wires (blue and Green) from the powerhead harness, at the plug connector under the powerhead cowling.

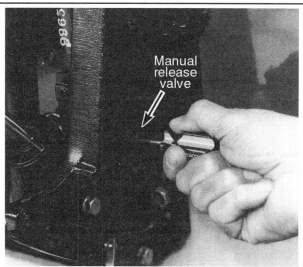

Fig. 17 The engine can be lifted if the trim system has failed by rotating the manual release valve counterclockwise about three complete turns and lifting the unit to the full up position

9-12 TRIM AND TILT

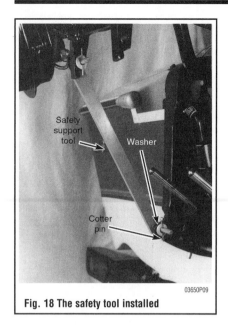

Fig. 18 The safety tool installed

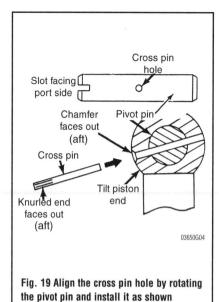

Fig. 19 Align the cross pin hole by rotating the pivot pin and install it as shown

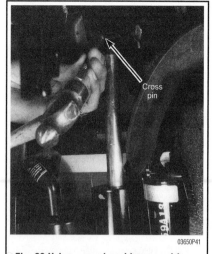

Fig. 20 Using a punch and hammer drive the cross pin into place flush with tilt piston end

5. Remove the harness retainer and any other clamps securing the wire harness run to the trim/tilt motor.
6. If the cross pin has a large head, use a suitable punch and remove the cross pin from the upper pivot pin.
7. If the cross pin is headless, use a large blunt nose punch and drive the upper pivot pin out from the clamp bracket and cylinder ram, this will shear-off the cross pin.
8. Use a suitable punch and remove the cross pin from the lower pivot pin. With a large blunt nose punch, drive the lower pivot pin out from the lower clamp brackets and end of the cylinder.
9. Swing the trim/tilt assembly upper end first clear of the transom, and then lift the unit from the clamp brackets.

To install:
10. Raise the outboard and install the safety support tool bar if not previously installed.
11. Position the trim/tilt unit between the clamp brackets, bottom end in first. Slide the pivot pin through the starboard clamp bracket, and then through the lower trim cylinder pivot. Using a suitable punch and hammer, drive the pivot pin through until it is flush with the port side transom bracket. Install a new cross pin through the pivot pin.
12. The pivot pin can be rotated with a large screwdriver until the hole in the pin is aligned with the hole in the end of the tilt piston. The chamfered hole in the end of the piston must face outward. If it does not, remove the pivot pin and rotate the piston in the cylinder ½ turn.
13. Insert the pump motor wiring harness up through the starboard transom bracket and into the powerhead cowling area.
14. Rotate the trim/tilt unit until the upper cylinder piston aligns with the outboard upper pivot point. Align the hole in the piston rod with the pivot, and then install the pivot pin. Using a suitable punch and hammer, drive the pivot pin through until it is flush at both ends. Use a flat blade screwdriver and align the pivot pin with the cylinder piston cross pin holes.
15. nsert the cross pin through the pivot pin and cylinder piston rod. Using a punch and hammer, drive the cross pin flush.
16. Route the trim/tilt harness to the starboard side of the powerhead cowling area. Insert the Green and Blue wire connectors into the same color coded connectors from the powerhead electrical harness. Secure the harness and wires with tie straps.
17. Close and secure the powerhead cowlings; remove the safety support tool; and connect the battery. Perform the Hydraulic Bleeding procedures outlined in this chapter.
18. Check to be sure the battery has a full charge, and then move the outboard unit through several cycles full down to full up for trailering. Observe the system for satisfactory operation electrical and hydraulic. Check the system for possible leaks.
19. If any problems are encountered, check the troubleshooting portion of this section to isolate the problem. If the service work has been performed with care according the instructions in this chapter, performance should be satisfactory.

HYDRAULIC SYSTEM BLEEDING

This trim/tilt system with one large cylinder has been designed with a "self bleeding" system. Fluid is pumped from one side of the cylinder to the opposite side as the piston extends or retracts. Therefore, any air trapped on one end of the cylinder, is pushed out back to the reservoir, when the piston bottoms out.

Usually two or three cycles of the system will remove all entrapped air from the cylinder. The following short sections outline the necessary steps to perform during the cycling sequence just mentioned. Two sets of instructions are included. One with the system installed on the outboard, and the other with the system on the workbench.

System Installed

1. Check to be sure the manual release valve is fully closed prior to activating the pump motor.
2. Press and hold the up trim switch until the powerhead is in the full up position. Release the up switch and engage the lever to hold the outboard in the upright position.
3. Slowly remove the fill plug from the reservoir. Fluid level should be to the edge of the fill plug opening. If necessary add Quicksilver Power Trim and Steering Fluid. If the Quicksilver product is not available, Automatic Transmission Fluid (ATF) Type F, FA or Dexron, may be used. Tighten the fill plug securely.
4. Press and hold the down trim switch until the powerhead is in the full down position. Release the down switch.
5. Press and hold the up trim switch until the powerhead is in the full up position. Release the up switch. Continue to cycle the system in this manner two or three times, and then check the fluid level in the reservoir. Add fluid if necessary, and tighten the fill plug.

System Removed

♦ See Figure 21

1. Check to be sure the manual release valve is fully closed prior to activating the pump motor.
2. Connect a 12-volt power source Red lead to the pump motor Green lead. Connect the power source negative Black lead to the pump Blue lead. Hold the connections until the piston has fully extended.
3. Slowly remove the fill plug from the reservoir. The fluid level should be to the edge of the fill plug opening. If necessary, add Quicksilver Power Trim and Steering Fluid. If the Quicksilver product is not available, Automatic Transmission Fluid (ATF) Type F, or Dexron, may be used. Tighten the fill plug securely.

TRIM AND TILT 9-13

shaft. Make contact with the other probe on the commutator. If the meter fails to indicate continuity, the armature is grounded and must be replaced.

6. Check the commutator for run out. Inspect the armature shaft and both bearings for scoring.

7. Turn the commutator in a lathe if it is out-of-round by more than 0.005" (0.13mm).

8. Never undercut the mica because the brushes are harder than the insulation. Undercut the insulation between the commutator bars $\frac{1}{32}$" (0.80mm) to the full width of the insulation and flat at the bottom. A triangular groove is not satisfactory. After the undercutting work is completed, clean out the slots carefully to remove dirt and copper dust. Sand the commutator lightly with No.600 sandpaper to remove any burrs left from the undercutting.

9. Make contact with the ohmmeter leads to any two commutator bars. The ohm meter should indicate continuity. Check the armature a second time on the growler for possible short circuits.

10. Test the brushes by connecting one meter lead to one positive brush and the other meter lead to the second positive brush. The ohmmeter must indicate continuity between the brushes. If the meter indicates any resistance, check the lead to the brush and the lead to the positive terminal solder connection. If the connection cannot be repaired, the brushes must be replaced.

11. Connect a meter lead to one of the positive brushes, and make contact with the other meter lead to the end cap. The ohmmeter must register no continuity.

12. Repeat this test for the other positive brush. If the meter registers continuity in either test, the positive brush is shorted to the end cap. Inspect the connections and repair or replace as necessary.

13. Slide the positive brushes from the end cap. Make contact with one meter lead to the motor frame on an unpainted surface. Make contact with the other meter lead to the positive brush. The meter should register no continuity. If the meter registers continuity, the positive brush is shorted to the motor frame. Inspect the connections and repair or replace as necessary.

14. Make contact with one meter lead to one of the negative brushes. Make contact with the other meter lead to the end cap. The meter should indicate continuity. If the meter registers no continuity, the negative brush has an open in the circuit. Inspect the connections and repair or replace as necessary.

15. Make contact with a meter lead to one of the positive insulated brushes. Make contact with the other meter lead to the positive terminal on the exterior of the motor housing. The meter should register continuity. Repeat the test for the other positive brush. If the meter registers no continuity in either test, the positive brush has an open in the circuit. Inspect the connections and repair or replace as necessary.

16. Make contact with a meter lead to an unpainted portion of the motor frame. Make contact with the other meter lead to the positive terminal on the exterior of the motor housing. The meter should register no continuity. If the meter registers continuity, the terminal stud, or one or both of the positive brushes is grounded to the motor frame. Inspect the connections and repair or replace as necessary.

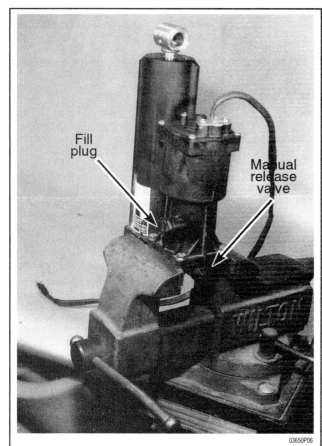

Fig. 21 With the trim/tilt unit clamped in a vise, the manual release valve may be opened and closed during the "bleeding" process, as described in the text

4. Reverse the 12-volt power source connections, Red lead to the pump Blue lead and Black lead to the pump Green lead. Hold the lead connections until the piston is fully retracted.

5. Continue to cycle the system in this manner two or three times, and then check the fluid level in the reservoir. Add fluid if necessary, and tighten the fill plug securely.

6. Disconnect the 12-volt power source leads from the pump motor.

Tilt/Trim Motor

TESTING

This simple procedure will quickly determine if the pump motor requires service.

1. Disconnect the pump motor wire harness from the system wire harness. Connect the pump motor Blue wire, to the battery positive terminal and the Green wire to the battery negative terminal. The motor should operate in one direction.

2. Reverse the two wires connected to the battery and pump. The pump should operate in the opposite direction.

3. If the motor fails to run in one, or both directions, disassemble the motor and check for defective components.

4. Check the armature for a short circuit by placing it on a growler and holding a hack saw blade over the armature core while the armature is rotated. If the saw blade vibrates, the armature is shorted. Clean between the armature bars, and then check again on the growler. If the saw blade still vibrates, the armature must be replaced. Occasionally carbon dust from the brushes will short the armature. Therefore, blow the slots in the armature clean with compressed air.

5. Make contact with one probe of an ohmmeter on the armature core or

REMOVAL & INSTALLATION

♦ See Figures 22, 23, 24, 25 and 26

1. Loosen and remove the four Phillips head screws from the pump motor end cap.

2. Using two flat blade screwdrivers, insert the blade of the screwdrivers into the slotted space between the end cap on opposite sides of the motor housing.

3. Push down on the screwdrivers and slowly separate the end cap, from the motor housing. Do not insert the screwdriver between the end cap and motor housing except at the slotted space. Damage to the end cap and/or motor housing will result and loosen the waterproof seal to the motor.

4. After the motor end cap has been separated from the motor housing, lift the end cap and wiring harness straight up and free of the motor armature.

5. Reach inside the motor housing and grasp the armature shaft. Due to the magnetic field of the motor frame, both the armature and frame will slide out of the motor housing as one unit. Lift up on the armature shaft, removing both the armature and motor frame from the motor housing. Take care not to lose the flat washer used on each end of the armature shaft. The washer on the lower end of the armature shaft will usually be in the bottom of the motor housing. Retrieve the washer from the motor housing and place it back on the lower armature shaft.

TRIM AND TILT

Fig. 22 Loosen and remove the four Phillips head screws from the pump motor end cap

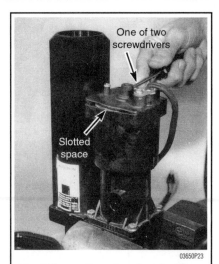

Fig. 23 Loosen the end cap by using two flat blade screwdrivers inserted in the slotted spaces between the end cap and the motor housing

Fig. 24 Lift the end cap and wiring harness straight up and free of the motor armature

Fig. 25 Reach inside the motor housing and grasp the armature shaft and lift it out of the reservoir housing

Fig. 26 Install a new O-ring onto the end cap and lubricate it with trim and tilt fluid

CLEANING & INSPECTION

♦ See Figure 27

Inspect the brushes in the end cap for wear. If the brushes are worn down to less than 3/8 in. inch, replace the brushes.

Inspect the ends of the pump driveshaft for worn or damaged splines. If the shaft is worn or damaged, replace the shaft.

Fig. 27 Measure the length of the brushes. If less than 3/8 in. remains, the brushes should be replaced

To install:

6. Insert the armature into the motor frame. Slide one flat washer onto each end of the armature shaft.

7. Insert the pump driveshaft into the lower end of the armature shaft. Lower the motor assembly, with the pump driveshaft going in first, down into the motor housing.

8. Install a new O-ring onto the end cap and lubricate the O-ring with fluid. Slide the brushes into the brush holders. Set the spring tensioner ends outside the holders to prevent the brushes from having any spring force applied.

9. Apply a small amount of marine lubricant, or equivalent, to the bushing in the end cap.

10. Lower the end cap down over the armature shaft. As soon as the brushes make contact with the commutator, release the brush spring tensioner against the brushes. The spring tensioner will now keep the brushes against the commutator.

11. Secure the end cap onto the housing with the four Phillips head screws.

TRIM AND TILT 9-15

Tilt/Trim Cylinder

REMOVAL & INSTALLATION

▶ See Figures 28 thru 39

1. Clamp the trim/tilt unit in a vise. Take care not to damage the manifold assembly.
2. Obtain a suitable container and be prepared to catch fluid escaping from the unit. Remove the manual release valve from the manifold assembly and the reservoir fill plug from the reservoir.
3. Extend the piston to the full up position by pulling up on the end of the piston. This may not be an easy task, a fair amount of muscle power and some rotation of the piston will probably be necessary. After the piston is fully extended, use a spanner wrench with 1/4in. diameter X 5/16in. inch long lugs, loosen and unscrew the cylinder end cap from the cylinder.
4. Slide the trim rod assembly upward until the piston is free of the cylinder.
5. Remove the fasteners securing the plate to the trim rod piston.
6. Lift off the plate, five check valve springs, seats and balls from inside the piston. Do not remove the check valve on the trim rod piston, retained with a roll pin.
7. Using a pair of snap ring pliers reach inside the cylinder and grasp the memory piston and pull it up and free of the cylinder.

➡ If the hydraulic system was not completely drained of all fluid, perform the next step over a drain pan to catch the hydraulic fluid behind the memory piston.

8. An alternate method to remove the piston is to first remove the trim/tilt unit from the vise. Turn the unit over and make a couple of quick solid "blows" against a piece of plywood or other hard surface. The memory piston will slide down and out of the cylinder.

To install:

9. Lubricate two new O-rings with fluid, and then install them into the recesses of the cylinder.
10. Install two new O-rings onto the pilot valve. Slide the pilot valve and spring into the manifold.
11. Align the pump and manifold assembly with the cylinder. Be sure the two O-rings and pilot valve are aligned.

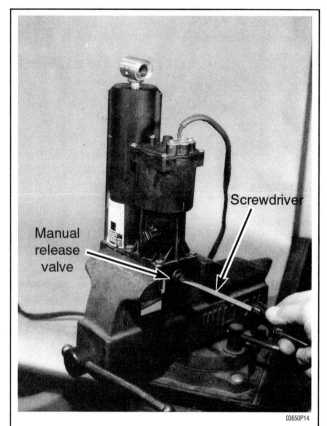

Fig. 29 To drain the unit remove the reservoir fill plug from the reservoir and the manual release valve from the manifold assembly

Fig. 28 Carefully clamp the trim/tilt unit in a vise as not to damage the plastic reservoir housing

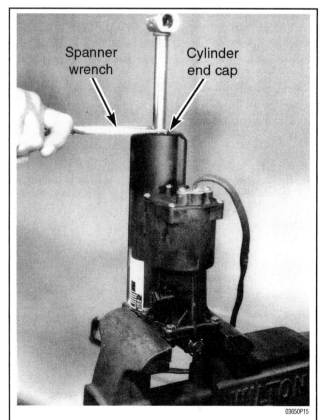

Fig. 30 Loosen and unscrew the cylinder end cap from the cylinder

9-16 TRIM AND TILT

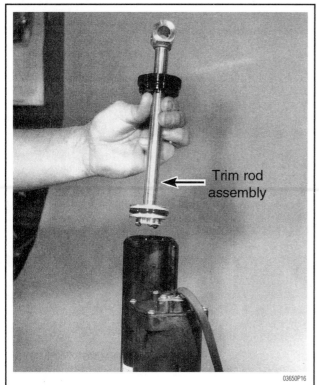

Fig. 31 Slide the trim rod assembly upward until the piston is free of the cylinder

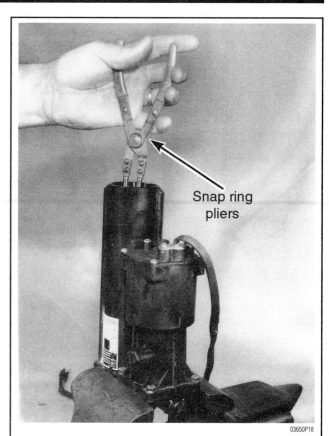

Fig. 33 Using a pair of snap ring pliers reach inside the cylinder and grasp the memory piston and pull it up and free of the cylinder

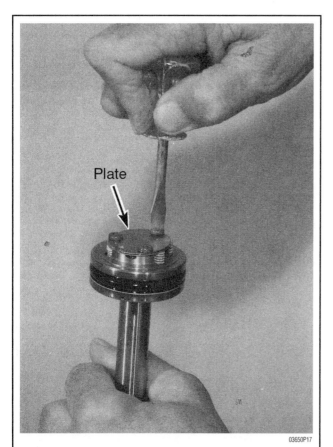

Fig. 32 Remove the fasteners securing the plate to the trim rod piston

Fig. 34 Turn the unit over and make a couple of quick solid blows against a piece of plywood to remove the memory piston if it can not be removed using the snapring pliers

TRIM AND TILT 9-17

Fig.35 Lubricate two new O-rings with fluid and install them into the recesses of the cylinder

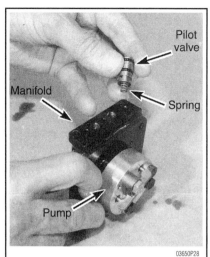

Fig. 36 Install two new O-rings onto the pilot valve and slide the pilot valve and spring into the manifold

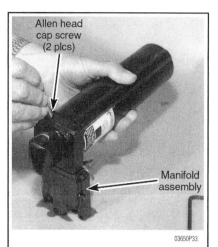

Fig. 37 Align the pump and manifold assembly with the cylinder making certain the two O-rings and pilot valve are in place

Fig. 38 Install a new O-ring into the groove in the bottom of the reservoir

Fig. 39 Install the reservoir over the pump and flush against the manifold with the fill plug hole facing aft

12. Install the two Allen head cap screws and tighten to a torque value of 100 inch lbs. (11Nm).
13. Install a new O-ring into the groove in the bottom of the reservoir. Lubricate the O-ring with fluid.
14. Slide the reservoir, with the fill plug hole facing aft, down over the pump and flush against the manifold.
15. Install four flat washers and bolts securing reservoir to manifold. Tighten the four bolts to a torque value of 70 inch lbs. (7.7Nm).

CLEANING & INSPECTION

♦ See Figure 40

➡ Never use an O-ring a second time. Always replace used O-rings with new.

During the assembling work, lubricate O-rings with either Quicksilver Power Trim and Steering Fluid or automatic transmission fluid.

Inspect the check valve on the back side of the trim rod piston. Clean the check valve with a parts cleaning solvent but do not remove the valve from the piston. If the valve is suspected to have failed, replace the trim rod piston.

Inspect the trim rod cap scraper seal. If the seal does not keep the rod clean, remove the trim rod piston from the end of the trim rod. Slide off the cap and replace the seal inside the cap.

Install a new O-ring on the trim rod cylinder piston. Lubricate the O-ring with fluid.

Examine the interior of the cylinder for scoring, roughness, or corrosion.

Inspect the edges of the memory piston and trim rod piston for scoring, roughness and corrosion.

Clean all parts with solvent and dry them with compressed air.

9-18 TRIM AND TILT

Fig. 40 Typical trim/tilt system completely cleaned and serviced with new O-rings ready to be installed between the clamp brackets

Tilt/Trim Switch

TESTING

1. Obtain a digital multimeter and set it for continuity reading. Open or remove the powerhead cowling and check the 20-Amp fuse for continuity. If the fuse is "blown", examine the wiring for breaks or damage which caused a direct short and the fuse to "blow". Repair the damaged wiring. If the fuse is satisfactory, proceed to the next step.
2. Set the digital multimeter for a 12-volt meter reading. Connect the Red meter lead to the starter solenoid positive (battery), terminal and the Black meter lead to a good powerhead ground. If there is no battery voltage indication, check the battery leads for proper connection and the battery charge. If the meter indicates battery voltage, proceed to the next step.
3. Connect the Red meter lead to the Blue/White wire from the remote control wiring harness. Connect the Black wire to a good powerhead ground. Depress the up trim button and check for battery voltage indication. If the meter indicates battery voltage, check the Black ground wires on the Up/Down trim relays for proper connection and continuity. If the continuity check of the relay ground wires is satisfactory, the pump motor is possibly defective. Perform the pump motor test covered in this chapter. If there is no battery voltage indication, proceed to the next step.
4. Connect the Red meter lead to the trim/tilt switch Red wire connection inside the remote control handle. Connect the Black meter lead to good powerhead ground. If the meter indicates battery voltage, the trim switch has failed or both wires Blue/White and Green/White from the remote control to the Up/Down relays has been damaged or severed. Inspect the wiring harness and connectors between the remote control and the powerhead. If there is no battery voltage indication, proceed to the next step.
5. Connect the Red meter lead to any instrument positive terminal and the Black meter lead to a good powerhead ground. Set the ignition switch to the on or run position. If battery voltage is indicated, the Red wire in the remote control handle between the ignition switch and the trim/tilt switch is open.
6. Repair the defective wiring. If there is no battery voltage indication, the Red wire between the cranking motor solenoid positive terminal and the ignition switch is open.
7. Repair the defective wiring.

Tilt/Trim Relay

TESTING

The single cylinder trim/tilt system uses a set of relays for the pump motor operation direction and control. Previous trim/tilt systems have used a set of solenoids to perform the same function. The relays are compact, sealed units and are easily changed with their plug-in feature.

If the trim system operates in one direction, but fails in the opposite direction, a quick and easy test can be performed to determine if one of the relays has failed and which one has failed.

1. Mark both relays, the top relay is the Up relay and the lower one is the Down relay.
2. Unplug both relays from the plug-in connectors. Plug each relay back into the opposite connector from which it was removed.
3. Depress the trim/tilt switch in the inoperative direction. If the trim/tilt system now operates, the relay is good.
4. Press the trim/tilt switch in the opposite direction. If the trim/tilt system fails to operate in the previously known good direction, the failed relay must be replaced.

Down Trim Inoperative; Up Trim Operates

1. Obtain a digital multimeter and set the meter for 12-volt DC reading. Connect the meter Red lead to the Green/White wire from the remote control wiring harness. Connect the Black wire to a good powerhead ground.
2. Depress the Down trim button and observe the battery voltage indication. If the meter indicates battery voltage, then the down relay is defective and must be replaced. If there is no indication, proceed to the next step.
3. Connect the Red meter lead to the Green/White wire (trim/tilt switches) inside the remote control handle. Connect the Black meter lead to good powerhead ground.
4. Depress the down trim button and check the meter for battery voltage indication. If the meter indicates battery voltage, the wiring between the remote control and the powerhead trim/tilt down relay is defective and must be repaired or replaced. If there was no voltage indication, proceed to the next step.
5. Connect the Red meter lead to the trim/tilt switch Red wire connection inside the remote control handle. Connect the Black meter lead to good powerhead ground.
6. Depress the down trim button and observe the meter for battery voltage indication. If the meter indicates battery voltage, the trim/tilt switch in the remote control handle is defective and must be replaced. If there is no battery voltage indication, check for voltage at the 20-Amp fuse and/or the wiring to the trim system from the main powerhead harness.

Up Trim Inoperative; Down Trim Operates

1. Obtain a digital multimeter and set the meter for 12-volt DC reading. Connect the Red meter lead to the Blue/White wire from the remote control wiring harness. Connect the Black wire to a good powerhead ground.
2. Depress the up trim button and observe for voltage indication. If the meter indicates battery voltage, the up relay is defective and must be replaced. If there was no voltage indication, proceed to the next step.
3. Connect the Red meter lead to the Blue/White wire (trim/tilt switches) connection inside the remote control handle. Connect the Black meter lead to good powerhead ground.
4. Depress the up trim button and check the meter for battery voltage indication. If the meter indicates battery voltage, the wiring between the remote control and the powerhead trim/tilt up relay is defective and must be repaired or replaced. If there was no voltage indication, proceed to the next step.
5. Connect the Red meter lead to the trim/tilt switch Red wire connection inside the remote control handle. Connect the Black meter lead to a good powerhead ground.
6. Depress the up trim button and check the meter for battery voltage indication. If the meter indicates battery voltage, the trim/tilt switch in the remote control handle is defective and must be replaced. If there is no battery voltage indication, check for voltage at the 20-Amp fuse and/or wiring to the trim/tilt system from the main powerhead harness.

REMOTE CONTROL BOX 10-2
DESCRIPTION AND OPERATION 10-2
TROUBLESHOOTING THE REMOTE
  CONTROLS 10-2
REMOTE CONTROL BOX 10-2
    REMOVAL & INSTALLATION 10-2
    DISASSEMBLY 10-2
    CLEANING AND INSPECTION 10-4
    ASSEMBLY 10-4
REMOTE CONTROL CABLES 10-5
    REMOVAL & INSTALLATION 10-5
    ADJUSTMENT 10-5
NEUTRAL START SWITCH 10-5
    TESTING 10-5
    REMOVAL & INSTALLATION 10-5
EMERGENCY STOP SWITCH
  (LANYARD) 10-6
    TESTING 10-6
    REMOVAL & INSTALLATION 10-6
IGNITION SWITCH 10-6
    TESTING 10-6
    REMOVAL & INSTALLATION 10-6
WARNING BUZZER 10-7
    TESTING 10-7
    REMOVAL & INSTALLATION 10-7
TILLER HANDLE 10-11
DESCRIPTION AND OPERATION 10-11
TROUBLESHOOTING THE TILLER
  HANDLE 10-11
TILLER HANDLE 10-11
    REMOVAL & INSTALLATION 10-11
ENGINE STOP SWITCH 10-13
    TESTING 10-13
    REMOVAL & INSTALLATION 10-14
EMERGENCY STOP SWITCH
  (LANYARD) 10-14
    TESTING 10-14
    REMOVAL & INSTALLATION 10-14

10
REMOTE CONTROLS

REMOTE CONTROL BOX 10-2
TILLER HANDLE 10-11

10-2 REMOTE CONTROLS

REMOTE CONTROL BOX

Description and Operation

The remote control box allows the person operating the boat to control the throttle operation and shift movements from a location other than where the outboard is mounted. In most cases, the remote control box is generally mounted at the helm station on the starboard side of the boat.

Several control box types are used, but all usually house a key switch, engine stop switch, choke switch, neutral safety switch, warning buzzer and the necessary wiring and cable hardware to connect the control box to the outboard unit.

→There are many different types of remote control assemblies that can be installed on your boat. The remote control assemblies covered in this manual are the ones most commonly used by the manufacturer.

Troubleshooting the Remote Controls

One of the things taken for granted on most boats is the engine controls and cables. Depending on how they are originally routed, they will either last the life of the vessel or can be easily damaged. These cables should be routinely maintained by careful inspection for kinks or other damage and lubrication with marine grade grease.

If the cables do not operate properly, have a helper operate the controls at the helm while you observe the cable and linkage operation at the powerhead. Make sure nothing is binding, bent or kinked. Check the hardware that secures the cables to the boat and powerhead to make sure they are tight. Inspect the clevis and cotter pins in the ends of the cables and also give some attention to the cable release hardware.

Another area of concern is the neutral start switch that is sometimes located in the control box and sometimes on the powerhead. This switch may fall out of adjustment and prevent the engine from being started. It is easily inspected using a multimeter by performing a continuity check.

Remote Control Box

REMOVAL & INSTALLATION

♦ See Figure 1

1. Disconnect both battery cables at their terminals.
2. Disconnect the main harness lead from the powerhead to the remote control box at the plug connector, if equipped.
3. Disconnect the trim/tilt harness at the connector plug, if equipped.
4. Remove the three screws securing the control box to the boat and remove any cable retainers securing the cables to the boat hull.
5. Move the control box to a convenient work surface.

To install:
6. Install the control box.
7. Install the three screws with flat washers and nuts. Tighten securely.
8. Connect the trim/tilt harness at the connector plug, if equipped.
9. Connect the main harness lead from the powerhead to the remote control box at the plug connector, if equipped.
10. Connect both battery cables at their terminals.

DISASSEMBLY

♦ See Figures 2 thru 10

1. Remove the retainer, if equipped, secured across the two cables.
2. Remove the two bolts and the retaining clamp across the throttle cable.
3. Remove the throttle arm securing bolt and the shift arm securing bolt.
4. Remove the small cotter pin at the throttle cable swivel joint and then remove the throttle arm from the cable.
5. Remove the small cotter pin at the shift cable swivel joint and then remove the shift arm from the shift cable.
6. Lift the throttle cable out of the box.
7. Remove the junction board and then lift the shift cable out of the control box.
8. Turn the control box over and loosen the Allen screw at the base of the remote control lever.
9. Work the lever up and free of the knurled shaft of the throttle gear.
10. Remove and discard the felt washer around the throttle gear shaft.
11. Remove the two retaining plate screws.
12. Lift the retaining plate off the shift box mating surface.
13. Place a shop towel over the throttle gear, detent ball and spring assembly before disturbing any component.

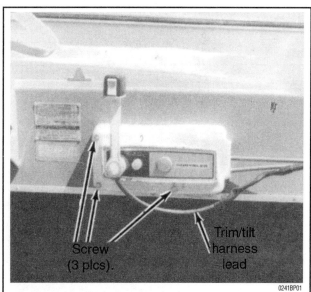

Fig. 1 Control box installed with the three mounting screws identified

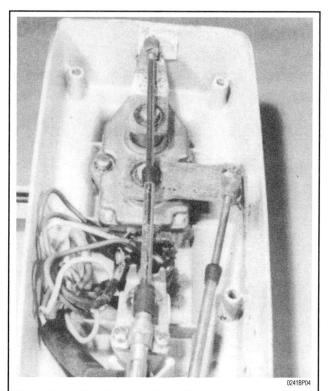

Fig. 2 Interior view of an opened control box showing the wires neatly routed and tucked away from moving parts

REMOTE CONTROLS 10-3

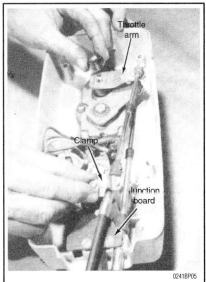

Fig. 3 Removing the throttle cable and arm from the control box

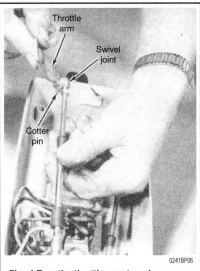

Fig. 4 Turn the throttle arm to gain access to remove the cotter pin holding the throttle cable swivel joint in place

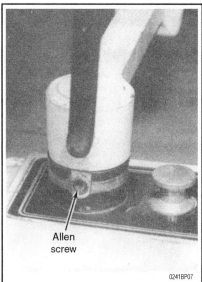

Fig. 5 Location of the Allen screw holding the control handle in place

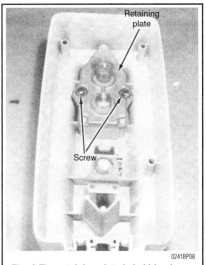

Fig. 6 The retaining plate is held in place with two screws

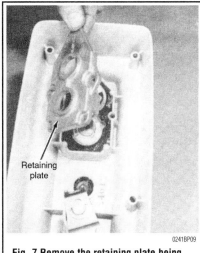

Fig. 7 Remove the retaining plate being careful not to loose the spring and ball detent just below it

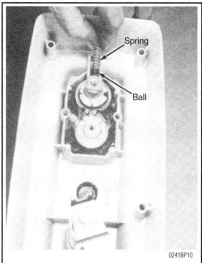

Fig. 8 Remove the throttle spring and ball detent

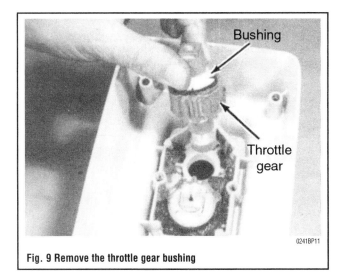

Fig. 9 Remove the throttle gear bushing

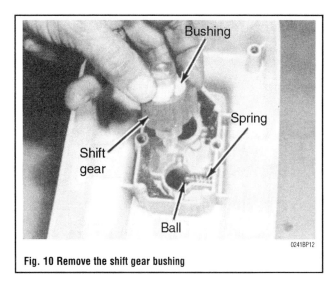

Fig. 10 Remove the shift gear bushing

10-4 REMOTE CONTROLS

> ### ✷✷ CAUTION
> The detent ball and spring are under pressure and are likely to fly out and cause personal injury or be lost.

14. Remove the spring and detent ball from the groove adjoining the throttle gear.

→Set these two small components apart so they will not be confused with the shift gear detent ball and spring, to be removed in a later step.

15. Lift out the throttle gear assembly.
16. Remove the plastic bushing and bowed washer from the top of the throttle gear shaft and then remove another plastic bushing from the bottom (knurled end) of the shaft.
17. Place a shop towel over the shift gear, detent ball and spring assembly before disturbing any component.

> ### ✷✷ CAUTION
> The detent ball and spring are under pressure and are likely to fly out and cause personal injury or be lost.

18. Remove the spring and detent ball from the groove adjoining the shift gear.

→Set these two small components apart so they will not be confused with the throttle gear detent ball and spring, removed in an earlier step.

19. Lift out the shift gear and remove the plastic bushing from the top of the gear.

CLEANING AND INSPECTION

♦ See Figure 11

Clean all metal parts with solvent, and then blow them dry with low pressure compressed air.

Never allow plastic bushings, nylon retainers, wiring harness retainers, and the like, to remain submerged in solvent more than just a few moments. The solvent will cause these type parts to expand slightly. They are already considered a "tight fit" and even the slightest amount of expansion would make them very difficult to install. If force is used, the part will most likely be distorted.

Inspect the control box housing for cracks or other damage allowing moisture to enter and cause problems with the mechanism.

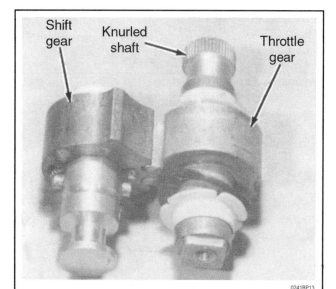

Fig. 11 Inspect the knurled shaft to be sure the splines have not become rounded. Any damage or excessive wear in this area will prevent the throttle gear from indexing properly and the shifting operation could be lost

Carefully check the teeth on the throttle and shift gear for signs of wear.

Closely inspect the condition of all wires and their protective insulation. Look for exposed wires caused by the insulation rubbing on a moving part, cuts and nicks in the insulation, and severe kinking which could cause internal breakage of the wires.

Inspect the detent ball for grooves. Any grooves or rough spots on the ball would impede smooth shifting.

ASSEMBLY

1. Applying a thin coat of water resistant multi-purpose lubricant to all moving parts as a precaution against moisture and corrosion. The lubricant will also help to ensure continued satisfactory operation of the mechanism and ease assembly.
2. Place the plastic bushing over the top of the shift gear.
3. Identify the detent ball and spring assembly removed from the shift mechanism. This assembly is smaller than the detent ball and spring removed from the throttle mechanism.
4. Place the shift spring into its groove.
5. Place the shift detent ball in the groove next to the spring.
6. Use a small screwdriver and hold back the detent ball against the spring while the shift gear is lowered into the housing with the gear teeth facing toward the throttle mechanism.
7. When the gear seats release pressure against the ball and allow it to spring back against the gear collar.
8. Install a plastic bushing onto the bottom (knurled) end of the throttle gear shaft.
9. Install the bowed washer and another plastic bushing onto the top of the throttle gear shaft.
10. Place the throttle spring into its groove.
11. Place the throttle detent ball in the groove next to the spring.
12. Use a small screwdriver and hold back the detent ball against the spring while the throttle gear is lowered into the housing with the gear teeth facing toward the shift mechanism.
13. When the gear seats release pressure against the ball and allow it to spring back against the gear collar.
14. Before installing the retaining plate be certain that the throttle detent ball is in the neutral recess of the throttle gear collar.
15. Position the retaining plate over the two gear shafts.
16. Install and tighten the two securing screws for the retaining plate.
17. Install a new felt washer over the throttle gear shaft.
18. Install the remote control lever onto the knurled end of the throttle shaft and secure it in place with the Allen screw.
19. Tighten the screw securely.
20. Loosen the locknut next to the shift cable terminal.
21. Rotate the cable terminal along the threaded portion of the shift cable until a cable length of 3/16" (4.75mm) extends beyond the terminal.
22. Tighten the locknut against the cable terminal to hold this position.
23. Insert the shift cable terminal into the rear hole of the shift arm and slide the cotter pin through the terminal to retain the cable.
24. Place the shift arm onto the shift gear shaft, install and tighten the bolt.
25. Install the junction board over the shift cable.
26. Adjust the position of the cable terminal on the throttle cable in the same manner as just described for the shift cable.
27. Insert the throttle cable end into the hole at the end of the throttle arm and slide the cotter pin through the terminal to retain the cable.
28. Place the throttle arm over the throttle gear shaft, install and tighten the bolt.
29. Place the throttle cable into the cable clamp and install and tighten the two retaining screws.

Install a retainer, if so equipped, across both cables before they exit the box.

30. Arrange the wires neatly inside the box. If possible, secure a "tie-wrap" around the bundle of wires to prevent loose wires from rubbing against and interfering with moving parts.
31. Align the mark on the wire harness plug with the mark on the coupler plug, and then connect the plugs, as required.
32. Connect the trim/tilt harness plug, as required.
33. Check operation of the remote control lever to and between each of the three positions—neutral, forward, and reverse.
34. The lever should move smoothly and with a definite action. The shift

REMOTE CONTROLS 10-5

cable and the throttle cable should move in and out of the sheathing without any indication of binding. The movement can be check at the powerhead end of the cable.

Remote Control Cables

REMOVAL & INSTALLATION

♦ See Figures 12, 13 and 14

1. Remove one bolt and loosen the other from each of the throttle and shift cable retainers mounted on the powerhead. This action will allow the retainer to swing down and clear the cable.
2. Use a slotted screwdriver and push the spring loaded barrel away from the sleeve at the end of the throttle cable. At the same time snap the throttle cable from the towershaft ball joint.
3. Remove the shift cable from the shift arm in the same manner.

To install:

4. Use a screwdriver to push back the spring loaded barrel from the end of the shift cable.
5. Snap the end of the cable over the ball joint on the shift arm.
6. Allow the barrel to spring back over the sleeve to lock it in place.
7. Secure the throttle cable to the towershaft ball joint in the same manner.

ADJUSTMENT

1. The groove in the cable should align with the groove in the retainer. If adjustment is necessary the cable adapter can be threaded in or out of the threaded area to achieve proper fit.
2. Slide the throttle cable retainer over the throttle cable and secure it with the two screws.
3. Secure the locknut on the cable against the adapter.
4. Secure the shift cable in the same manner.
5. Rotate the towershaft the full arc of its travel to make sure the throttle cable is free to slide through the retainer.
6. Test the operation of both cables they must move without binding for the full extent of travel of the remote control lever and the towershaft. The lever should move smoothly and with a definite action through all gears.

Neutral Start Switch

TESTING

♦ See Figure 15

1. The shift neutral start switch is located inside the control box, inside the lower cowling or mounted on the distributor, if equipped.

➡A neutral safety switch that is mounted in the control box requires the control box be partially disassembled to gain access to the switch.

2. Disconnect the two wires on the switch.
3. Using an ohmmeter set the Rx1000 scale connect one meter lead to each of the terminals of the neutral safety switch.
4. The switch should have good continuity with the engine shifted into neutral only. If the switch has continuity in forward or reverse gear the engine will be able to start in gear causing a safety problem. If there is no continuity in any shift position first be certain that the switch is being fully depressed in neutral.

A switch with no continuity will not allow the engine to be cranked over.

REMOVAL & INSTALLATION

1. A switch mounted inside the control box is removed by taking the backing plate off the control box. Disconnect the two wires and removing the fasteners holding the switch in place.
2. A switch mounted inside the lower cowling on the shift arm is removed by disconnect the two wires and removing the fasteners holding the switch in place.

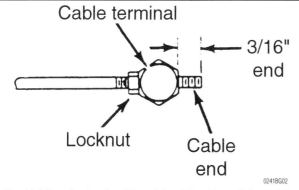

Fig. 12 Adjust the throttle cable and the shift cable until the end of each cable extends 3/16" beyond the cable terminal

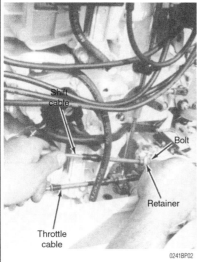

Fig. 13 Remove the screw from the each cable retainer and lift the control cable out of the retaining groove

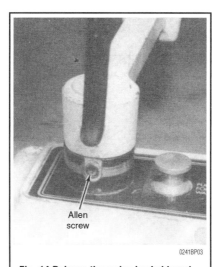

Fig. 14 Release the spring loaded barrel to remove the cable end from the ball retainer. Remember the longer cable adapter is on the throttle cable

Fig. 15 Typical location of the shift interlock switch

10-6 REMOTE CONTROLS

3. A switch mounted on the side of the distributor is removed by disconnect the two wires and removing the fasteners holding the switch in place.

To install:

4. The switch is installed by positioning it back into the control box, on the shift arm or on the distributor being certain the wires are on the proper terminals and are not in a position to be damaged during any additional assembly procedures.

Emergency Stop Switch (Lanyard)

TESTING

▶ See Figure 16

1. Disconnect the engine stop switch wiring harness.
2. Connect a multimeter between the switch harness leads.
3. With the switch engaged (stop switch lanyard in pulled), continuity should exist. With the switch released (stop switch lanyard in position), continuity should not exist.
4. If the switch does not function as specified, there is a fault with either the switch or switch wiring harness and the switch will need to be replaced.
5. If the switch functions properly, there may be a problem with the powerhead wiring harness.

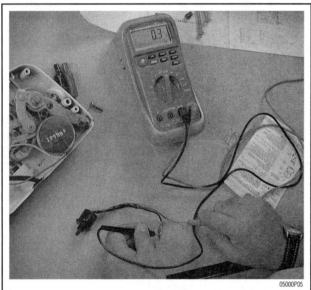

Fig. 16 Connect a multimeter between the switch harness leads. With the switch released (as shown), continuity should not exist

REMOVAL & INSTALLATION

1. Remove the control box from the boat and open the side covers to access the internal components.
2. Disconnect the emergency stop switch wiring harness.
3. Remove any wire straps that connect the switch to the control box.
4. Remove any retaining nuts/screws that secure the switch to the control box.
5. Remove the switch from the control box.

To install:

6. Install the switch into the control box.
7. As required, install the switch retaining screw/nut and tighten securely.
8. Install any wire straps that connect the switch to the control box or bracket.
9. Connect the emergency stop switch wiring harness.
10. Test the switch for proper operation.
11. Install the control box side covers and mount the control box in the boat.

Ignition Switch

▶ See Figures 17 and 18

TESTING

1. Remove the switch
2. Disconnect all leads from the switch terminals.
3. Keeping the switch in the off position use an ohmmeter on the Rx1 scale touching one meter lead to each of the two switch terminals marked with an "M". The meter should indicate continuity.
4. Connect one meter lead to the switch terminal marked with a "B" and the other meter lead to the switch terminal marked with an "I".
5. The meter should read constant continuity with the switch in the run, start and pushed in choke position.
6. Remove the meter lead from the terminal marked I and connect it to the terminal marked "C".
7. The meter should read constant continuity with the switch in the run position while the switch is pushed in and in the start position with the switch pushed in.
8. Remove the meter lead from the terminal marked "C" and connect it to the terminal marked "S".
9. The meter should read constant continuity with the switch in the start position.
10. Remove both test meter leads.
11. Install the switch and connect the proper leads to the correct switch terminal:
    - White lead to either "M" terminal.
    - Yellow lead to the "S" terminal.
    - Green lead to the "C" terminal.
    - Red lead to the "B" terminal.
    - Blue lead to the other "M" terminal.

REMOVAL & INSTALLATION

1. The switch is removed by loosening the retainer nut and sliding the switch out of the dash panel plate or the side of the control box with the back removed.

To install:

2. The switch is installed by sliding it back into its hole in the control box or dash panel plate being certain the wires are on the proper terminals and are not in a position to be damaged during any additional assembly procedures.

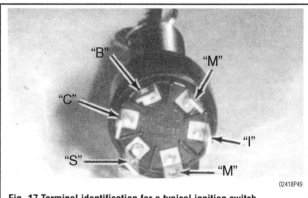

Fig. 17 Terminal identification for a typical ignition switch

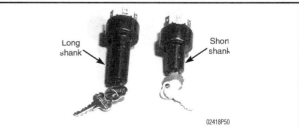

Fig. 18 Long shank and short shank ignition switches are not interchangeable but function and are tested in the same manner

REMOTE CONTROLS 10-7

Warning Buzzer

♦ See Figure 19

TESTING

1. Disconnect the two horn leads at their nearest quick disconnect fitting.
2. Connect one horn terminal to the negative post of a 12 volt battery
3. Connect the other terminal to the positive post of a battery using a jumper lead.
4. As soon as the lead makes contact with the positive battery terminal the horn should sound. If the horn does not sound or does not seem very loud, replace it.

REMOVAL & INSTALLATION

♦ See Figures 20, 21, 22, and 23

1. A warning horn mounted inside a control box is removed by removing the back cover plate from the control box. The horn wires are then disconnected and the two fasteners holding the horn in place are removed.

Fig. 19 A factory warning horn that is typically mounted inside the control box or behind the dash

2. A warning horn mounted behind the dash is removed by loosening the two screws or clamps holding the horn in place and the horn wires are then disconnected.

To install:

3. Installing either warning horn put it into place and properly fastens it.
4. Connect the wires to their proper terminals so they are not in a position to be damaged during any additional assembly procedures.

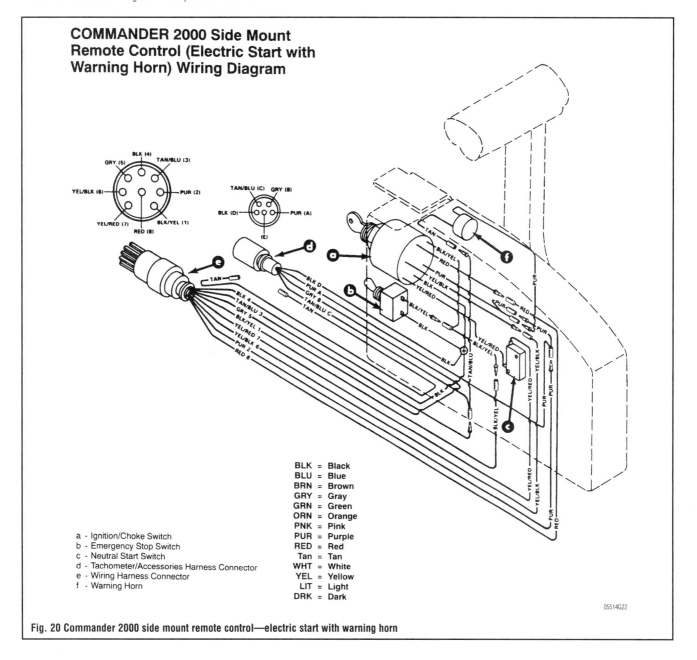

Fig. 20 Commander 2000 side mount remote control—electric start with warning horn

10-8 REMOTE CONTROLS

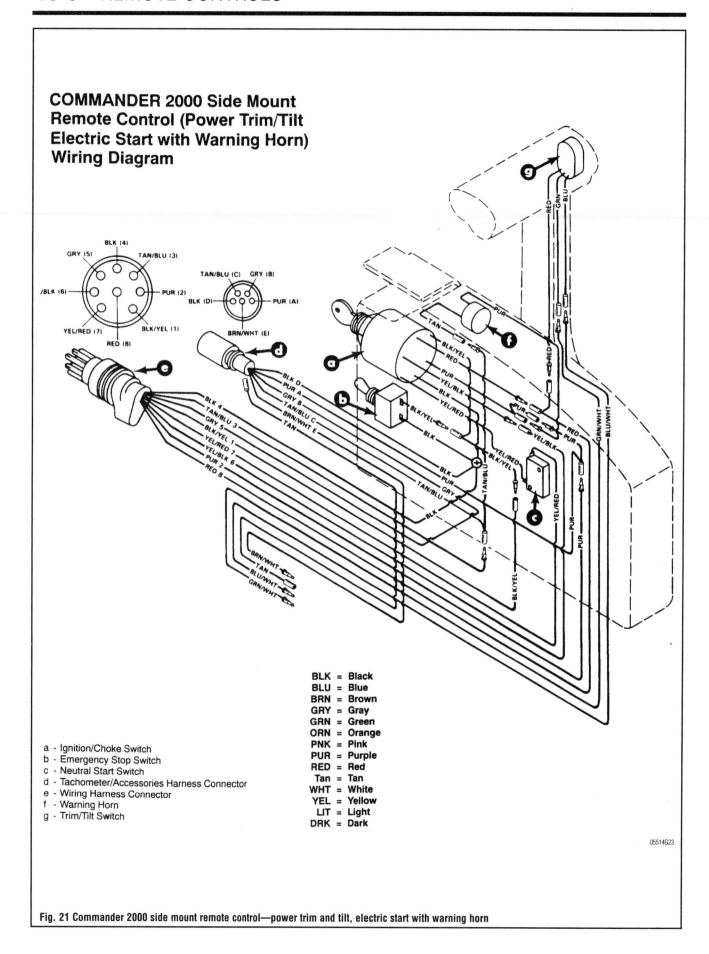

Fig. 21 Commander 2000 side mount remote control—power trim and tilt, electric start with warning horn

REMOTE CONTROLS

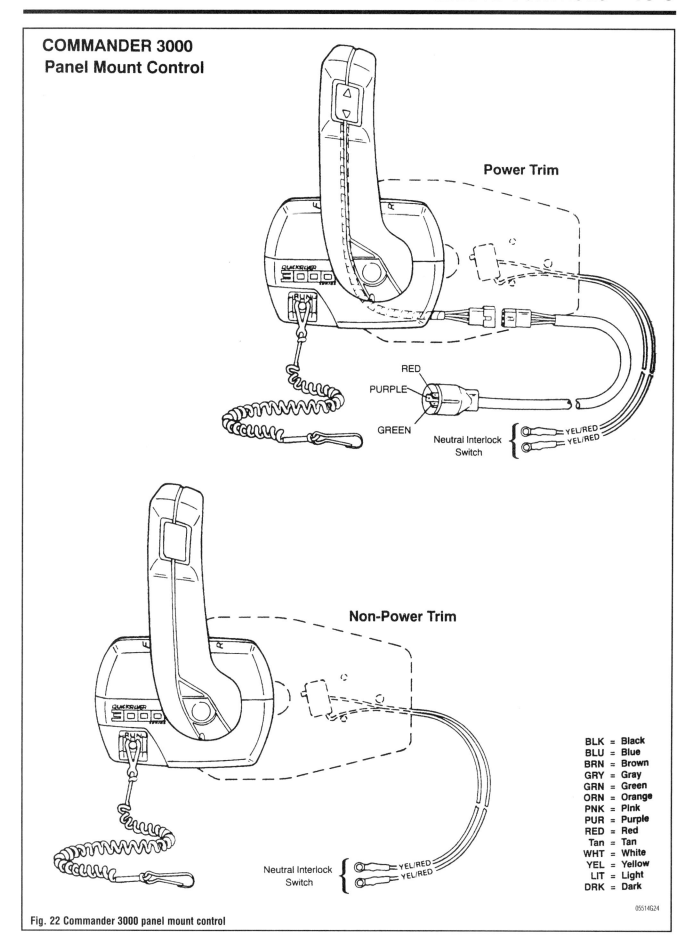

Fig. 22 Commander 3000 panel mount control

10-10 REMOTE CONTROLS

Commander 3000 Components

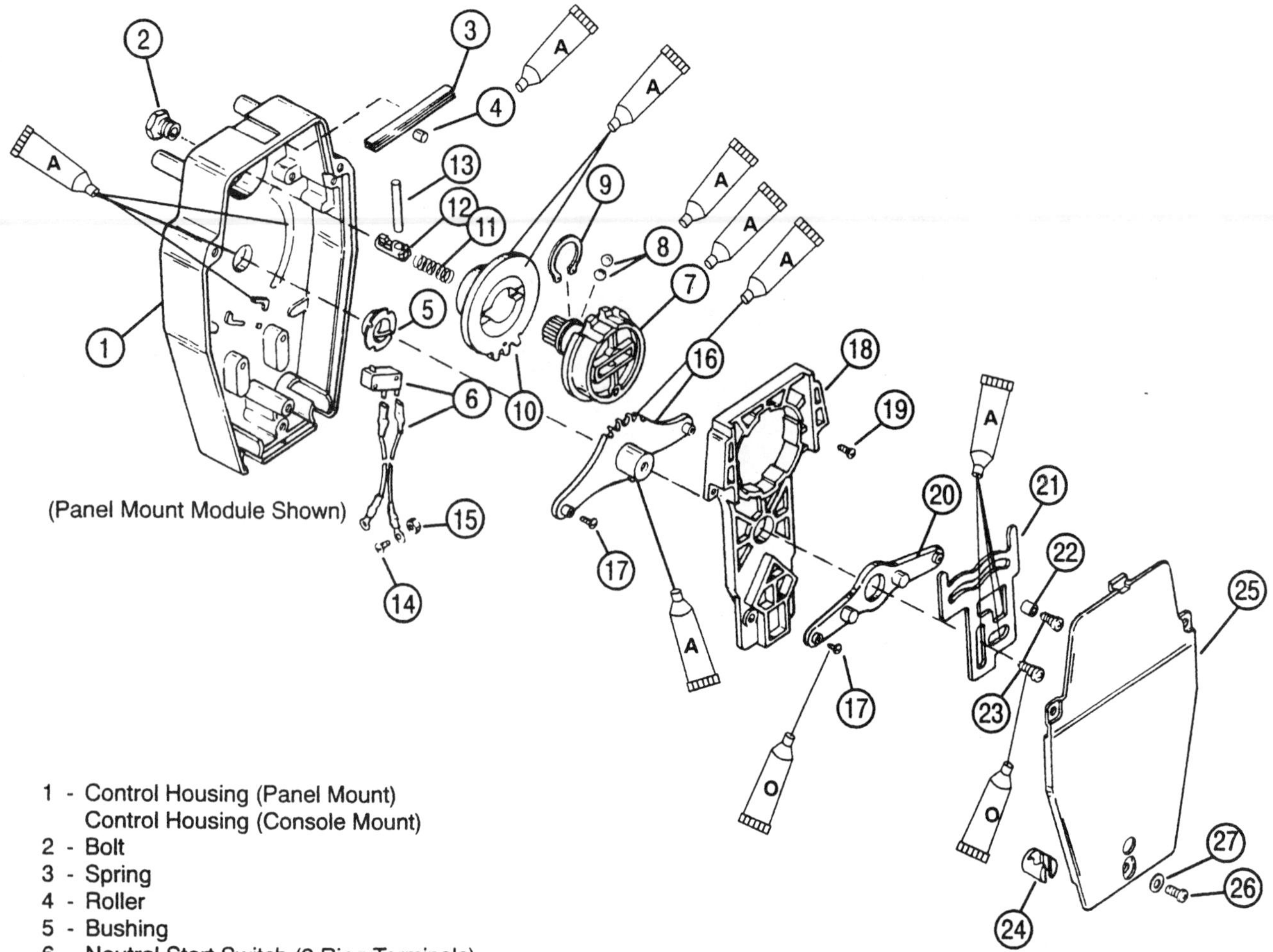

(Panel Mount Module Shown)

1 - Control Housing (Panel Mount)
    Control Housing (Console Mount)
2 - Bolt
3 - Spring
4 - Roller
5 - Bushing
6 - Neutral Start Switch (2 Ring Terminals)
    Neutral Start Switch (No Ring Terminals)
7 - Shaft Handle
8 - Steel Ball (2)
9 - Retaining Ring
10- Gear, Shift
11- Spring
12- Shaft (Throttle Only)
13- Pin (Gear Shift)
14- Screw
15- Nut
16- Shift Arm Assembly
17- Screw
18- Shaft Support Assembly
19- Screw
20- Throttle Arm Assembly
21- Plate
22- Roller (Throttle Plate)
23- Shoulder Bolt
24- Grommet
25- Back Plate
26- Screw
27- Washer

A — 2-4-C With Teflon
O — Loctite 242

Fig. 23 Commander 3000 panel mount control components

REMOTE CONTROLS   10-11

TILLER HANDLE

Description and Operation

Steering control for most outboards begins at the tiller handle and ends at the propeller. Tiller steering is the simplest form of small outboard control. All components are mounted directly to the engine and are easily serviceable.

Throttle control is performed via a throttle grip mounted to the tiller arm. As the grip is rotated a cable opens and closes the throttle lever on the engine. An adjustment thumbscrew is usually located near the throttle grip to allow adjustment of the turning resistance. In this way, the operator does not have to keep constant pressure on the grip to maintain engine speed.

An emergency engine stop switch is used on most outboards to prevent the engine from continuing to run without the operator in control. A small clip that keeps the switch open during normal engine operation controls this switch. When the clip is removed, a spring inside the switch closes it and completes a ground connection to stop the engine. The clip is connected to a lanyard that is worn around the helmsman's wrist.

Some tiller systems utilize a throttle stopper system that limits throttle opening when the shift lever is in neutral and reverse. This prevents over revving the engine under no-load conditions and also limits the engine speed when in reverse.

Troubleshooting the Tiller Handle

If the tiller steering system seems loose, first check the engine for proper mounting. Ensure the engine is fastened to the transom securely. Next, check the tiller hinge point where it attaches to the engine and tighten the hinge pivot bolt as necessary.

Excessively tight steering that cannot be adjusted using the tension adjustment is usually due to a lack of lubrication. Once the swivel case bushings run dry, the steering shaft will get progressively tighter and eventually seize. This condition is generally caused by a lack of periodic maintenance.

Correct this condition by lubricating the swivel case bushings and working the outboard back and forth to spread the lubricant. However, this may only be a temporary fix. In severe cases, the swivel case bushings may need to be replaced.

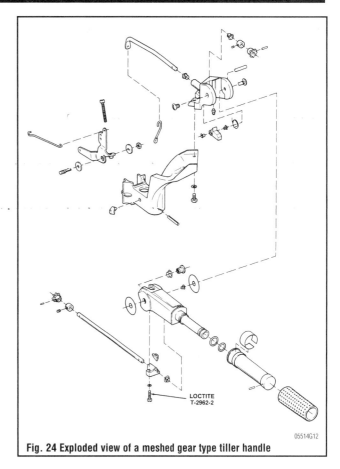

Fig. 24 Exploded view of a meshed gear type tiller handle

Tiller Handle

REMOVAL & INSTALLATION

Meshed Gear Type

♦ See Figure 24

1. Remove the flywheel.
2. Remove the stator plate assembly.
3. Remove the throttle cam link rod from the magneto control arm.
4. Remove the magneto control arm by taking the retaining nut off the pivot shaft and move the arm towards the full throttle position until enough clearance has been achieves to remove the arm.
5. Disconnect the magneto shaft link from the magneto control arm.
6. Remove the two fasteners holding the tiller handle to the lower cowl.
7. Remove the tiller handle assembly.
8. Using a pin punch drive the retaining pin through the end of the twist grip.
9. Pull the twist grip off the end of the tiller handle.
10. Remove the twist grip O-rings
11. Loosen the hinge pin screws on either side of the tiller handle.
12. Back the screws out approximately half way and hit the head of the screw to remove the pivot retainer on each side.
13. Remove the tiller handle from the tiller handle mounting bracket.
14. Loosen the Allen head set screw to free the throttle shaft.
15. Remove the screws holding the throttle shaft bearing support brace.
16. Remove the bearing support brace.
17. Remove the shaft and gear assembly.
18. Slide the throttle shaft support bushing off the shaft.
19. Remove the pivot bushings and throttle shaft bushings as necessary.
20. Secure the throttle shaft in a vise with the retaining pin facing up.
21. Using a pin punch drive the pin through the shaft and remove the throttle shaft gear.
22. Remove the throttle shaft from the vise.
23. Support the tiller handle bracket in a vise.
24. Loosen the Allen head set screw on the support bushing to free the magneto control shaft.
25. Using a pin punch drive the pin through the shaft and remove the magneto control shaft gear.

To install:

26. Place the magneto control shaft gear on the appropriate shaft and align the holes.
27. Using a pin punch drive the pin through the magneto control shaft gear and the shaft securing it to the shaft.
28. Slide the bushing support toward the gear leaving on more then .005in. (.127mm) endplay.
29. Tighten the bushing set screw to 45inlb (5.0Nm).
30. Remove the tiller handle bracket from the vise.
31. Secure the throttle shaft in a vise with the retaining pin hole facing up.
32. Place the throttle shaft gear on the appropriate shaft and align the holes.
33. Using a pin punch drive the pin through the throttle shaft gear and the shaft securing it to the shaft.
34. Remove the throttle shaft from the vise.
35. Install the pivot bushings and throttle shaft bushings as necessary.
36. Slide the throttle shaft support bushing on to the shaft.
37. Install the shaft and gear assembly.
38. Install the throttle shaft bearing support brace.
39. Install the screws holding the throttle shaft bearing support brace.
40. Install the handle into the mounting bracket.
41. Align the holes in the tiller handle with the holed in the tiller handle bracket.
42. Install the two pivot screws.

10-12 REMOTE CONTROLS

43. Install the twist grip O-rings.
44. Push the twist grip onto the end of the tiller handle.
45. Align the hole in the shaft with the hole in the twist grip.
46. Using a pin punch drive the retaining pin through the end of the twist grip to secure it to the shaft.
47. Hold the tiller handle assembly in place.
48. Install the two fasteners holding the tiller handle to the lower cowl.
49. Connect the magneto shaft link from the magneto control arm.
50. Install the magneto control arm by aligning the arm towards the full throttle position until it can be put in place on the pivot stud.
51. Install the retaining nut on to the pivot shaft.
52. Install the throttle cam link rod to the magneto control arm.
53. Install the stator plate assembly.
54. Install the flywheel.
55. Pull the twist grip out until the throttle shaft gear seats against the throttle shaft bearing support brace.
56. Slide the bushing support toward the throttle gear leaving on more then .005in. (.127mm) endplay.
57. Tighten the bushing set screw to 45inlb (5.0Nm).

✴✴ CAUTION

The tiller handle gear set must be aligned and set properly or loss of throttle control could result.

Single Cable Type

▶ See Figures 25, 26, 27, 28 and 29

1. Remove the manual starter.
2. Remove the flywheel.
3. Remove the screw connecting the cable end adapter to the magneto plate.
4. Disconnect the two wires for the emergency stop switch.

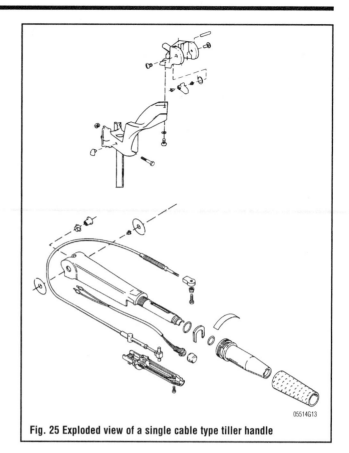

Fig. 25 Exploded view of a single cable type tiller handle

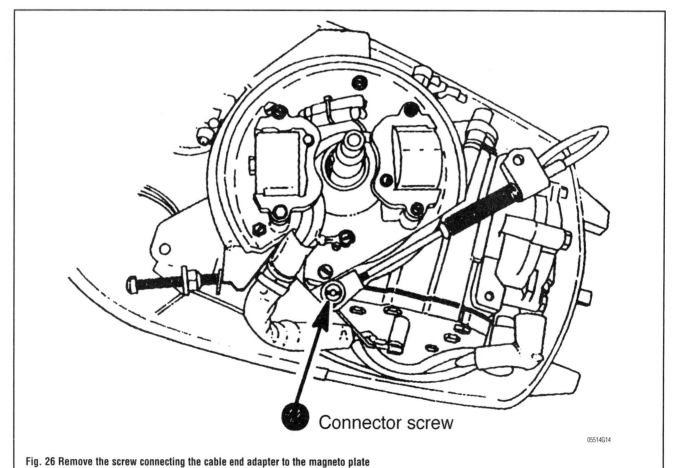

Fig. 26 Remove the screw connecting the cable end adapter to the magneto plate

REMOTE CONTROLS 10-13

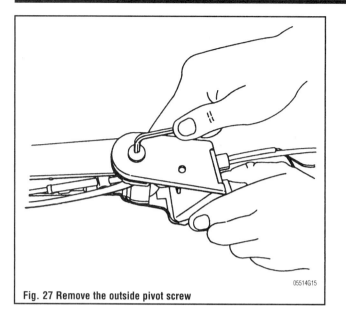

Fig. 27 Remove the outside pivot screw

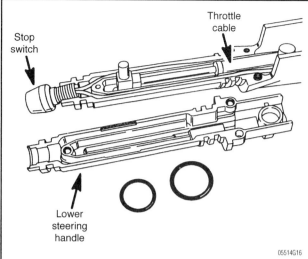

Fig. 28 Location of the throttle cable and stop switch with the proper wire routing in the upper handle housing

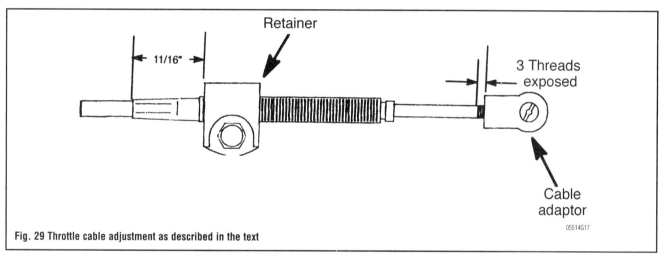

Fig. 29 Throttle cable adjustment as described in the text

5. Remove the two fasteners holding the tiller handle to the lower cowl.
6. Remove the tiller handle assembly.
7. Remove the twist grip retainer from the bottom of the handle.
8. Remove the twist grip.
9. Remove the outside pivot screw.
10. Loosen the inside screw approximately half way and hit the head of the screw to remove the pivot retainer.
11. Remove the inside screw.
12. Unthread and remove the cable adapter and cable retainer from the throttle cable.
13. Remove the tiller handle from the tiller handle mounting bracket.
14. Remove the O-rings from the tiller shaft that had been covered by the twist grip.
15. Remove the three screws holding the upper & lower handle assemblies together.
16. Remove the stop switch.
17. Remove the throttle cable.

To install:
18. Install the throttle cable.
19. Install the stop switch.
20. Install the three screws holding the upper & lower handle assemblies together.
21. Install the O-rings from the tiller shaft that will be covered by the twist grip.
22. Install the tiller handle into the tiller handle mounting bracket.
23. Thread the cable adapter and cable retainer on to the throttle cable.
24. Install the threaded bushing into the inside hole of the tiller handle.
25. Install the inside pivot screw.
26. Install the outside pivot screw.
27. Remove the twist grip.
28. Remove the twist grip retainer from the bottom of the handle.
29. Hold the tiller handle assembly in place.
30. Install the two fasteners holding the tiller handle to the lower cowl.
31. Disconnect the two wires for the emergency stop switch.
32. Thread the throttle cable adapter on the shaft until there are three threads showing on the cable.
33. Thread the cable retainer down the shaft until it reaches $11/16$ between it and the beginning of the swedged fitting.
34. Install the screw connecting the cable end adapter to the magneto plate.
35. Install the flywheel.
36. Install the manual starter.

Engine Stop Switch

TESTING

1. Disconnect the engine stop switch wiring harness.
2. Connect a multimeter between the switch harness leads.
3. With the switch engaged (button pushed), continuity should exist. With the switch released (button not pushed), continuity should not exist.
4. If the switch does not function as specified there is a short in either the switch or harness and the switch should be replaced.

10-14 REMOTE CONTROLS

5. If the switch functions properly, there may be a problem in the powerhead wiring harness.

REMOVAL & INSTALLATION

1. Disconnect the engine stop switch wiring harness.
2. Remove any wire straps that connect the switch to the tiller handle or bracket.
3. Remove any retaining nuts/screws that secure the switch to the tiller handle.

➡ **Some switches are screwed into the end of the tiller handle and simply unscrew from the handle.**

4. Remove the switch from the tiller handle.
To install:
5. Install the switch on the tiller handle.
6. As required, install the switch retaining nut/screw and tighten securely.
7. Install any wire straps that connect the switch to the tiller handle or bracket.
8. Connect the engine stop switch wiring harness.
9. Test the switch for proper operation.

Emergency Stop Switch (Lanyard)

TESTING

1. Disconnect the engine stop switch wiring harness.
2. Connect a multimeter between the switch harness leads.
3. With the switch engaged (stop switch lanyard pulled), continuity should exist. With the switch released (stop switch lanyard in position), continuity should not exist.
4. If the switch does not function as specified there is a short in either the switch or harness and the switch should be replaced.
5. If the switch functions properly, there may be a problem in the powerhead wiring harness.

REMOVAL & INSTALLATION

1. Disconnect the emergency stop switch wiring harness.
2. Remove any wire straps that connect the switch to the tiller handle or bracket.
3. Remove any retaining nuts/screws that secure the switch to the tiller handle.

➡ **Some switches are screwed into the end of the tiller handle and simply unscrew from the handle.**

4. Remove the switch from the tiller handle.
To install:
5. Install the switch on the tiller handle.
6. As required, install the switch retaining nut/screw and tighten securely.
7. Install any wire straps that connect the switch to the tiller handle or bracket.
8. Connect the emergency stop switch wiring harness.
9. Test the switch for proper operation.

HAND REWIND STARTER 11-2
DESCRIPTION AND OPERATION 11-2
TROUBLESHOOTING THE HAND REWIND
  STARTER 11-2
OVERHEAD TYPE STARTER 11-2
1-CYLINDER POWERHEAD 11-2
   REMOVAL & INSTALLATION 11-2
   DISASSEMBLY 11-3
   CLEANING & INSPECTION 11-5
   ASSEMBLY 11-5
2 AND 3-CYLINDER POWERHEADS 11-6
   REMOVAL & INSTALLATION 11-6
   DISASSEMBLY 11-7
   CLEANING & INSPECTION 11-9
   ASSEMBLY 11-10
BENDIX TYPE STARTER 11-11
   REMOVAL & INSTALLATION 11-11
   DISASSEMBLY 11-12
   CLEANING & INSPECTION 11-13
   ASSEMBLY 11-14
   INTERLOCK ADJUSTMENT 11-15

11
HAND REWIND STARTER

HAND REWIND STARTER 11-2
OVERHEAD TYPE STARTER 11-2
BENDIX TYPE STARTER 11-11

11-2 HAND REWIND STARTER

HAND REWIND STARTER

Description and Operation

The main components of a hand rewind starter (recoil starter) are the cover, rewind spring and pawl arrangement. Pulling the rope rotates the pulley, winds the spring and activates the pawl into engagement with the starter hub at the top of the flywheel. Once the pawl engages the hub, the powerhead is spun as the rope unwinds from the pulley.

Releasing the rope on rewind starter moves the pawl out of mesh with the hub. The powerful clock-type spring recoils the pulley in the reverse direction to rewind the rope to the original position.

Newer design manual starters employ the same principles of inertia found in electric starter motors. A nylon pinion gear slides upward and engages the flywheel ring gear as the starter rope is pulled. At the same time, the rewind spring is tightened so their starter rope can be recoiled.

A predetermined ratio designed into the starter provides maximum cranking speed for easy starts and easy pulls. Some models have a cam follower that prevents activation of the manual starter if the shift lever is not in the neutral position. This system is commonly referred to as a Neutral Start Interlock system.

Troubleshooting The Hand Rewind Starter

Repair on hand rewind starter units is generally confined to rope, pawl, nylon pinion gear (on inertia starters) and occasionally spring replacement.

✶✶ CAUTION

When replacing the recoil starter spring extreme caution must be used. The spring is under tension and can be dangerous if not released properly.

Starters that use friction springs to assist pawl action may suffer from bent springs. This will cause the amount of friction exerted to not be correct and the pawl will not be moved into engagement.

Models equipped with a neutral start interlock system may experience a no-start condition due to an interlock cable that is out on adjustment. The hand rewind starter should only function when the shift handle is in the **NEUTRAL** position.

OVERHEAD TYPE STARTER

1-Cylinder Powerhead

♦ See Figures 1, 2, and 3

REMOVAL & INSTALLATION

♦ See Figure 4

1. Remove the powerhead cover.
2. Remove fasteners securing the manual starter to the engine.
3. Lift the starter free of the powerhead.

To install:
4. Put the starter in place on the powerhead.
5. Install the fasteners securing the manual starter to the engine.
6. Install the upper powerhead cover.

Fig. 1 An overhead style starter inverted to show the friction shoes located in the center of the unit

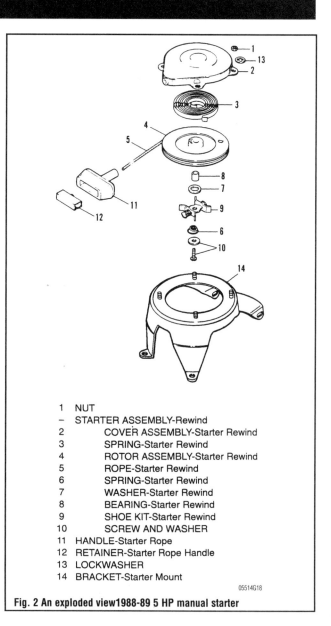

1		NUT
—		STARTER ASSEMBLY-Rewind
2		COVER ASSEMBLY-Starter Rewind
3		SPRING-Starter Rewind
4		ROTOR ASSEMBLY-Starter Rewind
5		ROPE-Starter Rewind
6		SPRING-Starter Rewind
7		WASHER-Starter Rewind
8		BEARING-Starter Rewind
9		SHOE KIT-Starter Rewind
10		SCREW AND WASHER
11		HANDLE-Starter Rope
12		RETAINER-Starter Rope Handle
13		LOCKWASHER
14		BRACKET-Starter Mount

Fig. 2 An exploded view 1988-89 5 HP manual starter

HAND REWIND STARTER 11-3

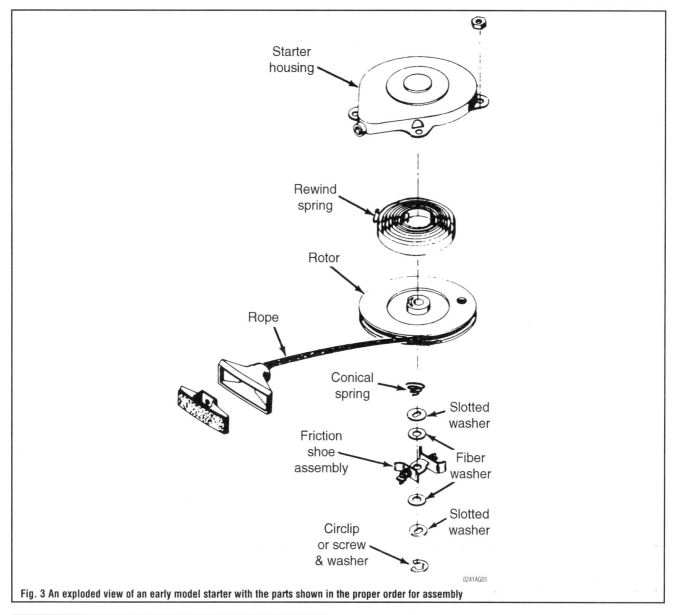

Fig. 3 An exploded view of an early model starter with the parts shown in the proper order for assembly

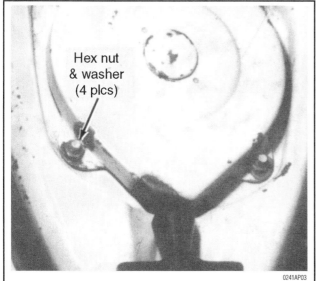

Fig. 4 Starter removal on an early model 4 HP

※ CAUTION

Be certain to remove the spark plug lead before testing the starter to avoid the possibly starting

7. Test the completed work by cranking the powerhead
8. Install the powerhead cover as required.

DISASSEMBLY

♦ See Figures 5 thru 9

1. Turn the starter housing upside down.
2. Pull the rope out as far as it will go and tie a knot to prevent the rotor and spring from rewinding.
3. Remove the center screw, washer and collar.

➡ **Models without the center screw: use a pair of needle nose pliers and compress the two small springs on the friction shoe assembly then remove the circlip from the shaft.**

4. Remove the slotted washer and the fiber washer.
5. Lift out the friction show assembly, the fiber washer, slotted washer and the conical brake spring.

HAND REWIND STARTER

> ⚠️ **CAUTION**
>
> The rewind spring is under pressure and can fly out during disassembly or attempted removal.

If the spring should accidentally be released, severe personal injury could result from being struck as the spring releases its tension. Therefore, the following step must be performed with care to prevent personal injury to self and others in the area.

➡ If the rewind spring is not the cause of the starter malfunction and is not to be replaced it is best to leave the spring in place clean and lubricate it as necessary.

6. Lift the bearing free of the rotor.
7. Untie the knot made in the rope and allow the rotor to unwind slowly.
8. Without disturbing the rewind spring, carefully separate the rotor from the housing.
9. If the spring is to be replaced, obtain two pieces of wood, a short 2" x 4" (5cm x 10cm) will work fine. Place the two pieces of wood approximately 8" (20 cm) apart on the floor.

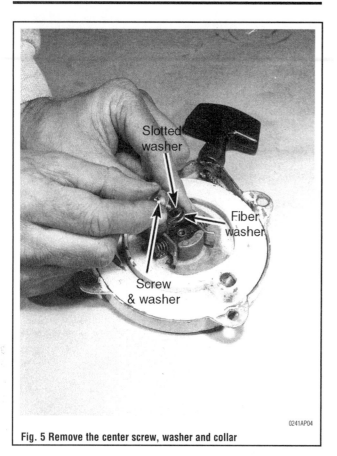

Fig. 5 Remove the center screw, washer and collar

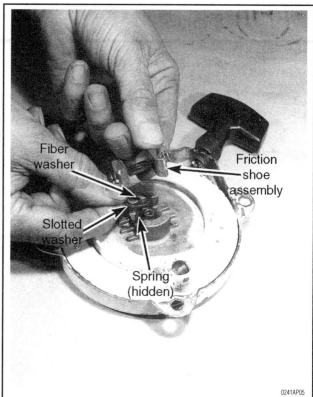

Fig. 6 Lift out the friction show assembly, the fiber washer, slotted washer and the conical brake spring

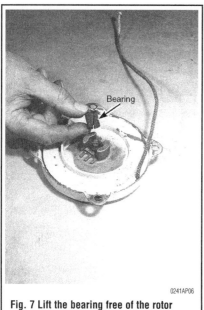

Fig. 7 Lift the bearing free of the rotor

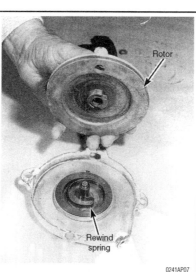

Fig. 8 Without disturbing the rewind spring, carefully separate the rotor from the housing

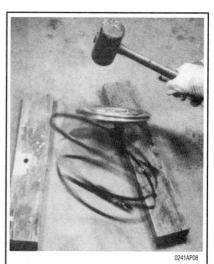

Fig. 9 Tap the housing with a soft mallet while Standing behind the wood keeping away from the openings as the spring releases

HAND REWIND STARTER 11-5

10. Center the housing on top of the wood with the spring side facing down making sure the wood is not touching the spring.
11. Stand behind the wood keeping away from the openings as the spring releases.
12. Tap the housing with a soft mallet.
13. The spring will fall and unwind almost instantly.
14. Untie the knot and pull the rope free of the rotor.

CLEANING & INSPECTION

Wash all parts except the rope and the handle in solvent and blow them dry with low pressure compressed air.

Remove any trace of corrosion and wipe all metal parts with an oil dampened cloth.

Inspect the rope. Replace the rope if it appears to be weak or frayed. If the rope is frayed, check the holes through which the rope passes for rough edges or burrs. Remove the rough edges or burrs with a file and polish the surface until it is smooth.

The starter rope length should be 48" (122cm). Check the length of a newly purchased rope. If the rope is not being replaced, check the length for evidence of stretching while in use.

Inspect the starter spring end hooks. Replace the spring if it is weak, corroded or cracked. Inspect the inside surface of the housing rewind recess for grooves or roughness. Grooves may cause erratic rewinding of the starter rope.

Coat the entire length of the used rewind spring (a new spring will be coated with lubricant from the package), with low-temperature lubricant.

Inspect the two curved shoe plates or the engaging dogs for wear. Replace as necessary.

If the starter is equipped with shoe plates the leading edge of the plates must be sharp. When the starter rope is first pulled the friction shoe assembly pivots and the sharp leading edge of the shoes dig into the flywheel. When the starter rope is pulled out further the friction shoe assembly begins to rotate causing the flywheel to also rotate.

Weak shoe springs will prevent the plates from fully retracting to clear the flywheel. If the plates do not clear the flywheel the shoes will drag on the flywheel while the powerhead is operating. This action will quickly round off the leading edge of the shoe. If the leading edge is rounded off through use or continual dragging efficiency of the road cranking mechanism will be severely impaired.

ASSEMBLY

◆ See Figures 10 thru 14

1. Tie a figure 8" knot about an inch (2.5cm) from one end of the rope.
2. Push the knot into the recess.
3. Pull the long end of the rope to lock the knot firmly in the recess.
4. Hold the rotor with the brake shoe cams facing upward.
5. Wind the starter rope evenly around the rotor counterclockwise.
6. If replacing the spring remove the retainers from the shipping container used to keep the spring from accidentally falling out of the container. Place the shipping container over the spring recess with the tabs resting on the outer edge of the recess.
7. Align the hook on the outer end of the spring with the housing anchor or notch.
8. Place two large blade screwdrivers into the holes of the shipping container over the tensioned spring.
9. Push on both screwdrivers at the same time to press the spring out of the shipping container and into the spring recess of the housing.

➥It is highly recommended that in the event the rewind spring had been removed that it be replaced. The old spring can be installed with the following directions but is a very difficult task and one could get injured while attempting it.

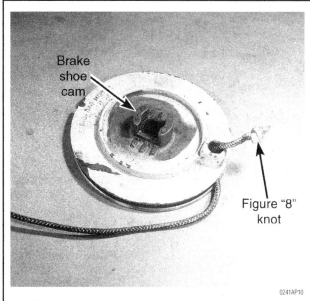

Fig. 10 Tie a figure 8" knot about an inch (2.5cm) from one end of the rope

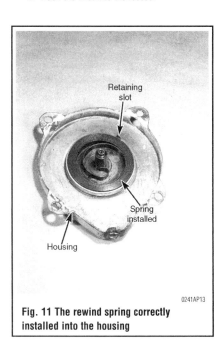

Fig. 11 The rewind spring correctly installed into the housing

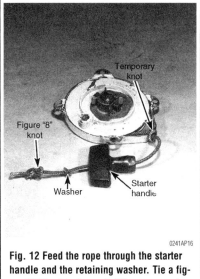

Fig. 12 Feed the rope through the starter handle and the retaining washer. Tie a figure "8" knot in the end of the rope, and pull the knot back into the handle

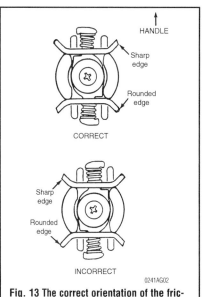

Fig. 13 The correct orientation of the friction shoe assembly

11-6 HAND REWIND STARTER

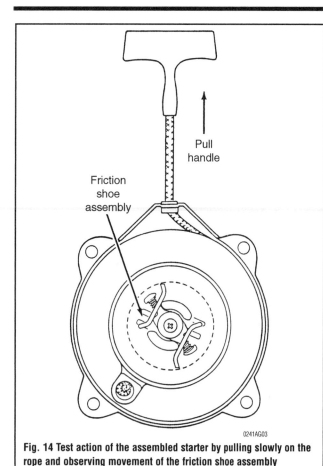

Fig. 14 Test action of the assembled starter by pulling slowly on the rope and observing movement of the friction shoe assembly

✱✱ CAUTION

Wear a good pair of gloves and a face shield while winding and installing the spring. The spring will develop tension and the edges of the spring steel are extremely sharp. The gloves will prevent cuts to the hands and fingers.

10. Loop the spring loosely into a coil to enable it to be fed into its recess.
11. Insert the hook on the end of the spring into the retaining slot on the inside perimeter of the spring compartment.
12. Feed the spring around the inner edge of the compartment and at the same time rotate the housing clockwise to help guide the spring into place.
13. Hold the rope at the notch cut into the face of the rotor.
The rope in the notch will permit the rotor to be wound later and not become wedged between the rotor and the housing after the rotor is installed.
14. Place the rotor over the center shaft of the housing and down on top of the spring.
15. Index one of the slots on the underside of the rotor with the loop at the end of the spring.
16. Slide the bearing down over the housing shaft with the shoulder of the bearing facing upward.
17. Hold the starter rope against the notch in the rotor and rotate the rotor counterclockwise for three full turns.
18. Hold the rotor against the housing to prevent the rotor rewinding.
19. Thread the end of the rope through the guide hole and tie a temporary knot in the rope to prevent the rotor from unwinding.
20. Feed the rope through the starter handle and the retaining washer.
Tie a figure "8" knot in the end of the rope, and pull the knot back into the handle.
21. Tuck the knot neatly in the handle recess.
22. Test the rewind spring tension by pulling the rope out as far as possible. Hold the rope out and turn the rotor 1/8 to 3/8 turn counterclockwise.

➡ If the rotor cannot be rotated the extra distance the rope is too short. Disassemble the starter and measure the entire unknotted rope length.

23. Untie the temporary knot and allow the rope to slowly feed into the guide hole.
24. After the spring has completely rewound the handle should be flush against the guide hole opening and be held under slight tension.
25. Place the conical spring over the shaft with the taper facing upward.
26. Index the slotted washer over the shaft followed by a new fiber washer.
27. Insert the friction shoe assembly between the stationary cams.

➡ The friction shoe assembly must be installed with the sharp edges of the shoe on the upper right and lower left positions.

28. Place a new fiber washer and slotted washer over the friction shoe assembly.
29. Install and tighten the center screw and washer.
30. Models with circlip: install the circlip to retain the components installed on the shaft.
Test action of the assembled starter by pulling slowly on the rope and observing movement of the friction shoe assembly. The rotor must rotate clockwise; the friction shoe assembly must pivot counterclockwise; and the shoe assembly must change shape. The two cams will push the sharp edge of each shoe plate outward. If the rounded edge of the plates are pushed outward the friction shoe assembly has been installed upside down.

2 and 3-Cylinder Powerheads

A black plastic housing identifies the manual starter covered in this section and once removed there are no dogs that contact the flywheel. The starter has a one way bearing that grabs a spindle that is part of the flywheel nut to crank the engine over.

REMOVAL & INSTALLATION

◆ See Figures 15, 16, 17 and 18

1. Unsnap the interlock link rod free of the lower lock lever.
2. Remove the three attaching bolts securing the hand rewind starter to the powerhead.
3. Lift the starter housing up and free of the powerhead.
To install:
4. Slide the starter assembly down over the crankshaft and into position on the powerhead.

Fig. 15 Remove the three attaching bolts securing the hand rewind starter to the powerhead

HAND REWIND STARTER 11-7

Fig. 16 Lift the starter housing up and free of the powerhead

Fig. 17 Secure the starter with the attaching hardware. Tighten the bolts alternately and evenly

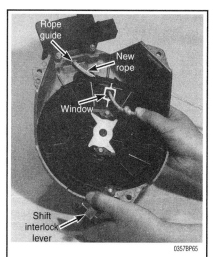

Fig. 18 This type of hand rewind starter has a window on the top of the housing. This window permits a broken pull rope to be replaced without removing the starter from the powerhead

5. Secure the starter with the attaching hardware. Tighten the bolts alternately and evenly.
6. Snap the interlock link rod into the lower lock lever.

DISASSEMBLY

▶ See Figures 19 thru 28

1. Pull about 1 foot (30cm) of rope out of the starter housing and tie a loose knot in the rope to prevent the rope from rewinding back into the housing.

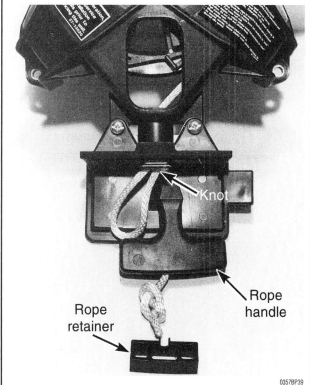

Fig. 19 Pull about 1 ft (30cm) of rope out of the starter housing and tie a loose knot in the rope to hold the starter rope for pull handle removal

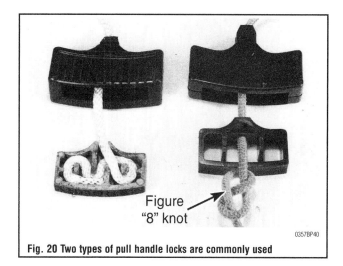

Fig. 20 Two types of pull handle locks are commonly used

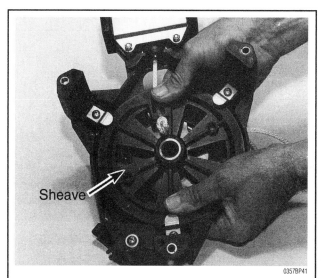

Fig. 21 Rotate the sheave until the knot in the end of the rope is aligned with the hole in the sheave

11-8 HAND REWIND STARTER

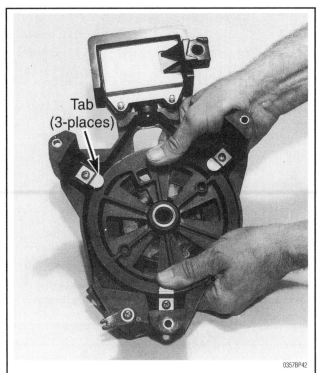

Fig. 22 Once the starter rope has been removed from the pull handle hold the sheave and let the spring tension slowly release

Do not tighten the knot because it will be necessary to untie the knot with one hand later.

2. Remove the handle retainer; feed the rope back through the handle; untie the knot; then pull the handle free of the rope. Two type handles are used, as shown.

→ If the only service on the hand starter is to replace a broken rope, the disassembling procedures may be stopped after the next step. However, if the rewind spring is to be replaced or other service work performed, it is best to leave the rope on the sheave.

3. Rotate the sheave until the knot in the end of the rope is aligned with the hole in the sheave. Hold the tension on the sheave with one hand, and with the other hand, untie the loose knot in the rope. Pull the rope free of the sheave and housing.

4. Ease your grasp on the sheave and allow the sheave to rotate until all tension on the spring is released. If the rope has not been removed, it will simply wind around the sheave.

5. Loosen the three screws securing the tabs holding the sheave in place. It is not necessary to remove the tabs. Rotate the tabs to permit the sheave to clear the housing.

✲✲ WARNING

The rewind spring is a potential hazard. The spring is under tremendous tension when it is wound. If the spring should accidentally be released from its container, severe personal injury could result from being struck by the spring with force. Therefore, the following two steps must be performed with care to prevent personal injury to self and others in the area.

6. Very carefully lift and rock the sheave. The spring will disengage from the sheave and remain in its container in the housing.

Fig. 23 Loosen the three screws securing the tabs holding the sheave in place

Fig. 24 Very carefully lift and rock the sheave to remove it from the housing

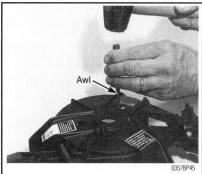

Fig. 25 Using a hammer and awl, pop the rewind spring container out of the starter housing

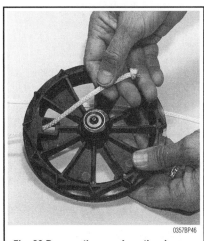

Fig. 26 Remove the rope from the sheave

Fig. 27 Drive the plastic cap free of the sheave using a punch and hammer

Fig. 28 Using a long thin blade screwdriver pry the oil seal out of the sheave

HAND REWIND STARTER 11-9

7. Place the starter housing on a flat surface in the upright position resting on its three legs. Using a hammer and awl, pop the rewind spring container out of the starter housing. There is no reason whatsoever to remove the spring from the container. If the spring is in satisfactory condition, it will be lubricated prior to installation. If the spring is broken, a new spring will come in a container, lubricated, and ready for installation.

8. If the rope was not removed, untie the knot and remove it from the sheave.

9. Drive the plastic cap free of the sheave using a punch and hammer.

10. Pry the oil seal out of the sheave, with a long thin blade screwdriver.

CLEANING & INSPECTION

▶ See Figures 29, 30 and 31

1. Wash all parts except the rope and the handle in solvent, and then blow them dry with compressed air.

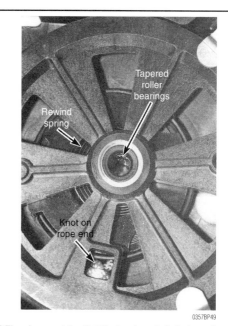

Fig. 30 The sheave of the 6–15hp hand rewind starter relies on a clutch arrangement containing tapered roller bearings located in the center of the sheave. If the clutch is defective, the sheave assembly must be replaced.

Fig. 29 A new rewind spring will arrive packed in a container, lubricated, and ready for installation.

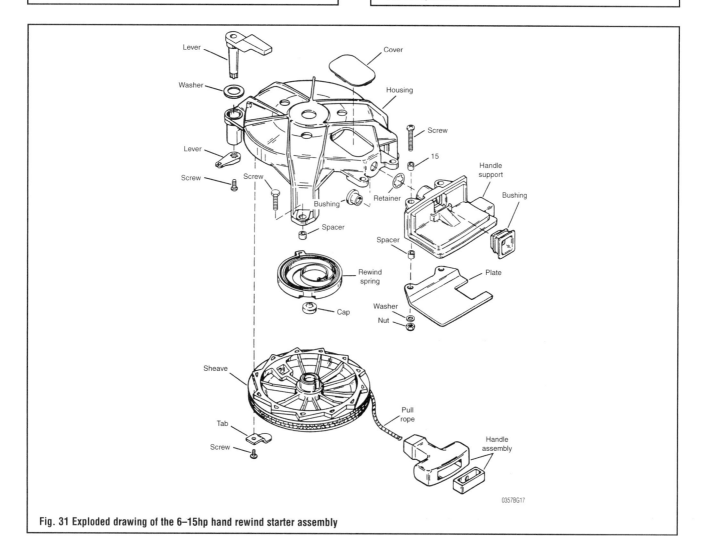

Fig. 31 Exploded drawing of the 6–15hp hand rewind starter assembly

11-10 HAND REWIND STARTER

2. Remove any trace of corrosion and wipe all metal parts with an oil dampened cloth.

3. Inspect the rope. Replace the rope if it appears to be weak or frayed. If the rope is frayed, check the holes through which the rope passes for rough edges or burrs. Remove the rough edges or burrs with a file and polish the surface until it is smooth.

4. Inspect the starter spring end hooks. Replace the spring if it is weak, corroded or cracked. Inspect the tab on the spring retainer plate. This tab is inserted into the inner loop of the spring. Therefore, be sure it is straight and solid. Inspect the inside surface of the sheave rewind recess for grooves or roughness. Grooves may cause erratic rewinding of the starter rope.

➡ **If starter operation was erratic or excessively noisy, check the starter clutch for damage from lack of lubrication. If necessary, replace the complete sheave assembly with a pre-lubricated starter clutch installed.**

ASSEMBLY

▶ See Figures 32, 33, 34 and 35

1. Install a new oil seal into the center of the sheave with metal part facing UP. This task may be accomplished using the proper size socket and a hammer.

Fig. 32 Install a new oil seal into the center of the sheave with metal part facing UP

Fig. 33 Secure the seal in place by taping the plastic cap into the center of the sheave using a soft head mallet

2. Tap the plastic cap into the center of the sheave using a soft head mallet. The cap secures the seal in place.

3. Snap the rewind spring container, with the spring inside, into the housing. A new spring will come with the container ready for installation.

4. Insert the sheave into the starter housing. Swing the three tabs around to lock the sheave in place. Tighten the tab screws securely. Hold the sheave to maintain tension and at the same time rotate the sheave counterclockwise as far as possible to wind the spring. Continue holding pressure on the sheave, but let it slip a little until the rope knot recess in the sheave is aligned

Fig. 34 Snap the rewind spring container, with the new spring inside, into the housing

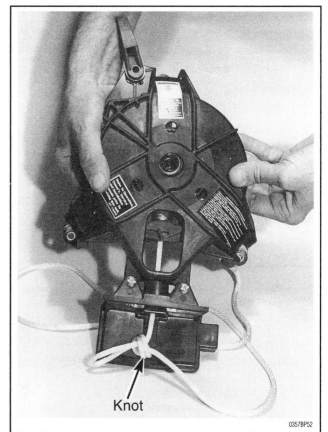

Fig. 35 Once the starter spring has been preloaded hold tension on the sheave and tie a knot in the rope to prevent the sheave from rewinding

HAND REWIND STARTER   11-11

with the rope hole in the starter housing. Hold tension on the sheave for the next step.

5. Feed the rope through the rope knot recess of the sheave and the hole in the starter housing. Hold tension on the sheave with one hand and tie a figure "8" knot close to the end of the rope. Pull the rope through until the knot is up tight in the rope recess of the sheave. Continue holding tension on the sheave for the next step.

6. Still holding tension on the sheave and at the same time holding the sheave in place within the starter housing, tie a knot in the rope as close as possible to the housing to prevent the sheave from rewinding. Now, release a bit of pressure on the sheave. The spring will rewind just a bit and pull the knot in the rope up tight against the housing.

7. Feed the free end of the rope through the handle and secure it with the retainer, or with a figure "8" knot and the retainer. After the handle is secured, untie the knot next to the housing and allow the sheave to rewind pulling the rope around the sheave.

BENDIX TYPE STARTER

♦ See Figures 36 and 37

REMOVAL & INSTALLATION

♦ See Figures 38, 39, 40 and 41

1. Remove the engine cover.
2. Pull the recoil rope out and knot the rope approximately one foot from the end.
3. Allow the rope to retract and be held by the knot.
4. Remove the pull handle by untying the knot or working the rope from the retainer.
5. Hold tension on the rope and untie the knot that was used to hold the tension during the handle removal procedure.
6. Slowly let the rope retract in through the lower cowl and secure it with a knot.
7. Disconnect the interlock rod between the shift mechanism and the interlock assembly on the starter.
8. Remove the fasteners holding the starter in place on the engine block.
9. Remove the starter.

To install:
10. Put the upper and lower brackets in place.
11. Align the interlock levers.
12. Lay the starter in place.
13. Install the fasteners that hold the starter in place on the engine block.
14. Connect the interlock rod between the shift mechanism and the interlock assembly on the starter.
15. Feed the rope around the idler pulley if equipped then through the nylon insert or rubber grommet at the cowling rope guide.

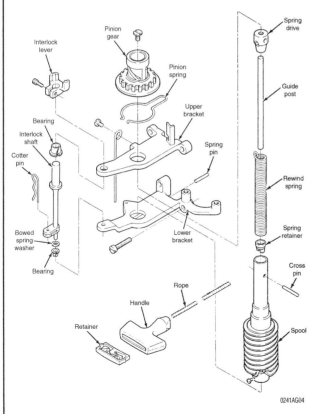

Fig. 36 Exploded drawing of a bendix type starter that would be found on a 7.5hp powerhead

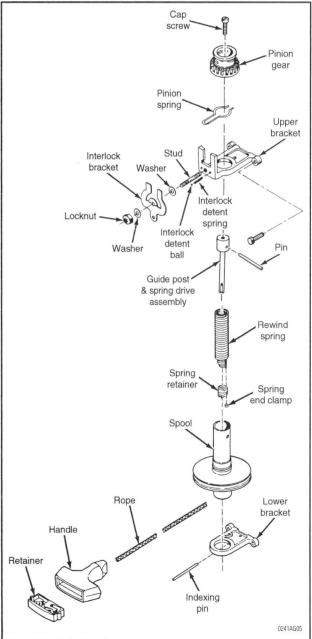

Fig. 37 Exploded drawing of a bendix type starter that would be found on a 9.9 or 15hp powerheads

11-12 HAND REWIND STARTER

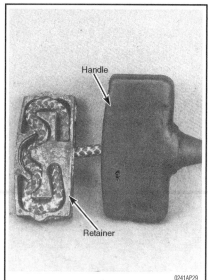

Fig. 38 Weave the rope through the retainer then push it firmly into the handle

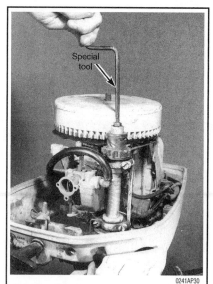

Fig. 39 Tightening the recoil spring using the special tool

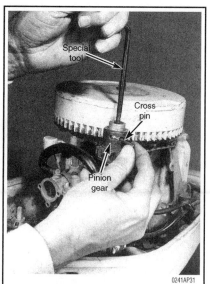

Fig. 40 With the recoil spring under tension place the pin through the starter shaft

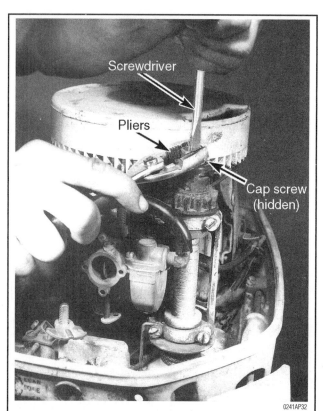

Fig. 41 Prevent the spring drive from rotating with a pair of pliers and tighten the cap screw securely

16. Slide the handle onto the rope end.
17. Tie a figure "8" knot at the starter handle leaving any excess rope in place.
18. Rotate the special angled tool counterclockwise until all slack is removed from the starter rope and the handle is lightly seated against the rope guide in the lower cowling.
19. With the handle lightly against the lower cowling continue rotating the tool 7½ to 8 turns.
20. Hold the tool firmly under tension and align the hole in the spring drive with the hole in the starter spool.

➡ It is suggested that a second person assist in holding the spring tension while the cross pin is being installed.

21. While holding this aligned position use a pin punch to drive the cross pin through the pinion gear, the spool bore, the spring guide, and finally flush with the other side of the pinion gear.
22. Position the cross pin ends flush with the diagonal slot in the pinion gear.
23. Remove the special tool.
24. Install the cap screw into the top of the spring drive.
25. Prevent the spring drive from rotating with a pair of pliers and tighten the cap screw securely.
26. Check action of the starter.
27. Check to be sure the interlock system prevents operation of the starter except when the unit is in neutral. Adjustment of the interlock system is described in the last portion of this section.

DISASSEMBLY

▶ See Figures 42 thru 47

1. Secure the starter in a vise.

➡ A special angled tool is now required, P/N T-3139-1 for the 7.5hp model or T-2985 for the 9.9 and 15hp models. If the special tool is not available a long bolt with the same threads as the cap screw removed may be used.

2. Thread the tool into the threads from which the cap screw was removed.
3. Have an assistant hold the special tool and use a pin punch to drive the cross pin out of the starter shaft and bendix area.
4. Once the cross pin has been removed relieve all remaining preload tension on the spring by slowly turning the special tool until the spring tension has been relieved.
5. Lift off the bendix gear and pinion spring.
6. Unwind the rope from the spool.
7. Remove the rope by either untying the knot, which may be on the top surface of the spool, or removing the screw that secures the rope end to the spool.
8. Remove the starter spring.
   a. On 7.5 hp pry the one or two spring ears clear of the posts on the spring drive and remove the spring retainer and guide post.
   b. On 9.9hp and 15hp the spring drive and the guide post are manufactured as one piece. Slide the spring end and drive post from the rewind spring. Pry the spring end open and remove the rewind spring from the retainer.

HAND REWIND STARTER 11-13

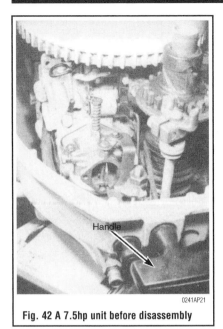

Fig. 42 A 7.5hp unit before disassembly

Fig. 43 A 9.9hp unit with the starter rope slacked to remove the pull handle

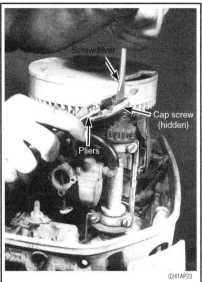

Fig. 44 Holding the shaft to remove the cap screw for special tool installation

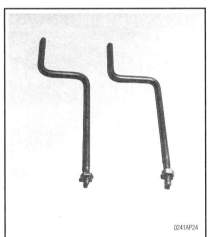

Fig. 45 Special angled tools used when servicing spool type hand rewind starters. Check with the local dealer for the correct number for the powerhead being serviced.

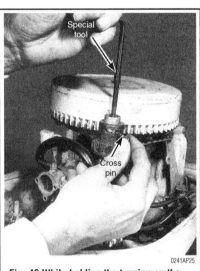

Fig. 46 While holding the tension on the starter spring remove the drive pin

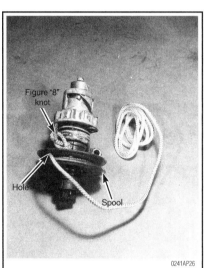

Fig. 47 A 9.9hp starter removed and ready for further disassembly

CLEANING & INSPECTION

▶ See Figures 48 and 49

Clean disassembled parts with solvent, and then dry them with compressed air.

Clean exposed parts of the rewind starter not disassembled.

Check to be sure that other visible areas are in satisfactory condition for further service.

Dampen a shop cloth with warm soapy water and clean the spiral groove of the starter spool.

Inspect the starter shaft bushing areas are not worn or scored beyond use.

Inspect the rope guides for sharp edges that could quickly wear or fray the rope.

Inspect the rope guide at the lower cowling. A nylon insert or a rubber grommet must be installed at this location to prevent wear on the rope. Many times the operator fails to pull the rope directly outward. Therefore, the opening in the cowling is a prime cause of excessive chafing and rope wear.

Throughly inspect the condition of the bendix gear. It should be replaced it shows signs of wear in the engagement tooth areas of the ramp area where it rotates on the starter shaft to engage the flywheel.

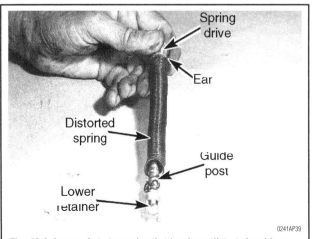

Fig. 48 A damaged starter spring that has been distorted and has a broken attaching ear

11-14 HAND REWIND STARTER

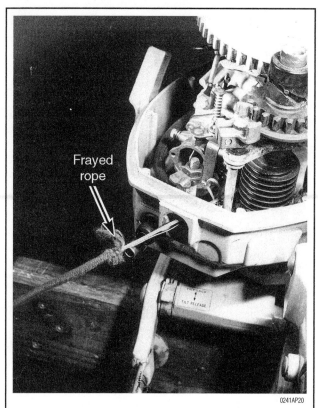

Fig. 49 The protective braided sleeve of this pull rope has been chewed away by sharp edges on the rope guide

Any broken or damaged parts should be taken to your parts dealer to be used as a guide to ensure the proper part is obtained for the unit being serviced.

ASSEMBLY

♦ See Figures 50 thru 55

1. Install the starter spring on the 7.5 hp powerhead as follows:
   a. Place the lower retainer onto the guide post.
   b. Insert the post and retainer into the rewind spring at the end with the ears.
   c. Slide the post and retainer down to the bottom of the spring.
   d. Guide the round post on the base of the retainer into the horizontal loop at the bottom of the spring.
   e. Insert the spring guide over the top of the guide post and hook the two spring ears over the posts on the spring guide.
2. Install the starter spring on a 9.9hp and 15hp powerhead as follows:
   a. Slide the spring end into the hole in the lower retainer. Install the small clamp over the spring end. Slide the spring drive and guide post assembly

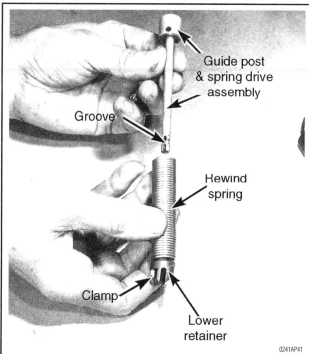

Fig. 51 Arrangement of parts for the guide post and spring drive assembly—9.9 and 15hp engine

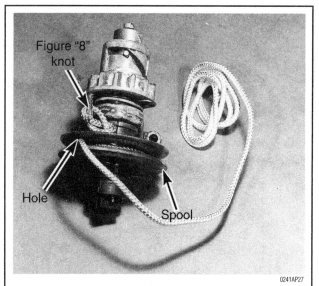

Fig. 50 Method used to secure the rope end to the spool—9.9 and 15hp

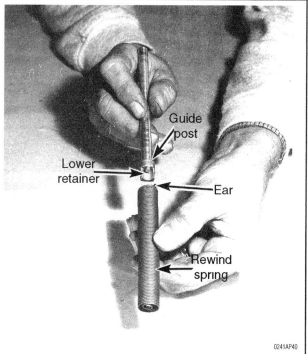

Fig. 52 Arrangement of parts for the guide post and spring drive assembly—7.5hp

HAND REWIND STARTER 11-15

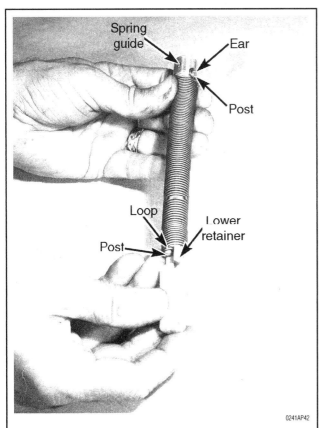

Fig. 53 Spring assembly with the spring tab aligned with the shaft

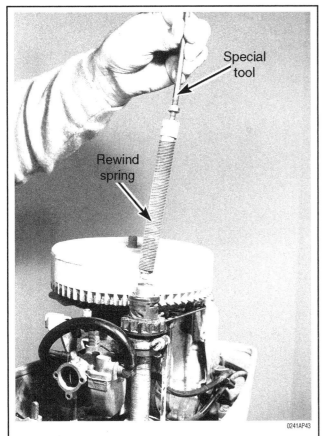

Fig. 54 Installing the starter spring

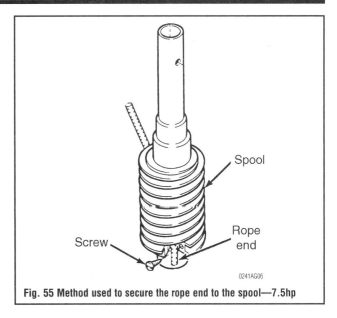

Fig. 55 Method used to secure the rope end to the spool—7.5hp

into the rewind spring. Index the groove at the end of the post into the lower retainer.

3. Lubricate the starter spring assembly with a generous coating of water resistant multi-purpose lubricant.
4. Thread the special tool into the spring guide.
5. Insert the assembly into the spool bore.
6. The lower slotted end of the retainer must index into the groove of the retainer extension.
7. Verify that the retainer has indexed into the groove by rotating the special tool. The starter spool should also rotate in the same direction. If the spool fails to rotate as the special tool is rotated rock the tool until the retainer and the extension do index.

➡ The bendix gear, pinion spring and brackets will be installed as part of the installation steps section.

8. Melt the end of the rope over a match flame to fuse the strands together.
9. Install the starter rope on a 7.5 hp powerhead as follows:
    a. Push the rope end into the hole at the base of the starter spool.
    b. Pull the rope through until approximately 3/16" (4.7mm) extends beyond the hole.
    c. Bend this end of the rope to the right and tuck it under the head of the retaining screw.
    d. Tighten the screw securely.
10. Install the starter rope on a 9.9hp and 15hp powerhead as follows:
    a. Pass the rope end through the hole in the top spool surface.
    b. Tie a figure "8" knot in the rope end above the surface, leaving approximately an inch (2.5cm), of rope beyond the knot.
11. Wind the rope around the spool or spiral counterclockwise until the turns fill the spool or the spiral.

INTERLOCK ADJUSTMENT

7.5 HP

♦ See Figure 56

➡ It is recommended that an assistant be used during this procedure.

1. Move the shift lever to the neutral position.
2. Hold the interlock lock shaft down against the lower starter mounting bracket.
3. Loosen the screw securing the interlock lever to the interlock shaft.
4. Rotate the pinion gear counterclockwise as far as possible while pulling up on the starter spool.
5. Rotate the interlock lever until the arms are an equal distance from the pinion gear teeth.
6. Using a feeler gauge check the clearance between the bottom of the interlock lever and the top of the pinion gear flange.

11-16 HAND REWIND STARTER

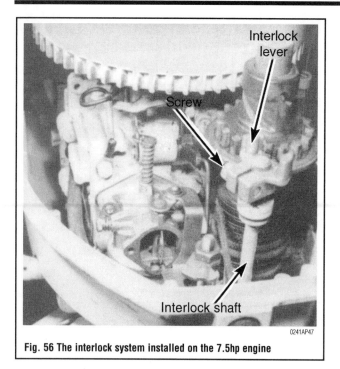

Fig. 56 The interlock system installed on the 7.5hp engine

7. The recommended clearance is 0.020" (0.51mm).
8. Tighten the screw to hold the interlock lever in this position.
9. When the shift lever is in the neutral position, and the starter rope is pulled, the interlock lever should clear the pinion gear flange and permit the gear to rise and engage the flywheel.
10. When the shift lever is in the forward or reverse position, the interlock lever should hold the pinion gear down and not allow the gear to rise and engage the flywheel.

9.9hp and 15hp

♦ See Figures 57, 58 and 59

1. To adjust pry the shift rod ball joint free of the gear shift lever.
2. Loosen the locknut at the ball joint and adjust the length of the shift rod until the interlock bracket swings far enough to block action of the pinion gear.
3. Tighten the locknut to hold the adjustment.

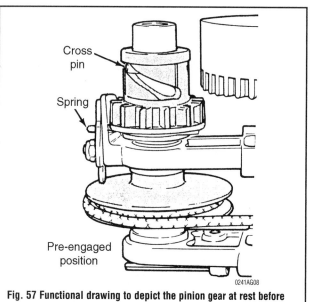

Fig. 57 Functional drawing to depict the pinion gear at rest before the rope is pulled

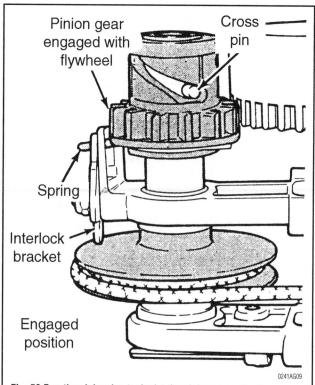

Fig. 58 Functional drawing to depict the pinion gear after the rope is pulled with the outboard unit in neutral

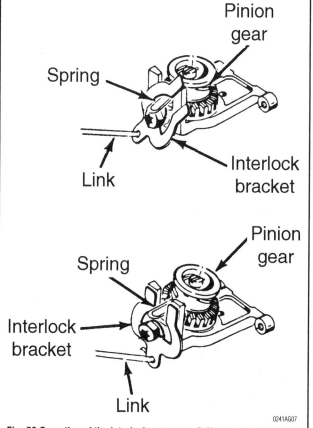

Fig. 59 Operation of the interlock system on 9.9hp and 15hp powerheads

GLOSSARY

GLOSSARY

Understanding your marine mechanic is as important as understanding your outboard. Most boaters know about their boats, but many boaters have difficulty understanding engine terminology. Talking the language of outboards makes it easier to effectively communicate with professional mechanics. It isn't necessary (or recommended) that you diagnose the problem for them, but it will save them time, and you money, if you can accurately describe what is happening. It will also help you to know why your boat does what it is doing, and what repairs were made.

AIR/FUEL RATIO: The ratio of air-to-fuel, by weight, drawn into the engine.

ALTERNATING CURRENT (AC): Electric current that flows first in one direction, then in the opposite direction, continually reversing flow.

AMP/HR. RATING (BATTERY): Measurement of the ability of a battery to deliver a stated amount of current for a stated period of time. The higher the amp/hr. rating, the better the battery.

ATOMIZATION: The breaking down of a liquid into a fine mist that can be suspended in air.

AXIAL PLAY: Movement parallel to a shaft or bearing bore.

BACKFIRE: The sudden combustion of gases in the intake or exhaust system that results in a loud explosion.

BACKLASH: The clearance or play between two parts, such as meshed gears.

BATTERY: A direct current electrical storage unit, consisting of the basic active materials of lead and sulfuric acid, which converts chemical energy into electrical energy. Used to provide current for the operation of the starter as well as other equipment, such as the radio, lighting, etc.

BEARING: A friction reducing, supportive device usually located between a stationary part and a moving part.

BEFORE TOP DEAD CENTER (BTDC): The point just before the piston reaches the top of its travel on the compression stroke.

BORE: Diameter of a cylinder.

BTDC: Before Top Dead Center.

BUSHING: A liner, usually removable, for a bearing; an anti-friction liner used in place of a bearing.

CARBON MONOXIDE (CO): A colorless, odorless gas given off as a normal byproduct of combustion. It is poisonous and extremely dangerous in confined areas, building up slowly to toxic levels without warning if adequate ventilation is not available.

CHECK VALVE: Any one-way valve installed to permit the flow of air, fuel or vacuum in one direction only.

CIRCLIP: A split steel snapring that fits into a groove to hold various parts in place.

CIRCUIT: Any unbroken path through which an electrical current can flow. Also used to describe fuel flow in some instances.

COMPRESSION CHECK: A test involving cranking the engine with a special high pressure gauge connected to an individual cylinder. Individual cylinder pressure as well as pressure variance across cylinders is used to determine general operating condition of the engine.

COMPRESSION RATIO: The ratio of the volume between the piston and cylinder head when the piston is at the bottom of its stroke (bottom dead center) and when the piston is at the top of its stroke (top dead center).

CONDUCTOR: Any material through which an electrical current can be transmitted easily.

CONNECTING ROD: The connecting link between the crankshaft and piston.

CONTINUITY: Continuous or complete circuit. Can be checked with an ohmmeter.

CRANKCASE: The lower part of an engine in which the crankshaft and related parts operate.

CRANKSHAFT: Engine component (connected to pistons by connecting rods) which converts the reciprocating (up and down) motion of pistons to rotary motion used to turn the driveshaft.

CYLINDER: In an engine, the round hole in the engine block in which the piston(s) ride.

DETONATION: An unwanted explosion of the air/fuel mixture in the combustion chamber caused by excess heat and compression, advanced timing, or an overly lean mixture. Also referred to as "ping".

DIAPHRAGM: A thin, flexible wall separating two cavities, such as in a vacuum advance unit.

DIODE: An electrical device that will allow current to flow in one direction only.

DISPLACEMENT: The total volume of air that is displaced by all pistons as the engine turns through one complete revolution.

DVOM: Digital volt ohmmeter

END-PLAY: The measured amount of axial movement in a shaft.

FEELER GAUGE: A blade, usually metal, of precisely predetermined thickness, used to measure the clearance between two parts.

FIRING ORDER: The order in which combustion occurs in the cylinders of an engine.

FLYWHEEL: A heavy disc of metal attached to the rear of the crankshaft. It smoothes the firing impulses of the engine and keeps the crankshaft turning during periods when no firing takes place. The starter also engages the flywheel to start the engine.

FOOT POUND (ft. lbs. or sometimes, ft. lb.): The amount of energy or work needed to raise an item weighing one pound, a distance of one foot.

FUEL FILTER: A component of the fuel system containing a porous paper element used to prevent any impurities from entering the engine through the fuel system. It usually takes the form of a canister-like housing, mounted in-line with the fuel hose, located anywhere on a vessel between the fuel tank and engine.

FUEL INJECTION: A system that sprays fuel into the cylinder through nozzles. The amount of fuel can be more precisely controlled with fuel injection.

FUSE: A protective device in a circuit which prevents circuit overload by breaking the circuit when a specific amperage is present. The device is constructed around a strip or wire of a lower amperage rating than the circuit it is designed to protect. When an amperage higher than that stamped on the fuse is present in the circuit, the strip or wire melts, opening the circuit.

HORSEPOWER: A measurement of the amount of work; one horsepower is the amount of work necessary to lift 33,000 lbs. one foot in one minute. Brake horsepower (bhp) is the horsepower delivered by an engine on a dynamometer. Net horsepower is the power remaining (measured at the flywheel of the engine) that can be used to power the vessel after power is consumed through friction and running the engine accessories (water pump, alternator, fan etc.)

HYDROMETER: An instrument used to measure the specific gravity of a solution.

IMPELLER: The portion of the water pump which provides the propulsion for the coolant to circulate it through the system

INCH POUND (inch lbs.; sometimes in. lb. or in. lbs.): One twelfth of a foot pound.

INJECTOR: A device which receives metered fuel under relatively low pressure and is activated to inject the fuel into the engine under relatively high pressure at a predetermined time.

JOURNAL: The bearing surface within which a shaft operates.

KNOCK: Noise which results from the spontaneous ignition of a portion of the air-fuel mixture in the engine cylinder.

MISFIRE: Condition occurring when the fuel mixture in a cylinder fails to ignite, causing the engine to run roughly.

MULTI-WEIGHT: Type of oil that provides adequate lubrication at both high and low temperatures.

NEEDLE BEARING: A bearing which consists of a number (usually a large number) of long, thin rollers.

OEM: Original Equipment Manufactured. OEM equipment is that furnished standard by the manufacturer.

PING: A metallic rattling sound produced by the engine during acceleration. It is usually due to incorrect timing or a poor grade of fuel.

POLARITY: Indication (positive or negative) of the two poles of a battery.

PRELOAD: A predetermined load placed on a bearing during assembly or by adjustment.

PRESS FIT: The mating of two parts under pressure, due to the inner diameter of one being smaller than the outer diameter of the other, or vice versa; an interference fit.

PSI: Pounds per square inch; a measurement of pressure.

RECTIFIER: A device (used primarily in alternators) that permits electrical current to flow in one direction only.

REGULATOR: A device which maintains the amperage and/or voltage levels of a circuit at predetermined values.

RESISTOR: A device, usually made of wire, which offers a preset amount of resistance in an electrical circuit.

RPM: Revolutions per minute (usually indicates engine speed).

SENSOR: Any device designed to measure engine operating conditions or ambient pressures and temperatures. Usually electronic in nature and designed to send a voltage signal to an on-board computer, some sensors may operate as a simple on/off switch or they may provide a variable voltage signal (like a potentiometer) as conditions or measured parameters change.

SHIM: Spacers of precise, predetermined thickness used between parts to establish a proper working relationship.

SOLENOID: An electrically operated, magnetic switching device.

SPLINES: Ridges machined or cast onto the outer diameter of a shaft or inner diameter of a bore to enable parts to mate without rotation.

STARTER: A high-torque electric motor used for the purpose of starting the engine, typically through a high ratio geared drive connected to the flywheel ring gear.

STROKE: The distance the piston travels from bottom dead center to top dead center.

TACHOMETER: A device used to measure the rotary speed of an engine, shaft, gear, etc., usually in rotations per minute.

THERMOSTAT: A valve, located in the cooling system of an engine, which is closed when cold and opens gradually in response to engine heating, controlling the temperature of the coolant and rate of coolant flow.

TOP DEAD CENTER (TDC): The point at which the piston reaches the top of its travel on the compression stroke.

TORQUE: Measurement of turning or twisting force, expressed as foot-pounds or inch-pounds.

TUNE-UP: A regular maintenance function, usually associated with the replacement and adjustment of parts and components in the electrical and fuel systems of a engine for the purpose of attaining optimum performance.

VOLTAGE REGULATOR: A device that controls the current output of the alternator or generator.

ANODES (ZINCS) 3-9
    INSPECTION 3-9
    SERVICING 3-9
AVOIDING THE MOST COMMON MISTAKES 1-3
AVOIDING TROUBLE 1-2
BASIC ELECTRICAL THEORY 5-2
    HOW ELECTRICITY WORKS: THE WATER ANALOGY 5-2
    OHM'S LAW 5-2
BATTERIES (BOAT MAINTENANCE) 3-10
    CLEANING 3-10
    MAINTENANCE 3-10
    STORAGE 3-11
    TESTING 3-11
BATTERY (CHARGING CIRCUIT) 5-48
    BATTERY CABLES 5-50
    BATTERY CHARGERS 5-49
    BATTERY CONSTRUCTION 5-48
    BATTERY LOCATION 5-49
    BATTERY RATINGS 5-49
    MARINE BATTERIES 5-48
BEARINGS 7-19
    GENERAL INFORMATION 7-19
    INSPECTION 7-20
BENDIX TYPE STARTER 11-11
    ASSEMBLY 11-14
    CLEANING & INSPECTION 11-13
    DISASSEMBLY 11-12
    INTERLOCK ADJUSTMENT 11-15
    REMOVAL & INSTALLATION 11-11
BOAT MAINTENANCE 3-10
BOATING SAFETY 1-3
BOLTS, NUTS AND OTHER THREADED RETAINERS 2-11
BREAKER POINTS 5-11
    POINT GAP ADJUSTMENT 5-11
    REMOVAL & INSTALLATION 5-12
    TESTING 5-12
BREAKER POINTS (MAGNETO) IGNITION 5-8
BUY OR REBUILD? 7-9
CAN YOU DO IT? 1-2
CAPACITIES 3-32
CAPACITOR DISCHARGE MODULE (CDM) SYSTEM 5-38
CARBURETED FUEL SYSTEM 4-3
CARBURETOR 4-7
    OVERHAUL 4-10
    REMOVAL & INSTALLATION 4-7
CARBURETOR SPECIFICATIONS 3-32
CDM IGNITION COIL MODULE 5-43
    REMOVAL & INSTALLATION 5-43
    TESTING 5-43
CENTER MOUNTED PISTON TRIM/TILT SYSTEM 9-3
CHARGING CIRCUIT 5-45
CHARGING COILS (CAPACITOR DISCHARGE MODULE SYSTEM) 5-41
    REMOVAL & INSTALLATION 5-42
    TESTING 5-41
CHARGING COILS (PRESTOLITE® ELECTRONIC IGNITION) 5-27
    REMOVAL & INSTALLATION 5-28
    TESTING 5-27
CHARGING COILS (THUNDERBOLT® ELECTRONIC IGNITION SYSTEM) 5-35
    REMOVAL & INSTALLATION 5-36
    TESTING 5-35
CHEMICALS 2-3
    CLEANERS 2-3
    LUBRICANTS & PENETRANTS 2-3
    SEALANTS 2-3
COMBUSTION 4-2
COMPRESSION TEST 3-13
    PRIMARY COMPRESSION 3-13
    SECONDARY COMPRESSION TEST 3-13
CONDENSER 5-12
    REMOVAL & INSTALLATION 5-12

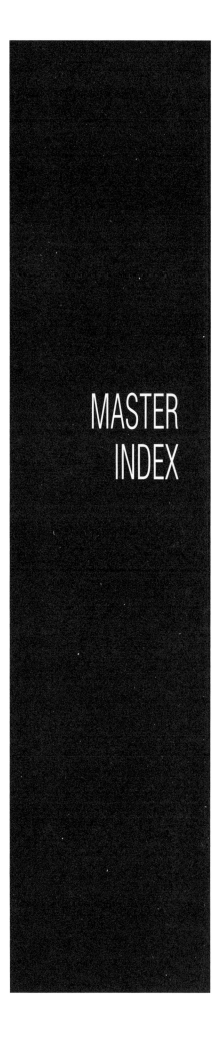

MASTER INDEX

MASTER INDEX

TESTING 5-12
CONNECTING RODS 7-17
    GENERAL INFORMATION 7-17
    INSPECTION 7-17
CONVERSION FACTORS 2-14
COOLING SYSTEM 6-2
COURTESY MARINE EXAMINATIONS 1-11
CRANKSHAFT 7-18
    GENERAL INFORMATION 7-18
    INSPECTION 7-18
CYLINDER BLOCK AND HEAD 7-11
    GENERAL INFORMATION 7-11
    INSPECTION 7-12
CYLINDER BORES 7-13
    GENERAL INFORMATION 7-13
    INSPECTION 7-13
    REFINISHING 7-13
CYLINDER HEAD TEMPERATURE SENSOR 6-10
    REMOVAL & INSTALLATION 6-10
    TESTING 6-10
DESCRIPTION AND OPERATION (CAPACITOR DISCHARGE MODULE SYSTEM) 5-38
DESCRIPTION AND OPERATION (CARBURETED FUEL SYSTEM) 4-3
    BASIC FUNCTIONS 4-3
    CARBURETOR CIRCUITS 4-4
    FUEL & AIR METERING 4-3
    FUEL PUMP 4-5
DESCRIPTION AND OPERATION (CENTER MOUNTED PISTON TRIM/TILT SYSTEM) 9-3
DESCRIPTION AND OPERATION (CHARGING CIRCUIT) 5-45
DESCRIPTION AND OPERATION (COOLING SYSTEM) 6-2
DESCRIPTION AND OPERATION (ELECTRONIC IGNITION SYSTEMS UP TO 25 HP) 5-22
DESCRIPTION AND OPERATION (HAND REWIND STARTER) 11-2
DESCRIPTION AND OPERATION (MANUAL TILT) 9-2
    SERVICING 9-2
DESCRIPTION AND OPERATION (MOTOROLA® DISTRIBUTOR IGNITION) 5-13
DESCRIPTION AND OPERATION (PRESTOLITE® ELECTRONIC IGNITION) 5-24
DESCRIPTION AND OPERATION (REMOTE CONTROL BOX) 10-2
DESCRIPTION AND OPERATION (SINGLE RAM INTEGRAL POWER TRIM/TILT SYSTEM) 9-11
DESCRIPTION AND OPERATION (STARTING CIRCUIT) 5-50
DESCRIPTION AND OPERATION (THUNDERBOLT® ELECTRONIC IGNITION SYSTEM) 5-32
DESCRIPTION AND OPERATION (TILLER HANDLE) 10-11
DESCRIPTION AND OPERATION (WARNING SYSTEMS) 6-10
DETERMINING POWERHEAD CONDITION 7-8
    COMPRESSION TEST 7-8
DIRECTIONS AND LOCATIONS 1-2
DISTRIBUTOR 5-15
    ASSEMBLY 5-19
    CLEANING & INSPECTION 5-19
    DISASSEMBLY 5-17
    REMOVAL & INSTALLATION 5-16
    TESTING 5-15
DO'S 1-11
DON'TS 1-12
ELECTRICAL COMPONENTS 5-2
    GROUND 5-2
    POWER SOURCE 5-2
ELECTRONIC IGNITION SYSTEMS UP TO 25 HP 5-22
EMERGENCY STOP SWITCH (LANYARD) (REMOTE CONTROL BOX) 10-6
    REMOVAL & INSTALLATION 10-6
    TESTING 10-6
EMERGENCY STOP SWITCH (LANYARD) (TILLER HANDLE) 10-14

    REMOVAL & INSTALLATION 10-14
    TESTING 10-14
ENGINE MAINTENANCE 3-2
ENGINE MECHANICAL 7-2
ENGINE STOP SWITCH 10-13
    REMOVAL & INSTALLATION 10-14
    TESTING 10-13
EQUIPMENT NOT REQUIRED BUT RECOMMENDED 1-10
    ANCHORS 1-10
    BAILING DEVICES 1-10
    FIRST AID KIT 1-10
    SECOND MEANS OF PROPULSION 1-10
    TOOLS & SPARE PARTS 1-11
    VHF-FM RADIO 1-11
FASTENERS, MEASUREMENTS AND CONVERSIONS 2-11
FIBERGLASS HULL 3-12
FLYWHEEL (BREAKER POINTS IGNITION) 5-10
    INSPECTION 5-11
    REMOVAL & INSTALLATION 5-10
FLYWHEEL (CAPACITOR DISCHARGE MODULE SYSTEM) 5-40
    INSPECTION 5-41
    REMOVAL & INSTALLATION 5-40
FLYWHEEL (ELECTRONIC IGNITION SYSTEMS UP TO 25 HP) 5-22
    INSPECTION 5-24
    REMOVAL & INSTALLATION 5-22
FLYWHEEL (MOTOROLA® DISTRIBUTOR IGNITION) 5-14
    INSPECTION 5-15
    REMOVAL & INSTALLATION 5-14
FLYWHEEL (PRESTOLITE® ELECTRONIC IGNITION) 5-25
    INSPECTION 5-27
    REMOVAL & INSTALLATION 5-25
FLYWHEEL (THUNDERBOLT® ELECTRONIC IGNITION SYSTEM) 5-34
    INSPECTION 5-35
    REMOVAL & INSTALLATION 5-34
FLYWHEEL TIMING MARKS 7-20
    ALTERNATE METHOD 7-21
    USING THE FACTORY TIMING GAUGE 7-20
FUEL 4-2
    ALCOHOL-BLENDED FUELS 4-2
    HIGH-ALTITUDE OPERATION 4-2
    OCTANE RATING 4-2
    RECOMMENDATIONS 4-2
    VAPOR PRESSURE 4-2
FUEL AND COMBUSTION 4-2
FUEL FILTER 3-5
    RELIEVING FUEL SYSTEM PRESSURE 3-6
    REMOVAL & INSTALLATION 3-6
FUEL LINES 4-26
FUEL PRIMER 4-22
    TESTING 4-23
FUEL PUMP 4-23
    OVERHAUL 4-24
    REMOVAL & INSTALLATION 4-23
    TESTING 4-23
FUEL/WATER SEPARATOR 3-7
    DRAINING 3-7
    SERVICE 3-7
GENERAL ENGINE SPECIFICATIONS 3-30
GENERAL INFORMATION (LOWER UNIT) 8-2
HAND REWIND STARTER 11-2
HAND TOOLS 2-5
    ELECTRONIC TOOLS 2-9
    GAUGES 2-10
    HAMMERS 2-8
    OTHER COMMON TOOLS 2-8
    PLIERS 2-8
    SCREWDRIVERS 2-8
    SOCKET SETS 2-5

MASTER INDEX

SPECIAL TOOLS 2-8
WRENCHES 2-6
HOW TO USE THIS MANUAL 1-2
IDLE SPEED AND MIXTURE ADJUSTMENT 3-18
ENGINES WITH A SINGLE CARBURETOR AND TILLER HANDLE 3-18
IDLE SPEED ADJUSTMENT 3-18
ENGINES WITH FACE PLATE MOUNTED CONTROLS 3-18
IGNITION AND ELECTRICAL WIRING DIAGRAMS 5-59
IGNITION COIL (BREAKER POINTS IGNITION) 5-13
REMOVAL & INSTALLATION 5-13
TESTING 5-13
IGNITION COILS (ELECTRONIC IGNITION SYSTEMS UP TO 25 HP) 5-24
REMOVAL & INSTALLATION 5-24
TESTING 5-24
IGNITION COILS (MOTOROLA® DISTRIBUTOR IGNITION) 5-21
REMOVAL & INSTALLATION 5-21
TESTING 5-21
IGNITION COILS (PRESTOLITE® ELECTRONIC IGNITION) 5-31
REMOVAL & INSTALLATION 5-31
TESTING 5-31
IGNITION COILS (THUNDERBOLT® ELECTRONIC IGNITION SYSTEM) 5-37
REMOVAL & INSTALLATION 5-38
TESTING 5-37
IGNITION MODULE (MOTOROLA® DISTRIBUTOR IGNITION) 5-21
REMOVAL & INSTALLATION 5-22
TESTING 5-21
IGNITION MODULE (PRESTOLITE® ELECTRONIC IGNITION) 5-32
TESTING 5-32
IGNITION SWITCH 10-6
REMOVAL & INSTALLATION 10-6
TESTING 10-6
IGNITION SYSTEM 3-17
IGNITION SYSTEM SPECIFICATIONS 5-44
INTRODUCTION (TUNE-UP) 3-12
LOWER UNIT 8-2
LOWER UNIT (ENGINE MAINTENANCE) 3-3
DRAINING LOWER UNIT 3-4
FILLING LOWER UNIT 3-4
OIL RECOMMENDATIONS 3-3
LOWER UNIT (LOWER UNIT) 8-4
REMOVAL & INSTALLATION 8-4
LOWER UNIT OVERHAUL 8-7
1984-91 70 TO 150 HP 8-19
ASSEMBLY & SHIMMING 8-24
CLEANING & INSPECTION 8-22
DISASSEMBLY 8-19
1991-95 70 TO 150 HP 8-27
ASSEMBLY 8-37
CLEANING & INSPECTION 8-33
DISASSEMBLY 8-27
SHIMMING 8-35
1995-99 70 TO 120 HP 8-42
ASSEMBLY 8-47
CLEANING & INSPECTION 8-45
DISASSEMBLY 8-42
SHIMMING 8-49
UNITS UP TO 7.5 HP 8-7
ASSEMBLY 8-12
CLEANING & INSPECTION 8-10
DISASSEMBLY 8-7
LUBRICATION POINT/FREQUENCY 3-5
LUBRICATION POINTS 3-7
INSPECTION & LUBRICATION 3-7
MAINTENANCE INTERVALS 3-29
MAINTENANCE OR REPAIR? 1-2
MANUAL TILT 9-2

MEASURING TOOLS 2-10
DEPTH GAUGES 2-11
DIAL INDICATORS 2-10
MICROMETERS & CALIPERS 2-10
TELESCOPING GAUGES 2-11
MOTOROLA® DISTRIBUTOR IGNITION 5-13
NEUTRAL START SWITCH 10-5
REMOVAL & INSTALLATION 10-5
TESTING 10-5
OVERHEAD TYPE STARTER 11-2
PISTON PINS 7-15
GENERAL INFORMATION 7-15
INSPECTION 7-15
PISTON RINGS 7-16
GENERAL INFORMATION 7-16
INSPECTION 7-16
PISTONS 7-14
GENERAL INFORMATION 7-14
INSPECTION 7-14
POWERHEAD 7-2
REMOVAL & INSTALLATION 7-2
POWERHEAD EXPLODED VIEWS 7-22
POWERHEAD OVERHAUL TIPS 7-9
CAUTIONS 7-10
CLEANING 7-10
REPAIRING DAMAGED THREADS 7-10
TOOLS 7-9
POWERHEAD PREPARATION 7-11
POWERHEAD RECONDITIONING 7-8
PRESTOLITE® ELECTRONIC IGNITION 5-24
PROFESSIONAL HELP 1-2
PROPELLER (ENGINE MAINTENANCE) 3-8
PROPELLER (LOWER UNIT) 8-3
REMOVAL & INSTALLATION 8-3
PROTECTIVE DEVICES 5-3
CONNECTORS 5-4
LOAD 5-4
SWITCHES & RELAYS 5-3
WIRING & HARNESSES 5-4
PURCHASING PARTS 1-3
RECIRCULATING SYSTEM 4-26
RECTIFIER (CHARGING CIRCUIT) 5-46
REMOVAL & INSTALLATION 5-47
TESTING 5-46
RECTIFIER/REGULATOR (CHARGING CIRCUIT) 5-47
REMOVAL & INSTALLATION 5-48
TESTING 5-47
REED VALVE 7-6
INSPECTION & CLEANING 7-8
REMOVAL & INSTALLATION 7-6
REGULATIONS FOR YOUR BOAT 1-3
CAPACITY INFORMATION 1-4
CERTIFICATE OF COMPLIANCE 1-4
DOCUMENTING OF VESSELS 1-4
HULL IDENTIFICATION NUMBER 1-4
LENGTH OF BOATS 1-4
NUMBERING OF VESSELS 1-4
REGISTRATION OF BOATS 1-4
SALES & TRANSFERS 1-4
VENTILATION 1-4
VENTILATION SYSTEMS 1-5
REMOTE CONTROL BOX 10-2
REMOTE CONTROL BOX 10-2
ASSEMBLY 10-4
CLEANING AND INSPECTION 10-4
DISASSEMBLY 10-2
REMOVAL & INSTALLATION 10-2
REMOTE CONTROL CABLES 10-5

ADJUSTMENT 10-5
REMOVAL & INSTALLATION 10-5
REQUIRED SAFETY EQUIPMENT 1-5
  FIRE EXTINGUISHERS 1-5
  PERSONAL FLOTATION DEVICES 1-6
  SOUND PRODUCING DEVICES 1-8
  TYPES OF FIRES 1-5
  VISUAL DISTRESS SIGNALS 1-8
  WARNING SYSTEM 1-6
SAFETY IN SERVICE 1-11
SAFETY TOOLS 2-2
  EYE & EAR PROTECTION 2-2
  WORK CLOTHES 2-3
  WORK GLOVES 2-2
SERIAL NUMBER IDENTIFICATION 3-2
SHIFTING PRINCIPLES 8-2
  STANDARD ROTATING UNIT 8-2
SHIM CHART—DRIVESHAFT 8-35
SINGLE RAM INTEGRAL POWER TRIM/TILT SYSTEM 9-11
SPARK PLUG WIRES 3-16
  REMOVAL & INSTALLATION 3-17
  TESTING 3-16
SPARK PLUGS 3-13
  INSPECTION & GAPPING 3-16
  READING SPARK PLUGS 3-15
  REMOVAL & INSTALLATION 3-14
  SPARK PLUG HEAT RANGE 3-14
  SPARK PLUG SERVICE 3-14
SPECIFICATIONS CHARTS
  CAPACITIES 3-32
  CARBURETOR SPECIFICATIONS 3-32
  CONVERSION FACTORS 2-14
  GENERAL ENGINE SPECIFICATIONS 3-30
  IGNITION SYSTEM SPECIFICATIONS 5-44
  LUBRICATION POINT/FREQUENCY 3-5
  MAINTENANCE INTERVALS 3-29
  SHIM CHART—DRIVESHAFT 8-35
  TUNE-UP SPECIFICATIONS 3-31
SPRING COMMISSIONING CHECKLIST 3-28
STANDARD AND METRIC MEASUREMENTS 2-13
STARTER (BENDIX TYPE) 11-11
STARTER (HAND REWIND) 11-2
STARTER (OVERHEAD TYPE) 11-2
  OVERHAUL 1-CYLINDER POWERHEAD 11-2
    ASSEMBLY 11-5
    CLEANING & INSPECTION 11-5
    DISASSEMBLY 11-3
    REMOVAL & INSTALLATION 11-2
  OVERHAUL 2 AND 3-CYLINDER POWERHEADS 11-6
    ASSEMBLY 11-10
    CLEANING & INSPECTION 11-9
    DISASSEMBLY 11-7
    REMOVAL & INSTALLATION 11-6
STARTER MOTOR 5-51
  ASSEMBLY 5-57
  CLEANING & INSPECTION 5-54
  DISASSEMBLY 5-53
  REMOVAL & INSTALLATION 5-52
  TESTING 5-55
STARTER RELAY 5-58
  REMOVAL & INSTALLATION 5-58
  TESTING 5-58
STARTING CIRCUIT 5-50
STATOR 5-45
  REMOVAL & INSTALLATION 5-46
  TESTING 5-45
SWITCH BOX 5-38
  REMOVAL & INSTALLATION 5-38

  TESTING 5-38
SYNCHRONIZATION AND TIMING 3-17
  MULTIPLE CARBURETOR ENGINES WITH MOTOROLA® DISTRIBUTOR IGNITION 3-21
    CHOKE ROD ADJUSTMENT 3-21
    IDLE SPEED ADJUSTMENT 3-24
    IGNITION TIMING ADJUSTMENT 3-24
    THROTTLE ROD ADJUSTMENT 3-22
    TIMING BELT ADJUSTMENT 3-23
  MULTIPLE CARBURETOR ENGINES WITH NO DISTRIBUTOR 3-24
    CHOKE ROD ADJUSTMENT 3-25
    IDLE SPEED ADJUSTMENT 3-27
    IGNITION TIMING 3-27
    THROTTLE ROD ADJUSTMENT 3-26
  9.9 HP AND LARGER ENGINES WITH A SPLIT GEARCASE 8-12
    ASSEMBLY 8-17
    CLEANING & INSPECTION 8-16
    DISASSEMBLY 8-13
    SHIMMING 8-19
  1984-89 ENGINES WITH A SINGLE CARBURETOR AND REMOTE CONTROL 3-19
    CHOKE ROD ADJUSTMENT 3-19
    IDLE SPEED & MIXTURE ADJUSTMENT 3-19
    IGNITION TIMING ADJUSTMENT 3-20
    THROTTLE ROD ADJUSTMENT 3-19
  1990-99 ENGINES WITH A SINGLE CARBURETOR AND REMOTE CONTROL 3-20
    CHOKE ROD ADJUSTMENT 3-20
    IDLE SPEED ADJUSTMENT 3-21
    IGNITION TIMING ADJUSTMENT 3-21
    THROTTLE ROD ADJUSTMENT 3-20
TEST EQUIPMENT (UNDERSTANDING AND TROUBLESHOOTING ELECTRICAL SYSTEMS) 5-5
  JUMPER WIRES 5-5
  MULTIMETERS 5-6
  TEST LIGHTS 5-5
THERMOSTAT 6-9
  REMOVAL & INSTALLATION 6-9
THUNDERBOLT® ELECTRONIC IGNITION SYSTEM 5-32
TILLER HANDLE 10-11
  REMOVAL & INSTALLATION 10-11
TILLER HANDLE 10-11
TILT/TRIM CYLINDER 9-15
  CLEANING & INSPECTION 9-17
  REMOVAL & INSTALLATION 9-15
TILT/TRIM MOTOR 9-13
  CLEANING & INSPECTION 9-14
  REMOVAL & INSTALLATION 9-13
  TESTING 9-13
TILT/TRIM RELAY 9-18
  TESTING 9-18
TILT/TRIM SWITCH 9-18
  TESTING 9-18
TOOLS 2-4
TOOLS AND EQUIPMENT 2-2
TORQUE (FASTENERS, MEASUREMENTS AND CONVERSIONS) 2-13
TORQUE SEQUENCE DIAGRAMS 7-29
TRIGGER COILS (CAPACITOR DISCHARGE MODULE SYSTEM) 5-42
  REMOVAL 5-42
  TESTING 5-42
TRIGGER COILS (PRESTOLITE® ELECTRONIC IGNITION) 5-29
  REMOVAL & INSTALLATION 5-30
  TESTING 5-29
TRIGGER COILS (THUNDERBOLT® ELECTRONIC IGNITION SYSTEM) 5-36
  REMOVAL & INSTALLATION 5-36
  TESTING 5-36
TRIM/TILT CYLINDER 9-8

MASTER INDEX

ASSEMBLY 9-10
CLEANING & INSPECTION 9-10
DISASSEMBLY 9-9
REMOVAL & INSTALLATION 9-8
TESTING 9-8
TRIM/TILT MOTOR 9-6
ASSEMBLY 9-7
CLEANING & INSPECTION 9-7
DISASSEMBLY 9-7
REMOVAL & INSTALLATION 9-7
TESTING 9-6
TRIM/TILT PUMP (CENTER MOUNTED PISTON TRIM/TILT SYSTEM) 9-4
ASSEMBLY 9-6
CLEANING & INSPECTION 9-6
DISASSEMBLY 9-5
REMOVAL & INSTALLATION 9-5
SYSTEM BLEEDING 9-6
TESTING 9-4
TRIM/TILT PUMP (SINGLE RAM INTEGRAL POWER TRIM/TILT SYSTEM) 9-11
HYDRAULIC SYSTEM BLEEDING 9-12
REMOVAL & INSTALLATION 9-11
TRIM/TILT RELAY 9-11
REMOVAL & INSTALLATION 9-11
TESTING 9-11
TRIM/TILT SWITCH 9-10
TESTING 9-10
TRIM/TILT SYSTEM (CENTER MOUNTED PISTON) 9-3
TROUBLESHOOTING ELECTRONIC IGNITION SYSTEMS UP TO 25 HP 5-22
TROUBLESHOOTING THE BREAKER POINTS (MAGNETO) IGNITION 5-9
TROUBLESHOOTING THE CDM SYSTEM 5-39
TROUBLESHOOTING THE CENTER MOUNTED PISTON TRIM/TILT SYSTEM 9-4
TROUBLESHOOTING THE CHARGING SYSTEM 5-45
TROUBLESHOOTING THE COOLING SYSTEM 6-2
TROUBLESHOOTING THE ELECTRICAL SYSTEM 5-6
OPEN CIRCUITS 5-7
RESISTANCE 5-7
SHORT CIRCUITS 5-8
VOLTAGE 5-6
VOLTAGE DROP 5-7
TROUBLESHOOTING THE FUEL SYSTEM 4-5
COMBUSTION RELATED PISTON FAILURES 4-7
COMMON PROBLEMS 4-6
TROUBLESHOOTING THE HAND REWIND STARTER 11-2
TROUBLESHOOTING THE LOWER UNIT 8-2
TROUBLESHOOTING THE MOTOROLA® DISTRIBUTOR IGNITION 5-14
TROUBLESHOOTING THE PRESTOLITE® ELECTRONIC IGNITION 5-25
TROUBLESHOOTING THE REMOTE CONTROLS 10-2
TROUBLESHOOTING THE STARTING SYSTEM 5-51
TROUBLESHOOTING THE THUNDERBOLT® ELECTRONIC IGNITION SYSTEM 5-32
TROUBLESHOOTING THE TILLER HANDLE 10-11
TROUBLESHOOTING THE WARNING SYSTEMS 6-10
TWO-STROKE CYCLE 7-2
TWO-STROKE OIL 3-2
OIL RECOMMENDATIONS 3-2
PRE-MIXING FUEL 3-2
TUNE-UP 3-12
TUNE-UP SPECIFICATIONS 3-31
UNDERSTANDING AND TROUBLESHOOTING ELECTRICAL SYSTEMS 5-2
WARNING BUZZER 10-7
REMOVAL & INSTALLATION 10-7
TESTING 10-7
WARNING SYSTEMS 6-10
WATER PUMP 6-3
CLEANING & INSPECTION 6-9
REMOVAL & INSTALLATION 6-3
SERVICE CAUTIONS 6-3
WHERE TO BEGIN 1-2
WINTER STORAGE CHECKLIST 3-28
WIRE AND CONNECTOR REPAIR 5-8

SELOC PUBLISHING'S FULL-LINE MASTER LIST

ISBN	PART NO	TITLE/DESCRIPTION	YEARS
OUTBOARDS			
089330018-7	018-7	1000 Chrysler Outboards, All Engines	1962-84
089330055-1	055-1	1100 Force Outboards, All Engines	1984-99
089330048-9	048-9	1200 Honda Outboards, All Engines	1978-99
089330007-1	007-1	1300 Johnson/Evinrude Outboards, 1-2 Cyl	1956-70
089330008-X	008-X	1302 Johnson/Evinrude Outboards, 1-2 Cyl	1971-89
089330026-8	026-8	1304 Johnson/Evinrude Outboards, 1-2 Cyl	1990-95
089330009-8	009-8	1306 Johnson/Evinrude Outboards, 3-4 Cyl	1958-72
089330010-1	010-1	1308 Johnson/Evinrude Outboards, 3, 4 & 6 Cyl	1973-91
089330040-3	040-3	1310 Johnson/Evinrude Outboards - All V Engines	1992-96
089330063-2		1310 Johnson/Evinrude Outboards - All V Engines	1992-01
089330052-7		1312 Johnson/Evinrude Outboards, All In-line engines/2 & 4 Stroke	1996-01
089330015-2	015-2	1400 Mariner Outboards, 1-2 Cyl	1977-89
089330016-0	016-0	1402 Mariner Outboards, 3, 4 & 6 Cyl	1977-89
089330012-8	012-8	1404 Mercury Outboards, 1-2 Cyl	1965-91
089330013-6	013-6	1406 Mercury Outboards, 3-4 Cyl	1965-89
089330014-4	014-4	1408 Mercury Outboards, 6 Cyl	1965-89
089330051-9	051-9	1416 Mercury/Mariner Outboards, All Engines	1990-00
089330050-0	050-0	1600 Suzuki Outboards, All Engines	1988-99
089330021-7	021-7	1700 Yamaha Outboards, 1-2 Cyl	1984-91
089330022-5	022-5	1702 Yamaha Outboards, 3 Cyl	1984-91
089330023-3	023-3	1704 Yamaha Outboards, 4 & 6 Cyl	1984-91
089330047-0	047-0	1706 Yamaha Outboards, All Engines	1992-98
STERN DRIVES			
089330029-2	029-2	3000 Marine Jet Drive	1961-96
089330005-5	005-5	3200 Mercruiser Stern Drive	1964-91
089330053-5		3206 Mercruiser Stern Drive - All	1992-01
089330004-7	004-8	3400 OMC Stern Drive	1964-86
089330025-X	025-X	3402 OMC Cobra Stern Drive	1985-95
089330056-X	025-X	3402 OMC Cobra Stern Drive	1985-99
089330011-X	011-X	3600 Volvo/Penta Stern Drives	1968-91
089330038-1	038-1	3602 Volvo/Penta Stern Drives	1992-93
089330041-1	041-1	3604 Volvo/Penta Stern Drives	1992-95
089330057-8	041-1	3604 Volvo/Penta Stern Drives	1992-02
INBOARDS			
089330049-7	049-7	7400 Yanmar Inboards	1975-98
PERSONAL WATERCRAFT			
089330032-2	032-2	9200 Kawasaki	1973-91
089330042-X	042-X	9202 Kawasaki	1992-97
089330045-4	045-4	9400 Polaris	1992-97
089330033-0	033-0	9000 Sea-Doo/Bombardier	1988-91
089330043-8	043-8	9002 Sea-Doo/Bombardier	1992-97
089330034-9	034-9	9600 Yamaha	1987-91
089330044-6	044-6	9602 Yamaha	1992-97
Seloc-On-Line (Internet Access)			
089330075-6	5000	One Mfg/Model - Subscription per user 3 years	1990-01
	5002	Master Pack -Case of 6 CD's including POP Display	
PROFESSIONAL TECHNICIANS MANUALS			
089330060-8		4500 Labor Guide - Johnson and Evinrude	1980-00
089330061-6		4550 Labor Guide - Yamaha	1980-00
089330062-4		4600 Labor Guide - Mercury	1980-00
BRIGGS AND STRATTON			
096376610-4	610-4	610 Briggs & Stratton 4 Stroke Horizontal Crankshaft	1950+
096376611-2	611-2	611 Briggs & Stratton 4 Stroke Vertical Crankshaft	1950+
096376612-0	612-0	612 Briggs & Stratton 4 Stroke Overhead Crankshaft	1950+

ENGINE FINDER

The following listings contain all engines covered in this manual

Model/Engine	Year
3 hp, 1 cyl, 2-stroke	1990 - 1994
4 hp, 1 cyl, 2-stroke	1984 - 1987
5 hp, 1 cyl, 2-stroke	1992 - 1999
7.5 hp, 2 cyl, 2-stroke	1985
9.9 hp, 2 cyl, 2-stroke	1984 - 1999
15 hp, 2 cyl, 2-stroke	1984 - 1999
25 hp, 3 cyl, 2-stroke	1994 - 1999
35 hp, 2 cyl, 2-stroke	1986 - 1991
40 hp, 2 cyl, 2-stroke	1992 - 1999
50 hp, 2 cyl, 2-stroke	1984 - 1999
60 hp, 2 cyl, 2-stroke	1985
70 hp, 3 cyl, 2-stroke	1991 - 1993
75 hp, 3 cyl, 2-stroke	1994 - 1999
85 hp, 3 cyl, 2-stroke	1984 - 1991
90 hp, 3 cyl, 2-stroke	1990 - 1999
120 hp, 4 cyl, 2-stroke	1990 - 1999
125 hp, 4 cyl, 2-stroke	1984 - 1989
150 hp, 5 cyl, 2-stroke	1990 - 1994

SelocOnLine

SelocOnLine is a maintenance and repair database accessed via the Internet. Always up-to-date, skill level and special tool icons, quick access buttons to wiring diagrams, specification charts, maintenance charts, and a parts database.
Contact your local marine dealer or see our demo at www.seloconline.com

SelocOnLine

Step 1
    Select your manufacturer
    Year / Model
    Engine

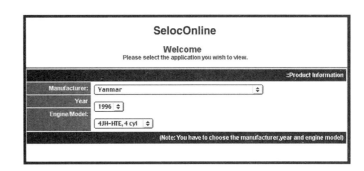

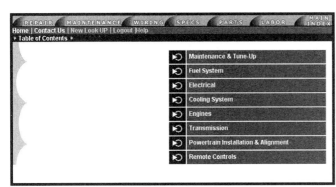

Step 2
    Select an Engine System

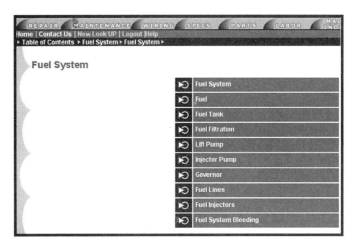

Step 3
    Select the repair

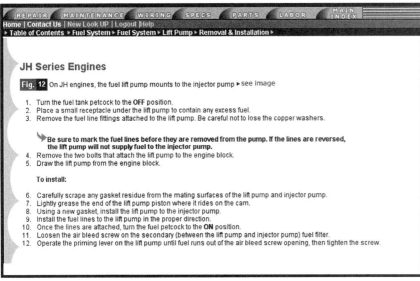

Step 4
    Print the repair procedure

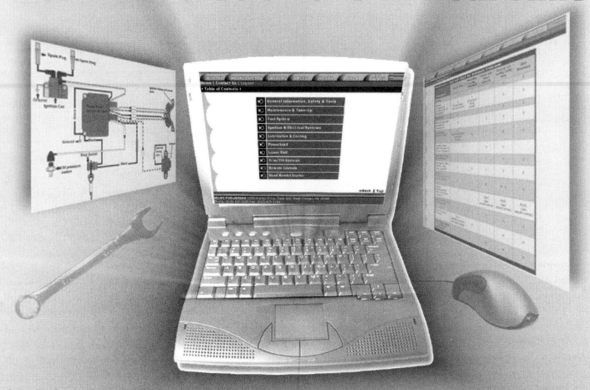

Seloc PRO
Real World Solutions In Real Time

Seloc Pro is a mechanical repair database that is accessed via the internet. There is no other tool in your shop that will save you as much time or provide you with as much productivity as Seloc Pro. Unlike manufacturer information you may presently use, Seloc Pro allows you to navigate the same way thru our database regardless of the manufacturer. Our database is written so that each procedure is 1-3 pages in length. New content or changes to content can be added in minutes.

Seloc Pro is always up-to-date.

Features Include:

Quick access buttons to databses for Wiring Diagrams, Specifications, Parts and Labor Times

Seloc Labor Times - Real world freshwater and saltwater times for hundreds of operations for Johnson, Evinrude, Yamaha, and Mercury from 1980 thru 2000

Hyper-linked index for quick access to unit repair sections

Mfgs covered include Force, Honda, Johnson, Evinrude, Mercruiser, Mercury, Suzuki, Yamaha and Yanmar.
Coming during the 4th quarter are Volvo Penta and OMC Stern Drive

Contact your local distributor for a demo
or Seloc at 866-735-6255

Seloc OnLine Solutions (SOS)

FEATURES AND BENEFITS

Seloc Publishing has been an innovator in helping boaters and technicians maintain and repair outboard and inboard engines and/or drive systems for over 30 years. From performing actual teardown procedures to digital photography, Seloc has been there all the way. Now Seloc wants to assist you in solving those difficult problems. We are as close as your fingertips or your phone. Not 100% sure about that procedure? Email or phone us, we can help you with your engine and boat related questions.

REAL WORLD SOLUTIONS IN REAL TIME!!

- Technical questions answered by factory trained technicians.
- Response to questions in less that 24 hours.
- Priority Response System (ability to have technician call you in 30 minutes or less).
- Technical documentation and diagrams
- Monthly newsletter containing helpful hints, and "How to" features.
- Future access to on-line repository of SelocSOS inquiries

One visit to SelocSOS could potentially save you hundreds of dollars. Above all, our priority is to get you back in the water. You have invested thousands of dollars on your boating investment. Now it is time to get a return on your investment. Visit us at www.selocsos.com or selocmarine.com and see what Seloc can do for you.

SELOC PUBLISHING PRESENTS

SELOC LABOR GUIDES

Johnson & Evinrude 1980-2000
Yamaha 1984-2000
Mercury 1980-2000

Seloc labor manuals cover all engines and thousands of operations. Our real world times include both freshwater and saltwater times.

Contact your local distributor or call us at 1-866-735-6255